THE TROPICAL RAIN FOREST

THE
TROPICAL RAIN FOREST
AN ECOLOGICAL STUDY

BY

P. W. RICHARDS, M.A., Ph.D.

Emeritus Professor of Botany, University of Wales,
formerly Fellow of Trinity College, Cambridge

CAMBRIDGE UNIVERSITY PRESS

CAMBRIDGE

LONDON   NEW YORK   NEW ROCHELLE
MELBOURNE   SYDNEY

Published by the Press Syndicate of the University of Cambridge
The Pitt Building, Trumpington Street, Cambridge CB2 1RP
32 East 57th Street, New York, NY 10022, USA
296 Beaconsfield Parade, Middle Park, Melbourne 3206, Australia

ISBN 0 521 06079 6 hard covers
ISBN 0 521 29658 7 paperback

First published 1952
Reprinted 1957 1964 1966 1972 1976
First paperback edition 1979
Reprinted 1981

Printed in Great Britain at the
University Press, Cambridge

TO
THE MEMORY OF
CARL SCHROETER
(1855–1939)
PROFESSOR OF BOTANY IN THE
FEDERAL POLYTECHNIC, ZÜRICH

Most studies of vegetation have been carried out in Europe, and I am of the opinion that owing to a paucity of material these investigations have begun with an inverted viewpoint. When studying the manifold types of vegetation, comparing them and relating them to each other, one ought logically to start with the richest and to derive from it the less complicated, impoverished types which have arisen from it by selection. The richest type of vegetation in number of species, volume and density, is found in the tropics. It is not the impoverished anthropogenic vegetation of Europe which should be the starting-point of one's investigations.

C. G. G. J. van Steenis (1937 *a*, transl.)

NOTE TO FIFTH REPRINT (1975)

As in the previous reprints the opportunity has been taken to correct a few misprints and other minor errors. No substantial revision has been attempted and no changes have been made in plant names.

P. W. R.

PREFACE

The scope of ecology is not easy to define—it has even been said that the only definition of ecology is that it is the subject-matter of the *Journal of Ecology*. In writing a book about the ecology of the Tropical Rain forest I have therefore had to decide for myself what was and what was not relevant to my theme; in this I have been influenced, no doubt, by my own particular whims and prejudices.

Because ecology is a synthetic science, embracing or touching many other disciplines, it has been my ambition to interest many who are neither botanists nor foresters—zoologists, geographers, in fact anyone who is concerned with the rain forest as a plant community or an environment. I have dealt scarcely at all with the economic aspects of my subject; my aim has been to provide a basis for future work, whether 'pure' or 'applied'. Because I hope the book will be of use to those not trained as professional botanists, I have tried to make the text as self-explanatory as possible and to avoid unnecessary technical terms.

No general account of the Tropical Rain forest has been written since A. F. W. Schimper published his great *Plant Geography* (1898; English edition, 1903), which has since been revised and expanded by Prof. F. C. von Faber (1935). My main qualification for such a formidable task is first-hand experience of rain-forest vegetation adding up to nearly two years. As this experience, though short, was very intensive, and as I had the unusual good fortune to visit each of the three chief tropical regions—South America, Africa and Malaysia—within a space of seven years, this qualification is perhaps not as painfully inadequate as it appears. The great development of interest in tropical vegetation during the last fifteen years has given rise to a voluminous and very scattered literature. In writing the book I have endeavoured to make full use of this, but much has had to be deliberately neglected and still more has probably been unintentionally overlooked. In a work of this kind it is inevitable that many statements will prove to be wrong, and in some places the facts may prove to have been misinterpreted. These shortcomings may not matter if the book stimulates further work. In my travels I have been impressed by the large amount of valuable ecological information which exists unpublished in the minds and notebooks of foresters and buried in departmental reports; I hope the publication of this book may coax some of these data from their hiding places. Much valuable information has been obtained from letters from various friends; the source of such data is indicated in the text by the name of my correspondent in brackets without date. In every chapter I have tried to point out the chief gaps in present knowledge and to suggest lines for future work. No better prospect for my work can be wished than that it may soon become out of date.

Many ecologists would agree that their science is not yet ripe for a rigid theoretical framework, but since a theoretical background of some kind is necessary, the general principles of the Anglo-American school of ecologists have been followed. The absence of a chapter on biotic factors is due, not to a failure to realize their importance, but to the lack of a sufficient body of suitable data.

Part of the matter in the book has appeared in a series of papers published from 1933 onwards. As might be expected, I have since modified some of the views and interpretations given in those papers.

With regard to the nomenclature of species, it is obvious that in a work of this kind, in which names are quoted from papers and books dealing with the flora or vegetation of many different countries, the author cannot answer for the correctness of every name used, though the nomenclature has been checked as far as time and opportunity have allowed. I am much indebted to various members of the staff of the Kew Herbarium for helping me in this part of the work. Where information has been taken from published books or papers the names given here are not always those used in the original, but the synonyms will be found in the Index of Plant Names (p. 425). In a few instances names of plants in the text have been placed in inverted commas; this indicates that the validity of the name or the correct citation is doubtful.

I could not have written a general account of the tropical rain forest without the help of many kind friends. Though it is impossible to acknowledge individually the help of all who have provided data, references to literature, or who have assisted in other ways, a word of special thanks is due to Dr Agnes Arber, F.R.S., who has given me much valuable advice on matters of presentation and has read and criticized a large part of the manuscript. Also to Sir Edward Salisbury, C.B.E., F.R.S., whose help in planning the book was invaluable; it was also a suggestion of his that gave rise to the 'profile-diagram' technique which has proved such a useful tool in the study of tropical vegetation. Special thanks for help of various kinds are also due to Dr J. S. Beard, Prof. H. G. Champion, Dr E. M. Chenery, Mr E. J. H. Corner, Mr T. A. W. Davis, Dr G. C. Evans, Mr P. J. Greenway, Prof. F. Hardy, the late Mr A. P. D. Jones, Mr R. W. J. Keay, Prof. J. Lebrun, Prof. G. Manley, Mr R. Ross, Dr C. G. G. J. van Steenis, Mr C. Swabey, Prof. J. S. Turner, Prof. T. G. Tutin, and Dr Frans Verdoorn. A word of gratitude is also due to the librarians of several libraries who have assisted me in searching for literature, especially the Librarian of the Imperial Forestry Institute, Oxford. For permission to reproduce figures and photographs, I have to thank Dr J. R. Baker, Dr J. S. Beard, Mr W. J. Eggeling, Dr G. C. Evans, Dr E. W Jones, Prof. F. W. Went, the Director of the Musée Royale d'Histoire Naturelle de Belgique, the Editor of the *Bulletin du Jardin Botanique de Buitenzorg*, the Forestry Department, Malayan Union and the Editor of the *Journal of Ecology*. Lastly, I am indebted to my wife for much help, especially in preparing the indexes.

P. W. RICHARDS

BOTANY SCHOOL, CAMBRIDGE
August 1948

CONTENTS

CHAPTER 1: INTRODUCTION

The biological spectrum of the Rain forest, p. 7. *Present and past distribution of Tropical Rain forest*, p. 10. *The significance of the Rain forest for ecology*, p. 15.

PART I. STRUCTURE AND PHYSIOGNOMY

CHAPTER 2: STRUCTURE: SYNUSIAE AND STRATIFICATION

The synusiae, p. 19. *Stratification*, p. 22. The profile diagram, p. 24. Stratification of Mixed Rain forest, p. 25. Stratification of Single-dominant forests, p. 34. Relation of stratification to floristic composition, p. 38.

CHAPTER 3: REGENERATION

Death of old trees and formation of gaps, p. 41. *Age-class representation*, p. 43. *Growth rate at various stages of development*, p. 46. *The Mosaic theory of regeneration*, p. 49.

CHAPTER 4: THE PHYSIOGNOMY OF THE TREES AND SHRUBS

Ecological morphology of the rain-forest flora, p. 54. *Habit of trees*, p. 55. *Bark*, p. 58. *Buttresses and root systems*, p. 59. Structure and development of buttresses, p. 62. Buttressing in relation to habitat, p. 65. Buttressing and soil conditions, p. 66. Theories of buttress formation, p. 69. Pneumatophores, p. 74. *Habit of shrubs*, p. 76. *Leaves of trees and shrubs*, p. 77. Buds, p. 77. Young leaves, p. 78. Leaf size and shape, p. 80. *Flowers and fruits*, p. 90. *Conclusions*, p. 94.

CHAPTER 5: THE GROUND HERBS AND THE DEPENDENT SYNUSIAE

Ground herbs, p. 96. *Climbers*, p. 102. Distribution and habit, p. 102. Part played in structure of forest, p. 106. Morphology and anatomy, p. 106. *Stranglers*, p. 109. *Epiphytes*, p. 110. Epiphytic flowering plants and pteridophytes, p. 112. Habitat and physiology, p. 114. Factors controlling the distribution of epiphytes within the forest, p. 116. Synusiae of epiphytes, p. 119. Vertical distribution of the synusiae, p. 121. Ecological morphology, p. 122. Non-vascular epiphytes, p. 126. *Saprophytes*, p. 128. *Parasites*, p. 130.

PART II. THE ENVIRONMENT

Chapter 6: CLIMATE

Chapter 7: MICROCLIMATES

Chapter 8: SEASONAL CHANGES

Chapter 9: SOIL CONDITIONS

PART III. FLORISTIC COMPOSITION OF CLIMAX COMMUNITIES

Chapter 10: COMPOSITION OF PRIMARY RAIN FOREST (I)

Chapter 11: COMPOSITION OF PRIMARY RAIN FOREST (II)

PART IV. PRIMARY SUCCESSIONS

CHAPTER 12: THE PRIMARY XEROSERE AND THE RECOLONIZATION OF KRAKATAU

CHAPTER 13: HYDROSERES

CHAPTER 14: COASTAL SUCCESSIONS

PART V. TROPICAL RAIN FOREST UNDER LIMITING CONDITIONS

CHAPTER 15: RAIN FOREST, DECIDUOUS FOREST AND SAVANNA

CHAPTER 16: THE TROPICAL RAIN FOREST AT ITS ALTITUDINAL AND LATITUDINAL LIMITS

PART VI. MAN AND THE TROPICAL
RAIN FOREST

Chapter 17: SECONDARY AND DEFLECTED
SUCCESSIONS

LIST OF TEXT-FIGURES

LIST OF PLATES

The plates are bound between pp. 300–301

CHAPTER 1

INTRODUCTION

'Its lands are high and there are in it very many sierras and very lofty mountains, beyond comparison with the island of Teneriffe. All are most beautiful, of a thousand shapes, and all are accessible and filled with trees of a thousand kinds and tall, and they seem to touch the sky. And I am told that they never lose their foliage, as I can understand, for I saw them as green and as lovely as they are in Spain in May and some of them were flowering, some bearing fruit and some in another stage, according to their nature.' In these words from Christopher Columbus's account[1] of the island of Española, written in 1493, we have a glimpse of a tropical landscape and perhaps the earliest description in literature of the kind of vegetation with which this book is concerned. Evergreen tropical forest, which in Columbus's time covered most of the West Indian islands, still stretches over mile after mile of the equatorial lowlands of South and Central America, Africa and south-eastern Asia. It is the characteristic vegetation of the wet tropics and occupies (or formerly occupied) all land surfaces with a sufficiently hot climate, and a sufficiently heavy and well-distributed rainfall, except for the small areas where the ground is too swampy or where, as on young volcanic lava, there has not been time for it to develop. In the ecological term, evergreen forest is the climax vegetation of the equatorial climate.

In scientific literature this evergreen forest is most often referred to as Tropical Rain forest (*tropische Regenwald*), a term coined for it by A. F. W. Schimper in his classical *Plant Geography* (1898, 1903); the Latin equivalent *pluviisylva* is also sometimes used. Humboldt, one of the first adequately to describe it, spoke of the great South American rain forest as the *Hylaea* (from ὕλη, a forest), and German writers often use the same word for the evergreen tropical forest of Africa. Germans also frequently use the word *Urwald*, meaning original or primitive forest, a term which has no necessary tropical implication, but which custom has attached to the Tropical Rain forest rather than to other types of primitive forest. To the Englishman in the tropics the Rain forest is usually the 'bush' or the 'jungle', in Australia 'brush' or 'scrub'.

The name 'Rain forest' is commonly given, not only to the evergreen forest of moist tropical lowlands—the plant-formation (or better, the formation-type) dealt with in this book—but also to the somewhat less luxuriant evergreen forest found at low and moderate altitudes on tropical mountains, and to the evergreen forests of oceanic subtropical climates, in south-western China, southern Chile, South Africa, New Zealand and eastern extra-tropical Australia. These other

---

[1] Jane's (1930) translation.

formation-types will here be called Montane Rain forest and Subtropical Rain forest respectively. In this book 'Rain forest' without qualification means Tropical Rain forest.

What is a Tropical Rain forest? Schimper (1903, p. 260) gave this brief diagnosis: 'Evergreen, hygrophilous in character, at least 30 m. high, but usually much taller, rich in thick-stemmed lianes and in woody as well as herbaceous epiphytes.' This definition fits the concept of Tropical Rain forest as most writers since Schimper have used it and is the one which will be adopted for our present purposes, but some modern workers wish to use the term in a less comprehensive sense or to abandon it altogether. In this narrower sense the term Rain forest would be reserved for the almost completely non-seasonal forest of tropical climates with the most evenly distributed rainfall; it would not be applied to the more seasonal types of evergreen forest found in areas with a marked dry season. To this question we shall return in Chapters 15 and 16, when attempt will be made to make clearer the differences between Tropical Rain forest on the one hand and, on the other, Montane Rain forest, Subtropical Rain forest and the various more or less deciduous tropical forest formations classified by Schimper as Monsoon forest, Savanna forest and Thorn forest.

A short definition, however accurate, will give but a poor idea of what a tropical rain forest is, especially to those who are familiar only with the vegetation of temperate countries such as Europe and North America. What are its chief characteristics? What kind of an impression does it make on an observer?

Most people who have no first-hand experience of tropical vegetation derive their ideas about it from books of travel, and these unfortunately are often biased or exaggerated, even when not positively inaccurate. Errors arise because travellers have been too quick to generalize from a single area, or because they have travelled mainly by river and have been misled into supposing that the interior of the forest is like the riverside fringe. The main source of inaccuracy, however, is that tropical vegetation has a fatal tendency to produce rhetorical exuberance in those who describe it. Few writers on the rain forest seem able to resist the temptation of the 'purple passage', and in the rush of superlatives they are apt to describe things they never saw or to misrepresent what was really there. In attempting to paint an accurate picture of a rain forest it will be necessary to point out and correct some of these common errors.

One of the outstanding features of the Tropical Rain forest, indeed, of all the vegetation of the humid tropics, is that the overwhelming majority of the plants are woody and of the dimensions of trees. Not only do trees form the dominants of the rain-forest community, but most of the climbing plants and some of the epiphytes are also woody. The undergrowth largely consists of woody plants— seedling and sapling trees, shrubs and young woody climbers. The only herbaceous plants are some of the epiphytes and a proportion, relatively small, of the undergrowth. Families of plants which in temperate regions are represented exclusively by low-growing herbs here assume the size and woodiness of trees. Spruce (1908, **1**, p. 256) says of the Amazon forests: 'Nearly every natural order of plants has here *trees* among its representatives. Here are grasses (bamboos) of 40, 60, or more feet in height, sometimes growing erect, sometimes tangled in thorny thickets, through which an elephant could not penetrate. Vervains form spreading trees with digitate leaves like the horse-chestnut. Milkworts, stout woody twiners ascending to the tops of the highest trees, and ornamenting them with festoons of fragrant flowers not their own. Instead of your periwinkles we have here handsome trees exuding a milk which is sometimes salutiferous, at others a most deadly poison, and bearing fruits of corresponding qualities. Violets of the size of apple trees. Daisies (or what might seem daisies) borne on trees like alders.' In the rain-forest region woody plants form not only the larger proportion of the mature forest vegetation, but play the chief part in its developmental stages, as in the colonization of bare rock or soil, and in the building up of vegetation in lakes and swamps.

The trees of the rain-forest community are extremely numerous in species and varied in size. The dimensions reached by tropical trees have sometimes been much exaggerated. The average height of the taller trees in a rain forest is rarely more than 150–180 ft. (46–55 m.), though individual trees over 200 ft. (60 m.) are not uncommon and some approaching 300 ft. (92 m.) have been reliably recorded. In Europe and North America, according to E. W. Jones (1945), dominant forest trees very commonly attain a height of 100 ft. (30 m.), and under exceptionally favourable conditions reach 150 ft. (46 m.). Thus though rain-forest trees are usually taller than those in temperate forests, they never reach the gigantic dimensions of the Californian redwoods or the huge gums (*Eucalyptus*) of Australia. In girth and diameter the largest rain-forest trees also fall far short of those of higher latitudes. The largest recorded girth measurement appears to be 56 ft. (17 m.), but trees of greater girth than a metre are uncommon, and rain-forest trees in general are more often remarkable for the slenderness than for the large girth of their trunks. Data on the dimensions of unusually large rain-forest and other trees are given in Table 1.

In a European or North American forest the dominant trees belong to a few, or often only a single, species; in extreme cases the forest may consist of twenty to twenty-five species (Braun, 1941, p. 236) or there may be as many as twenty-five to thirty species of tall trees per 0·5 hectare ('Atlantic' forest of North America, see Schimper, 1935, p. 831). In the Tropical Rain forest there are seldom less than forty species of trees over 4 in. (10 cm.) diameter per hectare and sometimes over a hundred species. A. R. Wallace (1878, p. 65) says: 'If the traveller notices a particular species and wishes to find more like it, he may often turn his eyes in vain in every direction. Trees of varied forms, dimensions and colours are around him, but he rarely sees any one of them repeated. Time after time he goes towards a tree which looks like the one he seeks, but a closer examination proves it to be distinct. He may at length, perhaps, meet with a second specimen half a mile off, or may fail altogether, till on another occasion he stumbles on one by accident.'

The richness of the tree flora is indeed the most important characteristic of the rain forest and on this many of its other features are directly dependent. Trees of different species are most commonly found mixed in fairly even proportions; more rarely one or two species are much more abundant than the rest.

TABLE 1. *Dimensions of the largest Tropical Rain forest trees*

The dimensions given are of the largest trees for which reliable measurements can be found in the literature. For comparison, measurements are given of exceptionally large extra-tropical trees.

Metres	Species	Locality	Authority
		(a) Height	
84	Koompassia excelsa	Sarawak (Borneo)	Foxworthy (1927, p. 84)
81	K. excelsa	Malay Peninsula	Foxworthy (1927, p. 84)
71	Eucalyptus deglupta	New Britain	Lane-Poole (1925a, p. 214)
70	Koompassia excelsa	Sarawak	Beccari (1904, p. 330)
70	Agathis alba	Celebes	van der Koppel (1926, p. 529)
59	Entandrophragma cylindricum	Nigeria	Kennedy (1936, p. 176)
56	Desbordesia glaucescens	Cameroons	Mildbraed (1922, p. 105)
111	Sequoia sempervirens	California	Tiemann (1935, p. 903)
107	Eucalyptus regnans	Victoria (Australia)	Tiemann (1935, p. 904)
75	Agathis australis	New Zealand	van der Koppel (1926, p. 530)
46	Fagus sylvatica	France	Elwes & Henry (1906, 1, p. 13)
40	Quercus robur	France	Elwes & Henry (1907, 2, p. 310)
		(b) Girth	
17	Entandrophragma cylindricum	Nigeria	Unwin (Kennedy, 1936, p. 176)
14	Entandrophragma cf. angolense var. macrophyllum	Nigeria	Richards
13	Bertholletia sp.	Brazil	Spruce (1908, 1, p. 18)
12*	Balanocarpus heimii	Malay Peninsula	Foxworthy (1927, p. 53)
c. 12†	Dryobalanops aromatica	Malay Peninsula	Foxworthy (1927, p. 45)
17	Tilia sp.	—	Kannegiesser (Büsgen & Münch, 1929, p. 38)
15	Castanea sativa	—	Kannegiesser (Büsgen & Münch, 1929, p. 38)
23‡	Sequoiadendron giganteum ('General Sherman')	California	Tiemann (1935, p. 909)
12	Quercus robur	Wales	Elwes & Henry (1907, 2, p. 309)
23§	Agathis australis	New Zealand	Tiemann (1935, p. 911)

    * 'This was the largest tree of which we have any record in the Peninsula' (Foxworthy).
    † Girth calculated from diameter.
    ‡ Calculated from diameter at 10 ft. from ground.
    § Calculated from diameter.

The Rain forest is thus usually a community with a large number of co-dominants, but sometimes there are only one or two dominants. Rain-forest communities with numerous dominants will be termed Mixed forests, those with a single, strongly dominant species Single-dominant forests (see Chapters 10–11).

Though so numerous in species rain-forest trees are on the whole remarkably uniform in their general appearance and physiognomy. The trunks are as a rule straight and slender and do not branch till near the top. The base is very commonly provided with plank buttresses, flange-like outgrowths which are a highly characteristic feature of rain-forest trees; such buttresses are very little developed in other plant formations. The bark is generally thin and smooth and rarely has deep fissures or conspicuous lenticels. The vast majority of the mature trees, as well as of the shrubs and saplings, have large, leathery, dark green leaves with entire or nearly entire margins; they resemble the leaves of the cherry laurel (*Prunus laurocerasus*) of English gardens in size, shape and texture. When the leaves are compound, as in leguminous trees, each leaflet tends to approximate in size and shape to the undivided leaves of the other trees. So uniform is the foliage that the non-botanical observer might easily be excused for supposing that the forest was predominantly composed of species of laurel. The similarity and sombre colouring of the majority of the leaves are mainly responsible for the monotonous appearance of the forest. Large and strikingly coloured flowers are uncommon; most of the trees and shrubs have inconspicuous, often greenish or whitish, flowers.

Uniform as are most of the woody species, there are a few members of the rain-forest flora which are widely divergent from the normal type in appearance, notably the palms and the species of *Dracaena* and *Pandanus*, but such bizarre and unfamiliar-looking plants are far less conspicuous in typical Rain forest than many descriptions would suggest. They are seldom common enough to affect greatly the physiognomy of the community as a whole and quite often they are entirely absent.

There is great variation in the height of the trees, and Rain forest is commonly several-layered in structure. Most often there are three tree layers (in addition to shrub and herb layers), but sometimes only two. There is thus, in Humboldt's often quoted phrase, 'a forest above a forest'. These tree layers are further discussed in Chapter 2.

The undergrowth of the Rain forest consists of shrubs, herbaceous plants and vast numbers of sapling and seedling trees. Books of travel often give a misleading impression of the density of tropical forests. Mature (primary) Rain forest is usually not difficult to penetrate. On river banks, or in clearings, where much light reaches the ground, there is a dense growth which is indeed often quite impenetrable, but in the interior of old undisturbed forest, if one's hands are free to bend back a twig here and there, it is not difficult to walk in any direction, though sometimes a detour has to be made to avoid a fallen log or a mass of lianes which have slithered down from above. When in the forest it is usual to carry a cutlass or parang, but more in order to mark a path to return by if necessary than to hack one's way. It is the slippery clay soil and the abundance of fallen logs and branches which make progress in the forest slow and laborious, rather than the thickness of the vegetation. Photographs give an exaggerated notion of the density of the undergrowth; it is usually possible to see another person at least 20 m. away. The forest interior is gloomy, but when the sun is shining the floor is dappled with sun-flecks. Giesenhagen (1910, p. 728) aptly compares the illumination in the rain forest to a European beechwood 'in its spring green'. The herbaceous ground flora is sparse; the number of species of ground herbs (per unit area) is usually less than in an English wood and the number of individuals fewer. The ground is only thinly covered with dead leaves and there are often patches of bare soil. Tropical forests in which the ground is 'covered with age-long accumulations of rotting vegetation' are quite exceptional or mere figments of the imagination.

Besides trees, shrubs and ground herbs, rain forest is composed of climbing plants of all shapes and sizes, and of epiphytes growing on the trunks, branches and even on the living leaves of the trees and shrubs.

The abundance of climbers is one of the most characteristic features of rain-forest vegetation. The majority of these climbers are woody (lianes) and have stems often of great length and thickness; stems as thick as a man's thigh are not uncommon. Some lianes cling closely to the trees that support them, but most ascend to the forest canopy like cables or hang down in loops or festoons. The number of species of climbing plant is enormous, and there is great variety of form and structure among them.

The epiphytic vegetation, as well as including algae, mosses, liverworts and lichens, as in temperate forests, consists of large numbers of orchids and other flowering plants and many ferns. The tree-top epiphytes (which are often not visible from the ground) include shrubby as well as herbaceous species, also the semi-parasitic Loranthaceae (mistletoes) and the curious 'stranglers' (mainly species of *Ficus*), which begin life as epiphytes and often develop afterwards into independent trees rooted in the ground. In no other plant community, except some types of Montane and Subtropical Rain forest, are epiphytes more abundant and luxuriant.

The Tropical Rain forest is thus extremely complex in structure and is built up of plants, mostly woody, but of the most varied life-forms, some of them quite unlike any found in temperate vegetation. Yet though it is in fact the most elaborate of all plant communities in structure and the richest in species, the first impression it makes is of sombreness and monotony. Though this is mainly due to the overwhelming predominance of woody plants and to the uniformity of foliage which prevails among nearly all the life-forms, it is increased by the absence of marked seasonal changes. In the rain forest there is no winter or spring, only a perpetual midsummer; the aspect of the vegetation is much the same at any time of year. There are, it is true, seasons of maximum flowering at which more species are in flower than at other times, and also seasons of maximum production of young leaves, but, for the most part, plant growth and reproduction are continuous and some flowers can be found at any time. This is a direct consequence of the climate, which is characterized by very slight seasonal variation, as well as by high temperature and humidity. No cold season or drought interrupts plant activities.

It has been said that the Tropical Rain forest is a *formation-type* or, in Clements' term, a pan-climax. This formation-type comprises three separate *formations*, occupying the chief regions of the world with an equatorial climate, the *American*

Rain forest in parts of Central and South America, the *African Rain forest* in tropical Africa and the *Indo-Malayan Rain forest* in Malaysia, the adjoining parts of south-east Asia, Australia and the Pacific islands. In each of these formations almost all the species and many of the genera and families are peculiar and not shared with the other two. The structure of each climax community, and the successional stages in its development, are, however, much alike. Each climax, as we shall see, also varies in a strikingly parallel manner in response to differences of climate and soil, in spite of the dissimilarity of the flora.

In any region where Rain forest is potentially the climatic climax only a part of the area is actually covered by mature forest. There may be large expanses where it has not as yet developed—swamps, lakes, river margins and estuarine mudflats occupied by seral communities in process of development towards the climax. There may also be bare rock and young volcanic lava where soil is gradually forming, and here, too, stages in the development of mature forest are found. In some tropical regions there are areas where local peculiarities of soil or topography make the development of typical Rain forest impossible, even though the climate is suitable. In such areas other kinds of vegetation, for instance, heath-like communities or savanna, may be found. Stable communities of this kind are termed edaphic climaxes.

In all but the most remote and least inhabited parts of the rain-forest zone there are large areas where the forest has been destroyed to make way for cultivation. Much land which has at one time been cultivated has been abandoned to the natural regrowth of vegetation. The vegetation occupying the sites where the primary forest has been destroyed soon takes on the aspect of a forest, but remains for many years easily recognizable; it is known as second-growth or secondary forest, in contrast to the virgin or primary forest. The terms primary and climax (or mature) are not synonymous, since an immature forest may not have been interfered with by man and is therefore primary though not climax. In Parts I, II, III and V we shall be concerned chiefly with mature or climax rain forest and in Parts IV and VI with primary and secondary successional communities.

THE BIOLOGICAL SPECTRUM OF THE RAIN FOREST

To define precisely the range of life-forms in the Rain forest, and the contribution made by each to its structure, it would be desirable to construct a 'biological spectrum' for a few typical samples of primary forest, using either the well-known 'epharmonic' life-form classification of Raunkiaer (1934) or one of the more purely 'physiognomic' systems which have been recently proposed, such as the 'Main life-form' classification of Du Rietz (1931). Unfortunately, however, there is not a single area of Rain forest for which complete information on the life-forms (particularly of the herbaceous and smaller woody plants) is available. In Tables 2 and 3 data are presented from which it is possible to arrive at a rough approximation to a 'biological spectrum'.

TABLE 2. *Life-forms of flowering plants in the Tropical Rain forest at Moraballi Creek, British Guiana.* (From the data of N. Y. Sandwith)

	No. of species	Life-forms according to Raunkiaer (1934)	Life-forms according to Du Rietz (1931)
Trees:			
Dicotyledons	163	Mega- and Mesophanero-phytes	Tall, High and Low trees
Tall palms	3		
	166		
'Shrubs' and ground herbs:			
Green herbs and 'shrubs'	80	Micro- and Nanophanero-phytes, Chamaephytes	Dwarf and Pigmy trees, Ctonophytic shrubs (?), Herbaceous plants
Saprophytes	13	—	—
Parasites	1		
Adventives (in clearings)	8	Therophytes	Herbaceous plants
	102		
Lianes and other climbers	83	Meso- and Megaphanero-phytes	Woody and Herbaceous lianes
Epiphytes:			
Shrubby (including hemi-epiphytes)	*c.* 10	Epiphytes	Epiphytoidic shrubs, Epi-phytoidic Dwarf shrubs, etc.
Herbaceous	*c.* 81	Epiphytes	Epiphytoidic Herbaceous plants
Semi-parasitic (Loran-thaceae).	7	Epiphytes	Parasitic Dwarf shrubs
	98		

The data in this table give no exact indication of the relative numbers of phanerophytes and chamae-phytes. If, however, we estimate the chamaephytes as half the 'green herbs and shrubs', plus the parasites and saprophytes (certainly an overestimate), and take the total flora, omitting the adventives, as 441, the Raunkiaer biological spectrum would be:

Stem succulents	Epiphytes	Phanero-phytes	Chamae-phytes	Hemicrypto-phytes	Geophytes	Helo- and Hydrophytes	Therophytes
0	22	66	12	0	0	0	0
		78					

TABLE 3. *Life-forms of flowering plants in sample plot* (400 × 400 ft.) *of Mixed Rain forest, Nikrowa, Nigeria.* (Data from Richards, 1939)

	No of species	Life-forms according to Raunkiaer (1934)	Life-forms according to Du Rietz (1931)
Trees over 4 in. (10 cm.) diameter:			
Dicotyledons	70	Mega- and Mesophanero-phytes	Tall, High and Low trees
Tall palms	0	—	—
	70		
'Shrubs'	9	Nanophanerophytes	Dwarf and Pigmy trees, Ctonophytic shrubs (?)
Ground herbs:			
Green herbs	16	Nanophanerophytes, Chamaephytes	Herbaceous plants
Saprophytes	0	—	—
Parasites	0		
Adventives	0	—	—
Lianes and other climbers	Numerous	Mega- and Mesophanero-phytes	Woody and Herbaceous lianes
Epiphytes	Numerous	Ephiphytes	Epiphytoidic Dwarf shrubs, Epiphytoidic Herbaceous plants

Table 2 is based on the collection of flowering plants made at Moraballi Creek, British Guiana, by Mr N. Y. Sandwith in 1929. The whole collection was obtained within a radius of about 5 miles (8 km.) in an area which was entirely forest-covered, except for some very small clearings (old lumbermen's camps, etc.). The Rain forest was partly primary (but not all of one type, see Chapter 10), and partly secondary, though the greater part of the collection was made in primary forest. The list, which includes a total of over 400 species of flowering plants, is certainly far from complete, and some constituents of the flora (tall trees, lianes and epiphytes) are less fully represented than others ('shrubs' and herbaceous ground flora).

The less complete data in Table 3 were obtained in the rain forest of Southern Nigeria and refer only to a single sample plot, measuring 400 × 400 ft. (122 × 122 m.).

Tables 2 and 3, imperfect as they are, illustrate well some of the salient characteristics of rain-forest vegetation and clearly demonstrate several important differences between it and temperate forest communities (see also Fig. 1). Though the data given are only for two small areas, there is no reason to believe that the conclusions drawn from them are not of general application.

The tables and Fig. 1 show, first of all, the enormous preponderance of woody phanerophytes ('Trees' in Du Rietz's classification) in the rain-forest flora. If comparative figures were available for numbers of individual plants, instead of merely for numbers of species,

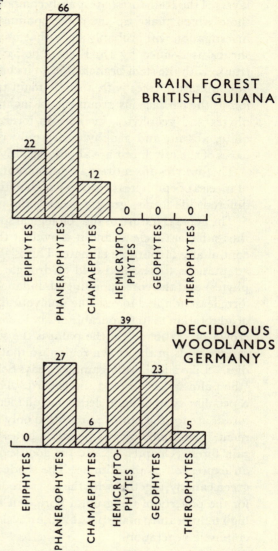

Fig. 1. Raunkiaer biological spectrum of rain-forest flora at Moraballi Creek, British Guiana, compared with that for deciduous woodlands, Germany. Each column represents the number of species of each life-form as a percentage of the total flora. The spectrum for Moraballi Creek is approximate only and was calculated as explained at the foot of Table 2. The spectrum for the German woodlands is the average of the spectra for six deciduous woodland associations in north-west Germany (Tüxen, 1929).

this preponderance would be overwhelming. Secondly, the 'shrub' and herb layers of the Rain forest are mainly composed of plants of different life-forms from those which make up the corresponding layers in temperate forests. Further investigation will probably show that most of the so-called shrubs are not true shrubs as defined by Du Rietz ('holoxyles...not developing a distinct main trunk, with the stem branched from its basal part, above or below the surface'), but miniature trees with a single main trunk (see p. 76). While in temperate forests the herbaceous ground flora includes a large proportion of hemicrypto-phytes and geophytes, in the rain forest hemicryptophytes are almost, if not quite, absent, and geophytes are only rarely present (see p. 98); the ground herbs of the rain forest are almost exclusively phanerophytes and chamaephytes.

Therophytes are entirely absent, except in clearings, by paths, etc. The number of species of epiphytes (a group which in the Rain forest includes plants of several different life-forms) greatly exceeds the number of ground herbs.

The general nature of the 'biological spectrum' for the Tropical Rain forest is thus sufficiently clear, and it is evident that it closely reflects the non-seasonal, continually favourable climate. This is shown in the absence of life-forms with adaptations to seasonal cold or drought (hemicryptophytes, geophytes, thero-phytes) and the corresponding abundance of evergreen phanerophytes, the life-form least adapted to resisting unfavourable conditions. An extreme case of lack of adaptation to unfavourable climatic conditions is furnished by the palms. The crown or 'cabbage' of the palm is the solitary perennating bud; there are no reserve buds in the crown region, so that if the apex is injured the whole shoot dies. 'Like a foolhardy gambler', says Schroeter (quoted by Rikli, 1943, p. 30), 'the palm stakes all on a single card.' The unbranched *Schopfbäumchen* (p. 57), a peculiar type of dicotyledonous phanerophyte very characteristic of the rain forest, are for the same reason fitted only for a permanently favourable environ-ment. The dominance of phanerophytes in most stages of successions leading to rain forest (Chapter 12, etc.) is good evidence that their abundance is in fact an expression of the climate. The shade cast throughout the year by the ever-green canopy, together with the severe root competition, are probably responsible for the poverty of the ground flora, but the constantly humid atmosphere and high temperature make possible the development of the rich, mainly herbaceous, epiphytic vegetation.

PRESENT AND PAST DISTRIBUTION OF
TROPICAL RAIN FOREST

The detailed distribution of the Rain forest as a formation-type is very imper-fectly known. The accompanying map (Fig. 2) is intended to show the area within which Rain forest is the climax vegetation under present climatic condi-tions; though only on a small scale, it cannot claim to be very accurate.

In some of the more densely populated regions of the tropics the natural vegetation has been so completely replaced by cultivation and anthropogenic plant communities that it is difficult or impossible to reconstruct with certainty the original climatic climax. Some remote areas are so little explored that the nature of the vegetation is almost unknown. Many of the published maps of the vegetation of tropical countries are of little value, and different authorities have given very different interpretations to 'Tropical Rain forest' and the other units mapped. Though the limits of the Rain forest will be further discussed in Chapters 15 and 16, it may be useful to give here a brief outline of its distribution, as far as it can be ascertained.

The Tropical Rain forest forms a belt round the whole earth, bisected somewhat unequally by the equator, so that rather more of its area lies in the northern than in the southern hemisphere. Owing to the presence of mountain ranges and plateaux, and to the irregular distribution of the controlling climatic factors (which in turn depends on the uneven distribution of land and sea), the belt is interrupted in several places. The northern and southern boundaries of the Rain forest do not coincide with any well-defined latitudinal limits; in some places they do not reach the geographical tropics, in others they extend slightly beyond them.

The largest continuous mass of Rain forest is found in the New World, in the basin of the Amazon. This extends west to the lower slopes of the Andes and east to the Guianas; it is broken only by relatively small areas of savanna and deciduous forest. The great South American hylaea or Rain forest extends south into the region of the Gran Chaco and north along the eastern side of Central America into southern Mexico and into the chain of the Antilles. In the extreme northwest of South America (Ecuador, Colombia) there is a narrow belt of Rain forest on the western side of the Andes. On the east coast of Brazil another narrow belt, separated from the Amazonian forest by a wide expanse of deciduous forest, runs from about lat. 6° S. to a little beyond the Tropic of Capricorn. The distribution of Rain forest in America is perhaps less well known than in any other major tropical region. The most recent map (Smith & Johnston, 1945) is no more detailed, and probably not very much more accurate, than that in Schimper's *Plant Geography* (1898, 1935).

In Africa the largest area lies in the Congo basin and extends westwards into French Equatorial Africa, the Gaboon and the Cameroons. As a narrow strip the evergreen forest continues still farther west, parallel to the coast of the Gulf of Guinea, through Nigeria and the Gold Coast to Liberia and French Guinea. This western extension is interrupted from western Nigeria to the eastern Gold Coast by a break in which the region of dry climate reaches southward to the coast. In the east the Rain forest scarcely reaches the region of the Great Lakes, and east of the Rift Valley typical Rain forest is absent. Southwards the forest extends, mainly as post-climax Gallery forest, towards Rhodesia.

The extent of the Central African evergreen forest and the degree to which it is interrupted by savanna areas have been the subject of much controversy. Early explorers such as Stanley believed that a continuous forest covered the greater part of the Congo basin. Later, however, De Wildeman (1913, 1926) suggested that this view was false and was due to most of the explorers having travelled by river. According to him the supposed forest massif of Central Africa consists chiefly of narrow strips of Gallery forest confined to the neighbourhood of rivers, the intervening country being mostly savanna. On this view most of the African Rain forest is not in equilibrium with present climatic conditions and is a post-climax of favourable soils and sites rather than a true climatic climax. De Wildeman's 'Gallery Forest theory' was vigorously challenged by Mildbraed (1914, 1923), and the older view that the forest is in fact continuous (though locally broken by patches of savanna, cultivation and secondary vegetation) is generally accepted by modern Belgian authorities. The subject has been critically reviewed by Lebrun (1936a), whose map of the Congo forest is probably the most accurate yet published.[1] A vegetation map of French West Africa has been given by Chevalier (1912, 1938a, etc.).

Rain forest is the natural climax in part at least of the Seychelles, Réunion, Mauritius and on the east coast of Madagascar, but very little primitive forest now remains in these islands.

In the eastern tropics Rain forest extends from Ceylon and western India to Thailand, Indo-China and the Philippines, as well as through the Malay Archipelago to New Guinea. The largest continuous areas are in the Malay Peninsula and the adjoining islands of Sumatra and Borneo, where the Indo-Malayan Rain forest reaches its greatest luxuriance and floristic wealth, and in New Guinea. In India the area of Rain forest is not large, but it is found locally in the Western and Eastern Ghats and, more extensively, in the lower part of the eastern Himalaya, the Khasia hills and Assam. In Burma, Thailand and Indo-China Rain forest is developed only locally, the chief climax formation being 'Monsoon forest'. In the eastern Sunda Islands from western Java to New Guinea the seasonal drought (due to the dry 'East Monsoon' from Australia) is too severe for the development of Rain forest, except as a post-climax in locally favourable situations. A good vegetation map of the Netherlands East Indies, showing the distribution of Rain forest in considerable detail, has been published by van Steenis (1935a).

In Australia the Tropical Rain forest of Indo-Malaya is continued south as a narrow strip along the eastern coast. Rain forest also extends into the islands of the western Pacific (Solomons, New Hebrides, Fiji, Samoa, etc.).

The areas indicated above are those in which rain forest appears to be the natural climax under the conditions of the present day. There are a few restricted regions where the forest seems now to be extending its area, e.g. the southern part of the Belgian Congo (Robyns, 1936), Uganda (Eggeling, 1947), but on the whole, owing to the activities of modern man, the area actually occupied by primary forest is rapidly diminishing. There are now large regions from which the forest has entirely disappeared, and others, like Southern Nigeria, in which

[1] Bernard (1945) has since shown conclusively that the belief that the Congo forest is not in equilibrium with the climate is without foundation.

the forest, once a continuous belt, has been broken up into isolated and widely separated blocks. Most of this destruction of the forest has taken place within the last fifty to a hundred years; thus, until very recently, virgin rain forest covered a large fraction of the earth's surface. From various kinds of evidence there are indications that climatic changes, as well as the destructive activities of man, have reduced the area of the Rain forest; large areas where its development is not now possible seem in earlier periods, when climate was differently distributed, to have borne forest similar to the Rain forest of to-day.

Indications of this kind are particularly numerous in Africa. Much has been written about the 'desiccation of Africa' (see especially Stebbing, 1937a, 1937b, and Lavauden, 1937) and about the supposed former greater extension of the forest-covered area. In West Africa, it is maintained, there has been a progressive drying of the climate and an advance of the Sahara southwards which is still in progress at the present time. This has caused a corresponding retreat of the evergreen forest (along with the Mixed Deciduous forest, Savanna and Thorn forest which form the Rain forest-Desert ecotone), so that localities which in historic times were surrounded by forest are now in savanna or desert. The evidence for this belief has been critically examined by Aubréville (1938).

Though the question is obviously of the utmost importance, it is highly controversial and can only be referred to briefly here. The problem is twofold. Can changes in climate and in the distribution of the climax plant formations be shown to have occurred within the historical period and to be still continuing, and secondly, have there been such changes during recent (Pleistocene and Quaternary) geological time? Aubréville shows that, although in the short historical period for which sufficient data are available, there may have been climatic fluctuations in West Africa, there is no evidence of a secular climatic trend. The description of Northern Nigeria, as it was in 1822–4, given by the travellers Denham and Clapperton, indicates conditions precisely like those of to-day and offers no support to the idea of a rapid deterioration such as Stebbing, for example, has postulated (Jones, 1947). If, in fact, there has been any southward movement of the desert, savanna and closed forest zones, it can be entirely attributed to human activities (shifting cultivation, burning, grazing), leading to the progressive replacement of climatic formations by biotic climaxes (see Chapter 15).

The question of changes in climate and vegetation between the Pleistocene and the present day is an entirely different one, and the evidence is of a different kind. Both geological data (plant remains, fossil soils, geomorphology, etc.) and bio-geographical data (relict species etc.) point to the existence in tropical Africa of wet (pluvial) periods when evergreen forest extended far beyond its present limits. It is tempting to suppose that some at least of these wet periods were contemporary with the Pleistocene glaciations of Europe, though as yet there is very little definite evidence in support of such a correlation (Zeuner, 1945).

Evidence for an extension of the evergreen tropical forest in East Africa during the Pluvial periods has been discussed by Moreau (1933). On the Usambara, Nguru and Uluguru Mountains there are small isolated patches (total area less than 300 sq. miles (777 sq.km.)), similar in character to Tropical Rain forest, but perhaps best described as Submontane Rain forest (see Chapter 16). Both the fauna and the flora of these forests, though largely consisting of endemic species, show a striking similarity to those of the West and Central African Rain forest. Moreau concludes, mainly from a study of the birds, that these East African Rain forests are relics of a large area which at one time joined them to the Central African forest. Such an extension of the forest area would demand an increase of 15–20 in. (38–50 cm.) in the average rainfall of East Africa. Evidence also from the distribution of birds suggesting a former greater extent of the African Rain forest, possibly great enough to connect it with that of the eastern tropics through Arabia and India, has been discussed by Chapin (1932), who gives a map showing the hypothetical distribution of forest in Africa during a humid period in the Pleistocene.

There is fossil evidence (see van Steenis, 1935a) of the existence in the lowlands of Java during the Tertiary of Dipterocarp forests such as are now characteristic of the rain-forest associations of Malaya, Sumatra and Borneo. The present lowland forests of Java (which have almost all been replaced by cultivation and teak plantations) are, except in the extreme west, 'Monsoon forests' rather than true Rain forests, and Dipterocarpaceae occur in them only sparingly. Though many dipterocarps are characteristic of seasonal types of forest, their former dominance may indicate a moister and less seasonal climate than at present and an extension of the Rain forest east of its modern boundary.

Still earlier, during the Cretaceous, Eocene and Oligocene periods, there was, even in countries as far from the tropics as Britain, vegetation which in floristic composition, and perhaps also in physiognomy, closely resembled the modern Tropical Rain forest, as the London Clay and other fossil floras bear witness. It would lead us too far to discuss this problem here.

It is likely that in some at least of the regions where Rain forest is now the climatic climax, similar vegetation has existed uninterruptedly since a very remote geological period. According to Posthumus (1931, see also discussion by van Steenis, 1935a), the Tertiary forests of Malaysia differed very little in their floristic composition, from the Rain forests of to-day.

THE SIGNIFICANCE OF THE RAIN FOREST FOR ECOLOGY

The Tropical Rain forest has been one of the last of the major formation-types of vegetation to be studied by plant ecologists, but it may well be that, as knowledge of its ecology increases, important principles will emerge which will throw much new light on the temperate plant communities with which ecologists have hitherto mainly concerned themselves. The vast wealth of species in the climax communities of the tropics imposes an obvious practical handicap on the plant ecologist, but has also important compensating advantages. The study of the highly complex relationships of the species with each other and with their environment in the floristically rich communities of the tropics cannot fail to reveal much that is new, especially concerning the laws governing the competition and dominance of species in mixed communities. The very large number of species in rain-forest associations offers an unrivalled opportunity, as will be seen in a later chapter, for disentangling the relation of plant form to environment: for example, the relation of leaf size and form to climate and microclimate. Though plant form is largely determined by genetical factors characteristic of the family, genus or species, it is also partly controlled by the environment, acting either indirectly through natural selection or directly during the development of the individual. In communities such as the rain forest, where there are numerous species of varied taxonomic affinity, but similar life-form, convergent (or epharmonic) resemblances between unrelated species are shown far more clearly than in communities consisting of only a few species. As van Steenis points out in the quotation at the beginning of this book, it would be more logical to begin ecological investigations with the unspecialized, floristically rich, plant communities of the equatorial region and to proceed from them to the specialized, impoverished and much modified communities of temperate regions, rather than vice versa. Tropical vegetation is still so little known that its significance for ecological science in general can be only dimly seen.

Botanists have tended to regard tropical vegetation as abnormal and temperate vegetation as normal because they are more familiar with the latter. W. H. Brown (1919) says: 'One raised in a temperate country will think of the deciduous broad-leaved forest as a usual one, and of the tropical forest as strange and bizarre. On the other hand, the inhabitants of the moist tropics look upon the deciduous forests in temperate regions as very strange. It is certainly reasonable to regard the broad-leaved evergreen forests of the tropics as the generalized type, and the deciduous forest of temperate regions as fitted to adverse winter conditions, just as the thorny vegetation of desert regions is suited to a dry climate. Nor is there any reason to regard the presence of palms, tree ferns and epiphytes as more peculiar than their absence, particularly as their absence is connected with unfavourable conditions. Likewise a dense forest of many species should not be regarded as more peculiar than a more open one composed of a few species.'

The Tropical Rain forest is the home *par excellence* of the broad-leaved evergreen tree, the plant form from which all or most other forms of flowering plants seem to have been derived. Converging evidence of many kinds indicates that our temperate floras have directly or indirectly a tropical origin, of which many temperate species still bear evident traces either in their phenology (see Diels, 1918) or in their structure. Bews (1927), for example, has found strong evidence that the moist tropical flora of Africa is older than that of the drier and cooler regions; his statistical comparison of fossil and modern floras suggests that from the beginning of the Cretaceous period until the end of the Tertiary a large part of the world had a 'phanerophyte climate', and that its vegetation was, at least in physiognomy, like the modern Tropical Rain forest. It may well be, as Bews believes, that the earliest angiosperms were similar ecologically and in their life-forms to the existing rain-forest flora, which may be regarded as an ancient type of vegetation from which the flora of the drier tropical and temperate regions has arisen relatively recently. The immense floristic richness of the Tropical Rain forest is no doubt largely due to its great antiquity; it has been a focus of plant evolution for an extremely long time.

Part I

STRUCTURE AND PHYSIOGNOMY

Man hat eine Pflanzengesellschaft noch nicht 'verstanden', wenn man weiss, unter welchen Bedingungen sie vorkommt. Viel wichtiger ist, zunächst zu ergründen, wie ein Pflanzenverein aufgebaut, wie seine Struktur ist. Die Vegetationskunde ist eine Formenlehre der Pflanzengesellschaft.

A plant community is not 'understood' if it is known merely under what conditions it is found. It is more important first to discover how it is built up, what is its structure. The science of vegetation is the study of the morphology of plant communities.

H. Meusel (1935, p. 269)

CHAPTER 2

STRUCTURE: SYNUSIAE AND STRATIFICATION

THE SYNUSIAE

A complex plant community is analogous (though admittedly only superficially) to a human society. The members of a human community form social classes, all the members of a given class standing in a similar relationship to the members of other classes and having a similar function in the society as a whole. Each human community thus has a characteristic social structure determined by the nature and the relative importance of the classes which compose it. In a like fashion the species in the more complex plant communities form ecological classes or groups. In the community as a whole the species are of varied stature and varied life-form, but the members of the same ecological group are similar in life-form and in their relation to the environment. These ecological groups, the analogues of the human social classes, will here be called synusiae, a term originally introduced by Gams (1918). A synusia is thus a group of plants of similar life-form, filling the same niche and playing a similar role, in the community of which it forms a part. In the words of Saxton (1924), who used the term in a slightly broader sense than Gams, it is an aggregation of species (or individuals) making similar demands on a similar habitat. The species of the same synusia, though often widely different taxonomically, are to a large extent ecologically equivalent.

In a plant community such as the temperate Summer forest the component synusiae are readily distinguished and have a relatively simple spatial arrangement. Commonly the woody plants form two layers or strata, one of tall trees, another of shrubs or smaller trees, while below these there are one or more layers of herbs and undershrubs (the ground or field layers) and a layer of mosses and liverworts. Besides these layers of self-supporting plants, each of which can be regarded as a separate synusia, there are climbers and epiphytes (the latter as a rule consisting entirely of non-vascular plants) and the synusia of saprophytes and parasites, which include a few flowering plants, as well as fungi and bacteria.

In the Tropical Rain forest the synusiae are more numerous and their spatial arrangement is far less obvious. Though tropical forests, like temperate, have a definite structure, they appear at first as a bewildering chaos of vegetation; in the often quoted phrase of Junghuhn, nature seems to show here a *horror vacui* and to be anxious to fill every available space with stems and leaves. A closer study shows, nevertheless, that in the Rain forest, as in other complex plant communities, the plants form a limited number of synusiae with a discernible, though complicated, arrangement in space. This arrangement is repeated as a pattern with only a small amount of variation throughout the Tropical Rain forest formation-type. The structure of primary Rain forest in America is essentially the same as in Africa or Asia. In all three continents the forest is composed of similar synusiae and, as will be seen later in this chapter, there is great similarity in their spatial arrangement. Species of corresponding synusiae in different geographical regions, as well as being alike in life-form, are to a considerable extent alike in physiognomy. The fundamental pattern of structure is thus the same through the whole extent of the rain forest.

As was seen in the last chapter, our knowledge of the rain-forest flora is still so imperfect that a complete biological spectrum for even a single limited area of rain forest cannot be constructed (whether the epharmonic life-form system of Raunkiaer is adopted or a purely physiognomic system such as that proposed by Du Rietz). Since the synusia is a group of plants of similar life-form, it follows that a final and complete classification of the rain-forest synusiae is not at present possible. Some kind of classification is, however, a practical necessity, and for the time being we must be content with an approximate and over-simplified scheme. That which will be adopted here is as follows; it is a modification of the scheme used by the author in previous works and has the merit of being convenient and easily applied.

Synusiae of the Tropical Rain forest

A. Autotrophic plants (with chlorophyll)
    1. Mechanically independent plants
        (a) Trees and 'shrubs'   arranged in a number of strata (layers)
        (b) Herbs

    2. Mechanically dependent plants
        (a) Climbers
        (b) Stranglers
        (c) Epiphytes (including semi-parasitic epiphytes)

B. Heterotrophic plants (without chlorophyll)
    1. Saprophytes
    2. Parasites

The basis of this scheme, as will be seen, is the means by which the plant satisfies its carbohydrate requirements. Each synusia represents a different method of succeeding in the struggle for food.

The autotrophic plants, which manufacture their own carbohydrates, are directly dependent on light for their existence and are divided into two groups according to their method of reaching it. The mechanically independent, or self-supporting, plants reach the light without assistance from other plants; the mechanically dependent plants (equivalent to the guilds or *Genossenschaften* of Schimper) cannot do so. The independent plants can be subdivided into several strata or layers, the number and structure of which are discussed below.

The dependent plants, on the other hand, are less clearly stratified; such tendency to stratification as they show is determined mainly by that of the independent plants which support them. One section of them, the synusiae of climbers, consists of plants which are rooted in the ground, but need support for their weak stems. Another, the epiphytes, are more or less short-stemmed plants which are intolerant of conditions at ground level and need to grow raised up and rooted on the stems and branches of other plants. One group of these, the semi-parasitic epiphytes (represented by the single family Loranthaceae), obtain water and mineral matter (though probably little or no organic food) from the trees on which they grow; the majority, however, depend on the supporting plant for mechanical reasons alone. The third section of dependent plants, here termed stranglers, are a transitional group. They begin life as epiphytes and later send roots to the ground; some of them reach a considerable size and may ultimately kill and supplant their 'host', as do, for example, the strangling figs (*Ficus* spp.). Stranglers thus begin life as dependent plants, but eventually become independent; they cannot be sharply separated from hemi-epiphytes, such as many Araceae, which send down roots to the ground, but never become self-supporting.

Lastly, there are the saprophytes and parasites, which together form the class of heterotrophic plants. They are not necessarily unresponsive to light, but as they obtain their organic food directly or indirectly from other plants, they are not dependent on it for their nutrition; they have, as it were, retired altogether from the struggle for light.

The subdivisions of these various groups will be further discussed in Chapters 4 and 5, where their physiognomic characters will be considered.

As far as is known there is no climax community to which the term Tropical Rain forest can properly be applied in which all of these ecological groups are not present. There is also no region of the Rain forest which possesses local synusiae not found elsewhere. The variations in the plan or pattern of structure which are met with consist chiefly of differences in the relative strength with which the various synusiae are represented. Thus in some rain-forest communities epiphytes are represented by more species and individuals, and show a wider range of form, than in others. Similarly, some Rain forests are richer than others in climbers. The spatial arrangement of the synusiae, which must now be considered, also varies to a greater or lesser extent.

STRATIFICATION

The plan of structure common to all climax Tropical Rain forest is most clearly manifested in the central feature of its architecture, the stratification of the trees, shrubs and ground herbs. It has long been a commonplace of the text-books that in tropical forests the trees form several superposed strata (the terms layer, story, canopy and tier are also used), while in temperate forests there are never more than two tree strata and sometimes only one, but the statement has been very variously interpreted and until recently its precise meaning has been far from clear. Sometimes it is stated categorically that there are three tree strata in the rain forest (according to a few authorities, more than three). Brown (1919, pp. 31–2), for example, describes the stratification of the Philippine Dipterocarp forest in these words: 'The trees are arranged in three rather definite stories. The first, or dominant, story forms a complete canopy; under this there is another story of large trees, which also form a complete canopy. Still lower there is a story of small scattered trees.' Some writers give the impression that the strata of the Rain forest are as well defined and as easy to recognize as in an English coppice-with-standards, but Brown continues: 'The presence of the three stories of different trees is not evident on casual observation, for the composition of all the stories is very complex and few of the trees present any striking peculiarities, while smaller trees of a higher story always occur in a lower story and between the different stories.'

There are also authors who state, or imply, that any grouping of the trees according to their height is arbitrary and that 'strata' have no objective reality. This is the point of view of Mildbraed (1922, pp. 103–4), who, with special reference to the forest of the southern Cameroons, says explicitly: 'It is often stated that the Rain forest is built up of several tiers or stories. These terms may easily give a wrong idea to anyone who does not know the facts at first hand. What is meant is merely that the woody plants can be grouped into 3, 4, 5 or perhaps more, height classes, according to taste. The space can indeed be thought of as consisting of height intervals of 5, 10 or 20 metres, and species are found which normally reach these height intervals when full grown. As, however, trees of all intermediate heights are present and the mixture of species is so great, these hypothetical height intervals never really appear as stories. It is truer to say that the whole space is more or less densely filled with greenery' (transl.). A similar opinion appears to be held by Chevalier (1917, pp. 354–5) and others.

In this chapter it will be shown that in most normal primary rain-forest communities tree strata actually exist and that they are usually three in number. The tree strata, though always present, are ill-defined and are seldom easy to recognize by casual observation. In addition to the tree strata, there is a layer of 'shrubs' and giant herbs (such as tall Scitamineae) and a layer of low herbs and undershrubs.

By a stratum or story is meant a layer of trees whose crowns vary in height between certain limits. In a several-layered forest each stratum will have a distinctive floristic composition, but since the forest is continually growing and regenerating, a proportion, perhaps the majority of the individual trees in the lower stories, will belong to species which will reach a higher stratum when mature. The crowns of young trees of a higher stratum and exceptionally tall ones of a lower will also be found between one stratum and the next; if such individuals are numerous, as is often the case, there will be little vertical discontinuity between neighbouring strata and the stratification will be much obscured. A tree stratum may form a continuous canopy or it may be discontinuous; that is to say, the crowns may be mostly in lateral contact with one another or they may be widely separated. The term canopy is sometimes used as a synonym for stratum; a canopy means a more or less continuous layer of tree crowns of approximately even height. The closed surface or roof of the forest is sometimes loosely referred to as the 'canopy'; in the Rain forest, as we shall see, this surface may be formed by the crowns of the highest tree stratum alone or (more frequently) by the highest and the second stories together.

The divergences of opinion as to the existence, nature and number of the rain-forest strata are mainly due to the difficulty of obtaining a clear view of the forest in profile. On the floor of the forest the observer is, as Mildbraed says, a prisoner. Visibility is very limited both vertically and horizontally. Above, the confusion of leaves, twigs and trunks only occasionally allows a glimpse of the crowns of the taller trees. Here and there, it is true, a gap in the canopy made by the death of a large tree will reveal rather more, or a river or a clearing will make it possible to see a whole cross-section of the forest from top to bottom. Such cross-sections, however, may be extremely misleading. When increased light is able to reach the lower levels of the forest, a vigorous outburst of growth takes place in the undergrowth, climbing plants grow down from the canopy and up from the ground, till the exposed surface becomes covered with a dense curtain of foliage which completely conceals the natural spacing of the tree crowns. The structure of the forest on a river bank is always different from that in the interior.

Attempts have been made to study the stratification of the Rain forest by plotting the frequency of trees in arbitrarily delimited height classes. Thus Booberg (1932) plotted the heights of all the numbered trees in the forest reserves of Java (using the measurements of Koorders); he found that the result gave a continuous curve, with no indication of modes of frequency at certain heights. From this he concluded that the several *étages* (stories) described by authors such as Diels and Rübel cannot be recognized in the forests of Java. A similar curve (Fig. 3), but showing a slight tendency to maximum frequency at a certain height class,[1] was obtained by Davis & Richards (1933-4, p. 367), using measurements of trees on clear-felled plots in Mixed Rain forest in British Guiana. That continuous curves do not disprove the existence of stratification is shown by Vaughan

[1] The 76-85 ft. (23-27 m.) class; this corresponds approximately with the B story (see below).

& Wiehe (1941, pp. 131, 137), who found that in the 'Upland Climax forest' of Mauritius the frequency of trees plotted against size (diameter) classes gave a continuous curve, though a very marked stratification was readily apparent to an observer (illustrated in their paper by an accurate perspective drawing).

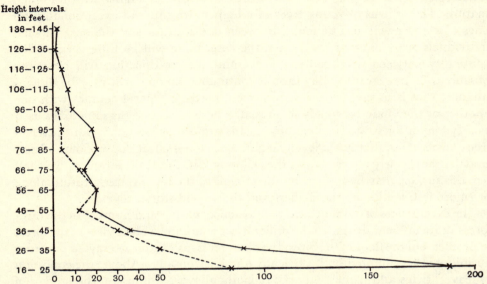

Fig. 3. Total heights and heights of lowest leaves of trees on clear-felled plots, Moraballi Creek, British Guiana. From Davis & Richards (1933–4, p. 367). The continuous line represents distribution of total heights among height intervals, the broken line the distribution of the heights of lowest leaves for all trees over 15 ft. (4·6 m.) on a total area of 400 × 50 ft. (122 × 15·3 m.).

The profile diagram

Because the direct observation of the stratification of the Rain forest usually offers insuperable difficulties, Davis & Richards (1933–4) adopted the device of constructing profile diagrams to scale from accurate measurements of the position, height, depth of crown and diameter of all the trees on narrow sample strips of forest. This technique, first applied to the forest of British Guiana, has since been used by various workers in several parts of the tropics and has proved a valuable means of studying and comparing the structure of tropical forest communities. The method may be briefly described as follows:

A narrow rectangular strip of forest is marked out with cords, the right angles being obtained with the help of a prismatic compass. In Rain forest the length of the strip should not usually be less than 200 ft. (61 m.); 25 ft. (7·6 m.) has proved a satisfactory width. All small undergrowth and trees less than an arbitrarily chosen lower limit of height are cleared away. The positions of the remaining trees are then mapped and their diameters noted. The total height, height to first (large) branch, lower limit of crown and width of crown of each tree are then measured. Often it is only possible to obtain these measurements by felling all the trees on the strip, and the trees must be felled in a carefully selected order, so that the heavier trees do not crush the smaller in falling, but it is sometimes possible to make measurements of sufficient accuracy from the ground by means of an Abney level. Felling has the advantage that herbarium material can be collected for the identification of the species.

In the last chapter it was pointed out that though Rain forest is most commonly mixed in composition, dominance being shared by a very large number of tree species, types of forest also exist in which a single species is dominant and forms a large proportion, or rarely the whole, of the highest stratum. The structures of Mixed and Single-dominant Rain forest differ in some respects, and the stratification of the two types will now be described separately.

Stratification of Mixed Rain forest

Figs. 4–6 are profile diagrams of typical Mixed Rain forests. Fig. 4 shows Mixed forest in British Guiana, Fig. 5 Mixed Dipterocarp forest in Borneo (without single-species dominance, but with 'family dominance' of the Dipterocarpaceae, see Chapter 10), Fig. 6 Mixed forest in Southern Nigeria. The British Guiana and Borneo examples are typical equatorial Rain forest (Rain forest in the narrow sense of Keay & Jones, Beard, etc., '*Forêt Équatoriale*' of the Belgians, see p. 338), while the Nigerian profile represents the slightly different climax type found in climates with a somewhat less evenly distributed rainfall ('Wet Evergreen forest', '*Forêt Tropicale*' of the Belgians). A careful study of the three diagrams will probably convey a clearer idea of the stratification of normal Rain forest than a description in words, but it will be desirable to describe the structure of each example briefly, supplementing the information obtained from the diagrams with notes made by direct observation in the respective areas.

Primary Mixed forest, Moraballi Creek, British Guiana (Fig. 4). Three strata of trees (which will be referred to as A, B and C, from above downwards) are shown in the diagram. The lowest (C) is continuous and fairly well defined; the upper two are more or less discontinuous and not clearly separated vertically from each other and in the original description (Davis & Richards, 1933–4, pp. 362–72) these were regarded as a single stratum of irregular profile. Most of the gaps in the highest stratum (A) are closed by trees in stratum B; strata A and B together thus form a complete canopy.

The height of the trees in A on the diagram is about 35 m. but elsewhere usually higher than this (to about 42 m.), that of B about 20 m., while stratum C includes trees between about 15 m. and the arbitrary lower limit of 4·6 m. (15 ft.), average height about 14 m. On the profile strip (135 ft. = 41 m. long) there are sixty-six trees over 15 ft. high; seven of these can be reckoned as belonging to the first stratum, twelve to the second and the remainder to the third.

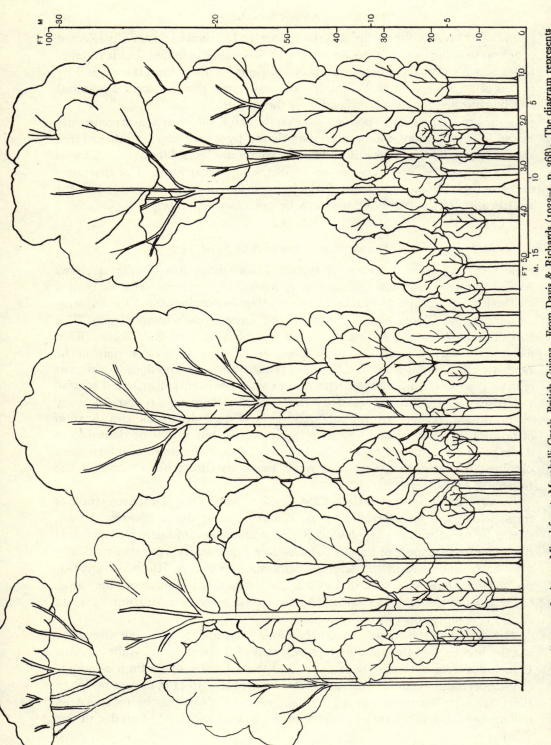

Fig. 4. Profile diagram of primary Mixed forest, Moraballi Creek, British Guiana. From Davis & Richards (1933–4, p. 368). The diagram represents a strip of forest 125 ft. (41 m.) long and 25 ft. (7·6 m.) wide. Only trees over 15 ft. (4·6 m.) high are shown.

The crowns of stratum A are only here and there in contact laterally, but it must be remembered that its canopy is more closed than it appears to be in the diagram because trees whose crowns overlap the sample strip, but whose bases are outside, are not shown. The trees of A belong to many species and families (Lecythidaceae, Lauraceae, Araliaceae, etc.). Their crowns are usually wider than deep and tend to be umbrella-shaped.

Stratum B is more continuous, but has occasional gaps. Like A it is composed of many species belonging to numerous families (mostly different from those of A). A considerable proportion of the trees are young individuals of species which reach the stratum A when mature. The crowns tend to be deeper than wide.

In stratum C there are very few gaps, and the density of foliage and branches is greater than at any other level of the forest, higher or lower. More than half the total number of individuals are young trees of species properly belonging to the higher strata; the remainder are small species peculiar to stratum C and mostly belong to families scarcely represented in strata A and B (especially Annonaceae and Violaceae). Both the young A and B trees and the true C species usually have long, tapering, conical crowns, much deeper than wide.

Below the three stories of trees represented on the profile diagram there are two other strata consisting chiefly or partly of woody plants; both are ill defined. The upper of these, the average height of which is about 1 m., may be called the shrub stratum (D), though the constituents include many young trees, small palms, tall herbs (Marantaceae etc.) and large ferns, as well as small woody plants of shrub or 'dwarf tree' life-form (see p. 76). The lowest stratum is the ground or field layer (E); it consists chiefly of tree seedlings, and herbaceous plants (dicotyledons, monocotyledons, ferns and *Selaginella*) form only an insignificant proportion of the total number of individuals. Like the shrub stratum, this layer is usually discontinuous, the constituents being very scattered, except in openings and occasionally where a social species forms a patch of closed vegetation.

There is no moss layer on the forest floor. Except for patches of mosses such as *Fissidens* spp. on disturbed soil (by overturned trees, armadillo holes, etc.), bryophytes are confined to the surface of living or dead trees (see p. 126).

As well as constructing a profile diagram, the forest stratification was examined by climbing a tall tree in forest similar to, but not actually adjoining, the profile strip. The following notes were made on the spot at a height of 110 ft. (33·6 m.); 'There is no flat-topped canopy; there are two more or less clear [upper] layers, but they are both discontinuous, so the general effect is very uneven. Any two tall trees may be separated from one another by one or more lower trees. The lower trees do not grow under, and are not much overshadowed by the higher. Practically all the lower trees of the canopy are covered and bound together with lianes. [These do not usually reach the trees of the highest stratum.] Apart from the ordinary upper canopy [i.e. stratum A] trees, there are rare ones which tower far above all others. Two such outstanding trees were seen whose whole crowns were well clear of all surrounding trees.' A view from this observation post is shown in Pl. II A; one of these 'outstanding trees' is visible on the horizon. No outstanding trees are shown in the profile diagram; such trees are probably not more frequent than 1 per sq.km. on the average, and their height is about 40–45 m. They belong to species rarely found in stratum A (e.g. *Hymenaea courbaril, Peltogyne pubescens*).

The results of direct observation from 110 ft. thus agree closely with the profile diagram. Both methods demonstrate that the forest is three-storied; the rare 'outstanding trees' do not constitute an independent stratum.

Primary Mixed Dipterocarp forest, Mt Dulit, Borneo (Fig. 5). The original description of the forest structure is given by Richards (1936). Here again three tree strata are recognizable, but instead of C being definite and A and B ill-defined, the highest stratum, though discontinuous, is well defined and sharply separated vertically from B, the division between B and C being difficult to draw. The average heights of the strata are about 35, 18 and 8 m. respectively. On the sample strip there are seven trees in stratum A and eighty-six in B and C (about forty-eight in B and thirty-eight in C); these numbers are not directly comparable with those for the Guiana profile, because the strip was longer (200 ft. = 61 m.) and because only trees 25 ft. (7·6 m.) high and over (instead of 15 ft. and over) were measured.

The crowns of A trees are mostly fairly well separated from each other and raised well above the very dense second stratum, hence, when the forest roof is looked at from above (e.g. from the crest of an escarpment), the individual crowns of the trees in A can be separately distinguished at a great distance. The majority of the trees in stratum A belong to the Dipterocarpaceae. The crowns are about 6 m. deep on the average and tend to be wider than deep. Stratum B is almost completely continuous, each crown usually being in lateral contact with several others; its canopy is almost as dense as that of C. The species here are very numerous and belong to a great variety of families; some immature dipterocarps are of course present, but the majority belong to other families. The crowns are about 4·5–6 m. deep and are mostly less wide than deep.

Stratum C is also continuous. It consists both of young individuals of species reaching strata A or B when mature and of small species which rarely exceed 15 m. in height; the former predominate. In both groups of species the crown tends to be very deep and narrow and conical in shape. The small species characteristic of this layer belong to many families; but it is interesting that, as in the Guiana forest, Annonaceae are abundant. The space between strata B and C is filled with trees of intermediate height. Observation over the whole district shows, however, that there are two quite separate groups of species, B species which average about 18 m. when mature and which do not flower when much smaller, and C species, about 8 m. high when mature, which begin to flower when very small. The absence of vertical discontinuity between the middle and lowest tree stories is therefore due partly to the abundance of immature B trees and partly to the presence of individuals of C species taller than the average.

Below the C tree stratum the vegetation becomes much less dense. No very clear stratification is evident in the smaller undergrowth, but it is convenient to divide it into a shrub stratum averaging 4 m. high, consisting of 'shrubs', palms and young trees, and a ground layer up to 1–2 m. high, consisting of tree seedlings and herbs, including ferns, *Selaginella* and rarely one or two species of very small palms. The number of species of herbaceous plants is considerable, but the tree seedlings outnumber them in individuals; on ten quadrats 1 m. square the total number of tree seedlings and other woody plants was 184 and of herbaceous plants 135 (shoots). The density of the ground layer is uneven; over large areas it is represented only by widely scattered plants; here and there it is fairly dense, though seldom as dense as in a deciduous wood in Europe.

There is no moss stratum on the ground in the Mixed Dipterocarp forest at low altitudes, but mosses are sometimes found growing on the ground on very steep slopes.

Primary[1] *Mixed forest* ('*Wet Evergreen forest*'), *Shasha Forest Reserve, Nigeria* (Fig. 6). The structure of this is described in detail by Richards (1939). Like the forest of Guiana and Borneo, the Nigerian forest has three tree strata. Only the lowest story (C), of trees up to 15 m. high (average about 10 m.), is continuous. Above it there is an irregular mass of trees of various heights, the tallest of which is 46 m. high. The crowns of these taller trees are sometimes in contact laterally, but there is no closed canopy above stratum C. Observation over the whole area of which the sample strip formed a part showed that this irregular mass could be separated into two strata, A of trees 37–46 m. (average about 42 m.), and B of trees 15–37 m. (average about 27 m.) high respectively. The former is represented by only one tree in the profile diagram.

The first stratum thus consists of very large trees scattered through the forest (though much more densely than the 'outstanding trees' of Guiana). The crowns of these trees are umbrella-shaped and extremely heavy; they are up to 25 m. or more wide. They are rarely in lateral contact and are raised well above those of stratum B. Story A consists of comparatively few species (in the area studied chiefly *Lophira procera* and *Erythrophleum ivorense*).

Stratum B is also open and the crowns are only occasionally in contact. In the diagram the gap in this layer beneath the crown of the large stratum A tree

[1] Recent work by the author (1948, unpublished) makes it seem likely that this forest has suffered disturbance in the past and is old secondary rather than truly primary; it is, however, probably sufficiently mature to differ little in structure from the primary (climax) forest of the region. The same is probably true of nearly all so-called virgin or primary forest in Nigeria, and perhaps in the whole of West Africa. Forest which has never at any time been cultivated exists in West Africa on swampy sites (Fresh-water Swamp forest, see Chapter 13), but elsewhere only on extremely limited areas, mostly on steep rocky slopes, etc .

is noteworthy. The trees of stratum B have small narrow crowns, usually under 10 m. wide. The species are numerous and belong to a wide range of families.

Stratum C is very dense and almost without gaps. The crowns are packed closely together and are usually tightly bound together by lianes. The majority of the trees in this layer are species which never reach a higher story; they belong to various families, but there is a tendency for a single species (the actual species varying from place to place) to be locally dominant. The remaining trees in the stratum are chiefly young B species; young A species seem to be strikingly rare. The C species, like those of the corresponding layer in Guiana and Borneo, mostly have small conical crowns, but old individuals sometimes have remarkably wide and heavy crowns.

The shrub stratum (D) is very indefinite. It consists largely of young trees belonging to strata B and C, so there is no clear division between this layer and the lowest tree story. Species properly belonging to stratum D (most of which are 'dwarf trees' rather than true shrubs, see p. 76) are few. The density of the shrub stratum is very variable; in undisturbed forest it is never so dense as to make progress difficult, and in some places both shrubs and ground layers are almost wanting.

The lowest layer of the forest is the ground layer (E), which consists of plants varying from a few centimetres to 1 m. or more high. The components are tree seedlings, dicotyledonous and monocotyledonous herbs, and ferns, the first generally predominating. This stratum is even more unevenly developed than D. Large stretches of the forest floor may be almost completely bare, but in places, especially in openings, the ground may be concealed by a dense growth of herbaceous plants and tree seedlings. There are no mosses on the ground.

A photograph taken at 78 ft. (24 m.) above ground in forest similar to the sample strip from which the profile was drawn is reproduced on Pl. III. Like the diagram this shows the dense mass of small trees with interwoven crowns about 9–12 m. high. Taller trees rise above this compact layer to various heights, but they do not form a closed canopy at any level, so that above 11 m. it is possible to see clearly for some distance in any direction. When climbing this tree one seemed to emerge into full daylight as soon as the dense C stratum was passed. The picture of the stratification obtained from this tree is thus very similar to that disclosed by the profile diagram.

The great discontinuity of the upper two tree strata is a remarkable feature of this forest. A very open A stratum is perhaps a consequence of the relatively severe seasonal drought, and may be a general feature of West African evergreen forest, since Aubréville (1933) describes the 'Closed forest' of the Ivory Coast as consisting of a dense mass of vegetation 20–30 m. high dominated by scattered taller trees; he compares its structure to a *taillis-sous-futaie* (coppice-with-standards). It is also noteworthy that in the 'Evergreen Seasonal forest' (*Carapa-Eschweilera* association) of Trinidad (Beard, 1944a, 1946b; see also Fig. 7), the canopy of the highest tree stratum is very discontinuous. On the other hand, since the forests of the Shasha Reserve have been much modified by selective felling and native cultivation it is possible that the openness of the upper strata is due to removal of a proportion of the larger trees, or that the community studied is an advanced stage in a secondary succession (see Chapter 17) and not in fact a true primary forest.

The three examples of Mixed Rain forest which have just been described are thus similar in the main features of their stratification, though they show some differences which may be of significance. From these data and from the other information available the following general statements about the structure of Mixed Rain forest seem justified:

(i) There are five strata of independent plants (in the sense of p. 20 above): the three tree layers, which we have termed the A, B and C strata respectively,[1] consisting entirely of trees, a D layer consisting mainly of woody, but often partly of herbaceous, species, which it is convenient to call the shrub stratum, though only a few of its components are shrubs in any exact sense of the word, and lastly the ground or field layer (E) of herbs and tree seedlings.

(ii) The height of each stratum varies from place to place, but not within wide limits. Thus the height of stratum A is about 30 m. or more in the Guiana forest described above, about 35 m. in the Borneo example and about 42 m. in the Nigerian example. Similarly, the height of the B strata is 20, 18 and 27 m. respectively, and the C 14, 8 and 10 m. respectively in the three examples.

(iii) Stratum A usually has a more or less discontinuous canopy, though there is considerable variation in this respect between the three profiles described; possibly in the tallest and most luxuriant Rain forest (e.g. in parts of the Malay Peninsula where the average height of the highest tree story is said to exceed 200 ft. (61 m.)), this layer may be practically continuous. There is some evidence that the A stratum becomes increasingly discontinuous as the climatic limits of rain forest are approached (see Chapter 15). The B stratum may be continuous or more or less discontinuous; C is always more or less continuous and is often the densest layer of the forest.

(iv) A vertical discontinuity between the canopies of neighbouring strata may or may not be apparent. Thus in the Guiana forest there is some discontinuity between the B and C strata, but little between A and B; in the Bornean forest there is a conspicuous gap between the canopies of A and B, but none between B and C; in Nigeria, as in Guiana, the main discontinuity is between B and C. The vertical limits of the shrub and ground layers are never very clearly defined.

(v) Each stratum in the forest has a different and characteristic floristic composition, but in all the strata except A and B young individuals of species which reach higher strata when mature form a large proportion of the total number of individuals.

[1] To avoid confusion with the ecological use of 'dominant' and 'subdominant', these terms are not applied here to strata, as is the common practice in forestry literature.

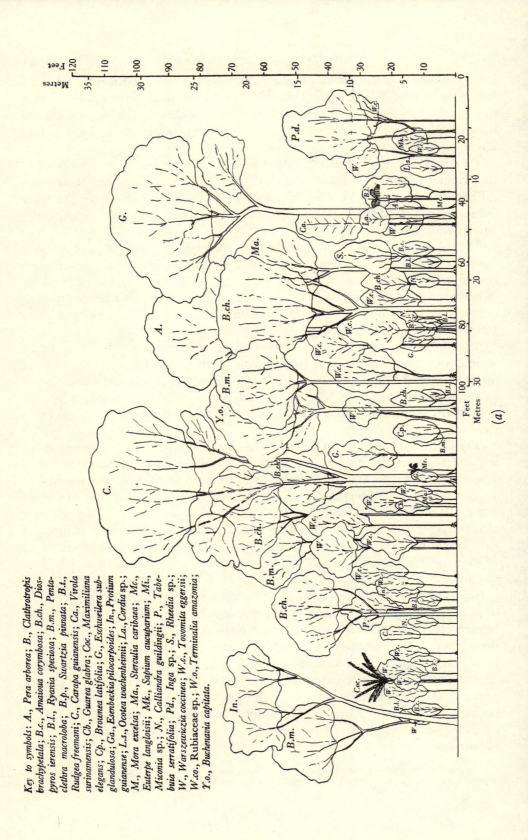

Key to symbols: A., Pera arborea; B., Clathrotropis brachypetala; B.c., Amaioua corymbosa; B.ch., Diospyros ierensis; B.l., Ryania speciosa; B.m., Pentaclethra macroloba; B.p., Swartzia pinnata; B.t., Rudgea freemani; C., Carapa guianensis; Ca., Virola surinamensis; Cb., Guarea glabra; Coc., Maximiliana elegans; Cp., Brownea latifolia; G., Eschweilera subglandulosa; Ga., Esembeckia pilocarpoides; In., Protium guianense; L.s., Ocotea wachenheimii; La., Cordia sp.; M., Mora excelsa; Ma., Sterculia caribaea; Mc., Euterpe langloisii; Mk., Sapium aucuparium; Mi., Miconia sp.; N., Calliandra guildingii; P., Tabebuia serratifolia; Pd., Inga sp.; S., Rheedia sp.; W., Warszewiczia coccinea; W.c., Tovomita eggersii; W.co., Rubiaceae sp.; W.o., Terminalia amazonia; Y.o., Buchenavia capitata.

(a)

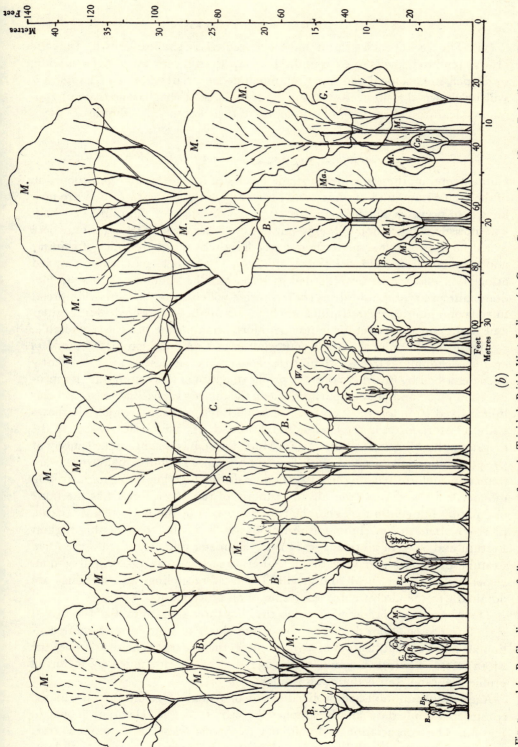

Fig. 7 a and b. Profile diagrams of climax evergreen forest in Trinidad, British West Indies. (a) Crappo-Guatecare forest (*Carapa-Eschweilera* association). (b) Mora forest (*Mora excelsa* consociation), Mayaro District. From Beard (1946a, figs. 2 & 3). Each diagram represents a strip of forest 200 ft. (61 m.) long and 25 ft. (7·6 m.) wide.

(vi) The trees of each stratum have a characteristic shape of crown. In A the crowns tend to be wide or even umbrella-shaped, in B they are as deep as wide, or deeper, in C conical and tapering, much deeper than wide. In Chapter 4 it will be shown that in each stratum the trees have other characteristic physiognomic features which are not a function of their taxonomic affinities.

Stratification of Single-dominant forests

Climax communities, in which a single species of tree forms a large proportion of the whole stand and in some cases over 80 % of the highest stratum, are now known to occur in all the main geographical divisions of the rain-forest belt. The floristic composition of these communities is discussed in Chapter 11, where it will be shown that some of them cover hundreds of square miles of country and may be termed consociations, while others are local patches in a mixed association and are better regarded as societies. The structure of these single-dominant forests probably always differs rather widely from the more widespread mixed communities, their stratification being, as might be expected, more regular and well defined. As yet, little precise information on the subject is available.

The stratification of the Mora consociation of Trinidad, dominated by *Mora excelsa*, a leguminous tree which may reach a height of at least 58 m., has recently been described by Beard (1946*a*). A profile diagram of this community is reproduced in Fig. 7 and, for comparison, one of the mixed *Carapa-Eschweilera* association (Mixed Rain forest, Evergreen Seasonal forest of Beard) which alternates with the Mora forest in the lowlands of Trinidad. As can be clearly seen in the diagram, the highest stratum of the Mora forest has a continuous canopy at 37–43 m. above the ground, with few or no gaps. The individual crowns 'are shaped in conformity with the adjacent ones, fitting together into a most striking mosaic....Viewed from the air the canopy of the Mora forest has the same undulating but continuous character as the waves of the sea' (Beard, 1946*a*, p. 173). *Mora excelsa* commonly forms 85–95 % of the trees in this highest stratum and 62 % of all trees 1 ft. (30 cm.) diameter and over. Below the *Mora* stratum there are two very discontinuous tree strata at 12–25 and 3–9 m. respectively. In the shrub and ground layers *Mora* saplings and seedlings are dominant to the almost entire exclusion of all other plants.

The Mora consociation of Trinidad thus has three tree strata, like the Mixed Rain forest associations described above, but the A stratum is much more continuous and the stratification as a whole altogether clearer. The structure of the Mora forest of British Guiana (see p. 237) has not been studied by means of profile diagrams, but is doubtless similar.

A previously unpublished profile diagram of Wallaba forest in British Guiana, constructed from data kindly supplied by Mr T. A. W. Davis, is shown in Fig. 8*a*. This consociation is dominated by *Eperua falcata*, a leguminous tree, and two other species of *Eperua* (*E. grandiflora*, *E. jenmani*) are also abundant.

The composition and characteristics of this very distinctive community are described on p. 240 (see also Davis & Richards, 1933–4). The diagram shows that, as in the Mora forest, the A stratum is much more even than that of Mixed forest, and is almost continuous; species of *Eperua* form a large majority of the trees in this layer. Below the highest stratum there is a sharply delimited layer of small trees about 8–15 m. high, representing the C stratum of Mixed forest, but a layer at an intermediate height, corresponding to the B stratum, is scarcely recognizable. The trees may thus be said to be practically two-layered and the forest to have an AC structure (as compared with the ABC structure of Mixed forest). Beneath the tree strata there is a rather dense shrub layer and a sparse ground layer of herbs and seedlings. Seedling and sapling *Eperua* are very abundant in the lower strata. An approach to an AC structure, very similar to that of the Guiana Wallaba forest is also shown by the *Dacryodes-Sloanea* forest of the Lesser Antilles (Fig. 8*b*), which also shows dominance (see Table 30, p. 261) of a single species.

Though in the Mora and Wallaba consociations the presence of a single strongly dominant species leads to development of a denser A stratum and the partial or complete suppression of the B stratum, in the Ironwood forest of Borneo and Sumatra (an apparently climax community dominated by the Borneo ironwood, *Eusideroxylon zwageri*, see p. 258) the B stratum, to which the dominant belongs, is extremely dense and A almost absent. Gresser (1919) described the forest as having a flat compact roof supported by trunks 15–20 m. high; this roof is formed chiefly by the crowns of *Eusideroxylon*. Only rarely is this canopy pierced by an occasional tall *Koompassia*, *Shorea* or *Intsia*, rising far above the general level. The undergrowth is quite open, but young individuals of the dominant are extremely abundant.

In the Rain forest of tropical Africa two widespread single-dominant climax communities are known, the *Cynometra alexandri* consociation and the *Macrolobium dewevrei* consociation. Both cover vast tracts of country in the eastern Belgian Congo, and the former also extends into Uganda. A detailed description of the structure of the *Cynometra* forest, illustrated by a profile diagram, has recently been published by Eggeling (1947), and Prof. J. Louis has kindly allowed me to examine unpublished profile diagrams of both consociations drawn by himself and Dr C. Donis in the Congo. The *Cynometra* consociation has a well-defined three-layered structure and differs from the Mixed forest chiefly in that in the latter the A story is composed of species mostly different from those of the B story, while in the former the majority of the trees in both the A and B stories are *Cynometra*. The average level of the top of the A story, which forms an almost continuous canopy, is 36 m., though occasional emergent trees rise considerably higher. Below this there is a middle layer of trees 11–21 m. high, which is fairly continuous except immediately below trees of the A stratum, and a C stratum, of regularly spaced trees up to about 11 m. high which does not form a closed canopy.

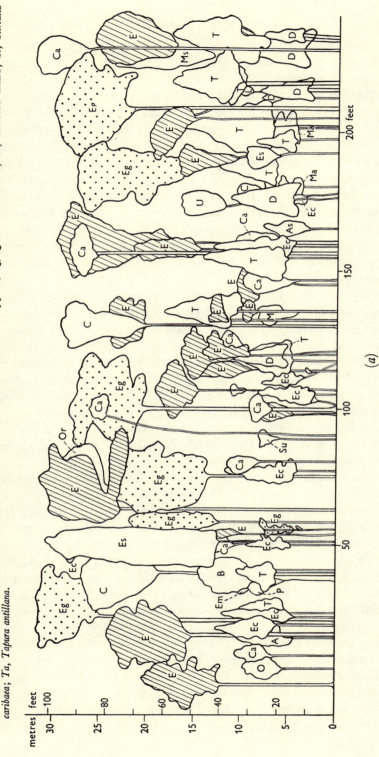

Key to symbols: (a) Am, Amaioua guianensis; As, Aspidosperma excelsum; B, Byrsonima sp.; C, Cassia pteridophylla; Ca, Catostemma sp.; D, Duguetia neglecta; E, Eperua falcata (soft wallaba); Eg, Eperua grandiflora (ituri wallaba); Ec, Ecclinusa psilophylla; Em, Emmotum fagifolium; Es, Eschweilera sp.; L, Licania heteromorpha; M, Matayba inelegans; Ma, Marlierea schomburgkiana; O, Ocotea sp.; Or, Ormosia coutinhoi; P, Pouteria sp.; Sw, Swartzia sp.; T, Tovomita cephalostigma; U, Unidentified. (b) Ab, Aniba bracteata; Bp, Beilschmiedia pendula; D, Dacryodes excelsa; Fo, Faramea occidentalis; Gm, Guarea macrophylla; Ii, Inga ingoides; L, Lauraceae (unidentified); Pm, Pouteria multiflora; Ps, Pouteria semecarpifolia; Q, Quararibea turbinata; Sa, Simaruba amara; Sc, Sterculia caribaea; Ta, Tapura antillana.

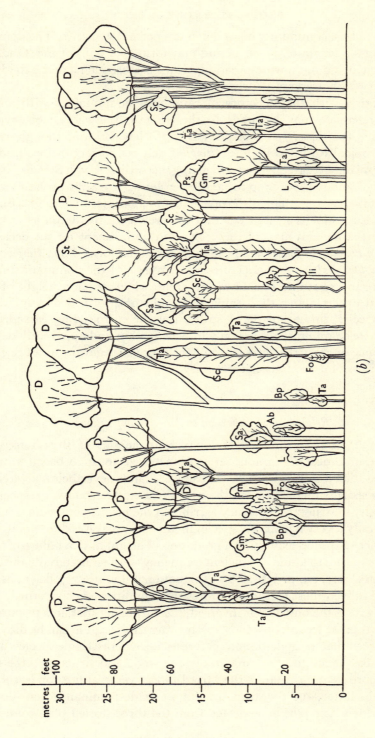

Fig. 8a and b. Profile diagrams of primary Rain forest. (a) Wallaba forest (*Eperua* consociation), Barabara Creek, Mazaruni River, British Guiana. After T. A. W. Davis (unpublished). (b) *Dacryodes–Sloanea* association, Dominica, British West Indies. After Beard (1949, fig. 8). Each diagram represents a strip of forest 25 ft. (7·6 m.) wide.

The *Macrolobium* community has a much less normal structure. The dominant species reaches a height of 35–40 m. and may form over 90% of the stand. It is much branched and has a widespreading crown casting such a dense shade that all the other strata are very poorly developed.

In connexion with the structure of rain-forest communities with a single dominant, reference may be made to the *Altingia excelsa* forest of Java and Sumatra and the *Eucalyptus deglupta* forest of New Britain, though neither is climax Tropical Rain forest. The *Altingia* forest is a society or localized consociation in mixed evergreen forest, occurring, according to van Steenis (1935 a), at altitudes of about 1000–1700 m. above sea-level; it is thus probably a type of Submontane, rather than of Tropical Rain forest (see Chapter 16). The remarkable feature of its structure is that the crowns of the dominant tree rise clear above the general mass of the vegetation and, except for an occasional individual of *Podocarpus imbricata*, the highest stratum is formed by *Altingia* alone. In the *Eucalyptus deglupta* forests (Lane-Poole, 1925 a, b) the dominant similarly forms a pure top stratum. The *Eucalyptus* trees, which reach a height of some 70 m., stand up far above the compact second story, formed of a mixture of evergreen species, similar to the A stratum of the surrounding Mixed forest. The *E. deglupta* community occurs only in New Britain and apparently depends for its existence on recurrent fires; it must therefore be regarded as a fire climax and not a true climatic climax community.

Relation of stratification to floristic composition

Salisbury (1925) has shown that the maximum heights of British trees and shrubs form an almost continuous series from the 36 m. high beech to undershrubs of less than a metre. Natural British woodlands nevertheless always have three rather sharply defined strata, the tree, shrub and herb layers respectively. This simplicity of structure Salisbury ascribes to the poverty of the flora and the presence of only a very few woody species in any one area; it is due 'not so much to the absence of species exhibiting a gradation of height as to a failure of all but a few woody species to attain a sufficient frequency to affect materially the woodland structure' (loc. cit. p. 336). This explanation is doubtless sufficient as far as British, and indeed most temperate,[1] forest communities are concerned, but if floristic poverty is the only factor involved, it is surprising that, in a tropical rain forest with its great wealth of woody species, any stratification can be discerned. It has been seen that in single-dominant forests, i.e. where a single species 'attains a sufficient frequency to affect materially the woodland structure', the stratification is more clear-cut than in the Mixed forest; yet stratification is still quite evident in the latter, and the structure of most single-dominant forests does not depart as widely as might be expected from the three-storied plan common to

---

[1] Some temperate forests have more than one tree stratum (see E. W. Jones, 1945).

the Mixed forests. It must be admitted, it seems, that the factors underlying the stratification of forest communities are as yet not fully understood.

The structure of a climax community in equilibrium with its environment is the expression of a complex balance, in which the struggle for light and probably other factors, such as root competition (see p. 190), of which we know very little, play important parts. The nature of the regeneration process, which will be considered in the next chapter, is also no doubt intimately connected with the *raison d'être* of forest structure.

CHAPTER 3

REGENERATION

Since the Tropical Rain forest is a climatic climax, it must, by definition, be in a state of equilibrium. When the trees die they are replaced by others of the same or different species; the floristic composition of a small area may vary in the course of years, but the average composition over a large area remains the same. Thus, though the individuals change, the community as a whole persists; indeed, in at least part of its area its structure and floristic composition have probably remained much as they now are for thousands, perhaps millions, of years.[1]

Although, apart from human interference, the Rain forest is stable, its equilibrium is dynamic and only relative. On the human time-scale its composition appears constant, but like all climax plant communities the Rain forest is subject to very slow and gradual change. As well as changes imposed by secular variations of climate, there are others due to the normal evolution and spread of species. New species evolve, others die out or come to occupy more restricted areas. Such changes are so slow that we can infer that they take place only from the facts of geographical distribution. Occasionally an aggressive species may find an opportunity for relatively rapid expansion into new areas. This appears to be happening, as Beard (1946a) and others believe, with *Mora excelsa* in Trinidad. This species is probably a comparatively recent arrival in the island and there is evidence that the Mora consociation is invading and replacing the surrounding Mixed forest at the present day. As far as is known, a change of this kind is uncommon; in general, rain-forest communities seem able to reproduce themselves, to all appearances indefinitely, without change.

The process of natural regeneration in tropical forests is no doubt exceedingly complex, and, though its practical importance to the forester is obvious, surprisingly little is known about it. Much of what has been written about the so-called 'natural regeneration' of Rain forest refers to the reproduction of a few economic species under conditions rendered more or less unnatural by the exploitation of timber. Before regeneration under these artificial conditions can be understood or controlled scientifically we need to know what happens under undisturbed conditions, and information about this is extremely scanty.

Among the first problems which suggest themselves are: What is the average age at death of trees in the different strata? What is the normal age-class representation of the chief dominants in a mature undisturbed Rain forest and how does the relative abundance of the species differ in different age-classes?

---

[1] In Trinidad, for example, the Miocene flora seemed to have been very little different from that of the present day (Hollick, 1924; Berry, 1925).

How does the growth rate of a large rain-forest tree vary during the successive stages of its development? At what stage does the heaviest mortality (and hence the most intense natural selection) occur? Finally, it may be asked, does the floristic composition of a *small* area of forest remain the same or is a Mixed Rain forest to be regarded rather as a large-scale mosaic, the species on a given area being succeeded not by the same but by a different combination of species? The solution of these and similar problems are among the most urgent tasks confronting the rain-forest ecologist at the present moment. Owing to the lack of data all the views expressed in this chapter must be considered tentative.

DEATH OF OLD TREES AND FORMATION OF GAPS

Very little is known of the average and maximum ages reached by rain-forest trees. In tropical forests with a marked seasonal drought the age of the trees can sometimes be determined approximately by means of the annual rings (this may even be possible with a few species in the 'Evergreen Seasonal' type of Rain forest), but in typical Rain forest with a relatively non-seasonal climate distinct annual growth rings are not formed (see p. 198). The age of the trees can therefore be found only by extrapolation from a series of careful measurements of the growth rate of individuals of different sizes over a period of several years—at best an unsatisfactory method. The observations of Brown (1919, p. 156) on the growth rate of *Parashorea malaanonan*, an important dominant of the Philippine Dipterocarp forest, show that an average individual of this species reaches a diameter of 80 cm. in 197 years; as no measurements are given of trees larger than 80 cm. it may be assumed that the greatest age reached was about 200 years. From measurements of growth rate in Malaya, Watson (1937) estimates the average maximum age of another dipterocarp, the relatively fast-growing *Shorea leprosula* (meranti tembaga), at 250 years. No other figures of this kind appear to exist. According to the measurements of Brown & Matthews (1914) second (B) story trees in the Philippines in general show a slower growth rate than the dominant dipterocarps, but no estimates of their maximum age have been made. The fast-growing, strongly light-demanding trees characteristic of young secondary forest have a much shorter life than primary rain-forest species (see p. 385).

When a large rain-forest tree dies it leaves a gap in the stratum to which it belongs, and such gaps play a very important part in the process of regeneration. When the tree dies slowly limb by limb, only a small gap is formed, but a large and apparently still vigorous tree is often overthrown by a tornado or sudden gust of wind; the gap may then be much larger. In undisturbed forest naturally formed gaps are numerous, and the remains of fallen trees are seen everywhere, though owing to the rapid rate of decay they disintegrate more quickly (unless the wood is exceptionally resistant) than in temperate forests.

The formation of a gap in one of the upper strata leads to the development of a dense patch of undergrowth, stimulated by the increased illumination and perhaps by the locally diminished root competition. Of the young trees in this patch of undergrowth the more light-demanding (intolerant) species respond more quickly than the shade-bearing (tolerant) species, which may be temporarily or permanently suppressed. In the larger gaps, besides species which will occupy a permanent place in one of the tree stories, there will be rapidly growing secondary forest species with good means of dispersal, which acquire a temporary foothold. Thus *Cecropia* spp. spring up in primary forest gaps in South America; *Musanga cecropioides* plays a similar part in Africa and species of *Macaranga* in Malaysia. In the gaps in the climax forest these species are soon replaced by the slower growing but more shade-tolerant primary forest dominants; in extensive clearings they become the dominants of the early stages in the secondary succession, and here, too, they eventually give place to primary forest species (see Chapter 17). A natural gap is thus the scene of a secondary succession on a very small scale.

The relationship between the normal process of regeneration in gaps and the secondary succession is illustrated by the interesting observations of Kramer (1933) on the upland rain forest of Mt Gedeh in Java. In artificial openings less than 10 ares in extent the existing regeneration of the primary forest dominants survived and made good growth. If the openings were 20–30 ares in extent, however, the regeneration was completely suppressed by the luxuriant growth of secondary forest species. Similarly Blanford (1929) noted that the best regeneration of dominants in the evergreen forests of Malaya was found in gaps not more than 20 ft. (6 m.) across; in larger gaps there was no regeneration at the centre.

The formation of gaps by the death of old trees is a normal part of natural regeneration and is probably necessary for the survival of the more light-demanding species playing a part in the composition of the climax forest, but the more extreme shade-tolerant species appear to be able to regenerate without the help of gaps. Thus among climax rain-forest trees in British Guiana, species which frequently overtop most of the other trees of the A story (p. 28), e.g. *Hymenaea* spp., are light-demanders and seem absolutely dependent on gaps for their regeneration; such species are rarely seen as young individuals in the normal shade of the forest. Others, e.g. *Ocotea rodiaei*, need at least a small gap to become established. Others again, such as *Licania venosa*, show a wide range of tolerance and appear to be able to regenerate both in gaps and under an unbroken canopy. The more extreme shade-tolerant species are intolerant of exposure to strong light (and fluctuating humidity?) and cannot survive in extensive clearings. Seedlings of *Ocotea rodiaei*, for example, make little growth in large clearings and appear stunted by drought (T. A. W. Davis).

AGE-CLASS REPRESENTATION

Whether helped by gaps or not, the regeneration of forest trees requires the presence of a sufficient number of seedlings, saplings and 'poles' to ensure replacement. Though the age-class representation of rain-forest trees cannot be directly studied (owing to the difficulty of determining their age), a superficial study of the size-class representation shows that among common dominants there is an enormous variation in the abundance of young stages from species to species (and sometimes in the same species from place to place). A species which forms a large proportion of the upper stories may be abundantly represented as a seedling and young tree, while another equally abundant species may be very poorly represented; sometimes a species abundant as a seedling is quite absent as an adult. In extreme cases a common dominant of the A or B strata may seem to lack regeneration altogether. In the evergreen forest of the Ivory Coast, for example, young individuals of certain species are so scarce that the natives say that these trees '*ne font jamais des petits*' (Aubréville, 1938). The possible significance of this puzzling state of affairs, which is characteristic, but by no means exclusively so, of Africa, is discussed below. In the 'Evergreen Seasonal forest' of Trinidad the species commonest in the tree stories appear to be also the commonest among the seedling population (Beard, 1946*b*, p. 63).

In Mixed Rain forests, as was pointed out in the last chapter, in the lower strata (C, D and E) young individuals of species belonging to the A and B strata outnumber those of species which never grow above these levels, but the evidence suggests that A species are in general much less well represented than B species. In Single-dominant forests, on the other hand, seedlings and other young stages of the dominant species are nearly always very abundant. Thus in the *Mora gonggrijpii* consociation of British Guiana and the *M. excelsa* consociation of Guiana and Trinidad the dense thickets of seedlings are a characteristic feature. Table 4 shows the size-class representation in sample plots of the four single-dominant forest types at Moraballi Creek, British Guiana. It will be seen that the dominant species is well represented in all the smaller size-classes except in the *Ocotea* consociation, in which, though seedlings are 'fairly abundant', established regeneration ('poles') is characteristically scarce. Even in these single-dominant communities there is no close relation between the abundance of seedlings and the abundance of mature trees. A great abundance of seedlings of the dominant is a feature of other Single-dominant forests, e.g. the *Eusideroxylon* consociation of Borneo and Sumatra and the *Cynometra alexandri* consociation of Uganda[1] (Eggeling, 1947). In the *Dryobalanops aromatica* societies of Sumatra (p. 259) the ground is everywhere covered with seedlings (van Zon, 1915).

---

[1] But in the *Parinari excelsa* consociation, a single-dominant community of lowland rain-forest type occurring in Uganda between 4500 and 5000 ft. (1400–1500 m.) (see p. 258), both seedlings and small saplings of the dominant are very scarce (Eggeling).

The differences in size- (and age-) class representation between different rain-forest trees are doubtless a matter of specific (hereditary) idiosyncrasy. They depend on the reproductive behaviour of the species and also on the ecological tolerance (temperament) of its young stages.

TABLE 4. *Size-class representation of dominant species on sample plots* 400 × 400 *ft.* (122 × 122 *m.*) *of four consociations at Moraballi Creek, British Guiana.* (After Davis & Richards, 1933–4)

Consociation	Average no. of seedlings under 2 m. high per m.²	No. of young trees under 4 in. (10 cm.) diam. and over 15 ft. (4·6 m.) high	No. of trees 4 in. (10 cm.) diameter and over Diameter class				
			4–8 in. (10– 20 cm.)	8–12 in. (20– 31 cm.)	12–16 in. (31– 41 cm.)	16–24 in. (41– 61 cm.)	24 in. (61 cm. and over)
Mora excelsa (Mora forest)	4·7	120*	28	19	17	24	21
M. gonggrijpii (Morabukea forest)	11·7	318	39	17	9	38	16
Ocotea rodiaei (Greenheart forest)	['Seedlings fairly abundant']	16*	2	5	10	38	18
Eperua falcata (Wallaba forest)	1·8	80*	47	43	37	64	3

* On these plots the trees of this size-class were counted on two strips together equal to one-eighth of the total area and the figure obtained multiplied by 8.

Different species of rain-forest trees differ widely in the frequency and abundance of seed production. Some species flower once a year, others at longer or shorter intervals (see Chapter 8). Mr T. A. W. Davis says of the British Guiana forest: 'Seed production is usually seasonal in the canopy, but in the undergrowth flowers and fruit are often borne almost continually in small quantities [cf. Chapter 8] or at irregular but frequent intervals. The majority of those which form the main canopy [strata A and B], however, flower either annually, biennially or even thrice a year, though seed is not always set, or if set, ripened; they probably produce good seed in fair quantity at least once in three years. The commonest species generally seed frequently; some, e.g. most species of *Eschweilera*, *Licania venosa* and *Ocotea rodiaei*, flower once or twice every year and usually ripen some fruit, though they may bear well only every second or third year; others, e.g. *Mora excelsa* and *M. gonggrijpii*, flower at intervals of roughly eighteen months or two years and normally have a good crop of seed each time they flower.... Flowering by no means always results in the ripening of even a little seed. Sometimes no fruit sets, presumably because the weather is unfavourable for fertilization. This seems to happen especially to species with a very short flowering season.... When seed is set wastage is, as a rule, not serious from such causes as weather and insects, sufficient being produced to allow for the loss of a considerable proportion of the crop. A number of trees, however, suffer heavy loss through parrots which feed principally on unripe fruit and are particularly fond of leguminous seeds. A flock of parakeets (*Brotogeris chryso-*

pterus) was seen feeding in a *Bombax surinamense* daily for weeks together on its soft and juicy unripe seeds; it is probably not a rare occurrence for certain trees to be stripped in this way of practically the whole of their crop.' Similar observations could doubtless be made in other regions of the Rain forest. Good and bad seed years are a common feature and are not necessarily dependent on intermittent flowering; they occur, for instance, in some Malayan dipterocarps, in the West African *Triplochiton scleroxylon*, and doubtless in many other rain-forest trees.

Germination seldom seems to be a critical stage in the establishment of regeneration. The seed of rain-forest trees usually shows a high percentage germination which may be of the 'simultaneous' type (Salisbury, 1929, p. 219), all the seeds of the same crop germinating together (usually immediately after falling), e.g. in *Mora excelsa*, or of the 'successive' type, germination being spread over a prolonged period, e.g. in *Ocotea rodiaei* (T. A. W. Davis). In either case, the forest floor is probably seldom (and only for short periods) too dry for successful germination and establishment of the seedlings. On the other hand, Aiyar (1932) states that in the Western Ghats (India) tree seeds falling just before or during the monsoon do not germinate, or if they germinate make little progress, until the wettest period is past. In *Mesua ferrea*, for example, the seeds fall before the monsoon, in May or June, but commonly do not germinate until September, the beginning of the dry season; *Palaquium ellipticum* drops its seeds during the monsoon and they germinate immediately, though the seedlings make little growth till the wet weather is over. The crops of seedlings are often extremely abundant, but always suffer a heavy mortality and only a very small proportion survive to become even established saplings. Eggeling (1947, p. 59) says of the *Cynometra alexandri* consociation in Uganda: 'Towards the end of the rainy season it is difficult to walk anywhere in the consociation without trampling on *Cynometra* seedlings. Four or five months later, at the end of the dry season, careful search is needed to find any young plants at all.' In the Dipterocarp forests of Malaya an enormous number of seedlings die in their first year from various causes (Blanford, 1929). Among the seedlings of the light-demanding shoreas only a very small proportion, according to Watson (1937, p. 147), survives longer than two years. Since seed in these trees is produced only at irregular intervals of about five years, 'we must visualize successive waves of short-lived seedlings, of which every single one is doomed unless there happens to be a suitable gap, an eventuality that may occur only once in a good many years, and not, as in a managed forest, at regular intervals as the parent trees reach economic maturity'. For more shade-tolerant trees gaps may be less necessary, but the rate of survival is probably no greater. Exceptional weather conditions may lead to the death of whole crops of seedlings; for instance, in one locality in British Guiana the drought of 1926 killed a whole thicket of *Mora gonggrijpii* seedlings (T. A. W. Davis). Such an occurrence is probably quite exceptional; the normal cause of death must be starvation due to root competition and lack of light.

When considering the relative abundance of different age-classes of rain-forest trees (and lianes), it is important to realize the great range of environmental conditions to which plants of the higher strata are exposed during different stages of their development. There is, as will be seen in Chapter 7, an extreme contrast, not only in illumination, but in atmospheric humidity and air movement, between the upper and lower levels of the forest. Thus while it may often happen that a tall-growing species is excluded from the A and B strata because the conditions in the undergrowth are unsuitable for its young stages and not because it would be unable to survive as an adult, it is also possible that the young stages can grow well in a habitat unsuited to the adult. In British Guiana some species of palm found as full-grown trees only in swamps are common as seedlings in all types of forest, to which their fruits are carried by birds. At Moraballi Creek young stemless plants of the palm *Jessenia oligocarpa* (which is not a swamp species) are abundant in the Wallaba consociation where the tree does not occur (or only rarely) as an adult. The often very striking differences in the form and structure of the leaves (and often in their arrangement) between young and mature individuals of trees and lianes of the A and B stories (see Chapters 4 and 5) reflect this great change in their environment. Such differences are developed in the most extreme degree in shade-bearing species.

GROWTH RATE AT VARIOUS STAGES OF DEVELOPMENT

Unless by a rare chance there is a suitable gap which it can quickly utilize, the young rain-forest tree must endure a long suppression period before there is an opportunity for it to take its place in the story to which it belongs. During this time it may produce a few new leaves but may make practically no growth in height at all or at best grow very slowly and struggle almost imperceptibly upwards through the undergrowth. Large numbers of young trees having survived successfully the dangerous period of seedling mortality probably die after a suppression period lasting for years or even decades, but the saplings of many rain-forest species evidently retain for a remarkably long time the ability to respond by rapid growth as soon as conditions become more favourable. This power of enduring a long suppression period is a specialized characteristic and is no doubt an important part of the biological equipment of a rain-forest tree. The large heavy seeds provided with abundant food reserves which are a common feature among tall-growing species (p. 94) may confer a considerable advantage in enabling the seedling to grow to a size at which it has some chance of surviving the suppression period. Herbert (1929) plausibly suggests that one reason why so few species of *Eucalyptus* have been able to invade the rain forest is that they lack the ability to survive suppression.

The incidence of a suppression period and the great variation in the growth rate at different stages of development are very clearly shown by the researches of Brown & Matthews (1914) and Brown (1919) on the growth of trees in the

Dipterocarp forests of the Philippines. These results, although concerned with a single area and very few species, will probably prove of wide application to rain-forest trees in general.

Careful measurements were made over a period of four years of the diameter of a long series of individual trees of three species in the evergreen forest at Mt Maquiling (altitude *c.* 100 m. above sea-level). One of these species, *Parashorea malaanonan*, a typical dipterocarp, was the dominant species in the A stratum; the others were *Diplodiscus paniculatus*, the commonest tree in the B stratum, and *Dillenia philippinensis*, another common B species. At the same time measurements were made of a group of *Parashorea* trees in an open situation at the edge of the forest. The diameters were measured once a year to the nearest 0·5 mm. The individuals of each species were grouped into diameter classes, and for each diameter class the average annual increment of diameter was determined; from this the average number of years spent in each diameter class was calculated. Brown considered that the individual variation in the growth rate of trees in the 0–5 cm. diameter class was so great that there was a large error in estimating the age of trees in that class, but there was probably not a large error in the estimates of the ages of trees in the larger classes.

The results are summarized in Table 5, and those for *Parashorea* and *Diplodiscus* are shown as graphs in Fig. 9. The higher points on the curve for *Parashorea* in the open have been calculated on the assumption that trees of large diameter grow equally fast in the forest and in the open.

TABLE 5. *Growth rate of three species of trees in the Dipterocarp forest at c.* 100 *m., Mount Maquiling, Philippine Islands.* (After Brown, 1919)

Species	Diameter class (cm.)	No. of individuals measured*	Average annual diameter increment (cm.)	Average no. of years in diameter class
Parashorea malaanonan	0– 5	4	0·07	71
(in forest)	5–10	10	0·27	19
	10–20	21	0·38	26
	20–30	19	0·49	20
	30–40	12	0·74	13·5
	40–50	28	0·82	12
	50–60	7	0·94	11
	60–70	10	0·75	13
	70–80	1	0·84	12
P. malaanonan (in open)	0– 5	13	0·42	12
	5–10	27	0·55	9·1
	10–15	7	0·73	6·8
Diplodiscus paniculatus	0–10	14	0·15	66·7
	10–20	20	0·31	32·3
	20–30	8	0·44	22·7
Dillenia philippinensis	0–10	1	0·46	22
	10–20	1	0·55	18
	20–30	6	0·39	26
	30–40	1	0·24	42

* Not all trees were measured every year throughout the four-year period.

The data for *Parashorea* show the effect of the suppression period in a very striking way. The trees in the forest grow exceedingly slowly until they reach 5 cm. diameter, then much faster. They take nearly as long to grow to 5 cm. diameter as from 5 to 40 cm. The trees in the open, on the other hand, grow quite rapidly even when small, and their growth rate is approximately constant throughout their development. The slow growth of the forest trees under 5 cm., for which root competition and competition for light must be largely responsible, is clearly due to their unfavourable environment. Trees of the 5–10 cm. class are

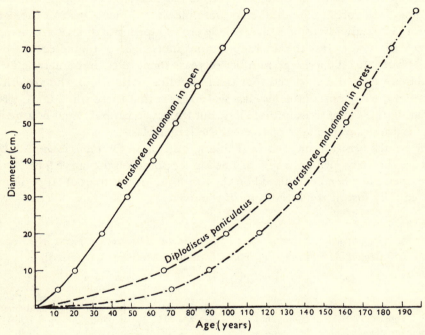

Fig. 9. Growth rate of *Parashorea malaanonan* and *Diplodiscus paniculatus* at 300 m., Mt Maquiling, Philippine Islands. After Brown (1919, fig. 2).

already 'high in the forest', while those over 20 cm. diameter are in the highest story and living under conditions which, at least as far as illumination is concerned, approximate to those in the open.

The figures for *Diplodiscus* show a slightly slower growth rate than that of *Parashorea*, but here, too, there is a marked suppression period. The calculated growth rates for the smaller diameter classes appear to be higher than those for *Parashorea*, because no trees of *Diplodiscus* less than 5 cm. in diameter were measured; if measurements of such trees had been included *Diplodiscus* would certainly have proved slower growing than *Parashorea* at all stages of its development. The figures for *Dillenia philippinensis* are based on very few measurements, but they also indicate a slower growth rate than that of *Parashorea*. Brown &

Matthews, as mentioned earlier, concluded that B story trees in general grew slower than the dominant dipterocarps.

That even very small *Parashorea* trees in the open grow rapidly shows, as might be expected, that the slow growth of young trees of this species in the forest is not due to inherent characteristics but is imposed on them by external conditions. The growth rate of a tree at all stages of its development is determined by two groups of factors, environmental and hereditary. The latter vary from species to species and, as we have seen, light-demanding species, given sufficient illumination, are usually faster growing than shade-tolerant species.

Very little is known about the mortality of rain-forest trees during the suppression period, but it is probably considerable. If this is so, these trees have two critical periods in their life history when mortality is heaviest; the first is certainly in the first or second year, in the seedling stage, the second perhaps towards the end of the suppression period. A young tree which has survived the suppression period, reached a height of 2 m. or so, and entered on the period of rapid growth would seem to have a high probability of becoming mature and reproducing. Salisbury (1930) has found that in many British herbaceous plants almost all the mortality of immature plants is in the seedling stage. Multitudes of seedlings die, but there is a high probability that those surviving beyond a certain, quite early, stage will grow to maturity; in rosette plants, for instance, individuals that survive long enough to form rosettes, usually live to reproduce. Phillips (1931) is also of the opinion that in the evergreen forest of Knysna in South Africa mortality is much heavier among the younger than among the older tree regeneration. Tropical rain-forest trees, on the other hand, though they also suffer their heaviest mortality in the seedling stage, are still uncertain of reaching the reproductive stage until they have survived the rigours of the suppression period.

THE MOSAIC THEORY OF REGENERATION

As we have seen, young individuals of rain-forest dominants may be scarce or apparently quite absent; the floristic composition of the upper stories is often very different from that of the population of seedlings and saplings destined to replace them. These facts have led Aubréville (1938) to put forward what may be termed the Mosaic or Cyclical theory of regeneration.

According to this view, the particular combination of species which form the dominants of a given small area of mixed tropical forest is constant neither in space nor in time. The dominant stories vary in composition from place to place (how far this is true will be discussed in Chapter 11) and, according to Aubréville, at the same place over a long period of years. The combination of dominant species at a given place and time is succeeded, not by the same combination, but by a different one. No combination of species is in permanent equilibrium with the environment. 'To borrow a comparison from algebra, one could say that the conditions of equilibrium between the determining factors of the habitat and the numerous independent variables constituted by the characteristic species of the community form a number of equations much fewer than the number of independent variables. All kinds of solutions are thus possible satisfying the equations of equilibrium' (Aubréville, *loc. cit.* p. 140, transl.). On this theory an extensive area of mixed forest may be regarded as a kind of mosaic, each unit of the patchwork being a different combination of dominant species. On any one small area different combinations will succeed one another more or less cyclically. If substantiated, this view demands considerable modification in our ideas of a climatic climax as a stable plant community of unvarying floristic composition.

Aubréville's theory, which certainly deserves serious consideration, is based mainly on studies of the floristic composition of the *forêt dense* (Rain forest and Mixed Deciduous forest) in the Ivory Coast. Aubréville quotes as an example a sample plot 1·4 hectares in extent in the interior of an 'untouched primitive forest' (Table 6). On this plot seventy-four species of tree were present, and *Piptadenia africana* was the commonest tree in the largest size-classes. There were seven individuals of this species of 0·5 m. diameter and over, but only two of 0·1–0·5 m., and no young plants of less than 0·1 m. A similar size-class distribution was shown, for example, by *Canarium schweinfurthii*, *Parkia bicolor* and *Parinari tenuifolia*. *Combretodendron africanum*, on the other hand, was represented by four individuals of 0·5 m. and over, but by ten of 0·1–0·5 m.; *Khaya ivorensis* by two large trees (over 0·5 m.), one individual of 0·1 m. and twelve young plants. From these figures it seems justifiable to conclude that as the present generation of dominant trees dies the composition of the stand will change and a new combination of dominants take the place of the present one.

A size-class distribution like that in Aubréville's sample plot appears to be a common feature of African forests and has also been noted elsewhere. Mildbraed (1930a), referring chiefly to the Cameroons, remarks that he has the general impression that the larger forest trees regenerate very feebly; the forest consists of relatively few large trees and innumerable saplings from the thickness of a walking stick to that of one's thigh, but of this mass of undergrowth young individuals of the larger species of trees form a very small part. Mildbraed attributes this scanty regeneration to the poor illumination in the lower levels of the forest which prevents the seedlings from germinating or soon kills those that do. On a sample plot of Mixed forest in the Shasha Reserve, Nigeria (Richards, 1939), *Erythrophleum ivorense* was the most abundant tree in the highest (A) stratum, but all the thirteen trees present were of 61 cm. diameter or more, and no seedlings, saplings or 'poles' were noticed, though they could scarcely have been overlooked. A similar, but less extreme, size-class distribution is characteristic of a large proportion of the dominant species in the Wet Evergreen forests of Benin Province. Watson (1937), speaking of Malaya, says that a poor representation in the underwood of species predominant in the overwood is a peculiarity of virgin, especially tropical, forests which is often commented on.

TABLE 6. *Composition of sample plot of primitive forest, Réserve forestière de la Massa Mé, Ivory Coast.* (After Aubréville, 1938, pp. 133–4)

Species	Young plants	Undergrowth and lower story Diameter class (m.)					Upper stories Diameter class (m.)						
		0·1	0·2	0·3	0·4	0·5	0·6	0·7	0·8	0·9	1·0	1·2	1·5
Piptadenia africana	—	1	1	—	—	1	1	—	1	—	1	1	2
Combretodendron africanum	—	4	3	2	1	1	2	—	1	—	1	—	—
Khaya ivorensis	12	1	—	—	—	—	1	—	1	—	—	—	—
Canarium schweinfurthii	—	—	—	—	—	—	—	—	1	—	—	1	—
Klainedoxa gabonensis	—	—	—	—	1	—	1	—	—	—	—	—	—
Parkia bicolor	—	—	—	—	—	—	1	—	—	—	—	—	—
Parinari tenuifolia	—	—	—	—	—	—	1	—	—	—	—	—	—
Pachylobus deliciosa	—	46	37	11	2	—	—	—	—	—	—	1	—
Funtumia africana	—	3	2	2	—	1	1	—	—	—	—	—	—
Diospyros sanza-minika	—	45	27	2	—	—	—	—	—	—	—	—	—
Strombosia pustulata	—	43	13	4	2	—	—	—	—	—	—	—	—
Scottellia coriacea	—	15	7	3	2	2	—	—	—	—	—	—	—
Monodora myristica	—	3	6	3	2	—	—	—	—	—	—	—	—
Parinari kerstingii	—	2	4	2	1	2	—	—	—	—	—	—	—
Hannoa klaineana	—	16	6	2	—	—	—	—	—	—	—	—	—
Allanblackia parviflora	—	15	3	3	1	—	—	—	—	—	—	—	—
Scytopetalum tieghemii	—	3	3	1	4	—	—	—	—	—	—	—	—
Calpocalyx brevibracteatus	—	16	5	2	—	—	—	—	—	—	—	—	—
Cola maclaudii	—	3	0	1	—	—	—	—	—	—	—	—	—
Daniella sp. aff. thurifera	—	3	2	1	1	2	—	—	—	—	—	—	—
Protomegabaria stapfiana	—	5	3	1	2	—	—	—	—	—	—	—	—
Trichoscypha arborea	—	2	5	—	—	—	—	—	—	—	—	—	—
Panda oleosa	—	—	3	—	2	—	—	—	—	—	—	—	—
Cola nitida	—	6	5	—	—	—	—	—	—	—	—	—	—
Irvingia gabonensis	—	2	3	1	1	—	—	—	—	—	—	—	—
Entandrophragma angolense	3	3	2	2	—	—	—	—	—	—	—	—	—
Diospyros kamerunensis	—	22	4	—	—	—	—	—	—	—	—	—	—
Guarea cedrata	7	7	2	1	—	—	—	—	—	—	—	—	—
Tarrietia utilis	53	6	2	1	—	—	—	—	—	—	—	—	—
Amphimas pterocarpoides	—	6	2	1	—	—	—	—	—	—	—	—	—
Phialodiscus plurijugatus	—	2	2	—	1	—	—	—	—	—	—	—	—
Baphia pubescens	—	10	3	—	—	—	—	—	—	—	—	—	—
Enantia polycarpa	—	3	3	—	—	—	—	—	—	—	—	—	—
Isolona campanulata	—	5	3	—	—	—	—	—	—	—	—	—	—
Aporrhiza sp.	—	3	3	—	—	—	—	—	—	—	—	—	—
Entandrophragma cylindricum	1	—	2	—	—	—	—	—	—	—	—	—	—
Turraeanthus africanus	48	4	—	1	1	—	—	—	—	—	—	—	—
Afzelia bella	—	3	—	2	—	—	—	—	—	—	—	—	—
Albizzia zygia	—	1	2	—	—	—	—	—	—	—	—	—	—
Guarea thompsonii	—	1	1	1	—	—	—	—	—	—	—	—	—
Coula edulis	—	1	2	—	—	—	—	—	—	—	—	—	—
Pleiocarpa mutica	—	7	2	—	—	—	—	—	—	—	—	—	—
Vitex micrantha	—	3	2	—	—	—	—	—	—	—	—	—	—
Garcinia polyantha	—	4	2	—	—	—	—	—	—	—	—	—	—
Macrolobium chrysophylloides	—	3	1	1	—	—	—	—	—	—	—	—	—
Discoglypremna caloneura	—	5	1	—	—	—	—	—	—	—	—	—	—
Ricinodendron africanum	—	—	1	—	—	—	—	—	—	—	—	—	—
Tylostemon mannii	—	2	1	—	—	—	—	—	—	—	—	—	—
Trichilia heudelotii	—	2	1	—	—	—	—	—	—	—	—	—	—
Lannea acidissima	—	—	—	—	—	1	—	—	—	—	—	—	—
Octoknema borealis	—	—	1	—	—	—	—	—	—	—	—	—	—
Xylopia elliotii	—	3	1	—	—	—	—	—	—	—	—	—	—
Pycnanthus angolensis	—	4	—	—	1	—	—	—	—	—	—	—	—
Cola chlamydantha	—	37	1	—	—	—	—	—	—	—	—	—	—
Myrianthus arboreus	—	12	—	1	—	—	—	—	—	—	—	—	—
Conopharyngia durissima	—	8	1	—	—	—	—	—	—	—	—	—	—
Phyllanthus discoideus	—	2	1	—	—	—	—	—	—	—	—	—	—
Bridelia micrantha	—	—	1	—	—	—	—	—	—	—	—	—	—
Napoleona sp.	—	9	—	—	—	—	—	—	—	—	—	—	—
Omphalocarpum sp.	—	5	—	—	—	—	—	—	—	—	—	—	—
Fagara macrophylla	—	4	—	—	—	—	—	—	—	—	—	—	—
Japaca guineensis	—	3	—	—	—	—	—	—	—	—	—	—	—
Mammea africana	—	1	—	—	—	—	—	—	—	—	—	—	—
Lophira procera	—	1	—	—	—	—	—	—	—	—	—	—	—
Antiaris welwitschii	—	1	—	—	—	—	—	—	—	—	—	—	—
Erythrophleum ivorense	—	1	—	—	—	—	—	—	—	—	—	—	—
Anopyxis ealensis	—	1	—	—	—	—	—	—	—	—	—	—	—
Anthocleista nobilis	—	1	—	—	—	—	—	—	—	—	—	—	—
Scaphopetalum amoenum	—	5	—	—	—	—	—	—	—	—	—	—	—
Microdesmis puberula	—	3	—	—	—	—	—	—	—	—	—	—	—
Baphia nitida	—	3	—	—	—	—	—	—	—	—	—	—	—
Carapa procera	—	1	—	—	—	—	—	—	—	—	—	—	—
Randia genipiflora	—	1	—	—	—	—	—	—	—	—	—	—	—
Chlorophora excelsa	1	—	—	—	—	—	—	—	—	—	—	—	—

Chengapa (1934), describing the regeneration of the Dipterocarp forests (a semi-evergreen rather than evergreen type) of the Andaman Islands, noticed a preponderance of mature and over-mature trees, a 'totally inadequate' representation of the younger age-classes and an almost complete absence of seedlings and saplings, except in tramway cuttings, on roadsides, abandoned camp sites and recently felled areas.

E. W. Jones (1945), in a paper on the regeneration of virgin forest in the North Temperate zone, has come to conclusions similar to those of Aubréville. In these forests, as in the forest of the Ivory Coast, the species regenerating abundantly are often different from those dominating the stand; seedlings of one or more species of the mature canopy are often completely absent. The abundance of seedlings is not necessarily a certain indication of which species are actually regenerating; seedlings which fail to become established may be very abundant and, on the other hand, where the average length of life of a dominant species is 300 years, an extremely small number of young individuals would suffice to perpetuate the community. Yet the available evidence suggests that in a North Temperate virgin forest the regeneration of a dominant species may require a rare combination of favourable circumstances. If such a combination arises at sufficiently long intervals of time, the floristic composition of the community would be continually changing and no one species or group of species would be permanently dominant. In a climax beech forest, for example, after one generation silver fir might become dominant in one part of the area and spruce in another. The climax would be a mosaic of communities in time and perhaps in space as well.

It would seem improbable that Aubréville's kaleidoscopic conception of a climax community can be applied to all Mixed Rain forest associations. In the Mixed forest of British Guiana, for example, all the abundant species in the larger size-classes are also abundant in the smaller, as can be seen in Table 7, though the relative abundance of the species in the smaller size-classes is different from that in the higher stories. The original data (Davis & Richards, 1933–4, table IV) also show that no species capable of reaching large dimensions is well represented in the smaller size-classes but not in the larger. In a mixed association such as this, it seems probable that the floristic composition will remain unchanged for an indefinite period.

Aubréville himself concedes that stable edaphic climax communities, such as mature mangrove or the association of *Cynometra megalophylla*, *Hymenostegia* spp., *Pterocarpus santalinoides*, etc., characteristic of river banks in the Ivory Coast, cannot be ever-varying in composition. In these associations one combination of dominants, consisting of species with specialized soil requirements or tolerances, seems able to maintain itself permanently. The same is certainly true of most single-dominant climax rain-forest communities, e.g. the *Mora excelsa* and *Eperua* consociations of Guiana, the *Cynometra alexandri* consociation of Africa. In these, as already mentioned, the younger age-classes of the dominant species are invariably well represented, and there is no reason to believe that the dominant will ever be supplanted by a different species as long as the environment remains unchanged.

Whether the Mosaic theory is a true interpretation of the facts or not and, if so, how widely it applies to mixed tropical forest associations must for the present be left undecided. The poor regeneration of the dominant species in African

TABLE 7. *Size-class representation of eight commonest tree species in a sample plot 400 × 400 ft. (122 × 122 m.) of Mixed Rain forest, Moraballi Creek, British Guiana.* (After Davis & Richards, 1933-4)

Species	Average no. of young trees under 4 in. (10 cm.) diam. and over 15 ft. (4·6 m.) high*	No. of trees 4 in. (10 cm.) diam. and over Diameter class				
		4–8 in. (10– 20 cm.)	8–12 in. (20– 31 cm.)	12–16 in. (31– 41 cm.)	16–24 in. (41– 64 cm.)	24 in. (64 cm.) and over
Eschweilera decolorans†} E. pallida†	128	19	14	7	1	1
E. sagotiana	48	5	6	14	13	1
Licania laxiflora	24	18	13	5	6	—
L. heteromorpha var. perplexans	176	29	12	—	—	—
L. venosa	128	26	16	9	10	—
Ocotea rodiaei	16	8	2	5	10	3
Pentaclethra macroloba	80	31	20	16	6	—

No data for seedlings available.

* The trees of this size-class were counted on two sample strips together equal to one-eighth of the total area and the figure obtained multiplied by 8.
† These species were not distinguished in the field.

forests seems in all probability to indicate that the composition of the community is changing. If the forest is in fact 'untouched and primitive', as Aubréville claims for the sample plot in Table 6 for example, the changes must be cyclical as the Mosaic theory implies. On the other hand, if the community has undergone disturbance in the past, the present combination of species may be a seral stage and the changes part of a normal (not cyclical) process of development towards a stable climax.

In a few cases the absence of regeneration may indicate that a species is in process of extinction. This is almost certainly true of the rare endemic tree *Canarium mauritianum* in the 'Upland Climax forest' of Mauritius (Vaughan & Wiehe, 1941, p. 136), but here the conditions are unusual in that the forest has been reduced in area to mere vestiges and natural conditions have been altered in many other ways by human interference.

CHAPTER 4

THE PHYSIOGNOMY OF THE TREES
AND SHRUBS

ECOLOGICAL MORPHOLOGY OF THE RAIN-FOREST FLORA

Anyone seeing an evergreen tropical forest for the first time cannot fail to be struck by the extraordinary prevalence of certain physiognomic characteristics in the vegetation, some of which are seldom or never met with in other plant formations. These characteristics recur throughout the Rain forest of the American and Old World tropics; in fact, they are typical of Rain forest wherever it exists and are not peculiar to any geographical region. As was pointed out in the introduction, this uniformity in the aspect of the rain-forest flora contrasts sharply with its taxonomic diversity.

A newcomer, for instance, would very soon notice the wing-like expansions at the base of a large proportion of the bigger tree trunks. Closer acquaintance with the forest shows that these buttressed trees do not belong to any one family or group of families; buttressing is a characteristic of species of the most diverse systematic position. Again, many rain-forest trees, especially the smaller ones, bear their flowers directly on the trunk or larger branches, the buds forcing their way through the bark. This unusual habit, known as cauliflory, is highly characteristic of the Tropical Rain forest and, like buttressing, occurs in species belonging to many quite unrelated families.

Besides such striking features as buttressing and cauliflory, various other peculiar morphological characters are widespread among rain-forest trees and, taken together, give them an easily recognizable physiognomy which is shared to some extent by the trees of the most nearly related formation-types, the Montane and Subtropical Rain forest. This is true not only of the trees, but in a smaller degree also of the other components of the community, the shrubs, ground herbs, epiphytes and climbers, all of which have their own special features, many of them no less remarkable than those of the trees.

Since the morphological characters in question are found in rain-forest trees throughout the tropics and do not depend on the systematic position of the species, they must be in some way related to the environmental conditions. It does not follow that they are necessarily adaptations in the ordinary sense of the word; indeed, it is questionable whether many of them have great survival value. Some at least are probably related to the environment 'causally', that is to say, they are, in some way not clearly understood, the inevitable result of the action of the habitat factors, any usefulness to the plant they may have being mainly incidental. The morphological characters we are concerned with may be said to be epharmonic, in the original sense of the term (Vesque, 1882), without making any assumptions as to the origin of the 'harmony' between plant and environment.

The physiognomy, or, as we may call it, the ecological morphology, of the rain-forest flora is a subject on which a vast number of observations have been made. Hitherto, in fact, it has been the aspect of rain-forest vegetation which has received more attention than any other. Countless writers, both travellers and professional botanists, have noted the occurrence of special characteristics in rain-forest plants and have suggested 'explanations' for them, sometimes plausible, but often far-fetched, usually intended to show how they are of use to the plants possessing them. In the space of a single chapter it is impossible to do justice to the extensive literature on the subject. A more comprehensive account of it, with a long list of references, will be found in the latest edition of Schimper's *Plant Geography* (1935).

It should be emphasized that, though all rain-forest trees have many features in common, each stratum has a distinctive physiognomy. For instance, the type of foliage characteristic of the tallest (A) stratum of trees is unlike that of the lowest tree and shrub (C and D) layers; buttressing is a feature found mainly in the upper and middle tree (A and B) strata and is absent in the lowest tree layer (C). On the other hand, cauliflory is characteristic chiefly of the lower tree and the shrub layers (C and D). Thus physiognomy is to a large extent a function of the stratum to which the plant belongs, and a description of the physiognomy of each stratum is complementary to the description of the stratification itself, given in Chapter 2.

HABIT OF TREES

The first characteristic of rain-forest trees to be considered is their form or habit. In the interior of the forest it is, of course, almost impossible to appreciate the general habit of the trees, but in clearings it can often be studied to advantage. It should, however, be remembered that trees which have been isolated or exposed for some time may undergo great changes of form and become very different from trees living in their natural surroundings.

In all stories the trunks of the trees are quite straight or very slightly sinuous. It is difficult to get exactly comparable data, but most tropical rain-forest trees give the impression of being thinner in proportion to their height than temperate trees. This is particularly striking in the case of the trees of story A and gives the forest as a whole an appearance of etiolation. When particular species of rain-forest trees are compared with allied species belonging to other plant communities, this straightness and thinness is again very apparent. The rain-forest species of *Eucalyptus*, for instance, differ from the rest of the genus in having more 'columnar' trunks (Herbert, 1929).

The impression of great height given by tall rain-forest trees depends partly on the relatively large proportion of the total height taken up by the unbranched bole. In mature trees the crown is usually smaller than in European trees, and the lowest branches arise comparatively near the top. The absence of branches on the lower part of the trunk depends, of course, on the competition between neighbouring trees and the low light intensity prevailing in the under-growth. The South American *Eperua falcata*, for example, under natural condi-tions has the long unbranched trunk characteristic of other tall rain-forest trees, but in the Botanic Garden at Port-of-Spain, Trinidad, planted specimens may be seen with branches bearing leaves and flowers to within a few feet of the ground, so that the habit recalls that of a tree in an English parkland. On river banks also trees branch nearly to the base of the trunk on the side towards the open water.

The shape of the crown, as was mentioned in Chapter 2, differs greatly in different strata. It depends on the relative height at which branching begins and on the length and angle of divergence of the main branches. These in their turn depend to a great extent on the immediate environment of the tree. Generally speaking, the taller the tree, the wider and flatter the crown. In the A story the crowns are usually wider than they are high (as seen in profile) and often tend to be umbrella-shaped, e.g. the South American *Peltogyne pubescens* and the West African *Lophira procera*. The umbrella shape is, however, never as strongly marked in rain-forest trees as in those of drier types of tropical vegeta-tion. In the B story the crowns are longer in proportion to their width; on the average they are about as deep as they are wide. Finally, in the C story tapering conical crowns, much longer than they are wide, are the rule. In the upper strata most of the main branches form an acute angle with the main trunk, while in the third story they tend to diverge at right angles.

In the course of its life an A story tree generally assumes successively the type of crown prevalent in the stratum in which it is at a given time. When young it will have the long tapering crown characteristic of the C story, when it is nearly adult it will have a more or less rounded crown, and when fully mature the crown will have the flattened form characteristic of the A story. The striking difference between the shape of the crown in young and old trees of the same species is particularly well shown in some Malayan dipterocarps, e.g. *Dryobalanops aromatica* (cf. Foxworthy, 1927, plate facing p. 46).

In some tropical trees, the 'pagoda trees' of Corner (1940), the branches form a series of tiers, like the eaves of a pagoda. This peculiar habit is chiefly due to the intermittent growth of the leading shoot, which gives rise to a whorled, or apparently whorled, arrangement of the main branches. The pagoda habit is strikingly illustrated by *Terminalia catappa* and the Asiatic *Alstonia scholaris*, but is not common in trees of typical mature Rain forest.

Tropical rain-forest trees differ considerably from those of the temperate zone in the degree of branching. The delicate tracery of twigs and branches characteristic of the majority of European trees is rare in the tropics, and many species have no branchings of a higher degree than third. According to Wiesner (1895, pp. 676–8), who paid special attention to the subject, branching to the fifth to eighth degree is usual in European trees, but in tropical rain-forest trees the highest degree of branching is generally the second or third or, at most, the fourth; branching of higher degrees than the fifth is never met with. According to Schimper (1935), however, observations by Koorders showed that higher degrees than the fifth occur, though very rarely. It is not uncommon for this restriction in the degree of branching to reach its extreme limit. In the second story in some Rain forests tall palms are found (e.g. *Oenocarpus* spp. and many others in South America), and these are, of course, entirely unbranched. In the lowest tree layer, besides palms and, in upland types of forest, tree ferns, small dicotyledonous trees with hardly any branches even of the first degree are not uncommon (*Schopfbäumchen* of Mildbraed, 1922, p. 112). Such trees often have exceptionally large leaves borne in a tuft at the top. Good examples are *Clavija* spp. in South America and *Phyllobotryum soyauxianum* in West Africa. A number of large rain-forest trees, which later have normally branched crowns, remain unbranched with their leaves crowded on to the last few centimetres of the stem until they are 6–7 m. or more high, e.g. the South American *Sterculia pruriens* and *S. rugosa*.

The tendency for the leaves to be crowded at the ends of the shoots is not limited to unbranched trees; it is equally well shown by many species with normal branching, e.g. many Malayan Anacardiaceae (*Campnosperma* and *Melanorrhoea* spp., etc.) and *Terminalia superba*, a common first-story tree of the West African Rain forest.

Tropical rain-forest trees often produce coppice-shoots very readily when the main trunk has fallen or decayed. One of the most remarkable examples is *Dicymbe corymbosa*, a leguminous tree dominant over large areas of Rain forest in the interior of British Guiana and also known from the Rio Negro region in the Amazon basin. In this species coppicing occurs not merely occasionally, but habitually. The primary main trunk, which reaches a height of 30 m. or more and a diameter of over half a metre, becomes surrounded by a group of coppice-shoots which spring from its base. When in the normal course of events the main trunk dies, one or more of the coppice-shoots takes its place and grows to similar dimensions. In time a new generation of coppice-shoots grows up round the secondary main trunk, and one or more of these become the tertiary main trunks and so on. The result of this extraordinary method of growth is that each individual has the form of a clump of trunks of various sizes, hence the vernacular name, clump wallaba. On a sample plot in British Guiana 800 sq.m. in area there were twenty-six individuals of this species and sixty-one trunks of 2 in. (5 cm.) diameter and over (T. G. Tutin).

In swampy rain forest in West Africa with very soft, yielding soil the tree *Grewia coriacea* frequently produces coppice-shoots. The trunk often slips into a horizontal position and a row of coppice-shoots grows out of it. These may all become large trunks and finally, when the parent trunk has rotted away, the old tree is replaced by a row of young independent individuals (Richards, 1939, p. 45). A similar habit is shown by *Dimorphandra conjugata* and the swamp forest *Cyrilla antillana* in British Guiana (T. A. W. Davis).

BARK

The bark of rain-forest trees is often remarkably thin and is generally smooth and light-coloured. Thick bark like that of a European oak or pine is uncommon, and even in large first-story trees the bark is often only a few millimetres thick. Foxworthy (1927) gives measurements of the thickness of the bark in a number of Malayan timber trees (mostly large A story species): the average is 0·4 in. (10 mm.), the maximum 'over 1 in.' (25 mm.) and the minimum 0·15 in. (4 mm.). At Moraballi Creek, British Guiana, the thinnest bark noted was 3 mm. thick. In trees of quite considerable diameter cork formation is sometimes so feeble that the chlorophyll in the cells beneath the bark is visible, e.g. in *Ficus cordifolia* (Schimper, 1935, p. 461). Trees of the B and C stories, as might be expected, usually have thinner bark than those of the highest story.

The smoothness which is such a common feature of the bark of rain-forest trees is no doubt a consequence of its thinness. The phellogen is generally superficial and continuous; the bark therefore does not tend to fissure and is often shed in very small flakes or granules which are barely noticeable.

Flaking and fissured barks are also found, however, but much less commonly. In *Distemonanthus benthamianus* and, according to Mildbraed (1922, p. 111), *Ochna calodendron*, both West African species, the bark comes off in large scales like that of a plane. That of *Planchonella obovata* of the Malayan region peels in large longitudinal flakes, while in *Tristania* spp., which are common in certain types of Malayan Rain forest, longitudinal flakes are formed which remain attached at the base of the tree. Furrowed or longitudinally fissured bark is much less uncommon than flaking bark; it occurs, for instance, in many Malayan dipterocarps and in *Mimusops djave* of West Africa. Many other exceptions to the general rule could be quoted. The thinness and smoothness of the bark of rain-forest trees is well illustrated by comparing *Lophira procera*, a tall tree common in the African Rain forest, with its very close ally (the two are often treated as one species) *L. alata*, which occurs in 'parkland' and savannas; the former has thin, slightly flaking bark, while in the latter the bark is thick and deeply furrowed and fissured.

Owing to a thin covering of whitish crustaceous lichens, the colour of the bark often seems to be lighter than it really is; but even when there is no lichen covering the bark is usually lighter coloured than in most temperate trees. Sometimes the bark of the upper branches of a tall tree is lighter coloured than that on the trunk; this seems to be due to the direct action of sunlight, which bleaches the pigments which give the bark its colour. Haberlandt (1926, p. 96) regarded the light-coloured bark of tropical trees as an adaptation to prevent excessive heating of the cambium beneath. Though light-coloured bark is very common, some rain-forest trees, e.g. *Lissocarpa guianensis* and many species of *Diospyros*, have dark, sometimes almost coal black, bark.

Thorny trunks, a characteristic of tropical forests in dry climates (see Chapter 15), are met with occasionally in Rain forest; examples are *Hylodendron gabunense* and species of *Fagara* in West Africa. In the small African tree *Citropsis articulata*, and in *Balanites wilsoniana* when young, the twigs and smaller branches are armed with sharp thorns. Many forest palms also have thorny stems.

BUTTRESSES AND ROOT SYSTEMS

Almost every writer on the Tropical Rain forest has commented on the buttresses of the trees, but until recently they have not been carefully investigated and have been dismissed with a casual reference to their presumed value as supports or 'anchors'. Some of the most useful papers on the subject are those of Chipp (1922), Senn (1923), Francis (1924), Navez (1930) and Petch (1930). Buttresses are of considerable ecological significance and are so important a physiognomic feature of rain-forest trees that it will be necessary to discuss in some detail their structure and development as well as their ecological aspects.

The trunks of rain-forest trees are frequently fluted or flanged for the greater part of their length; in extreme cases the trunk may be elaborately stellate in cross-section, e.g. in the South American *Aspidosperma excelsum* (Apocynaceae). Such flutings or flanges are, however, not usually included in the term buttresses, which is generally applied only to outgrowths confined to the lower part of the trunk. Local expansions of the lower part of the trunk (*empattements* of French writers), such as are common in the temperate trees, are also often seen in tropical trees; such expansions are broader and less well-defined than true buttresses, but grade into them imperceptibly. Such outgrowths can be divided into two main classes:

(i) *Stilt roots*. These are stout, woody, adventitious roots, which spring out from the main trunk at intervals up to 1 m. or more from the ground, bend downwards, often in a graceful curve, and enter the soil. Above ground they branch and sometimes anastomose; below ground they give rise to secondary and tertiary lateral roots and rootlets. They may be round in section and more or less concentric in structure or, owing to excessive secondary growth in the vertical plane, they may be flattened. The best-known examples of trees with the rounded, concentric type of stilt roots are the mangroves, *Rhizophora* (Combretaceae), but they also occur in many species of fresh-water swamp forest and of normal, well-drained Rain forest; e.g. species of *Clusia* (including the non-epiphytic species) and *Tovomita* (Guttiferae) in South America, *Macaranga barteri* and *Uapaca* spp. (Euphorbiaceae) in West Africa, *Elaeocarpus littoralis* (Tiliaceae), *Xylopia ferruginea* (Annonaceae), species of *Dillenia* (Dilleniaceae), *Eugenia* (Myrtaceae), etc. in Malaya. Similar, but straighter, stilt roots occur in some palms, e.g. the South American *Iriartea*, and in *Pandanus* (Pandanaceae). Flattened stilt roots are found in the South American *Virola surinamensis* (Myristicaceae), the African *Bridelia micrantha* (Euphorbiaceae), *Musanga cecropioides* (Moraceae) and *Santiriopsis trimera* (Burseraceae) and the Malayan *Casuarina sumatrana*. Some of the ground herbs of the rain forest also possess stilt roots, e.g. the African *Culcasia striolata* (Araceae), some Malayan Zingiberaceae and *Mapania* spp. Stilt roots are thus found in a wide variety of families and genera.

Most stilt-rooted trees are small to medium-sized and belong to the C story; comparatively few, e.g. the Malayan *Hopea mengarawan* and some species of *Dillenia*, are tall trees of the A or B story.

(ii) *Plank buttresses* or tabular roots. These are more or less flat, triangular plates subtended by the angle between the trunk and lateral roots running near the surface of the soil. They are produced by extremely epinastic secondary growth along the upper side of lateral roots, the original centre of growth (as seen in transverse section) always being along the base Rain-forest trees with plank buttresses are so numerous that it is unnecessary to quote examples. No exact figures can be given of the relative frequency of buttressing among trees of different size-classes or different stories, but observations in British Guiana, Borneo and Nigeria indicate that the highest proportion of buttressed trees is found in the A (tallest) story (trees over 30 m. high). In the B story buttresses are still very common, but the proportion of unbuttressed species is higher than in the A story. In the lowest tree story, except among young trees which when mature will belong to the higher stories, the buttressed habit is almost absent. There is thus a general (but far from exact) correlation between the height of the tree and the production of buttresses.

Stilt roots and plank buttresses resemble one another in being modifications of lateral roots. The distinction between them is that stilt roots are formed from adventitious roots springing from the main axis above the ground, while plank buttresses are developed on large lateral roots arising at or just above ground-level. This being so, it is not surprising that structures of an intermediate type are often met with. When stilt roots undergo a great amount of thickening in the vertical plane, they tend to resemble plank buttresses. On the other hand, though the lower margin of a plank buttress is generally sunk some centimetres below the surface of the ground, it may happen that there is a space between the lower margin and the ground at the proximal end, so that the buttress resembles a very wide stilt root. Occasionally the same individual may possess both stilt roots and buttresses, e.g. in the African *Grewia coriacea*.

From their method of origin, it is clear that plank buttresses can be formed only in trees with a superficial root system and well-developed laterals tending to run horizontally; stilt roots are more likely to occur in species in which the larger lateral roots bend vertically or obliquely downwards. Wilkinson (1939) classifies the root systems of Malayan forest trees into four groups: (i) laterals well developed, tap-root not persistent, (ii) laterals well developed, tap-root persistent, (iii) laterals and tap-root both well developed and persistent, (iv) laterals weak or absent. A similar classification is suggested by Coster (1932) for Javanese trees. Among the numerous examples of each group quoted by Wilkinson, it is noteworthy that all the species with conspicuous buttresses belong to groups (i), (ii) and (iii), and thus have strong lateral roots, while all the stilt-rooted species belong to group (iv). It may be added that there is some evidence, in general, that the larger rain-forest trees have superficial root systems, shallowly hemi-spherical or saucer-shaped and frequently lacking a tap-root, while the smaller trees have deeper and less wide-spreading roots, the system as a whole having the shape of a narrow inverted cone. If this is generally true, the frequency of buttressing in the upper, and of stilt roots in the lower, tree stories would be understandable.

Plank buttresses are very much commoner than stilt roots and of more ecological importance. Before dealing with the former in more detail, the latter may be briefly discussed.

Though trees of many families are habitually stilt-rooted, a given species may show a certain amount of plasticity in this respect; under some conditions stilt roots are present, under others they are not. Thus *Musanga cecropioides*, which in West Africa is invariably stilt-rooted, is said to be often not stilt-rooted in Uganda (Brasnett). In Malaya certain species which are normally stilt-rooted when growing in swamp forest do not pro-duce stilts on well-drained soil, e.g. *Eugenia longiflora*, *Archytaea vahlii*, *Myristica elliptica*, *M. maingayi*, *Maca-ranga* spp. (Corner). In Dominica (West Indies) *Amanoa caribaea* and *Oxythece pallida* are stilt-rooted in the *Amanoa* association of swampy podzolic soils but buttressed in the Lower Montane forest with normal drainage (Beard).

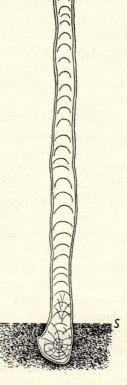

Fig. 10. Vertical transverse section through a buttress of *Argyrodendron trifoliolatum*. After Francis (1924). *S* shows the soil level.

The occurrence of stilt-rooted trees is to some extent correlated with soil conditions. Stilt-rooted trees are, as is well known, much more abundant in mangrove swamps than in normal Rain forest. They are also common in fresh-water swamp forests. In Southern Nigeria they were found to be definitely commoner in the Fresh-water Swamp forest, which had a soft, yielding soil, than in the Mixed forest on firm clay and loam. In a sample plot in the swamp forest stilt-rooted trees formed 7·3 % of all trees 4 in. (10 cm.) diameter and over; on a plot of Mixed forest on lateritic clay the corresponding figure was 1·3 %, and on another plot of Mixed forest on sandy loam it was 1·2 %

(Richards, 1939, p. 45). In Malaya stilt-rooted species are abundant in the swamp forests, and in Dominica (West Indies) a 'sub-type' of Rain forest occurs on flat, poorly drained ground which when well developed consists of an almost pure stand of two stilt-rooted species, *Symphonia globulifera* and *Tovomita plumieri* (Beard, 1944 c, p. 144). In rain forest on well-drained sites stilt-rooted trees are very rarely abundant enough to form a striking feature of the community, and are not conspicuously commoner on one kind of soil than another. In Malaya Mr E. J. H. Corner has noted only three stilt-rooted species (*Dillenia grandifolia, D. reticulata* and *Xylopia ferruginea*) in Rain forest on well-drained soil. At Moraballi Creek, British Guiana, where there were five types of primary forest on soils varying from a waterlogged silt liable to flooding through hard lateritic clay to sand, no difference was noted in the abundance of stilt-rooted trees in the five forest types.

Outside the tropics the stilt-rooted habit appears to be very rare, but there is little definite information on the point.

The obvious advantages of stilt roots to a tree growing in an unstable swampy soil would at first sight suggest that they are a clear case of an adaptation developed by the action of natural selection, but when the facts are closely examined there are difficulties in accepting any simple 'explanation'. If stilt roots are of great survival value for anchorage, it is surprising that they should be commoner in small trees mainly growing in the sheltered undergrowth than in tall trees with heavy crowns to support. Tall rain-forest trees, as will be mentioned later, are frequently blown over by the wind, even when quite healthy, so a more efficient system of anchorage would undoubtedly be an advantage to them. Among possible 'causal' factors which might explain the relative frequency of stilt roots in rain-forest trees is the continuously damp atmosphere in the lower levels of the forest which might tend to encourage the formation of aerial roots. A connexion between moisture and the production of stilt roots is suggested by the fact that in swamp forest in Malaya the height on the trunks to which stilt roots are found coincides with the highest level reached by floods in the district; in some localities it may be as high as 30 ft. (9 m.) (Corner). A similar observation has been made in British Guiana by Mr T. A. W. Davis. Possibly the production of stilt roots is also facilitated by the thinness of the bark, which, as has been shown, is a common feature in rain-forest trees.

We must now turn to the consideration of plank buttresses, and in the remainder of this chapter it will be convenient if the word 'buttress' is understood to refer only to them.

Structure and development of buttresses

The wood of which buttresses are composed is often harder than that of the trunk of the same tree; owing to this hardness the buttresses, when struck with an axe, give a peculiar hollow, ringing sound. According to Francis (1924, p. 31), who examined the buttresses in the Australian rain-forest tree, *Argyro-*

dendron trifoliolatum, the tissues composing the buttresses are similar to those of the trunk. The bark on the buttresses tends to be thinner than that on the trunk; in four Queensland rain-forest species the bark on the trunk was from 1·6 to 2·5 times as thick as that on the buttresses (Francis, 1924, p. 32).

As has been stated, the internal structure of a mature buttress is strongly epinastic, the centre of growth being near the ground (Fig. 10). When clear growth rings are visible, it can be seen that the early rings are normal and circular; the excessive thickening of the upper side only sets in after a certain time. The roots from which buttresses are formed thus begin with a normal structure and only acquire their peculiar modifications later on. This point is of some importance in connexion with possible 'causal' explanations of buttress formation.

The stage of growth of the tree as a whole at which buttress formation begins probably varies in different species, but in some cases at least it is quite early. In a number of Queensland species buttresses are already developed when the tree has attained from one-seventeenth to one-ninth of its maximum stem diameter and from two-ninths to one quarter of its maximum height (Francis, 1924, pp. 25–6). Similarly, Petch (1930, p. 278) found that in Ceylon (in a botanic garden?) buttresses were present in *Delonix regia* in trees 2–3 years old and not exceeding 12 ft. (3·7 m.) high. In plantations in Nigeria, MacGregor (1934, pp. 63–4) found that *Terminalia superba* began to produce buttresses when less than five years old and under 8 m. high. It is clear, therefore, that the assumption which is sometimes made, that buttresses do not develop until the tree is high in the forest and has acquired a heavy crown, is not justified.

It is a remarkable fact that in a tree with well-developed buttresses the trunk, instead of tapering steadily from the ground upwards, is always more or less markedly tapered from the top of the attachment of the buttresses *downwards* to the ground, as well as upwards. Thus Francis (1924, p. 31) found that in *Echinocarpus woollsii* in Queensland the diameter of the trunk above the buttresses was 60·4 cm., while at the surface of the ground it was only 22·6 cm. This downward taper is not shown in young trees and only becomes pronounced when the buttresses grow large. The reason for this reversal of the normal taper does not seem at all clear; one possible explanation is considered below.

In a few buttressed trees, e.g. *Canarium commune* and *Ceiba pentandra*, wing-like expansions similar to the buttresses are found in the angles between the branches and the main trunk and sometimes, perhaps, on the main trunk without relation to the branches.

The number of buttresses on one trunk varies from one to as many as ten. There are no exact data as to the relative frequency of different numbers, but from three to five is probably the commonest. Since three is the smallest number which can form mechanically efficient supports, the fact that it is comparatively rare for mature buttressed trees to have one or two buttresses is regarded as significant by those who look on buttresses primarily as an adaptation.

The size, thickness and shape of buttresses vary very widely. Though there is much variation between different individuals of the same species, these characters, as Chipp (1922) first pointed out, are to some extent constant within the species or larger taxonomic unit. Unfortunately, buttresses are rarely mentioned in systematic works and little exact information on the subject is available, but anyone familiar with tropical trees knows that, within limits, buttresses have a considerable value as diagnostic characters in the field.

The size of the buttresses in a given species of course depends largely on the age of the tree. In full-grown trees they may be of any size up to enormous plates extending up the trunk to a height of 9 m. or more and out from the trunk for an almost equal distance. The large buttresses of some species, e.g. the Malayan *Koompassia excelsa*, are occasionally used for making large dining tables. The size of the buttresses is sometimes of diagnostic value for distinguishing closely allied species; for instance, *Mora gonggrijpii*, an abundant tree in Guiana, constantly has smaller buttresses than the closely allied and more widely distributed *M. excelsa* (see Fig. 11).

The thickness of buttresses also varies greatly and sometimes has a diagnostic value. Some species produce buttresses which are only a few centimetres thick even when large, while other species always produce much thicker buttresses. Every transition is found from thin, almost knife-like, plates to broad indefinite flutings or ridges similar to those found on European trees such as *Taxus* and *Carpinus*.

The shape of buttresses as seen in profile varies greatly and is often highly characteristic of particular species. Chipp (1922) laid much emphasis on the diagnostic value of the ratio of height to base (or inclination of the outer edge). In doing so, however, he overlooked the fact that in many species, perhaps in all, this ratio changes, the buttresses becoming higher in proportion to their breadth as the tree grows older. Francis's observations on the change in this ratio in *Argyrodendron trifoliolatum* have already been mentioned, and a similar change has been noticed by the author in *Mora gonggrijpii* in British Guiana and in *Terminalia superba* in West Africa. Nevertheless, if trees in a similar stage of growth are compared, the inclination of the outer edge of buttresses has a certain value as a specific character. Thus *Couratari pulchra* can easily be recognized from other Lecythidaceae at Moraballi Creek, British Guiana, by its high, very steeply inclined buttresses. Similar very steep buttresses are a feature of the West African *Alstonia congensis*. Other instances could be quoted.

Buttresses are found in species belonging to a large number of unrelated families; nevertheless, among the families represented in the Tropical Rain forest, some show a much stronger tendency to produce buttresses than others. The Dipterocarpaceae, Leguminosae and Sterculiaceae, for instance, include many species which habitually have large buttresses, while in the tropical Annonaceae, Fagaceae (the Malayan *Lithocarpus bennettii* is an exception, according to E. J. H. Corner) and Lauraceae well-developed buttresses are usually absent.

Buttresses are also absent, at least in the Malayan species, in the Myristicaceae (Corner). Between different genera of one family, again, there are great differences; thus the Bombacaceae include trees with enormous buttresses, such as *Bombax* spp. and *Ceiba pentandra*, and also *Catostemma*, a genus including several abundant species of the Guiana Rain forest which have no buttresses at all. Similarly, though the majority of the leguminous trees of the Rain forest are more or less strongly buttressed, in the Malayan genus *Sindora* (which includes large timber trees) and a number of other Malayan genera, buttresses are never found (Corner). In the Dipterocarpaceae the genera *Dipterocarpus*, *Hopea* and *Shorea* include both buttressed and unbuttressed species. The large genera *Diospyros* (Ebenaceae) and *Eugenia* (Myrtaceae) include both buttressed and unbuttressed species in Malaya (Corner). Within the species buttresses appear to be a fairly fixed character, but some species show plasticity in this respect. Sometimes this plasticity is so great that some individuals have large buttresses and others of the same species none. Few definite observations have so far been made on this point, but Ghesquière (1925, p. 2) records that buttresses may or may not be present in the African species *Combretodendron africanum* and *Tetrapleura tetraptera*. *Eschweilera sagotiana* in British Guiana is normally almost unbuttressed, but occasionally has one or more large buttresses (Davis). Petch (1930, p. 278) found that in *Delonix regia* a seedling from an unbuttressed tree itself became buttressed.

From what has been said, it appears certain that the tendency to produce buttresses is an inherited character, dependent on the genetical constitution of the tree. The extent to which the tendency is realized, as will be seen later, depends largely on the environment.

Buttressing in relation to habitat

Though buttressing is a highly characteristic feature of tropical rain-forest trees, it is not entirely confined to them. The general distribution of the buttressed habit shows that the climatic conditions under which it is most strongly developed are a high rainfall (not necessarily at all times of year) combined with a high mean temperature.

Within the tropics buttressing is only prevalent at fairly low altitudes and where there is a high rainfall. It is common throughout the Rain forest, but also occurs in some types of Monsoon forest, e.g. in Burma. In Ceylon buttressed trees (e.g. *Bombax malabaricum*) occur in the open 'patanas' as well as in the forest (Petch, 1930, p. 277). In the West African Mixed Deciduous forest, which receives a rainfall of about 1200–1500 mm. (see p. 338), buttressed trees are as frequent, or nearly so, as in the typical Rain forest. In dry tropical regions buttressed trees are chiefly found in fringing (Gallery) forests by rivers.

In Montane Rain forest on tropical mountains buttressing is absent or uncommon. For example, in the 'Mossy forest' on Mt Dulit, Sarawak, which covered the top of the mountain down to about 970–1100 m., buttressed trees were quite absent. Shreve also found none in the Montane Rain forest on the Blue Mountains in Jamaica (Shreve, 1914a, p. 24). In Australia, where the rain forest extends without large interruptions from the tropical to the temperate zone, the number of buttressed species diminishes with increasing latitude. In the north, in Queensland, buttressed trees are abundant in the rain forest wherever the annual rainfall exceeds about 60 in. (1524 mm.) (Francis, 1924, p. 33). In the rain forest of northern New South Wales there are comparatively few buttressed trees (Fraser & Vickery, 1938, p. 155). Apparently even in the same species the production of buttresses decreases southwards, for *Laportea gigas* is buttressed in northern New South Wales, but unbuttressed in southern (N. A. Burges).

Generally speaking, the buttressed habit is absent in temperate forests, but even in some species of European trees buttresses are sometimes produced, though they are never as large or as thin as in tropical trees. Well-marked buttresses are found especially in *Populus italica* (cf. Senn, 1923) and *Ulmus* spp. (see photograph in Warming & Graebner, 1933, fig. 399); smaller buttresses are sometimes seen in *Fagus sylvatica* and other species.

Buttressing and soil conditions

The close correlation between the frequency of buttressed species, the size of the buttresses and the nature of the soil has been realized only recently, though it has often been observed that buttressed trees are, on the whole, more abundant in swamp forest than in rain forest on better drained land. De Wildeman (1930, p. 997) states generally that the production of buttresses is dependent, at least in part, on the nature of the soil, but Davis & Richards (1933–4, pp. 126 etc.) were able to show that in the Rain forest of British Guiana there is a close and definite correlation between the size of buttresses in the most abundant species of tree and the type of soil. They also found that the number of buttressed subordinate species varies in a parallel manner. At Moraballi Creek there were five types of forest in which the dominant or most abundant species of large trees were respectively, *Eperua falcata*, *Ocotea rodiaei*, *Eschweilera sagotiana*, *Mora gonggrijpii* and *M. excelsa*. The *Eperua* forest was found on coarse white sand on ridges and plateaux, the *Ocotea* forest on brown sand with a larger proportion of fine particles, the *Eschweilera* forest on loam, the *Mora gonggrijpii* forest on heavy clay and the *M. excelsa* forest on silt on the flood plains of streams. Through this series from the *Eperua* to the *Mora excelsa* type, the buttressing of the dominant or most abundant species of trees 16 in. (41 cm.) diameter and over increases gradually, but markedly, as can be seen in the scaled drawings reproduced in Fig. 11. At the same time the proportion of buttressed subordinate species increases. *Eperua falcata* is almost entirely unbuttressed, and relatively few strongly buttressed species are associated with it.

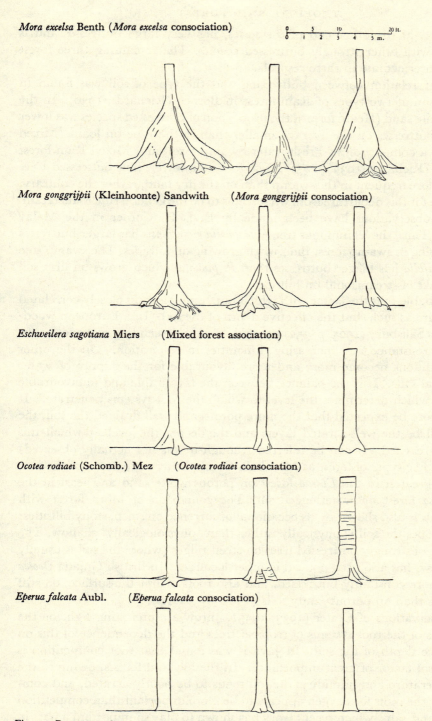

Mora excelsa Benth (*Mora excelsa* consociation)

Mora gonggrijpii (Kleinhoonte) Sandwith   (*Mora gonggrijpii* consociation)

Eschweilera sagotiana Miers   (Mixed forest association)

Ocotea rodiaei (Schomb.) Mez   (*Ocotea rodiaei* consociation)

Eperua falcata Aubl.   (*Eperua falcata* consociation)

Fig. 11. Buttresses of typical mature individuals of the most abundant species of tree in five forest types, Moraballi Creek, British Guiana. After Davis & Richards (1933–4, fig. 2, p. 127).

Mora excelsa, at the other end of the series, has very large buttresses and is associated with other strongly buttressed species. The remaining three forest types are intermediate in these respects.

A similar relation between buttressing and the type of soil was found in a comparison of two types of Rain forest in Borneo (Richards, 1936). In the type on white sand (Heath forest), the proportion of buttressed species was lower and their buttresses on the average smaller than in the type on loam (Mixed forest). In a comparison of Fresh-water Swamp forest and Mixed Rain forest in Nigeria (Richards, 1939, p. 45), it was found, however, that buttressed trees were not more frequent in the swamp than on the dry land, rather the contrary. The reason for this departure from what seems to be the general rule is uncertain.

Similar observations have been made by E. J. H. Corner in the Malay Peninsula. Thus, the leguminous tree *Koompassia malaccensis* has larger buttresses when growing in swamp forest than when growing on hillsides. The swamp tree *Pometia alnifolia* has bigger buttresses than *P. pinnata*, which grows on firm soil on the banks of streams and on hillsides.

It is probable that the factor with which buttressing is most closely correlated is not texture as such, but the effective depth of the soil. In a European wood-land soil, as Salisbury (1925, p. 352) has said, '...the descending root encounters increasing resistance and increasing difficulties in respiration. On the other hand, conditions become more and more favourable for the supply of water and mineral salts. It is the balance between the favourable and unfavourable conditions which determines the level to which the root systems penetrate.' It may therefore be expected that the more porous and well drained the soil, the thicker will be the well-aerated layers and the deeper the level at which this balance is reached. It can be inferred (though it was not actually observed) that in the five types of forest at Moraballi Creek the average depth of the root systems is greatest in the *Eperua* forest on porous white sand and least in the *Mora excelsa* forest on waterlogged silt. The connexion of Mora forest with shallow soils is also shown by its occasional occurrence on steep, stony hillsides; on these slopes the soil is physically rather than 'physiologically' shallow. The occurrence of strongly buttressed trees on steep ridges (where the soil is usually very shallow) has also been noticed in other localities. In British Guiana *Ocotea rodiaei* is more constantly buttressed, and also roots nearer the surface, on stiff clayey soils than on porous sandy soils (T. A. W. Davis).

The observations of Coster (1933, 1935 b) throw an interesting light on the shallowness of the root systems of tropical trees and the dependence of this on the effective depth of the soil. In Java it was found that root competition is an ecological factor of great importance. In the lowland forests, owing to the high temperature and humidity, the soil tends to be poorly aerated, and com-petition by the roots for oxygen appears to be more important than competition for water and salts, which recent work has shown to play so important a part in the ecology of temperate forest communities.

Since, as was pointed out earlier, buttresses can only develop in trees with large superficial lateral roots, the correlation of buttressing with the effective depth and aeration of the soil is easily understood. It is also not surprising that most, though not all, buttressed trees lack a tap-root. This was noticed long ago by Spruce (1908, **1**, p. 21) in South America, and more recently by Francis (1924) in Queensland and Petch (1930) in Ceylon. Opportunities for examining the root systems of buttressed trees are rare, but both in Borneo and Nigeria the author has always found that buttressed trees have no recognizable tap-root. For instance, an overturned, strongly buttressed tree of *Ceiba pentandra* on the bank of the Oni River in Nigeria, though of enormous size, had a root system consisting entirely of superficial lateral roots with small descending branches; the greatest depth of penetration was probably not more than half a metre. Exceptions are found—Foxworthy (quoted by Chipp, 1922) records a buttressed tree of *Dracontomelon dao* from the Philippines with a tap-root, and Francis (1924, p. 32) saw a specimen of *Argyrodendron trifoliolatum* in Queensland that had a tap-root although it had six buttresses. The trunk of the latter tree was, however, only 9·6 cm. in diameter, and in a mature individual a tap-root would probably not have been found. In Malaya *Eugenia grandis* and *Scaphium affine* have moderately buttressed trunks, yet have tap-roots (Corner).

Though buttressed trees usually have well-developed superficial root systems, the converse is not necessarily true. Some rain-forest trees have roots running horizontally for enormous distances above the surface of the ground, but such trees are not always buttressed. A tree of *Entandrophragma angolense* var. *macrophyllum* in the Okomu Forest Reserve, Nigeria (Pl. III B) had surface roots running out over 20 m. from the trunk, and was almost unbuttressed. Similar surface roots are a feature of the South American *Caryocar nuciferum*, a tree with only moderate-sized buttresses.

Theories of buttress formation

Four theories to 'explain' buttress formation have been put forward more or less explicitly.

(i) *The Adaptation theory.* In the last century buttresses were generally regarded simply as a special adaptation to help the tree to withstand stresses due to wind or gravity. The earlier supporters of this theory considered that the chief function was to resist *compressive* stresses, i.e. to act as props. Later writers, however, ascribe more importance to resistance to *tensile* stresses (Ghesquière, 1925). Where the buttresses are evenly distributed round the base of the trunk, as they often are, some of them must, of course, perform the one function and some the other, and in places where the wind does not blow constantly from one direction the same buttress will be subjected sometimes to compression and sometimes to tension. From the point of view of the adaptation theory, the distinction is therefore not of great importance.

If we accept the logical consequences of the adaptation theory we must

believe that the extra mechanical strength given by buttresses is of sufficient
value to give a species possessing them a selective advantage over a species
which does not. Buttresses (or rather the potentiality of producing them under
the appropriate environmental conditions) might arise by a single mutation or
by a series of mutations preserved and accumulated by natural selection.

Certainly, it can hardly be denied that buttresses are of value as mechanical
supports, and that a buttressed tree has an advantage over an otherwise identical
tree with none. There are, however, a number of reasons for doubting whether
the advantage is in fact great enough to be of much survival value.

The most important of these reasons is the difficulty of understanding why
they are so little developed in extra-tropical trees. The Californian sequoias and
the giant Australian eucalypts,[1] which are much taller than any tropical rain-
forest trees (see Chapter 1), have no buttresses. Though some parts of the
rain-forest belt are subject to tornadoes, in many tropical regions, e.g. British
Guiana, violent winds are less common than in temperate regions (see
Chapter 6). Besides this, rain-forest trees are often bound together by the
woody stems of lianes so that they support one another very effectively. Anyone
who has watched the felling of tropical trees knows that a large tree is often so
strongly bound to its neighbours by lianes that even when cut right through at
the base it will not fall. This is not usually true of the tallest trees, which are
often free of lianes, but such trees are not always the most strongly buttressed.
Mildbraed (1922), who believes in the adaptation theory, felt this difficulty; for
he says that, though the great height of tropical trees, the strength of tropical
tornadoes and the uneven canopy, which prevents rain-forest trees from shelter-
ing each other as much as those in European forests, make the presence of
buttresses understandable, there must be some factors peculiar to the tropical
rain forest which specially favour their formation. He notes that the low light intensity and high humidity in the undergrowth of the rain forest encourage the formation of aerial adventitious roots and suggests that buttresses have been evolved
from these. He looks on flattened stilt roots, such as those of *Santiriopsis trimera*,
as a stage in the evolution of plank buttresses from the rounded type of stilt root.

Another reason for doubting whether buttresses confer an important selective
advantage is that they are not most strongly developed where they are most
needed. For instance, they should be better developed in trees growing on soft,
yielding soils than in trees on hard, resistant soils. Though buttressing is usually
more pronounced in Fresh-water Swamp forest than in Rain forest on well-
drained ground, both in Borneo and in British Guiana, as was shown above (p. 66),
there is less buttressing on loose sandy soils than on firm clays and loams. Similarly,
it would be expected that buttressing would be particularly marked in trees growing in windy situations. At Moraballi Creek, British Guiana, however, buttresses
were best developed in the Mora forest which grew in the sheltered stream valleys
and least developed in the *Eperua* forest on the relatively exposed sandy ridges.

[1] According to W. M. Curtis' *Flora of Tasmania* (1962) *Eucalyptus regnans* in Tasmania has large buttresses.

On a mountain in the Philippines Whitford (1906) found that the buttressed habit disappeared altogether on the exposed ridges. From extensive experience in British Guiana Mr T. A. W. Davis believes that buttressed trees are in fact more often blown down by the wind than those without buttresses.

Obviously the adaptation theory cannot be ruled out all together; the reasons for doubting it are not conclusive, but at least they suggest that, like Mildbraed, we should look for some 'causal' explanation to supplement it. The usefulness of buttresses is perhaps incidental rather than a primary cause for their existence. It is also quite possible that other characters may compensate for the lack of buttresses, e.g. a root system well adapted to a poorly aerated soil which can grow deep and retain a tap-root under conditions which force the root systems of most plants to remain superficial. The other three theories are 'causal' explanations.

(ii) *The Negative Geotropism theory.* Francis (1924, p. 34) suggested that 'the upper part of the principal surface roots in buttressed species may be affected by negative geotropism and phototropism either directly or indirectly through the stem; and in this manner the perpendicular extension, which constitutes buttresses, may arise'. This theory, on the face of it, does not seem probable. Negative geotropism in roots is shown, for instance, by the pneumatophores of mangroves, but no instances are known of the gravity stimulus causing the vertical expansion of a horizontal organ (the normal reaction being, of course, a curvature). Most roots are insensitive to light and those which are sensitive usually react negatively; in any case it is hard to suppose that a fully formed root covered with cork could be sensitive to light.

(iii) *The Conduction Current theory.* This theory to account for buttress formation was suggested by Petch (1930), and is based chiefly on observations made in the Peradeniya Botanic Gardens and elsewhere in Ceylon. The starting point of the theory is the fact that buttressed trees usually lack a tap-root. Petch suggests that in a tree in which the greater part of the root system consists of superficial lateral roots, the transpiration stream bringing in dissolved nutrients from the soil is restricted in the trunk to limited tracts on the same radii as the lateral roots. This theory presupposes that lateral conduction in the trunk is relatively inefficient, an assumption supported by the experiments of Caldwell (1930) as well as by evidence brought forward by Petch himself. The currents of water and nutrients encourage growth in the sectors of the cambium on the same radii as the lateral roots at the expense of the intervening sectors, and this leads to the formation of ridges, which in time become buttresses. The frequency of buttressing in the wet tropics is due to the waterlogged condition of the soil, which prevents the development of tap-roots. All species are not buttressed, because in some species the tap-root may be tolerant of waterlogging.

Like the two theories already considered, the conduction current theory is open to objections. Some instances have been given (p. 69) of buttressed trees in which a tap-root is present. The theory does not explain why shallow-rooted trees without a tap-root so rarely develop buttresses in temperate regions. Again, although it gives a plausible explanation of why unequal growth should take place in the trunk, it is not clear why growth in the root itself should take place mainly along the upper side. It is at present uncertain what share the stem cambium normally has in buttress formation, but Petch himself (1930, p. 280) gives instances of roots in *Canarium commune* which have produced buttresses at some distance from the trunk and none on the trunk itself.

Petch's theory has one advantage not shared by other causal theories; it suggests a possible explanation of the downward taper of the lower part of the trunk which has been mentioned as a common feature of both stilt-rooted and plank-buttressed trees. We can imagine that if lateral conduction in the trunk is relatively inefficient, as Petch's theory demands, sectors of the cambium between the restricted tracts which are continuous with the lateral roots will be relatively poorly supplied with water and salts, but higher up the stem the ascending streams might gradually tend to spread out, so that above a certain level the whole cambium will be equally well supplied and will therefore grow evenly.[1] This level will be somewhere above the height at which the highest stilt roots or buttresses join the trunk. In some such way it might be possible to explain why, as Francis put it, 'the deficient portion of the lower part of the stem has passed into the buttresses'.

(iv) *The Strain theory.* According to Whitford (1906), Senn (1923), Navez (1924, 1930) and others, buttress formation is the direct response of the tree to the mechanical stimulation of strains set up by the wind. The most complete exposition of this view is that of Senn (1923), who, after his interest in the problem had been aroused by a visit to Java, made a detailed study of buttressing in the Lombardy poplar (*Populus italica*) in Switzerland. The buttresses of this species are smaller and thicker than those of most tropical trees, but otherwise they are similar and there seems no reason to suppose that the causes of the phenomenon are different in the two cases. Senn carefully examined the orientation of the buttresses in a large number of trees in different parts of the country. He could find no relation between the orientation of the buttresses and the distribution of water or nutrients in the soil; there was, on the other hand, an apparent relation to unilateral heat radiation, buttress formation being stimulated near surfaces reflecting heat, such as walls. Much more striking, however, was the correlation between the orientation of the buttresses and the direction of the prevailing wind (which is remarkably constant in many Swiss valleys). Out of the 461 trees examined, 443 (96·1 %) had more or less distinct buttresses and 407 (91·9 % of the total) had buttresses chiefly on the side of the tree *facing* the prevailing wind. Senn concluded from this that buttress formation is due to the stimulation of the cambium on the upper side of the proximal end of the lateral roots and in the adjoining part of the trunk by tensions set up by the wind.

[1] Harmon (1942) finds that in American trees total growth is greater on the radii continuous with the lateral roots than in between them.

Among the exceptions there were eight trees which had buttresses on the *leeward* side and also had a steep bank or rock abutting on that side. Senn provides an ingenious explanation for these cases. When the wind blows from the prevailing direction the lower part of the trunk tends to be bent slightly, owing to the resistance of the bank or rock; it therefore acts as a lever tending to pull on the roots on the leeward side and to push on the roots on the windward side. Thus the stresses are the opposite of those in a tree growing on level ground without obstructions, and hence the orientation of the buttresses is the opposite to the usual one. Three more exceptions are ascribed to the influence of unilateral heat radiation. In any case, as Senn points out, a mathematically exact correlation with wind direction cannot be expected, because buttresses are formed on pre-existing roots, the orientation of which is not influenced by the wind.

Senn also found that buttressing was most marked where the lower part of the trunk is shaded, that is, where it is surrounded by damp air. This would help to explain its frequency in the Tropical Rain forest.

A correlation between buttress orientation and the direction of the prevailing wind is certainly not very obvious in tropical trees. Petch definitely denies that such a correlation existed in the individuals of *Canarium commune* he examined at Peradeniya (1930, p. 280), but the question is one which can only be studied satisfactorily by statistical methods. Senn considered that photographs of tropical trees taken by himself and others indicated that there was a correlation with wind direction, but the only definite demonstration is that of Navez (1930), who made careful observations of a large number of trees of *Ceiba pentandra* in Cuba. These, like the Swiss poplars, showed a definite tendency for the buttresses to be on the side of the tree facing the prevailing wind. The data of Navez are statistically significant.

In trees leaning over the water on river banks in Malaya and British Guiana the buttressed species always have the largest buttress on the side away from the water, i.e. on the side receiving the greatest strain (Corner, T. A. W. Davis). It is also of interest that in *Canarium commune* and *Ceiba pentandra* small wing-like expansions are found in the axils of the larger branches; these, like the buttresses, might be induced by tension.

There is thus a certain amount of positive evidence in favour of the Strain theory, but several difficulties remain to be cleared up before it can be unreservedly accepted. Since buttressing appears to be partly a genetically determined character, found in certain families, genera and species and not in others, it must be assumed that there are hereditary factors controlling the sensitivity of the plant to stimulation by tension. Some plants would appear to be more sensitive than others to this stimulation. The theory does not fully explain why buttressing is so little developed in temperate trees, though Senn's observations that buttress formation in *Populus* is promoted by high temperature and humidity may give the clue to an explanation. It might perhaps also be suggested that the families and genera most sensitive to stimulation are not those which have proved adaptable to temperate conditions and hence have remained confined to the tropics. A further difficulty is that the Strain theory does not explain why buttresses are developed in many rain-forest trees while they are still young and are still in the undergrowth in a very sheltered environment. It is hard to believe that such trees are subjected to strains sufficient to induce buttress formation.

At present, therefore, there is no entirely satisfactory theory to explain the development of buttresses in rain-forest trees. The evidence on the whole supports some kind of 'causal' theory and suggests that the usefulness of the buttresses to the tree is incidental rather than primarily responsible for their development. To get more light on the problem a large number of exact observations are needed and, above all, experimental data. The possibility should be borne in mind that growth hormones may play an important part in buttress formation, and a knowledge of the production and translocation of such substances in tropical trees might contribute to a solution of the question.

The close relation which has been shown to exist between buttressing, the type of root system and the nature of the soil makes the phenomenon one of great interest to the ecologist. It is for this reason that what might otherwise seem to be a purely morphological problem has been treated here in some detail. It is hoped that this discussion will draw attention to a somewhat neglected subject. Incidentally, it may be noted that buttressing is a problem of economic importance. In felling tropical timber trees the cut is necessarily made above the buttresses, so the amount of waste is considerable. Any clue to the factors controlling buttress formation must therefore be of interest to the tropical sylviculturist.

Pneumatophores

In swampy types of Rain forest, though not in those where the soil is better drained, the so-called pneumatophores, or breathing roots, of some of the trees are often a striking feature. In some species these take the form of 'knees' or loop-like structures formed by short portions of superficial lateral roots which emerge into the air and immediately bend down and re-enter the soil; a series of such 'knees' may be formed on one root. In the tropical American *Pterocarpus officinalis* the pneumatophores are small, semicircular, wing-like upgrowths arising from lateral roots some distance farther from the trunk than the point at which the buttresses proper disappear into the ground. In other species they are short, tapering organs which grow up vertically into the air. These consist of side branches of superficial lateral roots; they are negatively geotropic and of limited growth. As is well known, several types of pneumatophore are common in mangroves, knee-roots occurring in *Bruguiera* spp., negatively geotropic aerial roots in *Avicennia* and *Sonneratia* spp., etc. It is less generally realized that they often occur in fresh-water swamp forest species.

Thorenaar (1927) drew attention to their occurrence in the peaty swamp forest (Moor forest, see p. 292) near Palembang in Sumatra, and according to van Steenis (1935 a) they are a general characteristic of swamp forest in the Netherlands East Indies. Pneumatophores similar to those of *Sonneratia* and *Avicennia* are mentioned and figured by E. Polak (1933, p. 27 and figs. 6 and 7) as occurring in *Ploiarium alternifolium*, *Cratoxylon* and *Tristania* spp. in peaty swamp forest in west Borneo. The author has seen both types of pneumato- phore in peaty swamp forest in Sarawak on various unidentified species of trees, and has observed knee-roots on *Mitragyna ciliata* (Rubiaceae) in non- peaty swamp forest in West Africa (Fig. 12). In *Symphonia globulifera*, a tree common in brackish and fresh-water Swamp forest in British Guiana, loop-like

Fig. 12. Loop-like aerial roots (pneumatophores) of *Mitragyna ciliata*, a tree of the West African Fresh- water Swamp forest. Shasha Forest Reserve, Nigeria. Drawn from a photograph. About one-third nat. size.

knee-roots are often produced in such abundance as to render walking difficult (T. A. W. Davis). Adamson (1910) has described pneumatophores in the south Indian riverside tree *Terminalia arjuna*, and Lam (1945, p. 48) in a rattan (climbing palm) in Fresh-water Swamp forest in New Guinea. Knee-roots and other types of pneumatophore are thus common in Fresh-water Swamp forests in all parts of the tropics. The knee-roots of *Taxodium distichum*, a tree of temperate and subtropical swamps, are, of course, well known.

The pneumatophores of mangroves, as their name suggests, are usually re- garded as ventilating organs which assist the tree in living on a soil almost completely deficient in oxygen. Their structure (numerous lenticels or con- stantly flaking bark, abundant intercellular spaces) certainly supports such an interpretation, and Karsten (1891) demonstrated that they do in fact give off carbon dioxide in considerable quantities (see also Troll, W. & Dragendorf, 1931; and Chapman, 1944), but W. Troll (1930) has recently shown that they have another function which may be of equal or greater importance. In mangrove swamps the soil level is continually rising—in the locality in Sumatra which Troll studied the yearly accretion was estimated at not less than 15–35 mm.— and the pneumatophores play an important part in enabling the tree to adapt itself to the rise. In *Sonneratia* and *Bruguiera* (and the same is true of *Avicennia*, according to Chapman) the great majority of the fine rootlets are borne by the underground part of the pneumatophores and not by the lateral roots directly. The absorptive part of the root system is thus situated near the surface of the soil, the deeper part of the root system presumably serving an anchoring function. As the soil level rises the pneumatophores grow upwards and produce fresh crops of rootlets at successively higher levels; in this way the active part of the root system is kept at a constant depth below the soil surface. No investigations have been made on the pneumatophores of fresh-water swamp trees in the tropics; but as they are similar in structure to those of mangroves their functions might be expected to be analogous, though it seems doubtful if the rate of accretion under fresh-water conditions ever equals that in mangrove swamps, especially in the peaty Moor forests.

It is perhaps of interest to note that in an area of swamp forest in Nigeria the author found that the knee-roots of *Mitragyna ciliata* were produced more abundantly in the wetter than in the drier parts. The same is true of *Symphonia globulifera* in British Guiana, which on drier ground produces hardly any knee-roots, except in occasional depressions (T. A. W. Davis).

HABIT OF SHRUBS

Besides the young individuals of trees belonging to higher strata, the 'shrub' layer of the rain forest includes plants of very varied habit. Some of the species, e.g. the tall Scitamineae, though belonging ecologically to the shrub layer, are not woody and are therefore morphologically herbs. Even among the woody members there is considerable diversity of habit. A few of them, e.g. various Rubiaceae, have no distinct main axis; they branch at or not far above the ground-level, and the plant is rounded or obconical in profile. Such species are true shrubs (or 'Dwarf shrubs') in the physiognomic sense as defined by Du Rietz (1931, p. 46), and only differ from the shrubs of European woodlands in being less branched. Other rain-forest shrubs—probably the majority— have a distinct main axis and resemble trees in miniature. These species belong to Du Rietz's life-forms 'Dwarf trees' and 'Pygmy trees' (1931, p. 45). In some of the latter group the stem is entirely unbranched and the leaves tend to be crowded at the top of the plant, e.g. in the West African *Angylocalyx oligophyllus* (Leguminosae) and *Sphenocentrum jollyanum* (Menispermaceae). Such 'shrubs' differ from the *Schopfbäumchen* already referred to (p. 57) only in their smaller size. The close resemblance in general form between these species and young individuals of true trees tempts one to regard them from an evolutionary point of view as precociously reproducing trees. However this may be, the presence of these miniature trees is a constant feature of the Tropical Rain forest and a striking difference from temperate forests.

Small palms are often abundant constituents of the shrub layer. In tropical America the group is represented by numerous species of *Bactris*, *Geonoma* and other genera, while in the Malayan region species of *Eugeissonia*, *Licuala*, *Pinanga*, etc., are common and sometimes locally dominant in the shrub layer. Since they are unbranched, these small palms, like the dwarf species of *Pandanus* and *Dracaena* sometimes found in the rain-forest undergrowth, are physiognomically very similar to the miniature trees just mentioned.

LEAVES OF TREES AND SHRUBS

The foliage of rain-forest trees and shrubs is very characteristic, not only in the shape, size and structure of the mature leaves, but in their arrangement, method of unfolding and protection (or lack of it) in the bud. Many of these features are undoubtedly connected, either as adaptations or 'causally', with the nature of the environment; most of them are of considerable diagnostic value in the definition of the tropical rain forest as a plant community and in distinguishing it from other plant formations. The evergreen habit, the most important characteristic of rain-forest foliage, will be considered in a later chapter in connexion with seasonal phenomena in general (Chapter 8).

Buds

The buds of rain-forest trees and shrubs, as might be expected, are less well protected than those of trees in other climates. Consequently, except in size, they often look very similar in the resting and the actively expanding condition. The connexion between bud protection and climate has been stressed particularly by Raunkiaer (1934), who divided his phanerophyte life-form into three classes: (i) evergreen phanerophytes without bud-covering, (ii) evergreen phanerophytes with bud-covering, and (iii) deciduous phanerophytes. In passing from the constantly hot and humid Tropical Rain forest to less constantly moist tropical and subtropical regions these three classes of phanerophyte become each in turn dominant, but one class nowhere prevails to the entire exclusion of the other two. Thus in the forests of the Congo basin the number of trees with protected buds increases from the humid centre to the periphery where there is a marked seasonal drought (Lebrun, 1936*b*, p. 174).

Even in the most humid Tropical Rain forest, though there is no cold season and no severe dry season, the leaf rudiments, at any rate of the taller trees, are to some extent liable to damage by desiccation. Thus, though the buds are undoubtedly less well protected than those of trees in other climates, a large number of species have buds protected in one way or another. Many, for instance, have hairy buds and others have buds covered with a mucilaginous or resinous secretion. There is also a considerable number of species with more specialized means of protection, such as wings on the petiole, stipules forming a hood (e.g. *Musanga cecropioides*) or other modifications. Some of these features have been described by Potter (1891), who showed that if the stipular hood which protects the buds of *Artocarpus incisa* was removed the young leaf became dwarfed and deformed. Even the most specialized of all types of bud-covering, namely, bud-scales (i.e. leaves entirely devoted to the protection of the buds), is found in some tropical trees. Holtermann (1907, p. 183), for instance, has described buds provided with scales in the purely tropical genus *Litsea* (Lauraceae) and other trees of the wet zone of Ceylon.

The bud-covering of tropical trees, when one is present, is usually of a soft and sappy consistency, not hard and dry, as in our familiar European trees, but Resvoll (1925) found that the resting buds of five species of *Quercus* in the rain forest of Java had dry bud-scales and differed very little in structure or appearance from those of the temperate *Q. robur*, except that the scales overlapped less closely. As such buds do not seem adapted to the continually moist climate of Java, Resvoll concluded that they are conditioned by internal factors common to the genus *Quercus* as a whole.

Young leaves

The newly unfolded young leaves of the trees are one of the most striking sights of the Rain forest. Their colour usually contrasts strongly with the deep green of the mature leaves, and until some time after they reach their full size the leaves are often almost entirely lacking in chlorophyll and may be brilliant red or crimson. Less frequently other colours are met with—in the African *Pancovia harmsiana* the young leaves are amethyst-coloured (Mildbraed, 1922, p. 110), and in various small unidentified trees in Borneo and the Malay Peninsula the young leaves are of a peculiar steel blue. Stahl (1893, p. 144) records white young leaves in *Cynometra cauliflora* and *Humboldtia laurifolia*, and the author has seen them in species of *Duguetia* in Guiana and in other tropical trees. Species of *Duguetia* with a fresh crop of leaves appear as if sprinkled with white paper. Leaves of an extremely pale green or slightly suffused with red are also found in some species.

When the production of young leaves is seasonal, as it often is, and many trees are in young leaf together, the beauty of the colours is not far short of that of a northern forest in the autumn. The brilliance of the colour often deceives an inexperienced observer into thinking that the trees are in flower.

Red-coloured young leaves are found in species of the most diverse families, but, like buttressing, they are to some extent a specific character, found in some species or genera and not in others closely allied. They are not confined to dicotyledons, but occur, for example, in some small palms. In the herbaceous undergrowth they are less frequent, but the author has seen red young leaves in the fern *Tectaria maingayi* in Borneo. This characteristic colouring of the young leaves is found in all geographical regions of the Rain forest. In Borneo it is a common feature of the trees of all strata; in Guiana it is also very frequent. In the West African rain forest a bright red colouring appears to be less common, but is well shown, for instance, by *Cynometra vogelii*.

Various teleological explanations, none of them very convincing, have been put forward to explain this feature of tropical trees; it has been suggested that the red colour (which is due to anthocyanin pigment in the cell sap) screens the chlorophyll from the destructive effects of very strong light, or that by promoting the absorption of heat it raises the temperature of the leaves and produces a supposedly beneficial increase in the rate of transpiration. These theories were critically discussed by A. M. Smith (1909), who showed that the thin young leaves of various tropical trees reach temperatures which may be either higher or lower than those of the thick mature leaves of the same species. He found that red young leaves in fact attain somewhat higher temperatures than green leaves of similar thickness and texture. Bünning (1947, pp. 101–2) has suggested that the red colouring may be important as a protection against ultra-violet light since this is strongly absorbed by the anthocyanin pigments.

Another peculiarity of the young foliage of rain-forest trees is that for some days after expanding to their full size the leaves are often quite limp and hang down as if wilted. Only later, as they become fully green, do they become stiff and assume an erect position. This also is a feature found in rain-forest trees in all parts of the tropics, though in West Africa it seems, like the red colouring, to be comparatively uncommon. It is best shown by various leguminous trees, but is also seen in many other families. In certain Caesalpinioideae, e.g. *Amherstia nobilis*, species of *Brownea* and *Saraca*, the young twigs as well as the leaves themselves are limp and hang downwards. In these trees the buds expand very rapidly, often overnight, so that Treub (quoted in Schimper, 1935, p. 484) spoke of them as 'pouring out' (*ausschütten*) their leaves. Funke (1929) has recorded interesting data on the time taken for the young leaves of various species to acquire greenness and rigidity.

The limpness of the young shoots of *Amherstia* has been shown to be due to a lack of mechanical tissues, not to the absence of turgor (Czapek, 1909), and in *Brownea grandiceps* Büsgen (1903, p. 438) found that there was an increase of dry as well as of fresh weight during expansion, indicating some true growth as well as the expansion of cells already formed. Probably in all these limply hanging young leaves and shoots there is a delay in differentiation; expansion and differentiation, instead of being more or less simultaneous as in most young leaves, are successive and separated by a considerable interval of time.

This peculiar sequence of events in leaf and shoot development occurs in trees of all stories. It is not quite unknown in other formations, e.g. it is shown by species of *Daniella* and *Isoberlinia* in the West African savannas, but is rare and much less pronounced outside the Rain forest. Various authors have regarded it as an adaptation to strong light, excessive heating or violent rain, or as a device for the better utilization of limited space (see Keeble, 1895; Czapek, 1909). The fact that these suggestions are so various strongly supports the view that the true explanation is a causal one connected with the nature of the climate; adaptive value, if any, may be merely incidental. This point of view is taken by Goebel (1924, pp. 230–6), who regards hanging young shoots (and the same interpretation can be applied to young leaves) as due to something in the fundamental nature of the plant, which in less favourable habitats would be quickly eliminated by natural selection. The shoots hang down because they must, not because it is advantageous for them to do so, and under the conditions of high humidity and temperature of the tropical forest the delay in the stiffening of the tissues which in a more rigorous environment would be harmful to the plant is here 'unpunished'. If it were found, says Goebel, that hedgehogs some-times impale apples on their prickles, need one conclude that prickles are specially adapted for carrying off apples? Many other supposedly adaptive characters of rain-forest plants may be regarded in the same way.

Leaf size and shape

Though in the Tropical Rain forest, as in other plant communities, there is a wide range in the type of foliage, the dominance of entire sclerophyllous leaves belonging to the 'mesophyll' size-class of Raunkiaer (1934, pp. 368-78) is a most striking feature. The type of leaf characteristic of the great majority of the trees and shrubs, regardless of systematic affinities, is the 'laurel type' of Warming; it might perhaps be termed the megasclerophyll to distinguish it from the microsclerophyll ('myrtle type' of Warming) characteristic of Mediterranean climates. Typically it is of a deep, sombre green and from oblong-lanceolate to elliptical in shape; the margin is entire or finely serrate, and there is often a long and distinct acumen, commonly forming a pronounced 'drip-tip'. The texture is hard, often more or less leathery, the upper surface glabrous and often highly polished, any tomentum or hairiness being confined to the lower surface. Where the leaves are compound the individual leaflets tend to approximate in size and shape to the prevailing type of simple leaf. Among the few conspicuous excep-tions are many of the Mimosoideae, which, especially in Africa and America, are often abundant components of the Rain forest; in these the leaflets are of nanophyll or leptophyll dimensions.

These facts are brought out clearly by a study of representative samples of the leaves of rain-forest trees and shrubs. Baker (1938) has published some striking photographs to show the extraordinary uniformity of size and shape in the leaves of the dicotyledons in the Sinharaja Forest in Ceylon. Out of forty-one species examined 90% (belonging to twenty families) had leaves with more or less marked 'drip-tips'. The same uniformity is shown by the samples of leaves from the rain forest of Borneo and the Shasha Forest Reserve ('Wet Evergreen forest') in Nigeria illustrated in Figs. 13 and 14.

An analysis of the sizes of the leaves, using Raunkiaer's classification, shows that in normal Tropical Rain forest at least 80% of the species (and individuals) of trees have leaves of the mesophyll size-class (area 2,025–18,225 sq.mm.). The macrophyll and microphyll classes are sparingly represented, while the nanophyll and leptophyll classes are usually scarce and often quite absent. Megaphylls, which are often regarded as especially characteristic of the wet

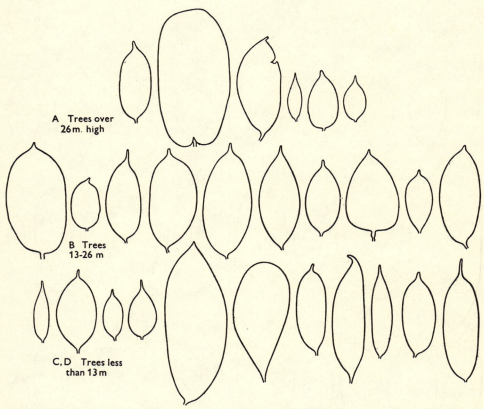

Fig. 13. Leaves of trees in primary Mixed Dipterocarp forest, Kapah River, Mt Dulit, Borneo. One-sixth natural size. Each is a typical mature leaf from a different species of tree. The stories are indicated by letters.

tropics, are comparatively rare in primary lowland Rain forest. The typical distribution among size-classes is illustrated by the figures in Table 8 for the Evergreen Seasonal forest of Trinidad and by the examples in Fig. 15. A statistical comparison of leaf sizes in Evergreen Rain forest and in the more continuously moist 'True Rain forest' would probably show that the average leaf size is somewhat greater in the latter than in the former.

The prevalence of entire margins is as striking as the uniformity of leaf size. In a large sample of leaves from two localities in the Rain forest (Wet Evergreen forest) of Nigeria 80% of the species had entire leaves, and of those with non-entire margins only a few (4% of the whole sample) were lobed or incised. In the Dipterocarp forest at 450 m. on Mt Maquiling in the Philippines, Brown

Fig. 14. Leaves of rain-forest plants from two localities in Southern Nigeria. One-sixth natural size. Each is a typical mature leaf from a different species. Only trees, 'shrubs' and ground herbs are shown. The stories are indicated by letters.

(1919, p. 381) found that 76% of the species and 80% of the individual trees had leaves with entire margins. Similar figures could be obtained for any mature rain-forest community. Exceptions to the normal type of leaf emphasize rather than break the general uniformity; no wonder that in the Amazonian forests Spruce (1908, **1**, p. 37) found the sight of a divided leaf 'a rare treat'!

It is remarkable how the rain-forest environment seems, as it were, to mould the foliage of all species coming under its influence to one particular form. This is well seen when the leaves of tropical species are compared with those of near

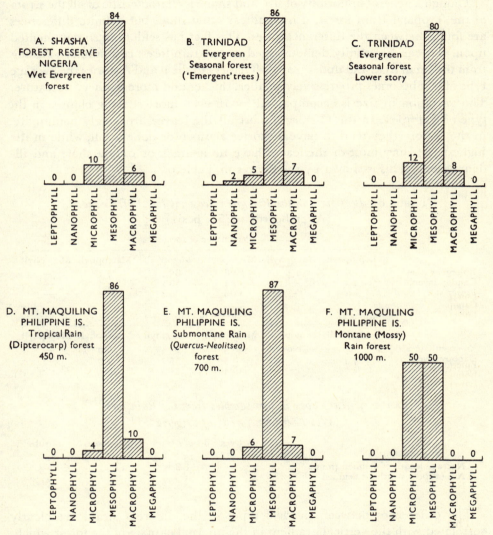

Fig. 15. Leaf size of rain-forest trees and shrubs according to Raunkiaer's classification. The columns represent the percentage of species in each class. B and C from Beard (1946*b*), D–F from Brown (1919).

relatives in temperate climates. In *Quercus*, for example, the leaves vary greatly in size, shape and texture, but in the nearly related, mainly tropical, genus *Lithocarpus* nearly all the species have entire, leathery, often longly acuminate leaves of mesophyll size (see Brenner, 1902). Some species of *Lithocarpus* conform to the normal rain-forest type of leaf with extreme faithfulness, e.g. *L. blancoi*. Another striking example is *Acer niveum*, a tree of the Malayan Rain forest, with entire, finely acuminate leaves, very different from those of most temperate members of the genus.

Though a general uniformity of size and shape is characteristic of all the strata of the Tropical Rain forest, it is significant that small but definite differences are found between the different stories. This fact has seldom been commented upon, but can be readily demonstrated in any rain-forest community. Passing from the small trees and undergrowth (C story) to the B and A stories the average type of leaf becomes progressively smaller, thicker and more leathery in texture. The variation in size is accompanied by an even more striking change in the type of leaf apex. In the C story almost all the leaves are longly acuminate; in the B story they tend to have a shorter acumen or none at all, while in the highest (A) story most of the leaves have no acumen or only a short and ill-defined one. This is shown by the figures in Table 9.

TABLE 8. *Leaf sizes in the Evergreen Seasonal forest of Trinidad.*
(After Beard, 1946 *b*, p. 61)

	Percentage of species					
	Leptophyll	Nanophyll	Microphyll	Mesophyll	Macrophyll	Megaphyll
Emergent trees'	—	2·4	4·8	85·7	7·1	—
'Canopy layer'	2·2	—	17·4	76·1	4·3	—
'Lower story'	—	—	11·8	80·4	7·8	—

	Percentage of individuals					
	Leptophyll	Nanophyll	Microphyll	Mesophyll	Macrophyll	Megaphyll
'Emergent trees'	—	0·0	20·5	83·5	16·0	—
'Canopy layer'	34·3	—	8·5	55·2	1·9	—
'Lower story'	—	—	16·4	71·5	12·1	—

TABLE 9. *Leaf apex in two samples from the Rain forest*
(*Wet Evergreen forest*) *of Nigeria*

	A story trees	B story trees	C story trees	Shrubs
No. of species	10	13	14	10
Average length of acumen (mm.)	2·9	8·8	12·9	12·4
No. of species without acumen	4	1	0	0

The differences between the leaf types in the different strata are clearly correlated with the vertical gradient in the internal climate of the forest which, as will be shown in Chapter 7, is very strongly marked. In each stratum there is a different microclimate, and the corresponding differences in leaf type show clearly the close connexion between leaf type and environmental conditions.

The internal climatic gradient is also reflected in the great differences which often exist between the shade and sun leaves and between the juvenile and adult leaves in the same species. The juvenile leaves may be many times as large as the adult leaves and often differ from them in outline and texture so widely that they would hardly be expected to belong to the same species. They almost always possess a 'drip-tip', even in species where the adult leaves are obtuse or emarginate, e.g. in the West African *Lophira procera* and in *Catostemma fragrans* of British Guiana (Fig. 16). Frequently the juvenile leaves are divided or compound and have non-entire margins, while the adult leaves are entire and undivided; thus the Malayan *Artocarpus elastica* has deeply incised leaves in the sapling, but entire leaves in the adult tree. In *Catostemma commune* of British Guiana the juvenile leaves have three to five leaflets and the adult leaves only a single leaflet. These striking changes in leaf form are found chiefly among trees of the highest story; in third-story trees the form changes very little, if at all.

No quantitative data are available as to differences in the average size or other characteristics of leaves in different types of Rain forest or in different habitats within the same type, but such differences probably exist and might be demonstrated by careful statistical studies. In swampy Rain forest, for instance, leaves of the larger size-classes would probably prove to be more strongly represented than in better drained types. In Borneo the leaves of the trees in the Heath forest, an edaphic climax on white sand soils (p. 244), appear to be on the average smaller, thicker, and harder in texture than in the Mixed forest on clays and loam (Richards, 1936, p. 28). A curious fact, particularly striking in the Malayan region, is that there is a marked tendency for species growing near the banks of rivers to have linear or lanceolate leaves, much narrower than those of allied species growing in the interior of the forest. This is shown in *Dipterocarpus oblongifolius*, species of *Fagraea* (Loganiaceae), *Garcinia* (Guttiferae), *Phoebe* (Lauraceae), *Syzygium* and *Eugenia* (Myrtaceae), etc., and by the leaflets of the pinnate leaves of *Dysoxylum* spp. It is not confined to particular families; it is even shown to some extent by the fronds of riverside ferns. This peculiarity was termed stenophyllism by Beccari (1904, p. 392), who first observed it in Sarawak; it was subsequently reported by Lam (1945, pp. 67-8) in New Guinea and by Strugnell & Mead (1937, p. 133) in the Malay Peninsula. Stenophyllous foliage is also a common feature among plants growing by swift streams in the Cameroons and elsewhere in West Africa; the author recently noted numerous stenophyllous trees and shrubs at Kwa Falls, near Calabar, including species of *Deinbollia*, *Psychotria* and other genera. According to Beccari, stenophyllism is 'due to the action of continuous currents of air, so constant along rivers, and secondly, to that of periodical floods'; but if it has any adaptive significance, it is more likely to be related to resistance to flowing water than to air currents.

As soon as we pass from the Tropical Rain forest to other formations, even those most closely related to it, such as the Submontane and Montane Rain forest, the prevailing type of leaf at once changes. This is well illustrated by Brown's (1919) very thorough study of the relation of leaf type to altitude on a mountain in the Philippines. Statistics were compiled of the leaves in virgin Dipterocarp forest at 450 m. (Tropical Rain forest), in *Quercus-Neolitsea* or

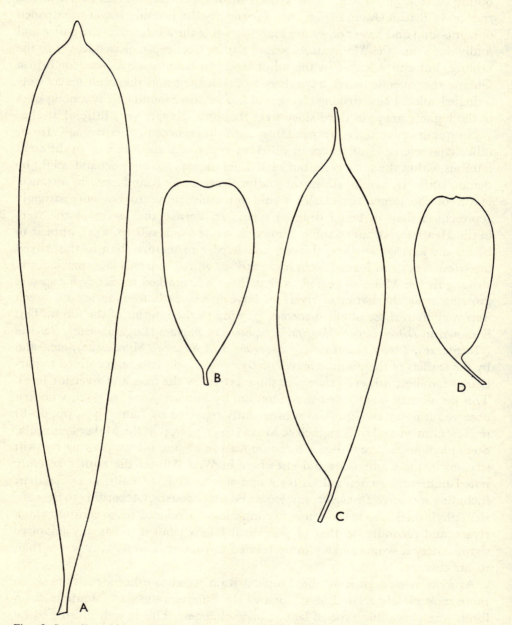

Fig. 16. Juvenile and mature foliage of rain-forest trees.   A. Juvenile leaf of *Lophira procera* from sapling
4 m. high.   B. Leaf from mature tree (over 30 m. high) of same species.   C. Leaf from seedling of
Catostemma fragrans.   D. Leaf from mature tree of same species. A and B from Shasha Forest Reserve,
Nigeria, C and D from British Guiana. All one-third natural size.

Mid-mountain forest (Submontane Rain forest) at 700 m. and in Mossy forest (Montane Rain forest) at 1000 m. As the altitude increases there is a marked increase in the percentage of small leaves and a decrease in the percentage of leaves with entire margins.

The extraordinary similarity of the foliage among rain-forest trees of the most varied systematic affinities cannot be due to chance. The fact that the leaflets of large compound leaves, such as those of the Leguminosae, often mimic simple leaves in both size and form shows still more clearly that the uniformity is due to the environment. To discuss fully why the entire, mesophyll, sclerophyllous leaf (megasclerophyll) is so strongly dominant in this particular environment would lead us far into physiology and causal anatomy. It must suffice to say that the crude teleological explanations put forward in the last century (adaptation to high humidity, violent rain, strong sunlight, etc.) no longer seem satisfactory, but our present knowledge does not allow a convincing alternative view to be suggested. Bews (1927, pp. 15–17) regards the rain-forest type of leaf as xeromorphic and connects its characters with the low specific conductivity of the wood for water which he believes to be characteristic of tropical and subtropical evergreen trees. Stocker (1935b) has made interesting observations on the transpiration of sun and shade leaves in the rain forest of Java.

The exaggerated acumen or drip-tip so often seen in rain-forest leaves has long attracted attention. It is prevalent throughout the tropical rain forest and occurs among ferns (Fig. 17) as well as among flowering plants. In any flora of a moist tropical region the phrases *folia longe acuminata*, and *folia caudato-acuminata* constantly recur. Drip-tips like those of modern plants are seen in the fossil leaves of some Tertiary floras. Outside the Tropical Rain forest well-marked drip-tips are a rare and sporadic feature. Thus they are uncommon in the 'Montane' Rain forest of Jamaica[1] (Shreve, 1914a and b) and even in the peculiar 'Upland Climax forest' of Mauritius they are rare (Vaughan & Wiehe, 1941, p. 152). It has also been seen that drip-tips are commoner and better developed in the lower than in the upper strata of the forest and in juvenile than in mature leaves of tall trees. The evidence therefore suggests a close connexion with the combination of wetness and heat characteristic of the rain-forest climate.

Jungner (1891) was the first to regard drip-tips as an adaptation for the rapid draining of the leaf surface. In the forests on Cameroons Mountain, which he visited with the object of observing the effects of extremely heavy rainfall on vegetation, he was much impressed by the frequency of drip-tips. According to his observations, leaves with a drip-tip dried quicker than those without one. Further, he found that leaves with a drip-tip were less frequently overgrown with algae, fungi, lichens and bryophytes than those without, and he believed that the presence of these epiphyllae interfered with assimilation to such an extent as to be a serious handicap to the plant. Jungner concluded that the

[1] See p. 368.

drip-tips, by allowing the rain water to run off quickly, kept the leaf surface cleansed from the spores of epiphyllae, as well as from the eggs and larvae of insects and soluble substances which might favour the growth of the epiphyllae. Stahl (1893) afterwards showed experimentally that the removal of the drip-tips greatly increased the time taken by the leaves to dry, but he was sceptical as to any correlation between drip-tips and the presence or absence of epiphyllae. The main importance of the drip-tip, he suggested, was to prevent the lingering of a surface film of water which would lower the leaf temperature and therefore depress the rate of transpiration. Stahl believed that a low rate of transpiration checked the uptake of mineral salts by the roots and was therefore disadvantageous, a belief for which there is little evidence (see p. 188).

Subsequently the views of Jungner and Stahl were much criticized, especially by Shreve (1914b). Working with undergrowth plants in the 'Montane' Rain forest of Jamaica, he was unable to show that the removal of drip-tips had much effect on the rate of drying. He found that epiphyllae were abundant on leaves of every type, including those with drip-tips; their occurrence depended solely on the moisture of the atmosphere surrounding the plant (see p. 84). The lowering of the temperature of the leaves due to a surface film of water was too small to affect their rate of transpiration. A water film, however, did decrease the water uptake of the leaves from the stem, probably partly owing to the stoppage of cuticular transpiration (which 'is slightly more than half the water loss in leaves of rain forest hygrophytes'), and partly owing to an actual absorption of water through the lightly cuticularized epidermis. McLean (1919, p. 31) held similar views.

It is unfortunate that the leaves Shreve worked with, as is shown by his photographs, were not typical megasclerophylls and had not the abruptly acuminate apex which is so common in the shrubs and smaller trees of the lowland rain forest. In harder and more glossy leaves the drip-tip might have a greater effect on the rate of drying. More experimental data on this subject are obviously needed.

Shreve's statements on the occurrence of epiphyllae can easily be confirmed by any careful observer; epiphyllae are abundant on any type of long-lived leaf regardless of the nature of the apex. Further, it is doubtful whether the epiphyllae have any serious effects on the host, as the covering of epiphyllae seldom becomes dense until the leaf is old and nearly ready to be shed. Shreve's results on transpiration apply a fortiori to more sclerophyllous leaves, in which cuticular transpiration is presumably negligible. It is therefore unlikely that Stahl's views can be upheld. On the other hand, other possible disadvantages of prolonged wetting are conceivable; for instance, according to unpublished results of Nakano, quoted by Schimper (1935, p. 126), leaves deprived of their drip-tips are said to photosynthesize less actively when wetted than undamaged leaves, the suggested reason being the amount of light absorbed by the water film which remains longer on the former.

A careful study of the effects of high temperature and humidity on the early development of leaves might give some suggestions as to a possible causal approach to the drip-tip problem. In this connexion the work of Holtermann (1907) is of great interest. This author found that in Ceylon the majority of the plants in the wet forests had longly acuminate leaves, while in the dry zone and on the mountain tops they had not. Often the leaves in the dry habitats were not only not acuminate, but were, in fact, emarginate. This difference was shown by different species of the same genus or even within a single species. Thus, the species of *Ficus* inhabiting the wet zone without exception had more or less long drip-tips, while the species in the dry zone had obtuse leaves: in *Memecylon varians* specimens from rainy localities had drip-tips 2–3 cm. long, but those from exposed peaks had none at all. In many species a drip-tip was present on the leaves when young, but later disappeared (see Fig. 16). A drip-tip, when one is present, develops much before the rest of the lamina. If the drip-tip is removed artificially when the leaf is young, the adult leaf becomes emarginate. Similarly, Holtermann found that in all cases investigated where the leaves were obtuse or emarginate when mature this was due to the withering away of the acumen in the young leaf. This withering appeared to be due to the relatively poor water supply of the acumen, the main vein being its only vascular supply. Further studies on the ontogeny of leaves with drip-tips are needed. W. Troll (1938, p. 1028) regards drip-tips as a type of *Nachläuferspitze*, i.e. due to prolonged growth of the leaf apex consequent on delayed differentiation. Such delayed differentiation would probably be encouraged by a uniformly moist atmosphere. On causal grounds it is easy to explain why leaves *do not* produce drip-tips in a dry environment.

Pulvini. Funke (1929, 1931) has recently drawn attention to the fact that 'joints' or pulvini (on the petiole or in compound leaves at the base of the leaflets) are an extremely common feature of the leaves of tropical trees and lianes. The author has also noted it in various ferns of the rain-forest undergrowth. Such 'joints' are comparatively rare in temperate plants, but Fraser & Vickery (1938, p. 154) mention them as 'almost universal' among trees of the Subtropical Rain forest in New South Wales. According to Funke, who worked in Java, the possession of one, two or even three 'joints' is much more typical of tropical leaves than a leathery texture, light-reflecting cuticle and other frequently mentioned attributes. The structure of the 'joints' varies and Funke has distinguished at least fifteen different types. There is no doubt that these structures serve as pulvini and assist in adjusting the position of the leaf lamina in relation to light, a function which, as is well known, is carried out by different plants in different ways. No suggestion can be offered as to why the pulvinus mechanism should be particularly common among tropical and not in temperate plants.

FLOWERS AND FRUITS

It is an accepted botanical principle that the reproductive organs of plants are more 'conservative', that is, less susceptible to the modifying influence of the environment, than the vegetative organs; we should therefore expect to find few common characteristics in the flowers and fruits of rain-forest trees and shrubs. In general this is true, but there are a few peculiarities which may perhaps have some ecological significance.

The commonest and most remarkable of these is cauliflory, the production of flowers on leafless woody stems. A familiar example of a cauliflorous species is the cacao, *Theobroma cacao*, a tree of the C story native in the rain forest of tropical America. To the visitor from a temperate country cauliflory seems one of the strangest features of tropical trees. In most European trees the flowers are usually produced only on the shoots of the current year, or occasionally on those of the previous year and they are thus borne only on small twigs, never on thick branches or on the trunk. In many tropical rain-forest trees, on the other hand, the flowers or inflorescences are borne, sessile or on very short leafless shoots, attached to the main trunk or larger branches, while the smaller branches, and sometimes the whole crown, remain purely vegetative.

Cauliflory may arise in two ways, but in either case the flowering shoots develop from buds originally subtended by foliage leaves. The commoner way is for the axillary buds on the twigs to produce flowers and to continue to do so year after year until the twig has become a thick branch or even the main trunk. Thus the tree may bear flowers anywhere from its crown to its base; in an old tree the older buds will have grown into short shoots with scale leaves and no visible internodes. Should accessory axillary buds arise on these short shoots, thick gall-like outgrowths may be formed. The second way in which cauliflory may arise is for the flower buds produced in the axils of the leaves on the twigs to remain dormant for a long period and to start to develop only after the shoot bearing them has reached a considerable diameter. The flower buds may then actually have to push their way through the bark. The developmental aspect of cauliflory has been discussed by Thompson (1943, 1946).

Several somewhat different phenomena are usually grouped together under the general heading of cauliflory. Mildbraed (1922, p. 115, reproduced at length in Warming & Graebner, 1933, pp. 741–8), in an excellent general account of the subject, has attempted to distinguish between and define them. *Simple cauliflory* is the production of flowers anywhere on the stem, on the trunk and branches as well as on the twigs; it is due to the buds remaining active for an indefinite period. This occurs in *Theobroma*, in species of *Angylocalyx*, *Diospyros*, *Drypetes*, and many other rain-forest trees. *Ramiflory* is the production of flowers on the larger twigs and branches, but not on the main trunk; in this case the buds remain active for a limited period only. Mildbraed gives as an example the African *Turraeanthus zenkeri* (Meliaceae), and the author has noted it in Borneo in *Polyalthia insignis* (Annonaceae) and other species. *Trunciflory* is the production of flowers on the main trunk only. This occurs in the African *Omphalocarpum procerum* (Sapotaceae), in species of *Diospyros* (Ebenaceae), *Cola* (Sterculiaceae), and many other trees. *Basiflory* is the production of flowers at the base of the trunk only, e.g. in the African *Uvariopsis sessiliflora* (Annonaceae) (Mildbraed, 1922) and in the Bornean *Polyalthia flagellaris* (Annonaceae). In another condition, termed *idiocladanthy*, the flowers are borne on comparatively long shoots, leafless or provided only with scale leaves, attached to the trunk or the lower part of the leafy branches, e.g. *Couroupita guianensis* (Lecythidaceae), *Annonidium mannii* (Annonaceae). (Species in which the inflorescence is a leafless panicle are, of course, not placed here, only those whose near allies bear solitary axillary flowers.) Finally, *flagelliflory* is the production of flowers on long, whip-like branches with scale leaves and long internodes which lie on the surface of the ground, e.g. *Paraphyadanthe flagelliflora* (Flacourtiaceae) (Mildbraed, 1922).

These terms of Mildbraed's are of some value in drawing attention to the variety of types of cauliflory which exist, but it should be remembered that every possible intermediate type can be found and that some kinds of cauliflory cannot be easily classified under any of the above headings. The commonest type is probably the production of flowers on the trunk and larger branches, but not on the twigs, e.g. species of *Polyalthia* and many other Annonaceae. Further, a single species may be variable in its behaviour; for instance, the author found that in *Diospyros 'confertiflora'* (Ebenaceae) in Nigeria some trees exhibited simple cauliflory (in Mildbraed's sense) and others ramiflory.

Mildbraed suggests that ramiflory is more primitive than the other conditions, and it is certainly tempting to suppose that ramiflory, simple cauliflory, trunciflory and basiflory, in this order, represent evolutionary stages of increasing specialization. In this connexion the genus *Ficus* is of great interest because it offers examples of a great variety of different types of cauliflory. In a detailed study of the Malayan species of the subgenus *Covellia*, Corner (1933) has shown that a complete series of types can be recognized. In *Ficus lepicarpa* the receptacles are borne in pairs in the axils of the leaves; in *F. fistulosa* and *F. scortechinii* a few receptacles may occur on the leafy shoots, but most are borne in compact clusters on the older branches or the trunk; in *F. obpyramidata* and *F. miquelii* all the receptacles are borne on the older leafless stem branches, and the main and the fruiting branches reach a much greater length, the clusters of receptacles looking like bunches of enormous grapes; in *F. hispida* the receptacles are mostly borne by leafless branches from the main stem, but some of these elongate up to 3 ft. or so and dangle, bearing the receptacles along their whole length. Finally, at the end of the series are the so-called 'geocarpic' species, e.g. *F. beccarii* and *F. geocarpa*. Here the fruiting branches arise only from the lower part of the main trunk up to 3 ft. from the ground and grow into little branched whip-like strands, up to 20 ft. (6 m.) long and an inch (2·5 cm.) or so thick at the base.

They bear caducous stipules near the growing point, but no leaves. As they grow, they hang down and then creep along the ground; when they meet an obstruction such as a bank of earth, they burrow into it like a root. According to Corner, this series might represent stages in an evolutionary process.

Cauliflory, in the broad sense, is quite common in rain-forest trees and shrubs and occurs in many unrelated families. Mildbraed (1922) lists some 278 cauliflorous species for Africa alone and estimates the number in the whole world at over a thousand. It is a striking fact that the great majority of these species are trees of small to moderate size or tall shrubs, belonging to the C story or the shrub layer. Among trees of the A story typical cauliflory is very uncommon, though ramiflory is shown by Malayan species of *Durio*, *Coelostegia*, *Myristica* and *Ficus*, and a species of *Polyalthia* has a habit similar to the 'geocarpic' figs. Some species of *Alexa* in British Guiana are ramiflorous (T. A. W. Davis). Some lianes are cauliflorous (see p. 108) and some ground herbs of the Rain forest show an analogous habit (see p. 102). The relative frequency of cauliflory in the lowest story must be taken into account in any attempt to explain its cause.

Ecologically, cauliflory is of interest chiefly because it is so characteristic of Tropical Rain forest, though in some regions, e.g. Southern Nigeria, it is comparatively uncommon. Outside the Rain forest the habit appears to be extremely rare, though the familiar judas tree, *Cercis siliquastrum*, and a few other temperate trees, show cauliflory of an unspecialized type.

The meaning of cauliflory has been the subject of much speculation, most of it entirely unsupported by experimental evidence. Mildbraed (1922) has discussed some of the more reasonable theories. The best known is that of Wallace (1878, pp. 34–5), who suggested that cauliflory was a device by which the smaller trees of the forest, unable to reach the abundant light and space of the upper stories, could display their flowers where they are more easily seen and reached by shade-loving butterflies. The lowest tree story, as was emphasized in Chapter 2, is normally the densest layer of the forest and the undergrowth is usually comparatively open, so the theory has some *a priori* probability. It is certainly true that many cauliflorous species require cross-pollination by insects —some of them are dioecious—though perhaps the insects concerned are ants rather than butterflies. If Wallace's theory of cauliflory is accepted, a similar interpretation can be given for the curious condition to which Mildbraed (1922) gives the name *penduliflory*. In this case the flowers are borne on long string-like peduncles which may reach a length of 2 m. and hang down from the branches so that the flowers are well clear of the leaves of the crown. This occurs in *Eperua falcata* (Leguminosae), a tall tree of South America, and *Mucuna flagellipes* (Leguminosae), an African liane. Mildbraed also mentions *Couepia longipendula* (Rosaceae), a South American tree, *Parkia pendula* (Leguminosae), an African tree and *Uragoga peduncularis* (Rubiaceae), an African shrub or small tree.

Other theories, for instance, those of Haberlandt (1926) and Klebs (1911, pp. 61–2), approach the problem of cauliflory from the physiological point of view. According to Haberlandt, evergreen trees, which unfold their leaves gradually and not in large numbers simultaneously, store in their trunks and large branches only such food materials as may be necessary for the growth of the flowers and fruits. If the latter are borne actually on the trunk and branches, the food need not be translocated back to the twigs, and there is thus a saving of time and energy. The most obvious objection to this theory is that the taller the tree, the more advantageous should cauliflory be to it, while it has already been seen that cauliflory is found chiefly in small trees. Klebs held that flower formation can only take place when there is a certain surplus of the products of assimilation in relation to the amount of available mineral salts. In tropical trees, in which assimilation can proceed without interruption all the year round, large quantities of assimilates accumulate in the trunk and branches. As at the same time the leaves keep up a constant drain on the supplies of mineral salts, conditions in the bark are always favourable to prolonged flower production from the buds. This is one of the few theories free from teleology.

The fact that tropical trees tend to have very thin bark (see p. 58) may of course explain to some extent how it is possible for flower buds to emerge through the bark, but it is in no sense an explanation of cauliflory. The whole subject is clearly in need of experimental investigation.

The floral biology of rain-forest trees and shrubs has been little investigated. It is well known that in the tropics birds (humming birds, Nectarinidae and many other families), as well as insects, play an important part in pollination. No doubt because flower-visiting insects and birds are abundant at all times of year, wind-pollinated flowers are scarce. Plants belonging to families which in temperate countries are usually wind-pollinated in the tropics are visited by insects and probably pollinated by them, though no nectar is produced. The sweet-scented male flowers of the Malayan oaks (*Quercus* and *Lithocarpus* spp.), for example, attract clouds of small bees, hoverflies and beetles; the female flowers are almost certainly pollinated by them, though they have no scent (Corner, 1940). In the Guiana forest the grass *Pariana radiciflora* is much visited and is probably pollinated by insects. Since the character of the bird and insect fauna is different and the extreme stillness of the air in the undergrowth is perhaps unfavourable to wind pollination, the proportion of species pollinated by various agencies is probably not the same in different stories, but definite information is lacking.

The seeds and fruits of rain-forest trees and shrubs, and their means of dispersal, are much in need of systematic statistical investigation. It is doubtful if there are any characteristics common to the dispersal units of rain-forest trees and shrubs as a whole. In young secondary forest (see Chapter 17) almost all the dominant species have seeds or fruits with efficient mechanisms for dispersal by wind or by animals (chiefly birds and bats); these mechanisms enable secondary forest trees to colonize bare ground and gaps in the primary forest with extraordinary rapidity. In primary Rain forest, wind and animal dispersal is less common, and many trees and shrubs appear to have no special adaptations for dispersal at all. The trees of different stories also seem to differ considerably in their dispersal mechanisms. Wind dispersal seems to be almost entirely confined to trees and lianes of the top story; this might be expected, since it is only this layer which experiences any considerable amount of air movement. In the Rain forest of Nigeria the author found that a large proportion of the taller trees have seeds or fruits apparently adapted to wind dispersal, while in the lower stories wind dispersal is quite absent, many of the species having heavy seeds which drop near the foot of the parent tree. (They may be subsequently dispersed by small mammals.) The same is probably true of the Malayan forest, where the dominant family of first-story trees, the Dipterocarpaceae, has winged (wind-dispersed) fruits. In the South American forest wind-dispersed seeds and fruits are uncommon in any story; in British Guiana the only common wind-dispersed trees of primary forest are *Aspidosperma excelsum* and *Bombax surinamense*, but bird, bat and mammal dispersal is frequent (T. A. W. Davis). In Guiana, and probably in other parts of the tropics, many trees, especially those of the top story, have very large seeds or fruits which drop heavily to the ground. No doubt heavy seeds are of some advantage to tall trees in that they do not tend to become lodged in the foliage of the lower stories, but they are also advantageous in providing the seedling with an ample food reserve so that it can grow quickly to a size large enough to withstand the long period of suppression which is usually in store for it (see Chapter 3). Large heavy seeds or fruits may actually prevent the species from occupying suitable habitats. Thus Witkamp (1925) believes that *Eusideroxylon zwageri* (Lauraceae), a tree which is the single dominant over large areas in Sumatra and Borneo (see p. 258), is absent from steep slopes because its heavy fruits roll down hill. The slow spread of *Mora excelsa*, which is extending its area in Trinidad, is attributed by Marshall (1934) partly to the weight and size of its seeds.

CONCLUSIONS

Our survey of the physiognomic characters of rain-forest trees and shrubs has shown that many of these characters are clearly connected, in some way or other, with the special environmental conditions of the community. Many vary significantly in frequency in different strata, which, in view of the great micro-climatic differences within the forest (Chapter 7), is not surprising. Though the physiognomic features in question are correlated with the habitat, occurring in large numbers of unrelated species wherever the same complex of environmental conditions is met with, for the most part they cannot be demonstrated to be 'adaptations' with a real survival value. Some of them may have a 'causal' origin, at present not understood. Many peculiarities seen among rain-forest plants can perhaps be interpreted in some such way as Goebel interpreted the 'hanging' young shoots of *Amherstia* etc.—they arise from obscure internal causes and are able to survive in the favourable conditions of the tropical forest, while any plant having such features would be quickly eliminated by selection in a more rigorous environment. In this connexion it is perhaps significant that the greatest variety of foliage in the rain forest is found in the undergrowth, where temperature and humidity are extraordinarily constant; the greatest uniformity of foliage is in the upper tree stories which have the least constant microclimate.

But here we are less interested in the more or less hypothetical 'adaptational' or 'causal' significance of the structural features of rain-forest trees and shrubs than in their diagnostic value in characterizing the community as a whole and its constituent strata. Enough has been said to show that the Tropical Rain forest as a formation-type, and the various associations and consociations of which it is composed, abound in highly characteristic structural features more or less independent of the systematic position of the species manifesting them. It is these which give the community its special physiognomy.

CHAPTER 5

THE GROUND HERBS AND THE DEPENDENT SYNUSIAE

The herbaceous ground flora of the Rain forest and the synusiae dependent on other plants for mechanical support or nutrition (the climbers, epiphytes, saprophytes, etc.) all live under specialized ecological conditions, and their constituent species are themselves highly specialized in structure and physiology. Climbers or lianes are a conspicuous feature of almost all rain forests and compete actively with the trees for light and space, but for the rest these very diverse groups of plants play only a minor part in the structure and general economy of the community. They are, nevertheless, of considerable interest to the ecologist, not only because of their wealth of real or supposed structural adaptations (which are for the most part very well known and will be dealt with only briefly here), but because of the closeness with which their distribution reflects variations in microclimatic conditions.

GROUND HERBS

The synusia of ground herbs is not synonymous with the 'field layer' or lowest stratum of the forest, because, as was pointed out in the last chapter, the majority of the plants in this layer are young individuals of trees, shrubs and woody climbers; on the other hand, not all the plants which can be classified morphologically as herbs form part of the field layer. Some grow as epiphytes, not rooted in the ground, others, e.g. some of the larger Scitamineae, though their stems are not woody, reach a height of 6 m. or more, and from the ecological point of view form part of the shrub or even the lowest tree story. In this section we are concerned only with the herbs which actually form part of the 'field layer', i.e. those not exceeding about 2 m. in height.

It has already been emphasized that in the Tropical Rain forest there is seldom a closed layer of herbaceous vegetation, as there often is in woodlands of the temperate zone. In lowland Rain forest, luxuriant herbaceous vegetation is found in small clearings, by paths and streams and in similar openings where the illumination is above the average. In the interior of the forest, herbs are found chiefly as scattered individuals widely separated from one another; sometimes they are almost absent, and social species forming patches of closed vegetation occur only occasionally. On steep slopes and in upland Rain forest, where for one reason or another the tree cover is less dense, herbaceous vegetation is generally more abundant.

Most of the herbaceous species abundant in openings can usually also be found in smaller numbers in the shade, except in the darkest parts of the forest, but some of the species living in the shade, e.g. *Trichomanes* spp. and some other ferns, never grow in openings. It seems, therefore, that there are two ecological elements among rain-forest herbs—they perhaps deserve to be regarded as separate synusiae—the shade-enduring and the shade-loving species. The first group reaches its maximum development under fairly strong illumination, the latter seem to be intolerant of strong light. The factor chiefly responsible for separating these groups is not necessarily light as such; the degree of tolerance for root competition and perhaps for an almost constantly saturated atmosphere may be also involved (see Chapter 7). The shade-enduring species often show a tendency to be social and grow in patches, while the shade-loving species are mostly solitary. The shade-enduring group includes the great majority of rain-forest ground herbs and is to some extent analogous to the 'wood-marginal' species in the ground flora of European Deciduous forest (Salisbury, 1925).

The synusia of ground herbs includes, in addition to dicotyledons and mono-cotyledons, numerous ferns, and in most localities species of *Selaginella* are very constant constituents. Even if the shade-tolerant and the shade-loving species are added together, the number of families and species represented is always restricted, though, of course, it varies very much from one region to another. In the richest regions it is probably never more than a fraction of the number of species of trees and shrubs on the same area. In an area of a few square kilometres, near Moraballi Creek in British Guiana, which is probably fairly typical in this respect, there were less than thirty common flowering plants, perhaps ten to twenty common ferns and one species of *Selaginella* in the 'field layer' of the primary forest (all types), as compared with some hundreds of species of trees and shrubs. In a similar area of primary forest in Nigeria the number of species of ground herbs would be not very different, but definite information is not available. The Malayan Rain forest, on the other hand (if the Mt Dulit district in Borneo is typical), is considerably richer in ground herbs than that of America or Africa. Here the floristic composition of the ground herb synusia varies greatly from place to place, often with no obvious relation to differences in soil or other conditions. Many species tend to occur on one or two neighbouring ridges and apparently nowhere else in the district.

Even in regions where the number of species is relatively large, the number of families of flowering plants represented in the ground herb synusia is small. In all Tropical Rain forests the Rubiaceae are well represented among the ground herbs. A few grasses (Gramineae) appear to be constantly present, but in un-disturbed forest they are not numerous in species or individuals—an important difference between the Rain forest and some of the drier types of tropical forest. In the rain forest of British Guiana in addition to these two families, Cyperaceae, Gesneriaceae, Marantaceae, Orchidaceae and Melastomaceae are usually present. In Southern Nigeria the commonest families include the Acanthaceae, Araceae, Commelinaceae and Liliaceae, though the Rubiaceae easily outnumber any other family of dicotyledons. In Borneo the families present are rather more numerous; those most abundantly represented are Araceae, Begoniaceae, Cyperaceae, Gesneriaceae, Melastomaceae and Zingiberaceae.

Many, but not all, of the ferns of the ground synusia belong to the shade-loving element and increase markedly in abundance in particularly damp and shady places, such as stream valleys and steep rocky slopes. The majority of the ground ferns belong to the Polypodiaceae, but species of *Trichomanes* are often present, though rarely abundant.

The limited number of species and families among the ground herbs is probably a consequence of the peculiar environmental conditions, which are so specialized that it is hardly surprising that comparatively few families have been able to produce species with the necessary equipment for succeeding under them.

The general characteristics and ecological morphology of the ground herbs are hard to describe in general terms because of the great variety of form found among them. In life-form they differ considerably from those of temperate woodlands. Plants of compact habit, such as rosette plants, are rare, and, as might be expected from the uniformity of the climate throughout the year, there are very few plants with resting buds on or below the soil surface. The majority of the herbs have comparatively elongated aerial shoots, which probably last for several years, often showing little tendency to become woody. No life-form statistics are available for the ground herb synusiae of tropical rain forests, but the majority of the herbs are probably, in Raunkiaer's terminology, herbaceous phanerophytes, and not, as in European Deciduous forest, hemicryptophytes or geophytes. Plants with underground rhizomes are frequent, but the rhizomes are adapted for multiplication and migration rather than for perennation. These rhizomatous species tend to be social and to form dense patches several square metres or more in extent, excluding other plants. Examples are *Leandra divaricata* (Melastomaceae) in British Guiana, the grass *Leptaspis cochleata* and species of *Dorstenia* and *Geophila* in West Africa, as well as *Susum malayanum* (Flagellariaceae) and many Zingiberaceae in the Malayan forest. In these, and probably most other ground herbs of the primary forest, reproduction appears to be chiefly vegetative.

Though plants with underground rhizomes, bulbs, tubers or corms are not uncommon, true geophytes in which the aerial shoots die down for some part of every year are rare and, as would be expected, seem to be found chiefly near the climatic limit of the rain forest where there is a relatively severe dry season. Thus they are probably quite absent in the typical rain forest of Borneo and British Guiana, but in the more seasonal evergreen forest of Southern Nigeria there are various Araceae (*Anchomanes* and *Amorphophallus* spp.) which are corm-geophytes and die down for several months every year, the inflorescence being produced in some species before, in others with or after, the new leaves. *Cyanastrum cordifolium* (Tecophilaceae), a common herb in the West African forest, was found to behave as a true geophyte in the Shasha Forest Reserve (Richards, 1939, p. 11), perennating by means of its corm, but in the Benin forests and in the Cameroons, perhaps because of the damper climate, the foliage appears to remain green throughout the dry season. In the Rain forest of the Malay Peninsula geophytes are represented by species of *Tacca* and by some Araceae (*Arisaema* and *Amorphophallus* spp.) with corms or tubers. In the seasonally flooded forests of the Congo there are herbaceous plants with bulbs or rhizomes which produce aerial shoots during the *dry* season only (Lebrun, 1936*a*, p. 26); these could be regarded either as geophytes or as hydrophytes.

The stems of rain-forest herbs tend to be sappy and brittle, no doubt because they have little lignified tissue, but sometimes they become woody, e.g. many Malayan species of *Didymocarpus* and *Sonerila*. In some species the stem is supported at the base by aerial 'prop' roots, e.g. in some Malayan Zingiberaceae.

The herbs of the rain forest are varied in their foliage and contrast with the monotonous uniformity of the trees and shrubs. The leaves of the majority are thin and soft in texture; they are hardly ever of the large, glossy, sclerophyll type. No doubt they are less long-lived than the leaves of the trees and shrubs—as is suggested by the rarity of epiphyllae on them. The greater variety among the leaves of the ground herbs as compared with the trees and shrubs is shown more in their colouring, surface and margin than in their size or general outline. No figures are available as to the relative numbers of leaves of different size-classes on Raunkiaer's scale (see p. 81), but 'mesophyll' leaves certainly predominate. Leaves of the smaller size-classes are, as might be expected, commoner than among the trees and shrubs, and leaves larger than 'mesophyll' are absent or very rare. The broadly elliptical to oblong-lanceolate acuminate shape is as prevalent as in the higher strata, but leaves with non-entire margins are more frequent. Though most of the leaves are to some extent acuminate, pronounced drip-tips are rarer than among the shrubs and smaller trees. The universal tendency towards broad leaves in the rain forest is shown very strikingly in the Gramineae and Cyperaceae of the ground layer; most of the species of these families normally occurring in primary rain forest have leaves quite different in shape from those of their allies in other habitats. The leaves of all typical rain-forest grasses (Fig. 18), e.g. *Ichnanthus panicoides* and *Pariana* spp. in South America, *Leptaspis cochleata* in West Africa, belong to what has been termed the 'Commelinaceae' type (Arber, 1934) and are very unlike the narrow leaves of tropical savanna or temperate grasses; the whole aspect of the plant suggests the Scitamineae or Commelinaceae rather than the Gramineae. Similarly, some of the Cyperaceae (Fig. 17) have remarkably broad, acuminate leaves. The leaves of these rain-forest grasses and sedges seem, as it were, to be trying to conform to the fashionable elliptical shape in spite of the difficulties imposed by their inherited type of construction.

The surface of the leaves in rain-forest herbs is seldom, if ever, as glossy as in most of the trees and shrubs. In general the surface is neither conspicuously glossy nor conspicuously matt. In some rain-forest herbs, as has often been remarked, the leaves have a velvety surface, e.g. in some Araceae and Orchidaceae, also *Neckia* sp. (Ochnaceae) in Borneo; such species, however, are not very numerous. The velvety appearance is produced by the cells of the upper epidermis which, as in petals with a velvety surface, project in the form of papillae. According to Stahl (1896) these papillae cause water falling on the

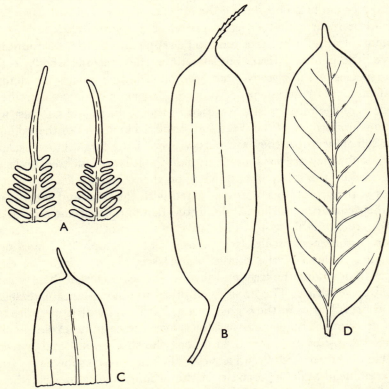

Fig. 17. Convergence in leaf form among rain-forest plants of widely different systematic affinity. A. Pinnae of *Pteris preussii*, a fern of the Nigerian Rain forest. B and C. Leaves of two unidentified Cyperaceae from the undergrowth of Dipterocarp forest, Mt Dulit, Borneo. D. For comparison, leaf of *Goniothalamus malayanus* (Annonaceae), a dicotyledonous shrub from the same locality as B and C. All two-thirds natural size.

leaf to spread out at once into a thin film over the surface, and also act as 'ray traps' increasing the absorption of the leaf for heat and light; he regarded them as primarily a device for increasing the rate of transpiration.

A much commoner feature of the leaves of rain-forest herbs is variegation. The leaves may be spotted or striped with white or paler green, e.g. species of *Begonia*, some South American species of *Cephaelis* (Rubiaceae), the African *Dracaena phrynioides* (Liliaceae), blotched or spotted with red, e.g. Malayan species of *Sonerila* (Melastomaceae). Variegated foliage is frequent among the ground herbs in Rain forest in most parts of the tropics and is very rare in the other synusiae, though a few epiphytes, e.g. South American *Peperomia* and *Vriesia* spp., also have variegated leaves. Stahl (1896) considered that variegation was, like velvet surfaces, an adaptation for increasing the rate of transpiration. Another peculiarity of colouring in the leaves of rain-forest herbs deserving mention is the metallic bluish green sheen well shown by some of the species of *Selaginella* commonly cultivated in hothouses. This also occurs in species of

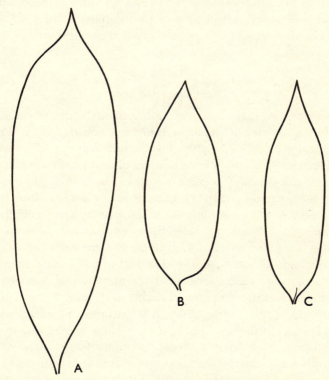

Fig. 18. Leaves of rain-forest grasses. A. Unidentified species from Mixed Dipterocarp forest, Mt Dulit, Borneo. B. *Leptaspis cochleata* from Shasha Forest Reserve, Nigeria. C. *Pariana campestris* from Surinam (from specimen in Cambridge University Herbarium). All one-third natural size.

Mapania (Cyperaceae), e.g. *M. monostachya* of Borneo, in ferns of the genus *Trichomanes* and probably in other genera. The bluish sheen is caused by granules of cutin in the epidermal cell walls; Gentner (1909) regarded it as an adaptation to the light conditions in the undergrowth, but its ecological significance, if any, is far from clear.

The leaves of rain-forest ground herbs thus have a number of peculiar features which have been supposed to be adaptations to the peculiar microclimate in which they exist; some of these features will be referred to again in Chapter 7.

The flowers of the ground herbs, like those of the woody plants, are seldom showy, but have few characteristics in common. A common feature, especially among the ground herbs of the Malayan Rain forest, is for the flowers to be borne at or just above the soil level on stalks which are leafless or bear only scale leaves; this peculiarity is perhaps comparable with cauliflory in trees and shrubs. In the flora of Borneo, among many other examples, are species of *Curculigo* (Amaryllidaceae) and *Forrestia* (Commelinaceae) in which the inflorescences arise from the base of the plant. In *Cyrtandra penduliflora* the inflorescence axis lies limply on the surface of the soil; in many Zingiberaceae it springs from the underground part of the plant at some distance from the leafy shoots. The meaning of this peculiarity is quite obscure.

CLIMBERS

Distribution and habit

Owing to their size and abundance, climbing plants, especially the large woody climbers or lianes, are one of the most impressive features of the Tropical Rain forest; their strange, often fantastic, forms contribute much to its unfamiliar aspect. The ivy, clematis and other climbers of temperate woodlands give but a faint notion of the diversity of such plants in the tropics. Lianes, sometimes called vines or bush-ropes,[1] are always present in normal rain forest; their stems, sometimes rope- or wire-like, sometimes as thick as an arm or thigh, vanish into the mass of foliage overhead and hang down here and there in gigantic loops. Often they pass from tree to tree and link the crowns so firmly that even if the trunk is cut through at the base, the tree will not fall. The stems of tropical climbers may attain incredible lengths. They may ascend one tree, descend to the ground and ascend another tree. Treub (1883, p. 175) measured the stem of a climbing palm 240 m. long. Such a length may be exceptional, and the majority of tropical climbers probably do not often exceed about 70 m.

Though climbers are a fairly well-defined plant form, weak-stemmed trees, such as the South American *Annona haematantha* (Annonaceae), and half-climbing shrubs, such as the African *Icacina trichantha* (Icacinaceae), are transitional between them and fully independent plants. Similarly, no sharp line can be drawn between climbers and epiphytes. Many epiphytes have climbing stems, and some root climbers, such as various Araceae, the American *Carludovica* (Cyclanthaceae) and the Malayan orchid, *Dipodium pictum*, start life rooted in the ground, but eventually lose their connexion with it and become epiphytes. Some hemiepiphytes (see p. 125) when full grown closely resemble large woody climbers.

Climbing plants are found in every climate and plant community where there are trees to support them, but they are far more abundant in the Tropical Rain forest than in any other plant formation. Schenck (1892–3, p. 57) estimated that more than ten-elevenths of all species of climbing plant occur within the

[1] The term 'climbing shrubs' used in some systematic works is misleading and hardly appropriate.

tropics; in the West Indies, where Rain forest and similar formations are the predominant vegetation, woody lianes form about 8 % of all the flowering plants (Grisebach, quoted by Schenck, 1892–3, p. 57). In Europe climbing plants form less than 2 % of the flora. According to Schenck the South American Rain forest is the richest in species of lianes, that of Asia comes next. The great abundance of lianes in the tropics is hardly surprising in view of the great advantage of the climbing habit in a closed and very tall community such as a rain forest. The climbing plant can reach the well-illuminated upper stories of the forest with great economy of stem material and can thus grow very rapidly to maturity.

Though in the early stages of their development rain-forest lianes must, of course, tolerate deep shade, the great majority of them are pronounced light-lovers. For this reason they always increase in abundance with increasing illumination. In undisturbed Rain forest of normal density, lianes, though always present, are never excessively common. They only become abundant enough to impede progress in well-lit places, such as clearings and river banks. By many tropical rivers, especially in South America, the lianes form thick curtains which completely screen the interior of the forest. When trees are felled, the climbers always increase temporarily in abundance and luxuriance. The gaps where large trees have been removed soon become filled with an impenetrable tangle of climbers through which young trees grow with difficulty, and an abnormal abundance of climbers is one of the characteristics of young secondary forest. In plantations the lianes are often a serious pest.

In the *Eperua* (Wallaba) forest at Moraballi Creek, British Guiana (p. 240), which is found on coarse white sand, lianes are markedly less common than in other types of forest on heavier soils in the same district (Davis & Richards, 1933–4, p. 125). A similar difference was found between the Heath forest on white sand in Borneo (see p. 244) and the Mixed forest on clay and loam (Richards, 1936, p. 32). The relative scarcity of lianes in these forest types is not connected with the light intensity, as in both the average illumination is higher than in normal Rain forest, but is possibly due to the slightly lower humidity of the air in the undergrowth; at Moraballi Creek the saturation deficit tends to be greater in the *Eperua* forest than in the other forest types (Davis & Richards, 1933–4, table VI, pp. 376–7). The humidity of the air is also probably lower in Borneo in the Heath forest than in the Mixed forest. It may be that lianes, on the whole, are rather intolerant of high rates of evaporation in the forest undergrowth.

Climbing plants are usually classified according to their means of attachment to their supports, into scramblers, twiners, root-climbers and tendril-climbers. Scramblers neither twine nor have specialized sensitive organs of attachment, though many of them have structures which passively assist them in climbing, such as branches diverging at right angles and recurved spines. Twiners also usually lack tendrils, but the tip of the young stem performs revolving movements in a constant direction, so that the plant becomes securely wound round its support. Root-climbers attach themselves by means of specially modified aerial roots which cling to the surface over which the plant grows. Tendril-climbers, the most specialized of the four classes, possess organs sensitive to the presence of a support which fix themselves actively, usually by curling round it. Tendrils are of very varied morphological nature. Some climbing plants can be classified under more than one of these headings, as they have more than one means of attachment; species of *Hoya* (Asclepiadaceae), for instance, twine and also bear roots at the nodes.

From the ecological point of view the method of climbing is less important than the size of the climber and the maximum height which it reaches. Rain-forest climbers fall into two, not always sharply separated, groups: the large, almost always woody, lianes which reach the crowns of trees in the B (less frequently the A) story and are therefore exposed to more or less full sunlight when adult, and the much smaller, mainly herbaceous, climbers which seldom emerge from the shade of the undergrowth. These two groups differ not only in their microclimatic environment, but in their relations to other plants in the community and are perhaps best considered as separate synusiae. Both groups include climbers of all four morphological types, but in different proportions, root-climbers being relatively less numerous in the synusia of large climbers than in that of small climbers.

The synusia of large climbers includes far more species than the other. The majority of these are dicotyledons of many different families and genera. Among families particularly well represented are the Annonaceae, Apocynaceae, Bignoniaceae, Combretaceae, Connaraceae, Convolvulaceae, Leguminosae, Menispermaceae and Sapindaceae. Among the monocotyledonous climbers the palms occupy the most important place. The rattans (rotangs) or climbing palms are found in all parts of the tropics, but are especially abundant in the forests of Malaya. Here they are represented by a wealth of species and play a large part in the physiognomy of the forest. Crowns of palms appear to be scattered everywhere in the canopy of the forest, but almost all these crowns belong not to tree palms, but to rattans. In the Rain forest of Queensland climbing palms (there known as lawyer vines) are also a conspicuous feature. In the South American and African forest they are much less common, and in West Africa they are usually absent from the normal primary forest and occur chiefly on river banks and in swamp forest. Other families of monocotyledons represented among the large climbers are the Orchidaceae (*Vanilla*) and Pandanaceae (*Freycinetia*); the Araceae should perhaps also be mentioned, though most of the climbing species appear to belong to the synusia of small climbers. Some species of *Gnetum* are tall climbers. A few climbing ferns, e.g. species of *Stenochlaena*, *Teratophyllum* and *Lygodium*, reach a sufficient height to be included in this synusia.

The seedlings and young plants of large woody climbers are at first erect and for some time may not differ in appearance from young trees or shrubs. Later the upper internodes begin to become much elongated and the flexible young stem seeks a support. The young stems often have extremely long internodes and very small leaves, the elongation of the former being accelerated and that of the latter retarded; lianes are thus plants in which the main axes are much etiolated. The nature of the support is probably of little importance; it may be a tree or shrub or the stem of another climber. Root-climbers can attach themselves to rocks as well as to trees. Freise (1939) states that in the Rain forest of south Brazil particular species of liane are always associated with particular species of supporting tree; this observation needs confirmation. A similar relationship has been claimed between the species of tree and its epiphytes (see p. 119). Twiners cannot climb trees of more than a certain diameter, as they are not able to encircle them. Should no suitable support be available in its immediate neighbourhood, the stem of the young climber bends downwards and creeps along the ground until it meets one. As soon as the climber finds something to which it can attach itself, it grows upwards very rapidly until, if the support is tall enough, it finds itself in the strong illumination of the upper levels of the forest. The scarcity of lianes intermediate in size between seedlings and mature plants is due to the short time taken in passing through the middle stages of growth. Some large climbers, if they happen to germinate in open well-illuminated situation with no near supports, remain as short erect shrubs and may flower when little more than a metre high. This behaviour was noticed by the author in *Agelaea trifolia* in Nigeria; inside the forest it reaches a height of 30 m. or more, but on waste land on the edge of native farms the same species is found as an upright shrub with no apparent tendency to climb.

The main stems of large woody climbers are often entirely unbranched, but once the canopy is reached numerous branches are produced and a large crown may be formed. In many species there is a sharp differentiation between the vegetative and flowering branches on the one hand and the greatly elongated climbing stems on the other. The latter are usually leafless in mature lianes, but may bear either scale leaves or normal foliage leaves when young. In many lianes the branches of the crown show no tendency to climb and lack the tendrils or other special means of climbing possessed by the main stem. The latter also often loses its special organs of attachment when it becomes mature. Doubtless the higher light intensity is the stimulus directly responsible for the change in the character of the shoots; the climbing stems are 'shade' shoots, the non-climbing stems of the crown are 'sun' shoots. The leaves of the two types of shoot are sometimes extraordinarily different (see below p. 107). Flowering does not generally begin until the canopy has been reached.

The attachment of old climbers is often a considerable height above the ground, so that great lengths of unsupported stem hang down from the tree tops. In such cases the climber must either have made use of a support which has since decayed and disappeared or it must have originally made contact with a young tree and have been carried up by its subsequent growth.

Part played in structure of forest

Woody climbers play a large part in the structure of the forest. The crowns of tropical trees, as has been stated (p. 56), tend to be less branched and less leafy than those of temperate trees of similar size. Lianes help to close the canopy and also, by covering over the gaps between neighbouring trees, greatly increase the shade cast. They compete with the trees for space and light and often cause their crowns to become one-sided or misshapen. Twiners occasionally have a marked constricting effect; when such climbers are removed from the trunks of felled trees, deep spiral grooves are sometimes seen on the bark. The weight of a climber may break the branches of the supporting tree; the mass of climber then slips down and may either die or begin to climb once more to the canopy.

Since climbers compete with the trees and interfere with their symmetry of growth the forester regards them as weeds; the cutting of their stems is thus an important operation in the routine of forest improvement. The lightening of the canopy brought about by thinning the trees usually greatly stimulates the growth of climbers, and under such conditions there is always a danger of their completely smothering seedlings and small saplings.

The synusia of small climbers, which never reach a great height above the ground and generally pass their entire life history in the shade, is much less important; the number of species composing it is small and their contribution to the forest structure negligible. For the most part they are twiners or root-climbers, or they may climb by both methods. In most types of rain forest this synusia is represented chiefly by monocotyledons and ferns, but in some well-illuminated types, such as the Heath forest of Borneo, small dicotyledonous climbers are rather numerous. In this forest type the synusia includes the insectivorous *Nepenthes*. Among the small monocotyledonous climbers the Araceae are the most important. In South America species of *Carludovica* (Cyclanthaceae) are sometimes a conspicuous feature; they climb by means of aerial roots to a height of some 5–10 m. and may subsequently lose their connexion with the ground, thus passing into the synusia of shade epiphytes. Ferns are very constant members of the synusia of small climbers; among the more important genera are *Lomariopsis*, *Lygodium*, *Stenochlaena* and *Trichomanes*.

Morphology and anatomy

The stems of climbers, especially the large woody species, show great diversity in both their external appearance and their internal structure. They vary greatly in thickness and may be rounded, angular or flattened in cross-section. Sometimes they are spirally twisted or cable-like, sometimes winged or almost like chains. Some of the South American *Bauhinia* (Leguminosae), the 'turtle-step bush-ropes' of the Guiana Indians, are ribbon-like and marked with deep undulations along both edges. In their internal structure, as is well known, the stems of lianes often show extraordinary anomalies in their method of secondary growth. Their anatomical and superficial peculiarities are often characteristic of the family or genus and assist in identification. According to Schenck (1892–3),[1] who has described the anatomy of lianes at length, the type of stem structure has little relation to the method of climbing, and the anomalous structure is often found only in the elongated climbing stems and not in the short 'sun shoots' of the crown. In the main stem the xylem commonly consists of a number of partly or completely separated strands embedded in softer tissue. The liane stem thus often resembles a rope or cable internally as well as externally and combines flexibility with great tensile strength. Both properties are necessary parts of the equipment of the plant for its special mode of life.

The subterranean roots of climbers show, as far as is known, no special characteristics, but the aerial roots of root-climbers often have peculiar features. In the Araceae, for instance, they may be of two kinds, adhering and nutritive. The former grow more or less horizontally away from the light and fix the plant to its support; the latter are strongly positively geotropic. The internal structure of the two kinds of root is strikingly different, the mechanical tissues being better developed in the adhering roots, the conducting in the nutritive roots.

The leaves of rain-forest climbers usually fall into the 'mesophyll' size-class and approximate in texture as well as size to those prevailing in the tree stratum which they reach. In shape, however, they tend to be different. Simple elliptical or oblong-lanceolate leaves are relatively uncommon, and there is a marked tendency, as in all climbing plants, towards short leaves broadest at the reniform or cordate base. Some climbers have peltate leaves. The main veins usually diverge in a palmate manner from the insertion of the petiole. The apex may or may not be acuminate, but well-defined drip-tips are not uncommon. The lamina is usually held at a wide angle to the petiole.

The type of leaf described is so strikingly prevalent among climbers, especially twiners, in the most different families, that it is difficult to believe it is due to chance. Even when the leaves are compound, as in leguminous and connaraceous climbers, the leaflets tend to conform to the prevalent shape. Schenck does not suggest any reasons for the dominance of this type of leaf, but Lindmann (1900) explained it teleologically, suggesting that a broad leaf widest at the base would have the advantage of providing a large transpiring area (which he considered to be necessary owing to the resistance to the transpiration current offered by the long slender stem), and would also allow the leaf to be easily inserted into a pre-existing leaf-mosaic.

The environment of a liane in its early life in the undergrowth is very different from that which it has when mature (see Chapter 7); the plant also, as we have mentioned, passes very quickly from one set of conditions to the other. It is therefore not surprising that in many lianes there are very striking differences between the shade leaves or 'bathyphylls' and the sun leaves or 'acrophylls'

---

[1] A good account of the external morphology of lianes is given by W. Troll (1938).

(terms suggested by Holttum, 1938*a*). This type of heterophylly is seen in an extreme form in certain root-climbers, e.g. in some Araceae (Goebel, 1913, pp. 388–91), in some species of *Ficus*, subgenus *Synoecia* (Corner, 1939), and in *Marcgravia*. In these plants the two types of leaf are so different that it is not surprising they have sometimes been thought to belong to different genera. In *Marcgravia* the young plant has small, shortly rectangular leaves of soft texture, with a velvet surface; they are always tightly pressed against the bark of the supporting tree at the edges. Later when the plant reaches the canopy it begins to branch, and leaves of entirely different form and texture—elongated, sclero-phyllous, glossy and freely projecting—are produced. The juvenile leaves no doubt act as a protection to the aerial roots, which are underneath them and help to retain water round them. In Guiana, the author noted that, as might be expected, the juvenile leaves wilt much more quickly when picked than the adult leaves.

Remarkable instances of heterophylly in climbing ferns are described by Holttum (1938*a*, pp. 427–9). In species of *Lomariopsis*, *Lomagramma* and *Terato-phyllum* the bathyphylls differ from the acrophylls almost as much as in *Marcgravia*, and it is interesting to note that the bathyphylls differ much more in different species than the acrophylls, just as the leaves of rain-forest ground herbs are more varied than those of the shrubs and trees. In Holttum's words, 'these climbing ferns make the best of both worlds; they have their roots in the moist bottom of the forest and develop there a frond system suited to such conditions; and by their strong climbing stems they climb to a lighter place, on the trunks of trees which may otherwise be without tenants, and there produce suitable larger and more resistant sterile and fertile fronds' (*loc. cit.* pp. 428–9).

Such extreme cases of heterophylly are not very common, but in lianes, as in trees, less remarkable differences between the young and old leaves are common. In some South American species of *Bauhinia*, for instance, the young leaves are bifid to the base and the two lobes are drawn out into long drip-tips, while the mature leaves are undivided and end in merely acute points.

The flowers and fruits of climbers offer few features of ecological interest, but it may be mentioned that some lianes, e.g. various Menispermaceae and West African species of *Pararistolochia* (Aristolochiaceae) (Pl. VI B), are cauliflorous, producing flowers only on the unbranched, leafless part of the stem. Cauliflorous species seem to form about the same proportion of the total among lianes as among trees, and whatever interpretation is offered in the one case must pre-sumably apply also to the other.

STRANGLERS

The stranglers are one of the most remarkable ecological groups in the Tropical Rain forest and belong to a life-form which has no parallel in European forests. By a strangler is meant a plant which begins its life as an epiphyte and later sends down roots to the soil, becoming an independent, or almost independent, plant and often killing the tree by which it was originally supported. Though no sharp line can be drawn between them and hemi-epiphytes (see p. 125), which send down roots to the soil, but never become mechanically self-supporting, the two groups are quite distinct in their importance in the general economy of the forest. The stranglers form a synusia which stands on the border-line between the dependent and the independent plants; most of them belong to the genera *Ficus* (Moraceae), *Schefflera* (Araliaceae) and *Clusia* (Guttiferae). A similar habit is shown by South American species of *Coussapoa* (Moraceae) and *Posoqueria* (Rubiaceae) and by some species of *Metrosideros* (Myrtaceae).

In the Rain forests of Africa, Indo-Malaya and Australia the 'strangling figs' (*Ficus* spp.) are abundant in both species and individuals and play a considerable part in the physiognomy of the Rain forest. The seeds germinate on tall trees far above the ground, often in a fork between the trunk and a large branch. The seedling becomes a stout bush which sends down positively geotropic aerial roots, some of which keep close to the trunk of the host tree, others descending vertically through the air like perpendicular cables; the roots eventually reach the ground and ramify in the soil. The descending roots multiply, branch and anastomose till the trunk of the supporting tree is encased in a network of extremely strong, woody meshes; the crown of the fig has meanwhile become large and heavy. After a time the 'host' tree usually dies and rots away, leaving the fig as a hollow, but quite independent, tree in its place. The exact cause of death in the 'host' is not clear, probably it is due to shading and partly, perhaps, to constriction and to competition by the roots of the fig. Some species of *Ficus* are more destructive to their 'hosts' than others (Corner, 1940, pp. 664–5). Some strangling figs become trees of great size, belonging to the highest story of the forest.

In the South American Rain forest, in the author's experience, strangling species of *Ficus* are not abundant, and those which do occur are rather hemi-epiphytic shrubs sending roots to the ground than true stranglers. By far the most important genus of stranglers in this region is *Clusia*, which, at least in Guiana and the West Indies, is represented by numerous species and is very abundant in individuals. The seedlings, like those of the strangling figs, germinate high up on first- or second-story trees. Aerial roots grow down to the ground and anastomose; short clasping roots are produced which fix the plant firmly to the supporting tree. These clusias develop large crowns, but do not form such a stout or close network of roots as the strangling figs. They seldom kill their host, though they often greatly deform and stunt its crown, and they do not usually become independent trees. An interesting detailed account of the life history of the West Indian *C. rosea* is given by Schimper (1888, pp. 56–60).

EPIPHYTES

Epiphytes grow attached to the trunks and branches of rain-forest trees, shrubs and lianes, some even on the surface of living leaves. In a closed forest the epiphytic habitat is the only 'niche' available for plants which combine small size with relatively high light demands; the advantages of good illumination are, however, offset by a precarious water supply and a lack of soil. A few epiphytes grow in the rain-forest undergrowth, where the conditions of illumination are little, if at all, more favourable than at ground-level; these may perhaps be plants which are intolerant of root competition or of the danger of smothering by the continual rain of dead leaves, or their roots may require better aeration than those of the ground herbs. The mode of life of epiphytes is highly specialized, and the epiphytic flowering plants and ferns of the rain forest consequently differ widely in physiognomy as well as in physiology from the herbaceous ground flora, to which, indeed, they are taxonomically not closely related. Schimper (1888) regarded tropical epiphytes in general as evolved from terrestrial plants growing in wet shady forests. The extreme intolerance of shade of most epiphytes is, however, against this view, and Pittendrigh (1948) has recently argued strongly in favour of the theory (originally suggested by Tietze) that the Bromeliaceae, the largest family of tropical American epiphytes, are derived from terrestrial ancestors which lived under semi-desert conditions. The same is probably true of the epiphytic Cactaceae, and the majority of epiphytic flowering plants may perhaps have had a similar origin.

In suitable open habitats (bare rocks, seashores, savannas) many epiphytes can grow successfully on the ground, but in the forest they are entirely dependent on their 'hosts', though usually for mechanical support only. Typical epiphytes do not require organic food obtained from other plants and are thus to be distinguished from the synusia of parasites. Only the Loranthaceae (mistletoes), which are abundant in all tropical forests, are semi-parasitic as well as epiphytic.

Though epiphytes are wholly dependent on other plants, they in their turn have little effect on their supports. A tree heavily laden with epiphytes may be somewhat more liable to be overturned in a high wind than a tree without them; by tending to retain water, epiphytes perhaps hasten the attacks of parasitic fungi, but in normal tropical rain forest they are seldom sufficiently abundant to be of vital importance in either of these ways. Thus, though epiphytes are a very characteristic element in the structure of the forest, their role in its economy is small. Their chief interest lies, on the one hand, in the clearness with which their distribution is correlated with ecological factors (microclimate, nature of the substratum), and on the other in their extraordinary structural specializations, which, more perhaps than those of any other group of plants, truly deserve to be called adaptations.

Epiphytes are interesting not only to the botanist; they are important as providing the chief, in some cases the only, habitat of certain animals, some of which play an important part in the rain-forest ecosystem. The roots of epiphytic flowering plants and ferns are one of the chief nesting places for arboreal ants. Many of the Bromeliaceae enclose large masses of humus and water in their rosettes of closely overlapping leaves, which are the home of a large and varied fauna. Terrestrial animals live in the humus above the water-level, and the water itself is the breeding place of frogs, large numbers of mosquito larvae and other aquatic insects. In tropical America the Bromeliaceae are so abundant that their 'tanks' form veritable 'aerial marshes' (Picado, 1912; 1913). Since the bromeliad-breeding mosquitoes include malaria carriers, in inhabited districts these 'marshes' are of great importance in the spread of disease.

In the Tropical Rain forest epiphytes are constantly present, but their abundance is very variable. There is often a greater variety of species and a larger number of individuals of epiphytes over a given area than of herbaceous plants on the ground. The abundance and variety of epiphytes is one of the most striking differences between the Tropical Rain forest and temperate forests. As well as algae, fungi and bryophytes, the epiphytic flora of the Rain forest includes a wealth of pteridophytes and flowering plants. It is the presence of these vascular epiphytes which especially distinguishes the Tropical Rain forest from temperate plant communities. In Britain the fern *Polypodium vulgare* is the only higher plant which is a true epiphyte, though various flowering plants occasionally flourish and may even reach maturity as epiphytes on trees. The normal habitat of these flowering plants is on the ground and their presence on trees is always more or less accidental—in Oliver's terminology (1930) they are 'occasional' or 'ephemeral', not 'typical', epiphytes. *Polypodium*, though it frequently grows on rocks and on the ground, grows to maturity and reproduces on trees and has some claim to be regarded as a 'typical' epiphyte. In the Tropical Rain forest the vast majority of the epiphytes are 'typical' and seldom or never grow on the ground; perhaps because of the poor and specialized nature of the ground flora 'occasional' and 'ephemeral' epiphytes are rarely seen. A few rain-forest epiphytes are shrubs of considerable size (e.g. some Ericaceae and Melastomaceae); the presence of these shrubby epiphytes was regarded by Schimper (1935, p. 403) as a feature distinguishing the Rain forest from other tropical forest formations.

An abundance of epiphytes, both vascular and non-vascular, is characteristic of the Montane and Subtropical Rain forest, the two plant formations most nearly resembling the Tropical Rain forest. In these cooler and perhaps even more constantly moist communities, for example, in the Montane forest of the eastern Himalaya and of Kinabalu and other mountains of the Malayan region, the epiphytic vegetation, though less rich in species, reaches a degree of luxuriance seldom approached in the lowland tropical forest. Accounts of forests in which the trees are weighed down and almost hidden by their load of epiphytes generally refer to Montane or Subtropical forests and not to true Tropical Rain forest.

Epiphytic flowering plants and pteridophytes

The epiphytic flowering plants of the Tropical Rain forest belong to a large number of genera, but to a very limited number of families. Schimper (1888) enumerated about thirty-three families and 232 genera of epiphytes, of which the majority are found in the tropics.[1] In addition, Schimper lists some twenty genera of pteridophytes. The number of epiphytic genera has been considerably increased since Schimper's time both by the subdivision of genera then known and by the discovery of new genera, but his list of families is still approximately complete. The families of flowering plants containing the largest number of genera and species of tropical rain-forest epiphytes are the Araceae, Bromeliaceae and Orchidaceae among the monocotyledons and the Asclepiadaceae, Cactaceae, Ericaceae (including the Vaccinioideae), Gesneriaceae, Melastomaceae and Rubiaceae among the dicotyledons. There are also a few epiphytic species of *Pandanus* (Pandanaceae) and *Solanum* (Solanaceae). The epiphytic pteridophytes of the rain forest include, besides numerous ferns, species of *Lycopodium* and *Psilotum*.

Most of the families of epiphytes are found in both the Old and New Worlds, but the Bromeliaceae, which in tropical America is one of the most abundant and well-represented families, is entirely absent in the other continents.[2] The epiphytic Asclepiadaceae are restricted to the eastern tropics. The Cactaceae is represented in tropical America by several genera and many species of epiphytes, but the only Old World genus is *Rhipsalis*, which occurs in tropical Africa, the Mascarene Islands and Ceylon.

Within the boundaries of the Tropical Rain forest the abundance and richness of the epiphytic flora varies considerably both from one region to another and from place to place within any one district. It should be noted that even under favourable conditions only a small proportion of the trees in normal Rain forest bear vascular epiphytes. The following figures for trees on clear-felled plots of Mixed Rain forest are probably typical: at Moraballi Creek, British Guiana, out of 193 trees 5 m. or more high on two plots, thirty (16%) bore vascular epiphytes, and on another plot, out of fifty-five trees 14 m. high or more twenty-one (38%) bore epiphytes (Davis & Richards, 1933–4, p. 378). At Mt Dulit in Sarawak 13% of the trees on one plot and 11% on another bore epiphytes (Richards, 1936, p. 16). Two plots in Nigeria gave figures of 15 and

---

[1] A few of the genera included in Schimper's list are hemi-epiphytes which become independent and are here regarded as stranglers (see p. 109 above).

[2] Chevalier (1938b) has recorded a terrestrial species of *Pitcairnia* from French Guinea.

24 % respectively (Richards, 1939, p. 27). The number of species of epiphytes on a single tree is seldom very large even in rich districts. The largest numbers of species recorded by the author are about fifteen in British Guiana and about thirteen in Nigeria. In the Budongo forest in Uganda, Eggeling (1947, p. 55) has recorded as many as forty to forty-five species of flowering plants and ferns on one tree; Longman & White (1917) listed sixteen species of vascular and thirty-four species of non-vascular epiphytes on a single, not exceptionally rich, tree in the Rain forest of Queensland.[1] No information is available as to the number of individuals of epiphytes for the tropics, but in New South Wales, Turner (quoted by Longman & White, 1917, p. 64) counted over 200 individuals of epiphytic orchids on one tree.

Of the great regions of the Tropical Rain forest, America is by far the richest in epiphytes and Africa the poorest. While it is, of course, dangerous to generalize from limited areas, the following data illustrate this point rather strikingly. In the Moraballi Creek district of British Guiana, an area of only a few square kilometres, the Oxford Expedition collected about ninety-eight species of epiphytic flowering plants and about thirty-four epiphytic ferns, and it was estimated that the actual number present was considerably greater. In a slightly larger area in Southern Nigeria the total number of vascular epiphytes found was only about twenty-one (including ferns), and it was estimated that the actual total present could not have been more than thirty-five.[2] Both districts were probably fairly typical of the lowland rain forest of their respective regions and were comparable in many respects. The investigation of a rain-forest area in tropical Asia or Australia would probably give an intermediate result.

These regional differences in the richness in species of the epiphytic flora are not correlated with differences in percentage of trees bearing epiphytes or with differences in the number of species per tree, as shown by the figures previously given. Local differences in the richness and abundance of the epiphytic flora are plainly related to climatic factors, but the cause of these regional differences is less clear. Owing to the uncertain nature of their water supply, the first requirement for a rich epiphytic vegetation is a well-distributed rainfall accompanied by a low average rate of evaporation. Such conditions are met with throughout the Rain forest, though it is true that in a large part of the African rain forest there is a longer and more intense dry season than in much of the American and Asiatic Rain forest. The relative poverty of the African forest in epiphytes may have a climatic cause, but this can hardly explain the great difference between the abundance of epiphytes in the Old and New World tropics generally, and some historical cause must probably be invoked. The

[1] In Tropical Montane and Submontane Rain forest much larger numbers of species are present. Bünning (1947) found over fifty species of ferns alone on one tree in a forest (probably Submontane) in the mountains of Sumatra.

[2] The Nigerian figures may give an exaggerated view of the poverty of the African forest in epiphytes. In Budongo forest, Uganda (area 352 sq. km.), Eggeling (1947, p. 55) has found nearly 100 species of vascular epiphytes.

greater richness of the New World is largely accounted for by the presence of the Bromeliaceae in addition to the pan-tropical families of epiphytes. For some reason a family able to play the role of the Bromeliaceae, which often act as pioneers and flourish in conditions which no other epiphytes could tolerate, has never evolved in the Old World.

Habitat and physiology

The habitat of epiphytes differs in several important respects from that of plants living on the ground. The fact that the substratum is raised above ground-level and is inclined at an angle varying from horizontal on large branches to vertical on main trunks in itself creates problems both for seed dispersal and successful maintenance. Secondly, the substratum is usually more or less smooth and therefore gives poor anchorage, though here and there hollows and fissures may offer easy sites for colonization.

Another important characteristic of the habitat is the shortage of soil. Epiphytes are necessarily dependent on the small quantities of debris in cracks and hollows on the tree or on such soil as they can themselves collect. Many epiphytes ('bracket epiphytes', 'nest epiphytes'), as is well known, are constructed so as to collect soil, and they always exploit this soil very intensively. The root systems of almost all epiphytes are inhabited by ants which gather dead leaves, seeds and debris of all kinds over a wide area and probably play a very important part in providing soil. Almost all the 'soil' available for epiphytes consists of humus derived from the dead remains of other plants. The humus on trees is important to epiphytes both as a source of mineral nutrients and because of its water-holding capacity. The scarcity of free inorganic ions for rain-forest epiphytes is accentuated by the leaching action of frequent heavy rain. The humus requirements of different species of epiphytes vary very greatly. In his study of the epiphytic vegetation of the Tjibodas (Submontane) forest in Java, Went (1940) has shown that some species, such as *Pholidota ventricosa*, in exposed situations always grow in large masses of humus, but in shady places can grow without much humus; for these epiphytes humus is probably more important as a reservoir of water than as a source of nutrients.

Apart from the mineral matter liberated from the decay of humus very small amounts may reach epiphytes as dust. According to Dixon (1882) the leaves of large epiphytic ferns in New South Wales have a remarkably high ash content (over 12%), and in the 'soil' collected by them there may be as much as 62% of 'sand'. For the luxuriant colonies of epiphytes on roadside trees (e.g. the rain tree, *Samanea saman*), the mineral nutrients conveyed in dust are no doubt important (see Pessin, 1925, pp. 32–3), but they can hardly play a large part in the interior of the forest where the atmosphere is singularly dust-free. Lausberg (1935) has shown that the leaves of many plants excrete considerable quantities of calcium, potassium and other ions. If rain-forest trees behave similarly, the mineral matter so excreted may form an important addition to the meagre nourishment of the epiphytes. Excretions from the trees must determine the remarkably specific relation between the species of tree and the composition of its epiphytic flora which sometimes exists (see p. 119).

Miehe (1911) examined the humus from the roots of epiphytes in Java and found no evidence that the microbiological processes taking place in it differed from those in normal soils. The existence of *Azotobacter* was not definitely demonstrated, but the data did not justify the conclusion that nitrogen fixation was not occurring. Nitrifying and cellulose-decomposing organisms are present and function as in other soils. According to Spanner (1939) all true epiphytes probably obtain their nitrogen as ammonium ions; in water culture the epiphyte *Myrmecodia* grew better in solutions containing ammonium salts than in nitrate-containing solutions of the same concentration.

The relatively small volume of soil which their roots can exploit, combined with excessive drainage and often with high rates of evaporation, renders epiphytes liable to frequent periods of severe water shortage—this is one of the fundamental differences between their habitat and that of the rain-forest ground flora. Even in a climate where rain falls almost daily, the epiphytes may lose large amounts of water in the intervals between showers. In most parts of the Rain forest drought periods lasting for days or weeks are not uncommon and unless the epiphytes can conserve their water supply or are drought-resistant they cannot survive. During the drought year 1923 in north Queensland large numbers of epiphytes died (Herbert, 1935, p. 350). An epiphyte must be able to absorb water quickly when it is available and not lose it too fast when it is not. The lack of vascular epiphytes in most temperate regions is doubtless due not directly to the low temperature but to the long 'physiological' drought of the winter. Only the smaller cryptogamic epiphytes, the mosses, hepatics and lichens, with their small water requirements and remarkable powers of surviving in an air-dry condition can flourish in such a climate.

The habitat of epiphytes is to some extent similar to that of plants growing on rocks. In both there is little soil and the water supply is usually intermittent; often the conditions of illumination and atmospheric humidity are similar. It is therefore not surprising that the flora of the two habitats is much alike and that some tropical epiphytes also occur on rocks.

The transpiration of epiphytic orchids has been investigated by Kamerling (1912) who found that the rates were very low; it has therefore been generally supposed that epiphytes as a group are characterized by low rates of transpiration. Spanner (1939), however, in his interesting study of the physiology of *Myrmecodia* and *Hydnophytum*, has shown that these two epiphytes, which are superficially very similar and often grow together in the same habitat, differ considerably in stomatal behaviour and rate of transpiration. Stomatal regulation is slow in *Hydnophytum*, rapid in *Myrmecodia*. The transpiration of a whole plant of *Hydnophytum* attains only about one-third of the maximum rate in *Myrmecodia*. Spanner believes that the view that most epiphytes have low rates of transpiration cannot be maintained and that among epiphytes, as in most other ecological groups of plants, there are wide physiological differences, some species having low, others high rates of transpiration. The osmotic pressure of epiphytes has been examined by Senn (1913) and Harris (1918) and more recently by Spanner (1939) in *Myrmecodia* and *Hydnophytum*. In general the osmotic values appear to be low. Blum (1933) found that in epiphytes in the Submontane Rain forest at Tjibodas (Java) the average suction pressure of the leaf cells in epiphytes with well-developed water-storage tissue was 8·4 atm., in those apparently without water-storage tissue 16·4 atm.

Factors controlling the distribution of epiphytes within the forest

The habitat of an epiphyte is almost always better illuminated and exposed to a lower and more variable atmospheric humidity than that of a plant living in the same climate on the ground; and in any closed plant community there is a gradient in microclimate from the ground upwards. In the rain forest the average illumination and the average amount of air movement increase from the herb layer to the highest tree story. In Chapter 7 it will be shown that there are parallel gradients in the range of temperature and in saturation deficit. The conditions in the lower levels of the forest are remarkably constant and differ widely from conditions in the upper levels, which approximate to those in the open.

Different species of epiphytes differ widely in their microclimatic requirements, as has been shown, for instance, by Wiesner's (1907, pp. 135–8) extensive studies of their light-requirements and by Went's (1940) recent work on the sociology of epiphytes at Tjibodas in Java (Fig. 19). Some species are restricted to very well-illuminated or to very shady habitats, while others are tolerant of a wide range of conditions; others again avoid both strong light and deep shade. These specific differences in requirements or tolerance are mainly responsible for the very marked variations in the composition of the epiphytic vegetation both with height above the ground and from tree to tree. The microclimatic gradients are by no means regular and vary from place to place owing to the irregular stratification of the trees and to specific and individual differences in the density of their crowns. Within the forest the epiphytes are therefore not regularly stratified, but form a series of superposed communities of very variable vertical distribution. Microclimatic differences other than those due to the vertical gradients sometimes have an important effect on the epiphytic vegetation. In Budongo forest, Uganda, the epiphytic flora of the swamp forest differs from that of the normal forest and some species, e.g. *Angraecum infundibulare*, are found only near water (Eggeling, 1947, p. 56).

Differences of inclination and aspect between different parts of the same tree affect the epiphytic vegetation not only by influencing the colonization of seeds and spores, but also by modifying the illumination and evaporation. Thus the

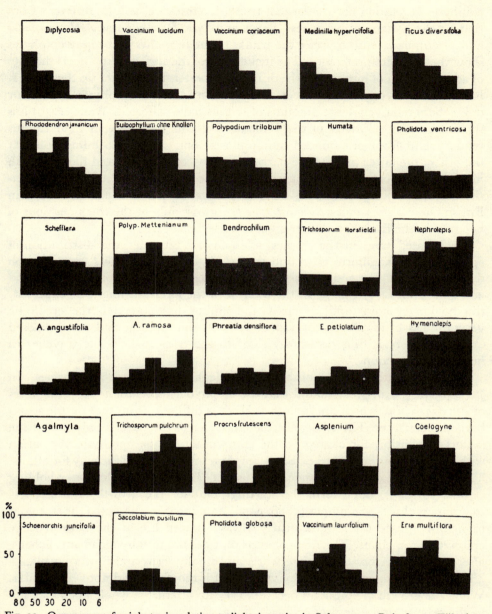

Fig. 19. Occurrence of epiphytes in relation to light intensity in Submontane Rain forest, Tjibodas, Java. Abscissae: light intensity on branches of 'host' tree as percentage of light intensity outside forest. Ordinates: percentage of trees at given light intensity bearing each species of epiphyte. *A. angustifolia, A. ramosa = Appendicula angustifolia* and *Appendicula ramosa. E. petiolatum = Elaphoglossum petiolatum. Trichosporum Horsfieldii = Aeschynanthus horsfieldii. Trichosporum pulchrum = Aeschynanthus pulchra.* From Went (1940).

epiphytic vegetation on the vertical trunk, for instance, will be different from that on the horizontal branches. Large epiphytic ferns such as *Asplenium nidus* and *A. africanum* tend to prefer the trunks to the branches, but many epiphytes show the opposite preference. The inclination of the surface also affects the rate of accumulation of humus and the drainage of water from higher up the tree. It has long been recognized that certain species are 'humus epiphytes' and occur only where relatively large amounts of humus have collected. Went (1940) has shown that the epiphytes of the Tjibodas forest in Java vary in their humus requirements from pronounced humiphiles at one extreme to humiphobes at the other; the latter prefer bare bark and seem actually to avoid habitats with much humus. A characteristic association of species occurs on humus accumulations. Some epiphytes have a marked preference for drainage channels; in British Guiana the small *Utricularia schimperi* grows only in the thick cushions of hepatics in the drainage channels below forks in the trees.

On isolated trees in the tropics, as in higher latitudes, the distribution of epiphytes is not uniform all round the trunks, different species occurring on different aspects, depending on the different degrees of exposure to sun, wind and rain (van Oye, 1921). In the interior of the forest, as might be expected, this no longer holds and, as far as the author has observed, the epiphytic vegetation is little affected by aspect even on the topmost and most exposed branches; there is thus no 'mossy side' to the trees to help the traveller in finding his direction.

The epiphytic vegetation on a tree also depends on its age and species. In Java, Went (1940, p. 22) found that the epiphytic 'association' on young trees of *Altingia excelsa*, which are smooth-barked, is different from that on the scaly-barked older trees. The change in the character of the bark is only one of the factors responsible for change in the composition of the epiphytic vegetation with increasing age of the tree. Some epiphytes may prepare the way for others. The Bromeliaceae, for instance, because of their remarkable water-holding powers (p. 125), often form the starting-point for the growth of other, less tolerant epiphytes. Schimper (1888, p. 58) noted that in Dominica the strangler or hemi-epiphyte *Clusia rosea* nearly always started life at the base of large epiphytic bromeliads. The first colonists on trees are usually algae and lichens, afterwards come mosses and liverworts and finally ferns and flowering plants. There is thus a succession of epiphytes, but it is indefinite and not very regular. When the growth of epiphytes becomes very luxuriant the whole mass may fall to the ground and colonization begin anew.

Differences in the abundance and floristic composition of the epiphytic vegetation depending on the species of tree are often striking. Thus, on the large bamboo *Dendrocalamus* and other trees with very smooth hard surfaces, the succession of epiphytes seldom passes the lichen stage. In Uganda shaggy-barked trees such as *Entandrophragma utile* are more heavily colonized than smooth-barked species, but the orchids *Ancistrochilus rothschildianus* and *Eulo-*

phiopsis lurida are exceptional in preferring the boles of smooth-barked trees (Eggeling, 1947, p. 56). The abundance of epiphytes on *Samanea saman*, compared with other trees commonly planted in tropical towns, is often very marked. Schimper (1888) remarked that in the southern United States *Epidendrum conopseum* occurred chiefly, though not exclusively, on *Magnolia*. In the Philippines the fern *Stenochlaena areolaris* is said to be epiphytic exclusively on *Pandanus utilissimus* (Copeland, 1907).

Until recently it has been generally assumed that the differences between the epiphytic vegetation of different species of tree are mainly due to purely physical factors such as differences in the roughness, water-absorbing power and method of peeling of the bark. The remarkable work of Went (1940) on the epiphytes of the Submontane Rain forest at Tjibodas, however, strongly suggests that chemical as well as physical factors play an important part. The epiphytes on a large number of trees of many species were examined by means of field-glasses and a telescope magnifying up to forty-two times. It was found that nearly all the species of vascular epiphytes could be readily identified in this way. The abundance of different epiphytes on each tree was recorded on an arbitrary scale. The composition of the epiphytic 'associations' was then tabulated in relation to species of tree, roughness of bark and other factors. Very great differences between the flora of different species of tree were apparent and, though the flora of each species of tree tended to change with age, the differences between one tree species and another were greater than between different individuals of any one species. In some instances the assemblage of epiphytes was so characteristic that the tree could be identified by its epiphytes; thus the different species of *Castanopsis* could be distinguished in this way. An analysis of the flora on different tree 'hosts' made it appear improbable that the specific differences were due only to physical differences in the bark, density of crown, etc.; Went felt compelled to assume that they depended largely on the chemical composition of the bark or of the drainage water on its surface. The presence of soluble substances in this water, e.g. tannin on *Castanopsis argentea*, could be readily demonstrated. It was noteworthy that humicolous epiphytes growing in large humus accumulations showed no preference for particular species of tree.

Such a specific relationship between the epiphytic vegetation and the 'host' tree, as Went has demonstrated, has not previously been suspected, but it may well prove to be quite general. There is evidence that the relations between the epiphytic mosses, hepatics and lichens of temperate regions and the species of tree are also to some extent specific.

Synusiae of epiphytes

The complicated interaction of the microclimatic gradient with the factors of the substratum which have just been discussed gives rise to a very great complexity in the distribution of epiphytes within a given area of forest, but in general, as has already been stated, they form a number of fairly well-defined superposed communities. These were recognized by Schimper (1888) who named them *étages* (stories), a somewhat misleading term. Here they will be regarded as synusiae and for practical purposes it will be sufficient to distinguish only three: (*a*) shade epiphytes, (*b*) sun epiphytes, (*c*) extreme xerophilous epiphytes.

Shade epiphytes are found mainly in the C story of the forest, on the trunks and branches of C-story trees, on young A- and B-story trees and on the stems of large lianes. Vascular epiphytes are found only rarely on the lower part of the trunk of large trees. On very large trees of the A story with very dense foliage the synusia of shade epiphytes occasionally reappears above the first fork in the centre of the crown. This community exists in deep shade and an atmosphere with a low, little-varying saturation deficit—a microclimate which differs little from that of the ground herbs. The shade-epiphyte community commonly consists chiefly of ferns, usually including the filmy ferns, *Hymeno-phyllum* and *Trichomanes* spp. Orchids and other flowering plants are seldom abundant, but in British Guiana, for instance, the minute *Peperomia emarginella* and one orchid, *Cheiradenia cuspidata* (always found clinging to large trunks at about 1–2 m. above the ground), are components of this synusia. In the Malay Peninsula, especially on trees of *Saraca* growing by streams, shade epiphytes, including *Hedychium longicornutum*, *Medinilla* spp., *Lycianthus parasitica* and many small orchids, are sometimes very plentiful (Corner).

Most shade epiphytes show no traces of xeromorphy, but, on the contrary, have a highly hygromorphic structure. About their actual water relations little is known. Schimper (1935, p. 24) states that in the thin-leaved epiphytic ferns growing in the shade, such as the filmy ferns, both losses and uptake of water are small and the daily march of transpiration follows the evaporation curve closely. The osmotic pressure of the root and leaf cells is low (not usually more than 4–6.5 atm.). In the filmy ferns, as is well known, the cuticle is permeable to water; the cells can imbibe large quantities of water through their walls and as rapidly lose it again; the fronds thus curl up in dry weather and expand again when moist conditions return. The water economy of these ferns is thus similar to that of mosses and liverworts.

The synusia of sun epiphytes is usually the richest of the epiphytic communities in both species and individuals. It occurs chiefly in the centre of the crowns and along the larger branches of A- and B-story trees. On very large trees sun epiphytes sometimes spread down the trunk below the first fork. They live in a microclimate intermediate between that in the undergrowth and that in the open and receive some shade, but much less than the shade epiphytes. They are exposed to a relatively high and very fluctuating saturation deficit. This synusia includes many ferns and a very large number of species of orchid and other flowering plants.

Sun epiphytes are extremely varied in mode of life (see pp. 110–11) and structure. Most of them are more or less xeromorphic, some markedly so. In their water relations they are probably as varied as in structure and it is difficult to generalize from the scanty information available (see Schimper, 1935, pp. 24–9). Many of them have been shown to be capable of withstanding large water losses; some species can lose over 50% of their fresh weight without harm. The osmotic pressure of the leaf cells is generally considerably higher than in shade epiphytes (von Faber, 1927; Senn, 1913).

Extreme xerophilous epiphytes live on the topmost branches and twigs of the taller trees; on exceptionally large outstanding trees they may occupy the greater part of the crown. They are almost fully exposed to sun and wind and live under conditions closely resembling those in the open. Since hardly any humus or dissolved salts can be washed down to them from above, they must also live under extreme conditions in respect of nutrition. This synusia includes comparatively few ferns and consists chiefly of flowering plants of highly specialized types—at Moraballi Creek, British Guiana, for instance, the chief representatives are narrow-leaved species of *Tillandsia* such as *T. bulbosa* (Bromeliaceae), *Aechmea* spp. (Bromeliaceae), *Codonanthe crassifolia* (Gesneriaceae), *Rhipsalis cassytha* (Cactaceae) and certain orchids. Several of these species are succulent and all of them show more or less marked xeromorphy. The water relations of the members of this synusia as such have not been studied, but some of the sun epiphytes and 'epiphytes growing in extreme habitats', whose water relations have been discussed by Schimper (1935, pp. 24–9),[1] should probably be included in this group.

Vertical distribution of the synusiae

Though there is no regular stratification of epiphytes and their vertical distribution varies from tree to tree, the average vertical range of the synusiae tends to be constant within any one type of forest and to differ from one type to another. In a community where the A and B stories are dense, the average vertical range of the epiphytes is high; in the more open types of forest it is always lower. This is strikingly shown in the five types of primary forest at Moraballi Creek, British Guiana. The average illumination in the undergrowth (as a percentage of full daylight)[2] is respectively: Mora forest, 1·33; Mora-bukea forest, 0·67; Mixed forest, 0·67; Greenheart forest, 0·81; Wallaba forest, 1·43. The vertical range of the epiphytes shows corresponding variations; in the three poorly illuminated types (Morabukea, Mixed and Greenheart) few epiphytes are visible from the ground; the shade synusia is scantily developed and typical sun epiphytes, such as *Tillandsia* spp., rarely occur below about 18 m. above ground-level, except for stray individuals which do not flower (lowest record for a mature plant, 9 m.). In the lighter types of forest (Mora and Wallaba) the epiphytes descend much lower and flowering individuals of

---

[1] See also Mez's work (1904) on the water economy of the 'extreme atmospheric' tillandsias.
[2] Determined at breast height using a photographic exposure meter (actinometer type).

Tillandsia spp. could be found as low as 3 m. from the ground. It was also noticed that in the dark types of forest when a plant of a sun epiphyte fell off a tree on to the ground, it nearly always soon died, while in the lighter types it often continued to flourish on the forest floor. In the Wallaba forest *Tillandsia* plants were sometimes met with which had apparently established themselves on the ground; when carefully examined these plants were always found to have the remains of a rotten branch beneath.

Similarly in Borneo epiphytes grow at lower levels in the well-illuminated Heath forest (p. 244) than in the darker Mixed forest. In the former they occasionally grow actually on the ground; thus in one locality *Asplenium nidus* was seen rooted in the sandy soil by a road through Heath forest (Richards, 1936, p. 32). In a sandy scrub area in west Borneo, B. Polak (1933, p. 26) found species of *Bulbophyllum* and other epiphytes growing on the ground.

In all parts of the tropics epiphytes grow on trees by rivers and isolated trees on cultivated land at much lower levels than in the interior of the forest.

The variation in the vertical distribution of epiphytes depends more on the light factor than on the average humidity of the air, which in a forest varies more or less inversely with the light intensity (Davis & Richards, 1933–4, p. 382). Pittendrigh (1948), who studied the vertical and geographical range of the Bromeliaceae in the forests and cacao plantations of Trinidad, also concluded that light is the chief controlling factor.

Ecological morphology

Unlike most other synusiae of rain-forest plants, epiphytes show little uniformity in foliage or morphology. Many of them, however, have very remarkable and varied structural specializations, obviously connected with their mode of life, which can certainly be regarded as 'adaptations' in the sense that they are a useful, and in many cases probably a necessary, part of the plant's equipment for its habitat. They may not be adaptations which have evolved by gradual modification in the habitat the plant now occupies; there is evidence that some at least of these specializations, e.g. the water-tanks and absorbing hairs of the Bromeliaceae, were inherited from terrestrial ancestors living under semi-arid conditions. The plants from which the epiphytes arose were thus 'pre-adapted' and possessed characters which made it possible for them to become rain-forest epiphytes (see above, p. 110).

The structural adaptations of epiphytes were first fully described in Schimper's classical *Die epiphytische Vegetation Amerikas* (1888). They need be discussed only briefly here, as the subject is dealt with in many text-books and in more detail by Goebel (1889) and Karsten (1895, 1925). Reference may be made also to the account of the ecology of tropical epiphytic ferns by Holttum (1938a) and to the list of literature in the latest edition of Schimper's *Pflanzengeographie* (1935).

The seeds and seedlings of epiphytes show many characteristic features related to their special problems of dispersal and establishment. Epiphytes, as Schimper first pointed out, almost without exception have wind-borne spores or fruits or seeds fitted for dispersal by animals or by wind. Fleshy fruits, often with sticky pulp which helps to attach the seeds to the bark, are found in *Rhipsalis* (Cactaceae) and many Bromeliaceae; the seeds of the latter are often deposited by birds in large numbers on telegraph wires, where they adhere and often germinate successfully. The spores of ferns and the very small seeds of *Aeschynanthus*, the Orchidaceae, etc. are well adapted to wind transport. The seeds of the epiphytic orchid *Stanhopea oculata* weigh only 0·000003 gm. (Ulbrich, 1928, p. 153) and those of species of *Aeschynanthus* 0·00002 gm. (Beccari, quoted by Goebel, 1889, p. 153). Seeds with parachute appendages are found in various Gesneriaceae, Rubiaceae, Bromeliaceae (sub-family Tillandsioideae), etc. A unique method of dispersal is found in *Tillandsia usneoides*, the Spanish moss of warm and tropical America;[1] seeds are rarely produced and roots are found only in young seedlings, but the beard-like masses of the plant are torn off and distributed by the wind, or used by birds for building their nests (Schimper, 1888).

Plants with no special mechanism for seed dispersal would probably be unable to maintain themselves as epiphytes, but no mechanism found in epiphytes is peculiar to them. As Schimper pointed out, the occurrence of the epiphytic habit in different families, at first sight random, becomes comprehensible when it is realized that all the families including epiphytes are those in which suitable dispersal mechanisms are common. Families which tend to have heavy seeds, such as the Leguminosae, have not given rise to epiphytes. On the other hand some families such as the grasses and the Compositae (with very few exceptions) have produced no epiphytes in spite of having suitable means of dispersal; in such cases some other item in the biological equipment of an epiphyte must be supposed to be lacking.

In *Aeschynanthus* and in the semi-parasitic *Loranthus* part of the young seedling forms a flattened disk, furnished with numerous root hairs, which assists it in becoming established. When adult, most epiphytes anchor themselves by very efficient root systems, though in some Bromeliaceae there are no roots and attachment is made in some other way. In many Araceae, Orchidaceae, etc. the root system is differentiated into nutritive and anchoring roots, the two kinds differing in tropisms, external appearance and anatomical structure, as well as in function. In the more specialized Bromeliaceae, in which absorption of water is carried out chiefly by peculiar multicellular hairs on the leaves, the string- or wire-like roots are very efficient organs of attachment, but probably absorb very little water; Schimper (1888, p. 68) showed that plants from which the roots had been cut off could maintain their water balance quite as well as an uninjured plant.

[1] It is uncommon in the Tropical Rain forest proper.

It is in relation to water and the accumulation of soil that epiphytes show the most striking and remarkable modifications of structure. Schimper distinguished four classes of epiphytes (in addition to the semi-parasitic epiphytes): proto-epiphytes, nest and bracket epiphytes, tank epiphytes and hemi-epiphytes.

Proto-epiphytes have no special structures for collecting water or soil and do not send down aerial roots to the ground. These are the least specialized group and are the least protected from the effects of intermittent drought and shortage of soil. Some, e.g. species of *Peperomia* and many ferns, have creeping stems and are thus able to exploit a large area of the substratum. Many proto-epiphytes have a xeromorphic structure and have water-storing organs of various types. In species of *Peperomia*, *Dischidia*, etc. the leaves are succulent; in many orchids there are swollen internodes known as pseudobulbs. Root tubers are found in Malayan species of *Vaccinium* and *Pachycentria*. The rapid absorption of water is assisted in some proto-epiphytes by a special tissue at the outside of the aerial roots, the velamen. The cells of the velamen, which are non-living, become full of water after rain, but in dry weather contain air and act as an insulating layer against excessive heating and loss of water. The velamen is especially charac-teristic of the roots of epiphytic orchids, but it is of interest that it also occurs in those of a few terrestrial herbaceous plants (Schimper, 1888; Goebel, 1922). It is possible that these species are epiphytes which have returned to living on the ground, but it is perhaps more likely that the velamen is a morphological feature which has evolved in a terrestrial environment and that, like the seed-dispersal mechanism, it is part of the equipment which has enabled certain species to adapt themselves to epiphytic life.

In nest and bracket epiphytes the plant is so constructed as to accumulate humus and debris, from which the roots derive water and mineral substances. In nest epiphytes, of which the large fern of the eastern tropics, *Asplenium nidus*, and many other ferns, aroids and orchids are examples, the roots form a dense interwoven mass, resembling a bird's nest; among these roots ants make their nests and humus gradually collects. In bracket epiphytes the leaves, or some of them, are bracket-like. The widely distributed genus of ferns, *Platycerium*, has two types of leaf, one of which, the mantle-leaves, form brackets. In species of the Malayan genus *Conchophyllum* (Asclepiadaceae) all the leaves are convex and appressed to the bark; debris accumulates in the space between the leaves and the bark. The extraordinary *Dischidia rafflesiana* resembles the bracket epi-phytes. It has two types of leaf, one having the form of a sack, the mouth of which is closed by an infolded flap. Ants make nests in these sacks and a soil gradually collects in them; the roots grow into this soil and absorb water and nutrients from it.

Ecologically similar to the bracket epiphytes are the extraordinary myrmeco-phytes *Myrmecodia* and *Hydnophytum*, both of which belong to the Rubiaceae. Several species of these genera are common members of the synusiae of sun and extreme xerophilous epiphytes in the Rain forest of the Malayan region. In these grotesque plants the stem forms a large tuber honeycombed by cavities which are invariably inhabited by ants. The cavities are not produced by the ants and are formed even when they are excluded. The tubers consist largely of aqueous tissue and there is little doubt that they function primarily as water-storing organs, though it is probable that nutrients can be absorbed from the ants' faeces and the humus which accumulates in the cavities. The plant is not dependent on nourishment obtained in this way, as it has been shown that it can grow normally in the complete absence of ants. These curious epiphytes have been much investigated (for literature, see Schimper, 1935, pp. 253–5 and Spanner, 1939). In some epiphytic ferns of the genera *Lecanopteris* and *Polypodium* the rhizome is flattened and much swollen, and is riddled with cavities inhabited by ants, as in *Myrmecodia* and *Hydnophytum*.

The tank or cistern epiphytes are represented by one family only, the Bromeliaceae, which are abundant throughout the tropical American forests and are one of their most characteristic features. In the epiphytic species, except in *Tillandsia usneoides* and those of similar habit, the long, narrow, stiff leaves form a rosette and their sheathing bases overlap so as to form a reservoir which in large plants may hold as much as 5 litres of water. In *T. bulbosa* the inner leaves meet above the tank, so that it holds water in any position; water enters through chinks between the leaves, and the plant is able even to grow upside down. Besides particles of humus, insects fall into the bromeliad tanks, and these together with the excreta and dead remains of the animals which breed there, add to the supply of nutrients in the water. The scale-like hairs on the leaves, which in many species are confined to the leaf bases lining the tank, absorb water and dissolved substances, but water cannot escape through them; they thus act as one-way valves (*Trichompompe* of Mez, 1904). The roots, as mentioned previously, often have only a mechanical function, so that the hairs are the only absorbing organs. This remarkable system of collecting and absorbing water and nutrients makes these Bromeliaceae largely independent of their substratum and is one of the reasons for their success in habitats in which few other epiphytes can establish themselves. That the system is efficient is shown by the great size reached by some Bromeliaceae—in *Glomeropitcairnia erectiflora* of Trinidad the inflorescence may be 9 ft. (3 m.) high (Pittendrigh, 1948, p. 67).

Schimper's last class of epiphytes, the hemi-epiphytes, develop long aerial roots which reach the ground. Once the connexion with the ground has been established these plants no longer have the difficulty in obtaining water and nutrients which confronts most other epiphytes. Some of them, e.g. various Araceae, are herbaceous and some, like the tropical American *Coussapoa fagifolia*, fairly large shrubs and as might be expected differ very little, except in their curious roots, from ground-living plants. They are linked by intermediate types, on the one hand with the stranglers (p. 109) which eventually become independent plants, and on the other with normal trees. Thus the Malayan *Pyrus granulosa* is sometimes an independent tree and sometimes an epiphyte with descending roots, half-way between a strangler and a true hemi-epiphyte (Corner, 1940, p. 529).

Non-vascular epiphytes

The epiphytic vegetation of the Tropical Rain forest includes lichens, algae, mosses and hepatics as well as vascular plants. Some of these non-vascular epiphytes grow in close association with ferns and flowering plants and may be regarded as forming part of the same synusiae; the larger mosses and hepatics, for example, are of the same order of size as the epiphytic filmy ferns and compete with them on equal terms. It will, however, be convenient to deal here with all the non-vascular epiphytes.

In the rain forest all the lichens and almost all the mosses, hepatics and algae are epiphytic. As was pointed out in Chapter 2, there is usually no moss stratum on the forest floor, though on very steep slopes and at high altitudes some bryophytes may be found on the ground. The chief reason for the absence of ground mosses is probably the smothering effect of the blanket of dead leaves; they are found on tree-trunks down to ground-level and are abundant on stones and fallen logs from which the leaves easily slide off. Freshly disturbed soil, on which a thick layer of leaves has not had time to collect, such as termites' nests and earth thrown up by burrowing animals or the roots of overturned trees, is also often colonized by small mosses (especially *Fissidens* spp.) and hepatics.

Besides the communities ('associules' of Clements, 1936) on stones, rotten logs and tree-stumps, non-vascular plants are found in the Rain forest as shade epiphytes, sun epiphytes and as epiphyllae on living leaves. The shade-epiphyte community occurs on trunks and twigs in the undergrowth and, like the synusia of vascular shade epiphytes, it often reappears in the crowns of large trees with thick foliage. The non-vascular sun epiphytes are found on the branches of the trees in the A and B stories and often also colonize the smaller twigs. Epiphyllae are found mainly in the shady undergrowth and seldom reach the higher tree stories. All these communities of bryophytes, lichens and algae undergo succession and can be found in seral and climax stages of development.

The shade community is dominated by bryophytes of many families (mosses, Lejeuneaceae, *Plagiochila* spp., etc.) and few algae or lichens play any part in it except in the early stages of succession. Unlike the corresponding community of flowering plants and ferns, it is much richer in species than the sun community. The twigs and branches of shrubs and C-story trees are often so thickly covered with mosses and hepatics that the bark is completely concealed; the trunks of large trees are seldom covered with a continuous carpet of bryophytes. Besides tufted and carpet-forming bryophytes of growth forms more or less familiar in temperate floras,[1] a characteristic feature of the shade community is the presence of pendent mosses, mostly belonging to the family Meteoriaceae; in these the

[1] The growth forms of mosses in the rain forest of Java have been described by Giesenhagen (1910).

primary stem creeps along a twig and gives off branches, in some species a metre or more long, which hang down into the air. The occurrence of these pendent mosses, which are found only in the tropical and other rain-forest formations, is no doubt related to the constantly high humidity, as the pendent habit offers the maximum exposure to evaporation. The bryophytes of the shade community have a typically hygromorphic structure; their leaves are usually large and the majority of the species have thin cell walls and very large cells.

The non-vascular sun epiphytes form a strong contrast to the shade epiphytes and very few species are common to both communities. Lichens are abundant (including large foliose forms such as species of *Parmelia*) and persist into the later stages of the succession. Mosses (e.g. species of *Macromitrium*) and hepatics (*Frullania* spp., Lejeuneaceae) are abundant but the number of species in any one district is usually restricted. Most of the bryophytes are of compact creeping or tufted habit and pendent species are absent or rare. Their structure is usually xeromorphic, the leaves being small and closely imbricated when dry, with thick cell walls and small cell lumens. The structural differences between the shade and sun bryophytes in the Tropical Rain forest are thus similar to those between sun and shade epiphytes in European woodlands (Olsen, 1917), but, if anything, more strongly marked.

The occurrence of epiphyllae (Busse, 1905) is a highly characteristic feature of rain forests, Tropical, Montane and Subtropical; outside the tropics and subtropics they are known in the evergreen forests of Japan and the Macaronesian islands (Richards, 1932; Allorge, 1939). Their presence demands long-lived evergreen leaves and constantly high humidity and temperature. In relatively dry Tropical Rain forests, such as the Wet Evergreen forest of Nigeria, they are absent except near streams and in swampy places. Epiphyllae, as has been mentioned, are found chiefly in the lower levels of the forest (C, D and E stories). Their chief habitat is on the leaves of shrubs and small trees.

The community of epiphyllae consists of algae (e.g. species of *Trentepohlia*, *Phycopeltis*, etc.), lichens (e.g. *Strigula* and many other genera) and leafy liverworts (chiefly Lejeuneaceae and species of *Radula*); epiphyllous mosses (e.g. *Crossomitrium* spp. in tropical America, *Taxithelium* spp. in Malaya) are relatively uncommon. Epiphyllous species rarely occur on bark, but bryophytes growing on bark sometimes spread on to the leaves, though they may not be capable of establishing themselves there in the first place. The lichens and algae are the earliest colonists on leaves and in comparatively dry situations the succession goes no further. Under more favourable conditions, however, liverworts then follow and usually eliminate the lichens and algae. In time the leaf may become covered with a dense felt of epiphyllous bryophytes, in which even seedling flowering plants (e.g. orchids, Bromeliaceae) may germinate. These flowering plants, of course, never grow to maturity. Ultimately the leaf falls off and the life of the little community comes to an end.

Epiphyllae are found only on fairly long-lived evergreen leaves of tough consistency; and, except occasionally on leaves hanging in a vertical position, they grow only on the upper surface. Their abundance varies greatly on leaves of different types. They prefer the leaves of some species and avoid those of others for reasons which are often not immediately apparent; this may perhaps depend on the nature of the cuticle or on substances excreted by the leaves. Epiphyllae do not avoid finely divided leaves (e.g. those of the Mimosoideae) or hairy leaves, but are unable to form extensive layers on them. As was seen in Chapter 4 there is no correlation between the presence of epiphyllae and the absence of a drip-tip. Epiphyllous bryophytes, like the epiphytic flowering plants so far considered, are not parasites and there is no evidence that they obtain organic food from the plants on which they grow. They may interfere with assimilation to a very slight extent and by tending to prevent the rapid drying of the leaf surface they probably encourage the attacks of fungal parasites, but, as stated in the last chapter, any adverse effects they may have are probably of negligible importance. With some of the epiphyllous algae and lichens the position is different. The alga *Cephaleuros virescens*, though green, is known to be partly parasitic and causes a virulent disease of tea and other plants. Other epiphyllous algae invade the tissues of the host and are certainly partial parasites, if only 'space parasites'. Fitting (1910) studied the relations of epiphyllous lichens to the leaves on which they grow and found that they fell into three groups; (*a*) those that penetrate more or less deeply into the tissues, e.g. *Strigula*, in which the algal component is *Cephaleuros virescens*; (*b*) those that penetrate the cuticle, but not the epidermis; and (*c*) those that do not penetrate the cuticle. Only group (*c*) is entirely non-parasitic. Group (*b*) is by far the most numerous.

Epiphyllae show many interesting structural modifications which appear to assist them in adhering to their substratum and many of the epiphyllous liverworts are constructed so as to hold large quantities of water by capillarity, either between their leaves and the substratum, or in 'water sacs'. Detailed accounts of the structure of epiphyllae have been given by Goebel (1888, 1889), Karsten (1895) and Massart (1898).

SAPROPHYTES

In the Tropical Rain forest, as in temperate woodlands, the vast majority of the saprophytes are lower organisms—bacteria and fungi. A synusia of small saprophytic flowering plants[1] is, however, usually present. These flowering plants, though widely distributed, are never, in the author's experience, very abundant and the statement sometimes made that in the Rain forest saprophytic fungi are replaced by flowering plants is certainly an exaggeration; the abundance of saprophytic fungi in the Rain forest is usually greatly underestimated, owing to the seasonal appearance of their fruit bodies (Corner, 1935). Though

---

[1] In British Guiana a small fern, *Schizaea fluminensis*, is often associated with these saprophytic flowering plants and may possibly be a partial saprophyte (Davis & Richards, 1933–4, p. 372).

the role of saprophytic flowering plants in the general economy of the forest is no doubt exceedingly small, they are of interest from several points of view.

Only a very limited number of families is represented in this community. Among the monocotyledons there are Burmanniaceae (*Gymnosiphon*, *Thismia* and other genera), Liliaceae (*Petrosavia*—only in the Malayan region), Orchidaceae (various genera) and Triuridaceae (*Sciaphila*, *Triuris*); among dicotyledons, Gentianaceae (*Leiphaimos*, Pl. VII A, *Voyria*, etc.) and Polygalaceae (*Epirrhizanthes*). Several of these genera are pan-tropical in distribution. Many of these little plants are of strange appearance and their morphology, anatomy and development have been much studied (Johow, 1885, 1889; Ernst, Bernard & others, 1910–14). The majority of them are holosaprophytes and nearly or entirely destitute of chlorophyll, but some at least of the orchids are partial saprophytes (containing some chlorophyll and probably capable of carrying out photosynthesis to a limited extent). In colour they are white (*Gymnosiphon*), bright yellow (*Leiphaimos aphylla*), pink, blue, or some shade of purple (*Sciaphila*).

The members of this synusia are found on the forest floor or occasionally on dead logs or stumps. They prefer intense shade and places where dead leaves tend to accumulate to a greater depth than usual, as in slight hollows. A favourite habitat for many of the species is in the corners between the buttresses of large trees. At Moraballi Creek, British Guiana, saprophytes were especially abundant in the Morabukea (*Mora gonggrijpii*) forest, the darkest of the five types of primary forest met with in the district. Whether light has any direct effect on the distribution of these plants is not known; their liking for very shady places may be due to an intolerance for an atmosphere even temporarily unsaturated. Their absence or scarcity in rain forest with a marked dry season, e.g. south-western Nigeria, suggests that they are not able to survive even a slight drying of the forest floor. The mode of occurrence of these saprophytes in districts where they are found is very characteristic. Over large areas they are quite absent or represented only by rare and isolated individuals of the commonest species; here and there, however, there are patches, often only a few square metres in extent, in which a number, sometimes considerable, of individuals of three or four species occur together. This type of distribution has been noticed in Borneo, British Guiana and elsewhere. It does not seem to depend entirely on the shade or the thickness of the dead-leaf litter; saprophytes are often absent where the conditions appear favourable for them. The underground organs of all these saprophytic flowering plants contain mycorrhizal fungi on which the plant may be dependent. It has been suggested (van der Pijl, 1934) that the patchy distribution of the saprophytes is due to factors affecting the mycorrhizal fungus. The relations between the distribution of the saprophytes and that of particular species of tree might be worth investigating.

It may be noted, incidentally, that in some of these saprophytes the root system is situated in the superficial humus layer, while in others it is in the underlying mineral soil, sometimes as deep as 15–20 cm. (Johow, 1889, pp. 486–7).

All the saprophytes so far mentioned are inconspicuous plants easily overlooked by the unpractised eye. Excepting some of the orchids, they are nearly all under 20 cm. high. The saprophytic orchid *Galeola altissima* is unique in being a root-climber attaining a height of some 40 m.

PARASITES

Apart from fungi and bacteria, there are two synusiae of parasites in the Tropical Rain forest, the root parasites growing on the ground and the semi-parasites growing epiphytically on the trees. No terrestrial semi-parasites are known from the rain forest.

The synusia of root parasites is of small importance, as the species are few and seldom represented by many individuals. Only two families occur, the Balanophoraceae and the Rafflesiaceae. The former is represented by several genera: *Helosis cayennensis* is widely distributed in tropical America and *Thonningia sanguinea* (Pl. VII B) is frequent in the evergreen forest of West Africa. The only genera of the Rafflesiaceae found in the Rain forest are the Malayan *Rafflesia*, famous for its gigantic flowers, and *Mycetanthe*.

Some of these root parasites have a restricted range of hosts—species of *Rafflesia*, for instance, occur chiefly on Vitaceae—others are fairly indifferent, e.g. *Thonningia sanguinea*. In any one district it is unusual to meet with more than one species of these root parasites. The scarcity of parasitic flowering plants in the rain forest is a surprising fact, for which there is no obvious explanation.

The epiphytic semi-parasites of the Rain forest consist of but one family, the Loranthaceae, though it is represented by numerous genera and species, and the individuals are often extremely abundant. Loranthaceae are met with throughout the entire area of the rain forest and in any one district a considerable number of species may often be found.

Most of the tropical Loranthaceae seem to have a very wide range of hosts and occur on trees of the most diverse genera and families; some species are occasionally hyperparasites, attacking members of their own family (Koernicke, 1910). Instances of hyperparasitism to the second degree (a parasite on a parasite on a parasite) have even been recorded. Though they show little limitation in their choice of hosts, the tropical Loranthaceae have a very well-defined vertical distribution. In rain forest of normal density and shadiness they always occur on the twigs and branches of the first-story and taller second-story trees; they never occur at lower levels, except in openings or on isolated trees. In more open types of forest they may descend almost to ground level. In the well-illuminated Heath forest in Borneo the author found *Macrosolen beccarii* growing on shrubs and small trees at 1–2 m. from the ground, while in the more shady Mixed Dipterocarp forest no Loranthaceae occurred below about 20 m., except in clearings and by rivers (Richards, 1936, p. 32). In Nigeria, *Loranthus leptolobus* is common on isolated cacao and cola trees on cultivated land at a height of only a few metres. The vertical distribution of the epiphytic semi-parasites thus corresponds exactly with that of the autotrophic sun epiphytes. In both, the distribution seems to be determined chiefly by an intolerance of shade.

The Loranthaceae of the Rain forest are more or less woody and vary in habit from erect shrubs to half-scramblers. The leaves are entire, coriaceous and, of course, evergreen; they usually conform closely to the type prevailing among the trees of the stratum where they occur. The resemblance of their leaves to those of their hosts is often quite deceptive and when not in flower their existence may be easily overlooked. Their brilliant colours often make them conspicuous when in flower. A tree laden with Loranthaceae is one of the most beautiful sights of the Malayan Rain forest.

Part II

THE ENVIRONMENT

Part II

THE ENVIRONMENT

CHAPTER 6

CLIMATE[1]

The dominating features of the tropical rain-forest climate are high and very even temperature and heavy rainfall spread over the greater part of the year. Within the general type of climate there are considerable variations, especially in the seasonal distribution of rainfall and temperature, yet throughout the rain-forest belt these main features of the climate remain essentially similar.

The distribution of the tropical rain-forest formation-type was outlined in Chapter 1. It covers an irregular area, with no simple latitudinal boundaries. Here and there, as in southern Brazil, it reaches or even passes the geographical tropics; in other places the rain-forest boundary extends only a few degrees north or south of the equator. This uneven distribution is due chiefly to the fact that the Tropics are not natural climatic boundaries, and ultimately depends on the uneven distribution of land and sea. In a climatological sense the Tropical zone is defined, not by limits of latitude, but by the isotherm of 20°C. mean annual temperature. Within the tropics in this sense there are great differences in both the amount and the seasonal distribution of the rainfall and correspondingly large variations in the plant covering, from luxuriant evergreen forest at one extreme to sterile desert at the other.

Rain forest can exist only where the annual total of rainfall exceeds a certain minimum and there is no prolonged seasonal drought; where there is a long and severe dry season evergreen forest is replaced by plant formations adapted to periodic drought. With increasing latitude and distance from the sea there is a change from the permanently moist equatorial climate to tropical climates with a less well distributed and usually lower rainfall which is reflected in a gradual change first from a relatively non-seasonal type of evergreen forest ('true rain forest' of Beard, 1944a) to the slightly less luxuriant, more seasonal, Evergreen Seasonal forest, then to Deciduous forest, Savanna woodland, Thorn forest and Desert (Chapter 15). Often at its border the tropical evergreen forest passes abruptly into open savanna, but this, as will be seen later, is a biotic or edaphic formation and not, as usually supposed, determined solely by climate. On mountains the lowland Tropical Rain forest gives place with increasing altitude to Submontane and Montane Rain forest (Chapter 16); for this change temperature rather than precipitation is mainly responsible. In general terms we may thus say that the northern and southern climatic boundaries of the tropical rain-forest formation-type are in most places determined by precipitation and its altitudinal boundaries chiefly by temperature.

[1] The important work of Bernard (1945) on the eco-climate of the central basin of the Congo unfortunately reached the author too late to be fully utilized in writing this chapter.

One of the outstanding features of all moist tropical climates is that seasonal changes of temperature are insignificant compared with seasonal variation in rainfall. The year has no summer and winter, only wet seasons and dry seasons. On these depend not only the seasonal rhythms of the vegetation, but also the seasonal activities of the human inhabitants. The relatively minute amplitude of seasonal change in the rain-forest climate is strikingly shown by the hythergraphs in Fig. 20, in which monthly mean temperatures are plotted against monthly mean totals of rainfall. In some parts of the Rain forest, for instance at Singapore and in some Pacific islands, even seasonal variations in rainfall are slight and the climate can be described as almost non-seasonal.

The annual cycle of wet and dry seasons in the Tropical zone depends directly on the altitude of the sun and usually repeats itself year after year with great regularity. The sudden, fluctuating and often unpredictable day-to-day changes of weather so characteristic of temperate countries with an oceanic climate, such as the British Isles, are seldom met with in the tropics. Variation from year to year, though probably less in equatorial than in temperate climates, is, however, by no means negligible and increases as the margin of the rain-forest belt is approached. In the Belgian Congo, for example, Goedert (1938) has shown that yearly variation in precipitation increases from the central Congo basin towards the coast. Though, as Moreau (1938) points out, the consequences of unusually dry or otherwise exceptional years may be less serious for natural plant communities in equilibrium with their environment than for agricultural crops, they are perhaps more important than is generally realized. Drought years are by no means unknown even far into the equatorial forest belt, but their effects are likely to be most strongly felt in the tension belt between one plant formation and another. Thus, the advance of the African savanna (a fire climax, see Chapter 15) into the evergreen forest is favoured by dry years (Swynnerton, 1917).

TEMPERATURE

The mean annual temperature in the tropical lowlands lies between about 20° and 28° C. In general, mean temperatures vary only very gradually from place to place; temperature gradients are therefore gentle and isotherms far apart. Near the equator the difference between the means for the hottest and coldest months is less than 5°, even in localities far in the interior of continents, such as Manaos in the Amazon valley; on small tropical islands it is sometimes less than 1° (Ocean Island in the Gilbert and Ellice group, 0·2°). The lowest mean temperatures usually occur in the wet season, the highest in the dry season. With increasing distance from the equator, seasonal variation in temperature increases, especially in regions with strongly marked wet and dry seasons, but even in the least equable tropical climates it seldom amounts to more than 13°. The mean daily range of temperature varies from 3° to 16° in different parts of the Tropical zone. Though minimum temperatures are high, maxima near the equator are lower than in southern Europe and rarely exceed 33–4°; towards the Tropics average annual maxima are much higher and may be over 50°.

In the area occupied by the Tropical Rain forest the mean temperature averages about 26° and the mean for the coldest month is rarely below 25°, but where rain forest of a more or less tropical type extends outside the actual Tropics, as in south Brazil, the annual mean may be only about 20°. Köppen (see p. 149) regarded a mean of 18° for the coldest month as the lower temperature limit for his 'Tropical Rain climate'. Temperature conditions in representative lowland rain forest localities are shown in Table 10.

TABLE 10. *Temperatures in the Tropical zone*

					Temp. (° C.)			
					Average extremes*		Recorded extremes	
Locality	Latitude	Mean annual	Mean of hottest month	Mean of coldest month	Max.	Min.	Max.	Min.
Rain forest:								
1. Apia, Samoa	13° 48′ S.	25·8	26·3	25·1	32·7	18·3	35·5	16·1
2. Singapore, Malaya	1° 18′ N.	27·2	27·7	26·4	33·8	20·5	36·1	18·8
3. Sandakan, North Borneo	5° 49′ N.	27·4	28·1	26·5	34·4	21·6	37·2	21·1
4. Mazaruni Station, British Guiana	6° 50′ N.	26·0	27·2	25·1	—	—	—	—
5. Manaos, Brazil	3° 8′ S.	27·2	28·2	26·5	36·6	21·1	38·3	18·8
6. Yangambi, Belgian Congo	0° 45′ N.	24·5	25·2	23·9	—	—	—	—
7. Eala, Belgian Congo	0° 3′ N.	25·6	26·3	24·6	—	—	37·2	15·2
8. Duala, Cameroons	4° 3′ N.	25·4	26·9	23·8	32·7	20·0	35·0	18·3
Tropical Semi-deciduous, Deciduous forest, etc.:								
9. Lagos, Nigeria	6° 27′ N.	26·9	28·5	25·3	35·0	19·4	40·0	15·5
10. Sokoto, Nigeria	13° 2′ N.	28·0	32·7	23·7	43·3	10·0	46·1	3·8
11. Rangoon, Burma	16° 47′ N.	27·3	30·6	24·8	38·3	15·0	41·6	12·7
12. Mandalay, Burma	21° 59′ N.	27·6	32·2	21·3	43·3	10·5	44·4	7·2
Rain forest (near altitudinal or latitudinal limit):								
13. Bambesa, Belgian Congo	3° 26′ N.	24·4	25·1	23·5	—	—	—	—
14. Masindi (1146 M.), Uganda	1° 35′ N.	21·8	22·6	20·7	—	—	36·7	9·4
15. Cairns, Queensland	16° 55′ S.	24·8	27·2	21·0	—	—	—	—
16. Rio de Janeiro, Brazil	22° 54′ S.	22·7	25·6	20·0	35·5	13·8	38·8	10·0

Sources: 4 from figures communicated by the Conservator of Forests, British Guiana; 6, 7, 13 from Bernard (1945); 14 from Eggeling (1947); 15 from data communicated by Prof. J. S. Turner; the remainder from Brooks (1932).

* I.e. the average of the highest (or lowest) temperatures recorded during each of a series of years.

Comparatively little is known of temperatures in the various zones of vegetation on tropical mountains. In the tropics the mean temperature decreases by about 0·4–0·7°/100 m. ascent. Thus in the Netherlands Indies, according to Mohr (1944, p. 113), the average temperature in the lowlands (0–200 m. Tropical Rain forest and 'Monsoon forest') is about 25–27°, in the foothill belt (200–1000 m. Tropical and Submontane Rain forest) about 19–24°, in the mountain belt (1000–1800 m. Submontane and Montane Rain forest) about 13–18°. Brown (1919) gives the maxima and minima for the different belts of vegetation on Mt Maquiling in the Philippines shown in Table 11.

The absolute minima of temperature in the rain-forest region are probably of little significance for the vegetation. Temperatures at or below freezing point have been recorded at low altitudes in some localities within the tropics, e.g. in the interior of South America, but probably nowhere within the rain-forest belt. The whole of the Tropical Rain forest proper (excluding the Subtropical Rain forest, e.g. that of eastern Australia) thus lies outside the limits of frost. It is noteworthy, however, that while some tropical plants are said to be killed by temperatures somewhat above freezing-point (Molisch, 1896) a few rain-forest species (e.g. *Eucalyptus deglupta*, Lane-Poole, 1938) show some degree of frost tolerance.

Table 11. *Average maximum and minimum temperatures in tops of tall trees at different altitudes, Mt Maquiling, Philippine Islands, October* 1912 *to January* 1915. (After Brown, 1919, p. 227)

Altitude (m.)	Vegetation	Average maximum (° C.)	Average minimum (° C.)
80	Parang (secondary communities)	38·2	24·0
300	Dipterocarp forest (Tropical Rain forest)	32·4	20·0
450	Do.	31·3	19·6
740	Mid-mountain forest (Submontane Rain forest)	30·4	18·1
1050	Mossy forest (Montane Rain forest)	28·1	17·3

The small seasonal variation of temperature which is characteristic, to a greater or lesser extent, of the whole Tropical zone depends partly on the small annual variation in the length of day. Even at the Tropics themselves the shortest day is approximately $10\frac{1}{2}$ hours and the longest only $13\frac{1}{2}$ hours. Other important factors are the thermostatic effect of the oceans, which form about three-quarters of the whole Tropical zone, and of the soil. In Indonesia (Mohr, 1944, p. 45) the soil temperature under forest at sea-level is 25–7° C. and at a depth of a metre it is quite constant; even at 60 cm. it is 'essentially constant'. At Yangambi in the Belgian Congo the mean temperature of the soil at 50 cm. below the surface is 26·2° and the annual range 1·5°; at a depth of 50–75 cm. all traces of a diurnal variation of temperature disappear (Bernard, 1945).

RAINFALL

Rainfall within the Tropical zone varies greatly in amount and in seasonal distribution. Over a large area there are annual totals of between 250 and 400 cm. Very locally enormous totals, sometimes exceeding 10 m., are found (Cherrapunji[1] in Assam, 1161 cm., Debundja at the foot of Cameroons Peak, 1017 cm.), while in the desert regions towards the northern edge of the zone very low rainfalls are found. The seasonal distribution of the rainfall is not necessarily related to the total amount. Very large totals may be strongly seasonal in distribution (at Cherrapunji there are four consecutive months each with a rainfall of considerably less than 10 cm.), on the other hand much lower totals may be relatively evenly distributed; but in general, in tropical regions where

[1] In fact just outside the Tropical zone (26° N.).

the total rainfall is low, it is also unevenly distributed (seasonal or irregular). For the vegetation, the distribution of the rainfall through the year is usually of far more importance than the annual total.

Stated in the most general terms, the seasonal distribution of rainfall within the tropics is a function of latitude: at the equator rain falls at all seasons (though there are probably no land surfaces within the tropics with a completely non-seasonal rainfall), in the zone from about 3 to 10°, north or south, there are two wet and two dry seasons, while still further from the equator there

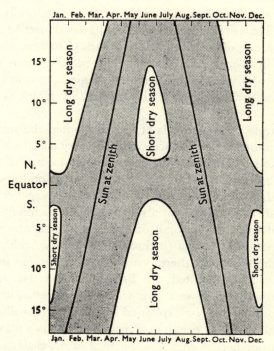

Fig. 21. Wet seasons and dry seasons in the tropics in relation to latitude. Diagrammatic. After E. de Martonne. Wet seasons shaded.

is a single wet and a single dry season in the year (Fig. 21). This variation of seasonal distribution with latitude depends chiefly on the annual passages of the sun overhead; the periods when the sun is in the zenith tend to be periods of heavy rainfall and the intervening periods are comparatively dry, or even quite rainless.

The zone of latitude where the sun is directly overhead at any given time of year becomes more strongly heated than the zones to north and south and therefore becomes a belt of ascending air currents. Air flows in to take the place of the rising air and the zone where the sun is overhead thus becomes a region of relatively little wind (the doldrums) and adjoins two belts of strong and constant winds (trade winds). Over the Indian Ocean the system is complicated by the monsoons, seasonal winds produced by the low pressure which develops over central Asia in summer. To a slighter extent the Sahara and the Australian desert give rise to monsoons.

By adiabatic expansion the ascending air currents over the latitude of greatest heating become cooled and produce rain. Thus rainfall in the Tropical zone is for the most part convectional and is the result of the cooling of air rising over heated land surfaces, though it may be produced in other ways, for instance by the upward deflexion of air currents by mountain ranges and by the convergence of air streams differing slightly in density, generally as a consequence of small differences of temperature. Owing to its usual method of origin, tropical rain is seldom gentle or of long duration; it tends to have a similar character to thunder showers in temperate countries, indeed it is often accompanied by thunder. As may also be expected from its origin, the rain in many tropical districts is remarkably constant in its time of falling, which is most commonly afternoon or early evening.

In a given latitude, if the periods of the year when the sun is overhead are far apart, there will be two wet seasons, while if they fall close together the two wet periods tend to become one, though the seasonal curve of rainfall may still show two distinct maxima. In this way arises the relation between the seasonal distribution of rainfall and latitude, which has already been indicated. In practice the actual relations are not always as simple as might be expected and there are local anomalies in rainfall distribution which are not always easily explicable. In south-eastern Asia (including the Malay Archipelago), owing to the incidence of the monsoon, there is a single rainy season in latitudes where a double one would be expected. In British Guiana (lat. *c.* 6° N.) the two maxima coincide with the lowest, not with the highest, altitude of the sun. Some of these anomalies can be explained by the irregular distribution of land and sea and by the effects of ocean currents. Such causes may account for the local dry areas in northern Venezuela, parts of north-eastern Brazil, the Gold Coast, etc., where both the total rainfall and its season of incidence differ from what might be expected from the latitude. Local factors sometimes give rise to considerable differences in rainfall between tropical localities very close together, even between different parts of the same town.

Since the Tropical Rain forest is, as has been said, dependent on abundant rain at all times of year and the absence of long droughts, it is the climatic climax in the equatorial zone of high and evenly distributed rainfall and in most of the zone with two wet seasons; in the zone with a single wet season it extends only so far as orographic or other local conditions increase the rainfall in the relatively dry season or where there are other compensating factors.

A study of the figures in Table 12 shows that in typical rain-forest localities the annual rainfall is at least 200 cm., but there is a considerable range in both the

TABLE 12. *Rainfall in the Tropical zone*

Monthly totals less than 10 cm. (4 in.) in *italics*. Months with highest rainfall in **heavy type**.

	Latitude	Jan.	Feb.	Mar.	Apr.	May	June	July	Aug.	Sept.	Oct.	Nov.	Dec.	Mean annual rainfall (cm.)	Mean no. of rainy days
Tropical Rain forest:															
1. Mazaruni Station, British Guiana	6° 50′ N.	21·0	11·5	13·1	18·4	32·2	**33·3**	27·5	19·6	14·4	12·8	15·0	27·2	246·9	212
2. Victoria, British Cameroons	4° 1′ N.	*2·7*	*6·7*	15·2	18·8	34·5	70·6	**106·9**	80·0	44·6	22·8	10·3	*4·3*	417·5	194
3. Yangambi, Belgian Congo	0° 45′ N.	*8·0*	10·0	15·5	15·3	14·7	11·3	12·3	14·2	18·6	**20·7**	17·4	11·1	169·1	—
4. Eala, Belgian Congo	0° 3′ N.	*8·6*	13·3	12·1	**16·5**	16·4	13·5	*7·6*	16·2	17·6	22·4	19·1	16·5	179·8	—
5. Singapore, Malaya	1° 18′ N.	25·2	21·6	18·8	19·3	17·0	17·3	17·3	19·1	17·3	25·2	25·2	**26·9**	240·8	177
6. Pontianak, Dutch Borneo	0° 6′ S.	26·9	24·4	28·2	28·2	27·2	22·6	16·5	21·8	21·8	37·3	**40·9**	33·3	322·5	182
7. Sandakan, British North Borneo	5° 49′ N.	**46·7**	24·4	20·3	10·4	15·0	18·5	16·5	20·6	23·9	25·4	37·3	45·0	304·0	181
8. Apia, Samoa	13° 48′ S.	**42·4**	39·6	34·0	25·9	14·2	13·2	*6·9*	*8·9*	12·7	16·0	24·4	34·5	272·7	210
Tropical Rain forest (near climatic limit):															
9. Port-of-Spain, Trinidad	10° 42′ N.	*6·9*	*3·8*	*4·6*	*4·8*	*8·9*	19·3	22·1	**24·1**	18·5	17·0	18·0	12·2	160·2	192
10. Martinique, French West Indies ('Mesophytic forest')	14° 40′ N.	19·0	11·4	*8·4*	*7·3*	20·0	23·3	25·4	28·1	28·1	34·9	**40·6**	21·2	267·7	290
11. Manaos, Brazil	3° 15′ S.	21·1	20·3	20·6	**21·3**	16·8	*9·9*	*4·6*	*3·3*	*3·6*	11·7	11·4	20·8	165·4	172
12. Akilla, Nigeria	6° 40′ N.	*2·8*	*3·5*	*9·6*	18·0	19·6	40·7	**41·9**	13·4	25·2	24·2	*8·0*	*1·1*	208·0	125
13. Budongo, Uganda*	1° 35′ N.	*2·3*	*7·7*	12·9	18·8	17·3	13·4	*9·8*	15·5	15·3	**16·0**	13·2	*7·3*	149·5	167
14. Batavia, Java	6° 22′ S.	30·2	**34·0**	20·3	14·2	10·4	*9·4*	*6·6*	*4·1*	*7·1*	11·4	14·5	19·3	181·5	138
15. Cairns, Queensland	16° 55′ S.	42·2	39·4	**46·0**	29·5	10·9	*7·1*	*4·1*	*4·3*	*4·3*	*5·3*	*9·7*	21·8	224·6	—
Tropical Deciduous forest, Savanna, etc.:															
16. Martinique, French West Indies ('Xerophytic forest')	14° 40′ N.	10·7	*5·2*	*5·2*	*5·2*	14·1	13·1	21·7	**23·4**	23·3	21·9	21·4	10·8	176·0	250
17. Kumasi, Gold Coast	6° 50′ N.	*1·6*	*5·6*	13·6	14·2	18·6	**22·4**	12·8	17·3	18·6	20·2	*9·9*	*3·2*	148·0	133
18. Sokoto, Nigeria	13° 2′ N.	*0·0*	*0·0*	*0·3*	*0·5*	*5·2*	10·4	16·3	**20·3**	10·7	*1·0*	*0·0*	*0·0*	64·6	55
19. Rangoon, Burma	6° 47′ N.	*0·5*	*0·0*	*0·8*	*3·6*	30·7	46·7	**54·6**	50·0	39·1	18·5	*7·1*	*0·8*	252·9	122
20. Mandalay, Burma	21° 59′ N.	*0·3*	*0·3*	*0·5*	*2·8*	**15·0**	14·0	*8·4*	11·7	14·5	11·9	*4·1*	*1·0*	84·5	51
21. Pasoeroean, Java	7° 41′ S.	24·4	**30·0**	20·3	13·2	*8·9*	*5·8*	*2·8*	*0·5*	*0·8*	*2·0*	*6·4*	16·0	130·9	94

* 1146 m. above sea-level.

SOURCES: 1, figures communicated by the Conservator of Forests, British Guiana; 2, Forestry Department, Nigeria; 3, 4, Bernard (1945); 6, 21, Braak (1936); 9, Beard (1946); 12, Richards (1939); 13, Eggeling (1947); 15, data communicated by Prof. J. S. Turner; 10, 16, Stehlé (1945, pp. 302, 359); 17, Foggie (1947); remainder from Brooks (1932).

total and the seasonal distribution   A few localities, such as Debundja in the Cameroons, owing to special local features of topography, have totals much exceeding 200 cm. There are also localities in which the annual total is considerably less than 200 cm. In some places, e.g. Mazaruni Station, British Guiana, and Singapore, no month has less than 10 cm.; on the other hand at Akilla in Southern Nigeria where, though the annual rainfall reaches the considerable figure of 208 cm., there are five successive months with less than 10 cm. and three of these have less than 5 cm. Probably the existence of rain forest in a climate with such a severe dry season is possible only where there is some compensating factor, in Akilla perhaps relatively high atmospheric humidity during the dry months (see p. 146). Another example of compensation for low rainfall by some other factor is afforded by the evergreen riverine or Gallery forests (see p. 316) which are similar to Tropical Rain forest in structure and physiognomy, but extend far beyond its normal boundaries as narrow strips along watercourses in regions of deciduous forest or savanna. In gallery forest edaphic moisture compensates for low annual or seasonal rainfall; such communities must be regarded as post-climaxes.

Within the rain-forest belt itself differences in the length and severity of the seasonal drought are expressed in distinct, though not very large, differences in the structure and physiognomy of the vegetation. The forest of regions with a long period (4 or 5 months) of little or no rain differs little in floristic composition from the (relatively) non-seasonal forests ('true Rain forests') of regions where the rainfall is more evenly distributed, but in the former the less continually favourable climate is reflected in a diminished luxuriance and in the presence of a proportion of plants such as deciduous trees and tuber-geophytes which have a resting period during the dry season. Provided the annual rainfall exceeds a certain minimum (for the Tropical Rain forest perhaps about 160 cm.) the nature of the climax vegetation is related to the seasonal distribution rather than to the total. Regions with an annual total far above the minimum for the existence of rain forest may bear drought-enduring types of vegetation such as deciduous forest. The determining factor is no doubt always the length and severity of the drought period which depends mainly but not entirely on the seasonal distribution of rainfall.

Not very much is known about the variation of rainfall with altitude on tropical mountains, though it is generally true that as in other parts of the world mountains receive more rain than the adjacent lowlands and that rainfall usually increases with the altitude. In some parts of the tropics, for example in east Java and the eastern Sunda Islands, the heavier rainfall and shorter dry season of the mountains permit the development of evergreen rain forest where the lowland climax is deciduous forest. In Java rainfall increases from the base of the mountains to a maximum at between 610 and 1220 m. (Junghuhn, quoted by Schimper, 1935, p. 1261) and then decreases towards the summits; according to the more recent data of Braak (1936, p. 20) the maximum rainfall does not occur at a constant level, but at about half the height of any given mountain. On Mt Maquiling in the Philippines rainfall is somewhat greater at 740 m. than at 1000 m. (Brown, 1919, p. 256). Such a rise to a maximum followed by a decrease with further increase of elevation, if at all general, is perhaps a feature only of rather high mountains.

ATMOSPHERIC HUMIDITY

For the plant ecologist the most useful index of atmospheric humidity is the saturation deficit of the air, which is a measure of its evaporating power. Figures for the saturation deficit are, however, not often available and usually one must be content with measurements of the relative humidity. Since the saturation deficit is a function of the temperature as well as of the relative humidity of the air, the same relative humidity will correspond to very different saturation deficits at different temperatures. If there are large differences of temperature in time or space in a region under consideration, relative humidity may thus be a very misleading index to the evaporating power of the air, the vitally important factor from the point of view of plant physiology. In the Rain forest, however, variation in temperature (daily, seasonal and from place to place) is sufficiently small for it to be usually permissible to regard the relative humidity as an approximate measure of evaporating power. A brief discussion of relative humidity data will thus help to give a clear picture of the rain-forest environment.

In open situations in the rain-forest region the relative humidity tends to be high during the greater part of the day, even in the dry season. It varies considerably during the hours of daylight and shows a seasonal variation parallel to that of rainfall. The daily and seasonal variations will be further considered in the next chapter, in connexion with microclimates. Most of the published data on relative humidity in the Tropical zone are not sufficiently detailed to allow more than rather vague general statements; often they are given as monthly means, which from the ecological point of view are not very illuminating, unless they refer to a definite hour in the middle of the day.

Table 13 gives the monthly mean values for relative humidity in a few selected rain-forest localities; figures are given for the early morning (when relative humidity is near its daily maximum) and for the middle of the day (when it is usually near its daily minimum).

In general, in the tropical rain-forest region, even in the open, relative humidity throughout the night is always at or near saturation: during daylight on dry days, it falls to values of the order of 65%, momentarily it may touch 55%, or even lower. At a locality as moist as Buitenzorg (Java) (mean annual rainfall 436·7 cm., driest month 22·7 cm.) an absolute minimum relative humidity of 28% has been recorded (Braak, 1924). The mean annual minimum relative humidity

TABLE 13. *Relative humidity in the Tropical zone*

Minimum monthly figures in **heavy type**

	Relative humidity (%)												Time of day	Mean for year (%)
	Jan.	Feb.	Mar.	Apr.	May	June	July	Aug.	Sept.	Oct.	Nov.	Dec.		
Tropical Rain forest:														
1. Mazaruni Station, British Guiana	94	**93**	**93**	**93**	**93**	94	94	94	**93**	**93**	**93**	94	07.00 L.T.	94
	79	76	**73**	75	81	82	81	77	73	72	74	81	13.00 L.T.	73
2. Eala, Belgian Congo	74	**70**	72	73	75	77	76	76	75	74	76	77	12.00 L.T.	75
3. Duala, French Cameroons	95	96	95	**94**	95	95	96	96	95	95	95	95	07.00 L.T.	95
	74	75	76	76	79	82	**82**	84	85	82	80	78	14.00 L.T.	80
4. Singapore, Malaya	84	83	**82**	83	85	**82**	**82**	**82**	**82**	**82**	83	83	09.00 L.T.	83
	72	**69**	75	76	76	74	73	75	75	76	77	77	15.00 L.T.	75
5. Sandakan, British North Borneo	**91**	**91**	**91**	92	92	92	**91**	**91**	**91**	92	93	92	06.00 L.T.	92
	77	75	74	70	69	69	69	76	**67**	70	75	76	14.00 L.T.	72
6. Apia, Samoa	82	81	81	79	78	77	77	76	**75**	77	78	79	09.00 L.T.	78
	79	78	78	76	76	**73**	75	**73**	75	76	75	77	15.00 L.T.	76
Tropical Rain forest (near climatic limit):														
7. Port-of-Spain, Trinidad	**92**	93	93	92	**91**	92	93	95	94	94	94	93	07.00 L.T.	93
	68	63	62	**60**	**60**	66	73	73	74	72	74	73	15.00 L.T.	68
8. Lagos, Nigeria	84	83	82	**81**	83	87	87	85	86	86	85	86	09.00 L.T.	85
	65	69	72	72	76	80	80	76	77	76	72	68	15.00 L.T.	74
9. Masindi, Uganda*	70	**66**	77	80	82	82	88	85	84	81	80	77	08.30 L.T.	79
	43	**42**	57	60	66	67	67	65	66	62	56	54	14.30 L.T.	59
10. Batavia, Java	90	92	90	89	89	89	87	84	82	**81**	84	87	08.00 L.T.	87
	75	76	73	71	69	68	65	**62**	**62**	65	69	73	14.00 L.T.	69
Tropical Deciduous forest, Savanna, etc.:														
11. Martinique, French West Indies ('Xerophytic forest')	87	85	85	**79**	81	80	83	84	85	89	90	90	07.00 L.T.	85
	74	70	**70**	72	73	72	75	78	78	81	83	79	17.00 L.T.	75
12. Kumasi, Gold Coast	**95**	95	96	96	96	96	96	96	96	96	96	97	07.00 G.M.T.	96
	55	**50**	58	65	59	67	79	85	79	63	66	59	14.00 G.M.T.	63
13. Sokoto, Nigeria	30	29	**24**	40	59	67	79	85	79	63	38	31	07.00 L.T.	52
	13	13	13	18	31	42	54	59	55	31	20	14	14.00 L.T.	30
14. Rangoon, Burma	**77**	83	85	83	83	89	92	91	90	84	83	78	08.00 L.T.	85
	51	54	54	64	75	86	88	88	86	67	71	61	18.00 L.T.	71
15. Pasoeroean, Java	84	86	85	81	78	77	73	69	**66**	67	71	79	08.00 L.T.	76
	72	73	71	67	65	63	58	57	**55**	57	63	70	14.00 L.T.	64

* 1146 m. above sea-level.

SOURCES: 1, from figures communicated by Conservator of Forests, British Guiana; 2, Bernard (1945); 7, Bain (1934); 9, Eggeling (1947); 10, 15, from data communicated by Dr C. G. G. J. van Steenis; 11, Stehlé (1945); 12, Foggie (1947); remainder, by courtesy of Meteorological Office, London.

at Singapore is 51% and at Pontianak (Borneo) 35% (Braak, 1931, Table 5). At Eala in the very moist central area of the Congo basin the mean humidity for the driest month (March) is 56% (Bernard, 1945, p. 114).

The nature of the vegetation, the continual moisture of the surface soil and the way in which books, clothes, leather, etc. become mouldy, unless frequently exposed to wind and bright sunshine, are clear indications of the high average relative humidity in the rain-forest region. It is therefore difficult to realize how great is the amplitude of the daily variation and how low are the values reached in open situations for short periods in the middle of dry days. These recurring periods of low humidity, which are found in the most humid parts of the rain forest region and may occur even in the wet season, though not of long duration, must have a considerable influence on the vegetation and their importance has probably been much underestimated in the past. Since the daily minimum of relative humidity occurs at, or shortly before, the time of day when the temperature reaches its daily maximum, the corresponding saturation deficit is considerable. The author made observations in a large clearing in the Rain forest of Borneo and found that on ten days in October (beginning of the rainy season) the mean minimum humidity was 61%, which, at the temperatures prevailing in the middle of the day, represents a saturation deficit of about 12 mm. (Richards, 1936, p. 8). At 15°, the temperature of an average day in a temperate climate, this would be equivalent to a relative humidity of 6%. Few extensive series of data on saturation deficit in tropical localities are as yet available. Baker & Harrisson (1936, pp. 459–60) give monthly mean saturation deficits at 2 p.m. for a locality in the New Hebrides. Bernard (1945) has recently published saturation deficit figures for a number of localities in the Belgian Congo. The observations of Evans (1939) on saturation deficit in the Rain forest of Nigeria are considered in Chapter 7.

In the drier parts of the Tropical zone the mean relative humidity is in general lower than in the rain-forest region; very low humidities are more frequent and last for longer periods. The decrease of atmospheric humidity from the coast inland is one of the outstanding features of the climatic gradient in West Africa, as is shown by the figures compiled by A. P. D. Jones & R. W. J. Keay in a recent (unpublished) study of vegetation zones and climate in Nigeria (Table 14).

TABLE 14. *Rainfall and relative humidity in vegetation zones of Nigeria.*
(After A. P. D. Jones & R. W. J. Keay, unpubl.)

Zone	Annual rainfall (cm.)	No. of months with less than 3 cm. rainfall	Lowest mean monthly relative humidity	
			Average for several years (%)	Lowest recorded (%)
1	Over 230	0–1·3	78–85	70–80
2	140–230	1–2·3	71–84	60–70
3	110–230	1·9–4·3	58–81	40–65
4	110–150	3·8–5·0	44–63	30–55
5	95–145	4·2–5·8	26–35	22–30

In Zone 1, which lies nearest the sea, the climax is typical rain forest; in Zone 2 it is Wet Evergreen forest (see Chapter 15) and in Zone 3 Dry Evergreen forest (Mixed Deciduous forest, see p. 338). In the remaining zones (4–5) the plant cover is savanna woodland, etc., mostly much modified by fire and biotic influences. There is evidence that the occurrence of evergreen and semi-evergreen forest in the coastal Zones 1–3 depends on the high minimum humidity during the dry season which modifies the effects of the long drought. The observations of Buxton & Lewis (1934, pp. 228–31) on relative humidity and saturation deficit at Gadau in Northern Nigeria show how different are the conditions during the dry season in the inland savannas. The present writer (1939, pp. 9–10) has previously shown that the distribution of rain forest and Mixed Deciduous forest in Southern Nigeria is not closely correlated with the length of the dry season and suggested that the determining factor might be the distance inland reached by moist air from the sea. The observations in Senegal quoted by Angot (1928, p. 148) show strikingly the remarkable effects of an intermittently blowing sea wind in a tropical climate with a severe dry season; during the afternoon in the dry season the sea wind may cause a rise of 60% or more in the relative humidity as well as a large drop in temperature.

On mountains in the tropics humidity rises with increasing elevation and an altitudinal zone is reached where the perpetual cloud and drizzle prevent it from ever falling far below saturation. The effect of altitude on humidity is illustrated by Brown's (1919, pp. 272–5) data for Mt Maquiling in the Philippines; at the base of the mountain (80 m.) the mean relative humidity under 'second-growth' trees was 82·2 % and the average daily minimum 68·6 % while in the open at 450 m. the mean was 87·9 % and the average daily minimum 75·5 %. In the 'Mossy forest' at 1000 m. the air remains at or close to saturation for months on end and the lowest relative humidity ever recorded was 91 %. The average daily evaporation shows a corresponding decrease with increasing altitude. On the highest tropical mountains drier conditions are found above this zone of maximum humidity.

WIND

The wind is important as an ecological factor from several points of view. The amount of air movement affects the rate of evaporation and is therefore one of the chief factors controlling the water relations of vegetation. Apart from the normal effects of wind on evaporation, desiccating winds are of great significance for the vegetation in some parts of the tropics, e.g. the 'east monsoon' in the Sunda Islands and the harmattan in West Africa, which for a short period each year reduces the humidity to very low values even in the rain-forest belt. The mechanical effects of strong winds are also ecologically important and may even prevent the development of climax vegetation.

In general wind velocities are lower and violent winds less frequent in tropical than in temperate regions, except locally as in the 'hurricane belt' of the West Indies. Mean annual wind velocities in rain-forest localities are commonly less than 5 km./hr. and seldom exceed 12 km./hr. In many parts of the tropics thunderstorms are very common and are often preceded by squalls of strong wind ('tornadoes') lasting for a few minutes only. In these the wind may momentarily reach very high velocities. Groups of trees, and even areas of forest several hectares in extent, felled by such squalls are not an uncommon sight in Nigeria and Guiana. The effects of hurricanes and cyclones, in the limited areas where they occur, are much greater. Vaughan & Wiehe (1937, pp. 301–2) contrast the devastating effects of cyclones on crops and plantations in Mauritius with the small amount of damage in the indigenous plant communities which must have become adapted to them by long selection. Beard (1945) has given the name 'Hurricane forest' to a type of subclimax forest deficient in large trees which is found on the slopes of the Soufriere in St Vincent (West Indies) above 500 m. The frequent hurricanes, together with the steepness of the slopes, make the development of climax rain forest impossible. Both the structure and the floristic composition of the vegetation are partly determined by the wind.

On exposed sea coasts strong winds may deform the crowns of the trees and bring about the development of a strip of characteristic woodland or scrub. Beard (1946b, pp. 108–10) describes a strip of Littoral woodland 20–400 m. wide on the windward coast of Trinidad; he attributes the peculiar physiognomy and floristic composition of this community to the salt-laden trade wind. The extreme dwarfing of the vegetation on isolated peaks and ridges on tropical mountains may be partly due to exposure to the wind, though there are probably other causes for the general reduction in the height of the trees with increasing altitude (see p. 153).

SUNSHINE AND CLOUDINESS

The length of day at the equator is 12 hours with only an insignificant seasonal variation in the length of twilight; with increasing latitude there is an increasing seasonal variation in the length of day, which at the two Tropics amounts to about 2 hours. Though the possible daily amount of sunshine is thus never less than 10 hours in any part of the Tropical zone, the actual amount of bright sunshine is always much less than this, owing to the high degree of cloudiness. Contrary to the popular belief, the Tropical zone is not characterized by a perennially cloudless sky; this is especially true of the equatorial belt, whose cloudiness has earned it the name *pot au noir* from French sailors. The neighbourhood of the equator is in fact a more or less permanent cloud belt where cloudless days are rare; the average cloudiness over the land is 5·2 (tenths) in the zone 0°–10° N. and 5·6 in the zone 0°–10° S. North and south of the equator cloudiness decreases, reaching its minimum near the two Tropics (over the land 3·4 at 20°–30° N. and 3·8 at 20°–30° S.) (Brooks, 1927a). The percentage cloudiness in the Tropical zone varies greatly during the year; usually it is greatest during the wet season, but in some tropical countries it is greatest during the dry season.

Since light clouds may not cut down the illumination appreciably, figures for the hours of bright sunshine, as recorded by a sunshine recorder, are of more ecological interest than the percentage cloudiness. Unfortunately, such data appear to be available only for comparatively few stations in the rain-forest region. The following figures (1920–9) for Mazaruni Station, British Guiana, may probably be taken as fairly typical:

Average daily sunshine	5·5 hr.
Average for sunniest month (September, dry season)	6·3 hr.
Average for least sunny month (June, wet season)	4·4 hr.

The data of Baker & Harrisson (1936, p. 458) for the New Hebrides indicate a similar average daily duration of sunshine, but a slightly greater range in the monthly means. In the Rain forest of Panama, according to Allee's data (1926a, p. 276), the hours of sunshine are considerably greater than in either of the two stations just mentioned (70% possible sunshine in dry season, 40% in wet season). Bernard (1945) estimates that at Yangambi in the Congo forest the average daily sunshine is about 5·1 hr., i.e. 42·3% of the possible insolation. The corresponding mean value for a number of lowland tropical stations with a maritime climate (data extracted by Bernard from Köppen & Geiger's *Handbuch der Klimatologie*) is 58%.

Anyone who has attempted to take photographs in a Tropical Rain forest will realize that the amount of bright sunshine is surprisingly small. In the author's experience, clouds are continually passing in front of the sun even in the dry season and it is often difficult to find periods of uninterrupted sunshine in which to give long exposures.

There is only scanty information on the light intensity and total solar radiation in tropical climates, but the available data indicate that, in spite of the high altitude of the tropical sun, radiation in the tropical lowlands (at ground level) is less rather than more intense than in temperate regions. Boerema (1919) found from pyroheliometer observations that when the altitude of the sun was comparable total radiation was considerably less at Buitenzorg in Java (6° 35' S.) than at Washington, D.C. (38° 52' N.). Bünning (1948, pp. 96–8), after reviewing his own and other measurements of light intensity in the tropics, quotes with approval Orth's (1939) conclusion that the light is in general less bright than in Europe. Bernard (1945) made a detailed study of radiation in the central Congo basin. Both the mean and the maximum (momentary) values for total radiation here are probably no higher than those recorded in temperate regions during the summer months and there is also a marked deficiency in the blue and ultra-violet wave lengths. The relatively low total radiation in this part of Africa is ascribed by Bernard to the high content of water vapour and impurities (desert and volcanic dust, smoke from bush fires, etc.) in the lower atmosphere. Whether a similar explanation can be applied to the Tropical zone generally is an open question.

CLIMATIC BOUNDARIES OF THE RAIN FOREST

The preceding sections of this chapter may have sufficed to give a general picture of the tropical rain-forest climate, the main features of which are well understood, but when an attempt is made to define this climate in precise terms, and to draw climatic boundaries between the Tropical Rain forest on the one hand, and Deciduous forest, Submontane Rain forest and Subtropical Rain forest on the other, great difficulties are at once encountered. At present such an attempt is often handicapped both by lack of knowledge of the exact position of the boundary of the Tropical Rain forest and by lack of climatic data for localities near where the boundary is supposed to lie.

In Chapter 1 it was pointed out that the rain forest is the potential climax over a much larger area than that which it actually occupies. There are large unforested areas where the development of rain forest is climatically possible: in some of these it has been destroyed by man and replaced by cultivation or secondary communities; in others, owing to various causes such as immaturity, and perhaps senescence, of soil or local peculiarities of topography, the climax is savanna or evergreen scrub. When defining the climatic limits of the tropical rain forest formation-type the whole area in which it is potentially or actually the climax vegetation must be taken into account. At the same time it must be remembered that the presence of a large mass of forest in itself modifies the climate to some extent, so that the climate of a potential rain forest area is not necessarily the same as that of the same area if it were in fact forest-covered.

Various climatologists and botanists, often with a view to determining the limits within which particular agricultural crops could be successfully grown, have attempted to classify climate by means of empirical 'climatic indices' integrating several climatic factors. Such indices have been used to define the climate of the rain forest and of other important world types of vegetation. The best known, and probably the most useful, classification of climate for ecological purposes is that of Köppen (1918; 1936) who defines 'Tropical Rain climates' (A climates) as follows: mean temperature of coldest month (t) above 18°, rainfall (in mm.) greater than $2t + 14$. This large class of climates he subdivides into Af (hot damp forest climates) with a mean rainfall for the driest month of not less than 60 mm., Am (forest climates, monsoon type) and Aw (periodically dry savanna climates). In both the Am and Aw climates there is a seasonal drought and the mean rainfall of the driest month is less than 60 mm. The Am climates are intermediate between the Af and the Aw and in them the total annual rainfall is sufficient to 'compensate' for the seasonal drought. (Thus if the driest month has 40 mm. the annual total must exceed 1500 mm., if it has 20 mm. the annual total must exceed 2000 mm. and so on.) The *Aw* climates have a severe dry season not 'compensated' by the annual total rainfall. In general the distribution of the *Af* climates coincides broadly with that of the Tropical Rain forest, that of the *Am* climates with Evergreen Seasonal forest, Monsoon forest, etc., and that of the *Aw* climates with Savanna. The correlation has been worked out in some detail by Bernard (1945) for the central Congo basin. The *Af* climates here form a narrow belt parallel to the equator and extending from it about 1° further to the south than to the north. This belt is flanked on both sides by *Am* climates, and these in turn by *Aw* climates. The *Af* region corresponds fairly closely with the area of the *véritable forêt ombrophile*, the *Am* belts with the *forêt tropophile* (probably equivalent to the Mixed Deciduous forest of West Africa, see Chapter 15) and the *Aw* regions with the Savanna.

A large number of other 'climatic indices' have been proposed and used with varying success to correlate climate with natural vegetation, among them Lang's 'rain factor' (precipitation/mean annual temperature), De Martonne's 'index of aridity' $\left(\dfrac{\text{precipitation}}{\text{temperature} + 10}\right)$, the Meyer ratio (precipitation/mean saturation deficit) and Thornthwaite's (1931) calculation of 'temperature efficiency' and 'precipitation effectiveness'.

Moreau (1938), who has given a very useful review of these classifications of climate as seen from the biological point of view, has pointed out a number of their disadvantages. Some climatic indices demand data, such as mean saturation deficits, which are available for very few places; others make no allowance for factors such as the 'occult' or 'horizontal' precipitation due to mists and the influence of moist winds, which may be locally of great importance in modifying the effects of rainfall and temperature and therefore in determining biological boundaries. All of these indices are based on 'normals' and give little weight to abnormal or extreme meteorological conditions which may well be decisive in ecology. It is a general experience that though some of these climatic indices give a fairly good broad agreement between the proposed climatic boundaries and the distribution of major plant communities, such as the rain forest, the correlation is seldom satisfactory in detail. To the field ecologist, who is aware that the distribution of plant communities (or rather, ecosystems) is governed by many factors, climatic and edaphic, all attempts to correlate biological limits closely with simple integrations of climatic factors appear foredoomed to failure. The ecological factors interact in such a complex manner that we cannot expect to define climatic limits by simple quantitative expressions, or indeed to define them at all, with our present very limited knowledge of climatic factors other than temperature and rainfall.

At its northern and southern boundaries, as we have already said, the Rain forest most often passes, usually by a gradual transition, into Deciduous forest or more abruptly into Savanna. In Chapter 15 it will be shown that Savanna is

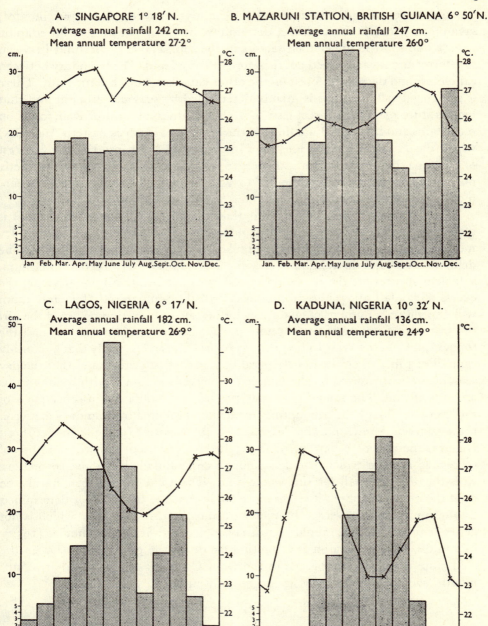

Fig. 22. Seasonal distribution of rainfall and temperature in tropical climates. A. Singapore. Equatorial climate; rainfall relatively evenly distributed. Climax vegetation: Tropical Rain forest. B. Mazaruni Station, British Guiana. Double rainy season; Climax vegetation: Tropical Rain forest. C. Lagos, Nigeria. Double rainy season; relatively severe seasonal drought. Climax vegetation: Tropical Rain forest (near climatic limit). D. Kaduna, Nigeria. Single rainy season; prolonged seasonal drought. Climax vegetation: Savanna. Data for B communicated by the Conservator of Forests, British Guiana, A, C and D from Brooks (1932).

probably never a climatic climax and that there is therefore no 'savanna climate'; savanna is either a biotic climax determined by recurrent fires or an edaphic climax due to little understood factors in the soil. The climatic limit between rain forest and savanna need not therefore be considered. The boundary between rain forest and deciduous forest on the other hand is a true climatic limit. There is good evidence that this is mainly determined by seasonal drought and that temperature plays little direct part in it. The altitudinal limit of Rain forest on mountains and the latitudinal limit in the few places (such as eastern Australia) where the Tropical Rain forest passes imperceptibly, not into Deciduous forest or Savanna, but into sub-tropical evergreen forest are almost certainly directly due partly to temperature, though to what element of temperature (mean, monthly minima, absolute minima, etc.) we are quite ignorant.

There is fairly general agreement that the climatic limit of rain forest is dependent on the length of the dry season expressed as the number of 'dry months', but various views have been taken as to how a 'dry month' should be defined. Provided the rainfall over a given period exceeds a certain minimum, sufficient to prevent the depletion of soil moisture below a critical value, the excess of rainfall above the minimum is probably immaterial. This minimum will depend on a number of factors, including the evaporating power of the air, the transpiring power of the vegetation, the characteristics of the soil and the topography. In the West Indies, Beard (1946 b) finds empirically that a monthly rainfall of 4 in. (10 cm.) can in general be regarded as critical and that the dry season can be measured by the number of successive months with less than this amount of rain. The same figure may possibly be significant in other parts of the tropics, though in attempting to define the climatic boundary between the evergreen forest and the Mixed Deciduous forest (p. 338) in Southern Nigeria, the author (1939, p. 9) found little correlation with the number of successive months which had less than 10 cm. rainfall, but some correlation with the total rainfall for the whole year. The total rainfall can hardly be itself the effective factor; it seems more likely that the boundary is determined by the minimum or average humidity during the driest months, which is not dependent on the total rainfall, but is correlated with it. A. S. Thomas (1932), in a study of the dry season as a limiting factor for the growth of cacao (in the wild state a rain-forest species) in the Gold Coast, has suggested that a 'dry month' should be defined by its 'degree of wetness', i.e.

$$\frac{\text{inches of rain} \times \text{number of rainy days}}{100}$$

The correlation between 'dry months', defined in this and various other ways, and the distribution of tropical plant formations needs to be further tested in as many areas as possible.

Since for plant growth the critical factor must be the water-supplying power of the soil in relation to water losses from transpiration, topographic and edaphic factors as well as climate must play a large part in determining the drought limit of Rain forest. Thus Beard (1946*b*) finds that in Trinidad the Semi-evergreen Seasonal forest becomes the climax on limestone in climatic regions where Evergreen Seasonal forest is the climax on less porous soils. Stamp & Lord (1923) and Stamp (1925) emphasize the importance of edaphic as well as climatic factors in determining the distribution of semi-evergreen and deciduous forest climaxes in Burma. A plant community, like an individual species, will necessarily be more sensitive to edaphic factors near the edge of its area of distribution than well within it. It is thus significant that in Australia, according to Francis (1929, p. 6), edaphic factors are increasingly important in limiting the distribution of Rain forest with increasing distance from the equator. Thus we should not expect the drought limit of Rain forest to give a perfect correlation with climatic factors alone, whether a single factor or a combination of factors is considered.

The factors limiting the upward extension of Rain forest on mountains and its poleward extension where it reaches a latitudinal boundary have been little investigated, apart from Brown's (1919) very thorough study of the vegetation on Mt Maquiling in the Philippines (see Chapter 16). On this mountain the Dipterocarp forest (Tropical Rain forest) gives place to 'Mid-mountain forest' (Submontane Rain forest) at about 600 m. and this in turn to 'Mossy forest' at about 900 m. The most conspicuous feature of the altitudinal zonation is the decrease in the height of the trees from the tall Dipterocarp forest (average height of A story trees 27·2 m.) to the dwarf 'Mossy forest' (average height of canopy about 5·9 m.). After a careful study of all the climatic factors and of soil conditions at different elevations Brown concluded that the most important climatic changes with increased altitude were an increase of atmospheric humidity accompanied by increased cloudiness and consequently decreased average light intensity, and decreased temperature. The increasing humidity and decreased temperature led to a rapid decrease in the rate of evaporation with rising elevation. A study of the growth rates of the trees suggested that the dwarfing of the trees could not be attributed to temperature alone, but was due to the combined effects of temperature and light intensity. The height of the trees agreed closely with the 'light-temperature indices' (the product of light intensity multiplied by 'temperature indices' for plant growth),[1] except at the lower elevations where the height is less than would be expected, probably owing to the effects of seasonal drought. Other striking features of the zonation are the increase in the abundance of both the epiphytes and the herbaceous ground vegetation with increasing altitude. Brown attributes the

[1] According to B. E. & G. J. Livingston (1913) and B. E. Livingston (1916). The 'exponential temperature index' is $\dfrac{2t-40}{18}$, where t is the 'normal daily mean temperature', taking the rate of growth at 40° F. (4·4° C.) as unity and assuming that the rate of growth doubles for every rise of 10° C. A 'physiological temperature index' based on experimental work with maize seedlings was also utilized.

former mainly to the lower evaporation and the latter to increased soil moisture combined with decreased evaporation.

Though it is obviously impossible to generalize from Brown's results alone, it is likely that temperature is usually not the sole factor responsible for the upper altitudinal and latitudinal limits of rain forest. The well-defined 'mist belt' or 'cloud belt' which exists on some tropical mountains may account for the transition from Submontane to Montane Rain forest, which is often abrupt (see Chapter 16), but can hardly be the cause of the upper limit of Tropical Rain forest which is usually at a lower altitude. The lowering of the vegetation belts, including the upper limit of Tropical Rain Forest, on isolated mountains, especially those near the sea, as compared with large mountain ranges (p. 347) is doubtless due to a corresponding lowering of the climatic belts.

The climatic limits of the Tropical Rain forest is clearly a subject in which there is much room for further research. A more exact definition of the limits than is at present possible would be of great value not only to the ecologist interested in the existing plant communities of the world, but also to the palaeo-botanist attempting to use fossil floras as indices of past climates. As an example Reid & Chandler's (1933) discussion of the climate of the London Clay (Eocene) flora may be mentioned. From the very careful analysis of the data it is impossible to escape the conclusion that this fossil flora, which includes species of *Nipa* (p. 300) and many other plants nearly related to members of the present Indo-Malayan flora, was like the flora of a modern tropical (or at least sub-tropical) estuarine region and must have existed under a climate very considerably warmer than that of any part of Europe to-day. Reid & Chandler's conclusion that the London Clay flora grew under conditions of heavy and well distributed rainfall, a frost-free winter and a mean annual temperature which could not have been much less than 70° F. (21° C.) can hardly be questioned, but further investigation of the climatic limits of Rain forest and other related plant communities might make possible a more precise definition of the 'London Clay' climate than this.[1]

NOTE ON SOME PHYSIOLOGICAL EFFECTS OF THE RAIN-FOREST CLIMATE

Since the climate of the Tropical Rain forest differs in so many respects from that of temperate regions, it is clearly of great importance to the ecologist to know something of the general effects of tropical conditions on the fundamental physiological processes of plants. It is impossible to enter here into a full discussion of the physiology of tropical plants, but a few brief comments on the

[1] It may be said, incidentally, that conclusions as to the climates of Tertiary floras are perhaps more firmly based when they are drawn from a statistical study of leaf sizes and similar features (cf. Bews, 1927) than when, like those of Reid & Chandler, they are derived from fruits and seeds and therefore rest on the *taxonomic affinities* of the fossil flora rather than its physiognomy, which appears, at least as far as modern vegetation is concerned, to be a very sensitive index of environmental conditions.

subject may help to make clearer the relations between the rain-forest climate and the vegetation, or at least to indicate the problems which require solution.

The newcomer to the equatorial region is invariably impressed by the overwhelming luxuriance of the vegetation, a feature which is clearly a reflexion of the moist, warm climate. The higher plants especially grow in unaccustomed abundance, size and variety. The impression of luxuriance is partly due to the very rapid growth of certain elements in the flora, particularly the species dominating the early stages of secondary successions (see Chapter 17). Waste ground, roadsides and land left uncultivated do not, as in temperate climates, remain practically bare for months or years, but become covered in a few weeks with dense vegetation several metres high. Some species grow astonishingly fast. In the bamboo *Dendrocalamus giganteus* an average rate of stem elongation as great as 22·9 cm. per day over a period of about 2 months has been observed; in one individual it reached 57 cm./day (Kraus, 1895). Many of the dicotyledonous trees of the secondary forest are also very quick growers. The African *Musanga cecropioides* becomes a substantial tree some 16 m. high in 14 years. In Java *Albizzia falcata* has been observed to reach a height of 25 m. in 6 years and 35 m. in 10 (Koorders & Valeton, 1894; Haberlandt, 1926, p. 111). Very rapid growth is much less conspicuous in the climax communities. Apart from the forest dominants which, as we saw in Chapter 3, may exist for years, making hardly any visible growth, there are many species which grow very slowly. Klebs (1911) has given numerous examples. Brown (1919) found that in the Philippines the dipterocarp, *Parashorea malaanonan*, grew more slowly than typical North American trees such as yellow poplar and white pine in the earlier part of its development, though later it grew faster. An extreme instance of a slow-growing tropical plant is the epiphytic orchid, *Taenophyllum obtusum*, in which the chief photosynthetic organs are the flattened aerial roots. In a series of measurements of the growth of these roots at different light intensities Wiesner (1897) found a maximum rate of 0·283 mm. per day, that is about 1/1000 of the rate in the bamboo mentioned above. According to Lebrun (1936*b*, p. 176) the mean annual increase of diameter of primary forest trees in the equatorial forest of the Congo is 0·5–2 cm.; the average increase is about 1·4 cm.

The rate of growth of the majority of tropical plants—at least those of the primary forest—may thus well lie within what would usually be regarded as 'normal' limits, but there are also some exceptionally fast growers; these raise physiological problems which are as yet largely unsolved. Three possible explanations of these very high growth rates suggest themselves: the average net assimilation rate (increase in dry weight of the whole plant per unit leaf area per day) may be higher than in any temperate plants, or a larger proportion of the total net assimilation may be expended in increasing the leaf area than in temperate plants (leading to a more rapid increase in size of the plant as a whole), or the explanation may simply lie in the long unbroken period of active growth which in some cases may not be interrupted by any resting period.

Data by which these three possibilities might be tested are still very scanty. Of recent work neither that of Guttenberg (1931) nor that of Stocker (1935*a*) indicates values for the assimilation rate in excess of what has been observed in temperate plants. Stocker's values are in fact rather low, probably owing to the nature of the gasometric technique used. Guttenberg's use of the 'half leaf' method, though it has the advantage that the plant material is not exposed to an unnatural external environment, leaves us in doubt both as to the amount and the destination of the substances translocated from the leaves; moreover the plants used are not among the most rapidly growing tropical species. We cannot at present distinguish between, or eliminate, the first two possibilities above mentioned. Detailed studies are required which would give a picture not only of the assimilatory activity of the leaves, but also of the growth pattern of the whole plant, in some of the very fast-growing species.

It seems probable that, in the early stages of their growth, rapidly growing tropical plants will not be found to differ widely in behaviour from temperate annuals such as *Helianthus annuus*, which under favourable conditions doubles its dry weight every week till the time of flowering approaches. After flowering and setting seed *Helianthus* dies, having reached within its span of some three months dimensions comparable with those of rapidly-growing tropical plants; the latter, on the other hand, generally continue their growth after flowering. In this connexion it is interesting that many fast growing tropical trees (see p. 385) are short-lived and die after a few years. It is not unlikely that the vigorous annuals of the temperate regions in which the life cycle is completed within a few months are, as it were, an extreme development of the fast-growing short-lived type of plant which dominates second-growth communities in the tropics.

The respiration of the leaves in tropical trees has been investigated by Stocker (1935*a*), who found that the average respiration per unit area of the sun leaves in three species, one of them the rain-forest tree *Stelechocarpus burahol*, was 1·3 mg. CO_2/sq.dm./hr. at 30° C., a value not very different from those obtained in Denmark for *Fagus sylvatica* (1·0 mg.) and *Fraxinus excelsior* (1·2 mg.). Stocker claims, however, to have shown that the respiration of tropical trees is in fact lower than that of European trees at the same temperature, the tropical plant at 30° C. respiring no faster than the Arctic plant at 10°. Since respiration rates expressed in these terms differ so widely from one species to another in the same climate, Stocker's theory of a 'physiological adaptation' of respiration to climate cannot at present be regarded as securely founded. In the absence of trustworthy evidence as to the rates of assimilation and respiration in tropical plants it is not surprising that comparative data on the compensation point (at which assimilation balances respiration) in tropical and temperate species are also at present too scanty to warrant definite conclusions; such information would, however, be of great value for understanding the relations between tropical shade plants and their environment.

Some plant physiologists have attempted to make generalizations as to the transpiration rates of tropical plants. Haberlandt (1898, 1899), for example, claimed that the transpiration rate of plants in the wet tropics was on the average two to three times less than in central Europe. Variation between different species is, however, so great that such statements have very little meaning. In tropical plants exposed to very high saturation deficits, such as the trees of the upper stories of the rain forest and plants of cultivated land and other open habitats, rates as high or higher than any found in European plants can be observed; on the other hand the shade flora of the rain-forest under-growth may show lower rates than typical European shade plants. The great differences between the humidity of the air in different tropical microclimates (see Chapter 7) are only partly responsible for the wide range of behaviour in different species. The rate of transpiration is a function not only of the difference in the partial pressure of water vapour in the air immediately surrounding the mesophyll cells of the leaf and that in the atmosphere at some distance from the plant, but also of the resistances to diffusion along the path by which the water vapour escapes from the leaf. It is in these diffusion resistances, if anywhere, that systematic differences in transpiration behaviour between tropical and temperate plants are likely to exist. This aspect of the matter does not yet appear to have been investigated and might throw some light on the wide range of transpiration rates which has been shown to exist among tropical plants inhabiting different microclimates. Stocker (1935 b) also finds reason for thinking that by adapting their 'surface development' (i.e. the ratio of surface to bulk), and the ratio of water-content to surface area, plants in all climates are able to keep their water expenditure within certain limits; in other words, their rate of transpiration relative to their total water-content shows only a moderate amount of variation. This view is hardly borne out by Stocker's own data, for his tropical shade plants *in the shade* had a much lower rate of transpiration than his sun plants *in the sun*, but at the same time a considerably higher water-content per unit area of surface.

An indirect effect of the tropical climate on plant growth which may be mentioned in conclusion concerns the supply of mineral nutrients. As we shall see in Chapter 9, the heavy and well distributed rainfall of the wet tropics tends continually to remove soluble salts from the soil. As the soil itself is in-variably acid and frequently porous in texture, the mineral supply to the plant always tends to be deficient. The physiological consequences of a very limited mineral nutrient supply for natural vegetation (as distinct from cultivated crops) in the tropics are as yet unexplored. We may at least expect that competition by the roots for the meagre supply of nutrients would be still more severe among tropical than among temperate plants (cf. p. 190), a fact which may have an important bearing on their growth rates and possibly on other aspects of their physiology.

CHAPTER 7

MICROCLIMATES

The last chapter was concerned mainly with climatic conditions for the rain-forest community as a whole, that is to say with the standard or general climate within regions occupied by rain-forest formations. This standard climate is but one among a considerable number of climates which exist together in different situations in the same locality; it is the climate recorded by instruments about 1 m. above the surface of the ground, exposed, according to the rules adopted by meteorologists, in more or less extensive clearings. At 100 m. above the ground, or at the soil level, the climate will be different from that at 1 m. and in the undergrowth of a dense forest it will be quite unlike that in a large open clearing. The standard climate is in fact arbitrarily selected from a constellation of partial climates or microclimates mainly because it approximates to the conditions normally experienced by a human being out of doors. Within a complex plant community, such as a tropical forest, many microclimates can exist together, some of them differing very widely from the standard climate of the same locality.

Plants are, of course, directly influenced only by their immediate surroundings. Hence, though the foliage of mature dominants of a forest community is living under conditions closely approximating to the standard climate, that of the young trees on which the regeneration of the community depends, and many of the subordinate species, exists in a widely different environment. A knowledge of the microclimates within the community is thus indispensable for a true understanding of the relationship between its members and the climate. The importance of this fact has been realized only comparatively recently and the study of microclimates is still in a very early stage. The microclimates of certain types of temperate forest have been systematically explored to a limited extent. Information on the microclimates of Tropical Rain forest is still scanty, though animal ecologists have already given a considerable amount of attention to the subject, chiefly in connexion with the smaller forest inhabitants, such as mosquitoes; it has become increasingly clear that the same area of forest can shelter two quite different faunas, that of the tree-tops and that of the undergrowth (see Allee, 1926 b; Hingston, 1932; Paulian, 1946, 1947) with very different ecological relationships. The chief aim of this chapter is to correlate the scattered data as far as possible and to try to give a connected account of the vertical and horizontal microclimatic variations within rain-forest communities, and thus to portray the conditions of life of their various synusiae.

In comparing the interior climates of a forest with its standard climate, we may assume that at a few metres above the uppermost layer of tree-crowns conditions are practically identical with those at 1 m. above the ground in an open clearing. A layer of crowns forms a barrier impeding, but not preventing, convectional air currents between the free atmosphere and the space below the crowns; it also offers a strong resistance to lateral air movements. A considerable fraction of the rainfall is intercepted by the trees; some of this runs down the trunks and some evaporates without reaching the soil. The trees also interfere with radiation both to and from the surface of the ground. During the day a large proportion of the incident light and heat is absorbed by the leaves. At night radiation of heat takes place mainly from the crowns of the tallest tree layer, though the interior of the forest becomes cooled by the sinking of the relatively dense, cold air from above. If dew forms, it will be on the crowns of the trees and not at ground-level. The effect of the tree-crowns will thus be to give the space below them a more equable climate than that in the open. It has a smaller range of temperature and relative humidity (or saturation deficit) and all changes in the outside atmosphere will be 'damped down' and followed by corresponding changes inside the forest only after a considerable time lag. The microclimates of the forest interior are, in fact, intermediate in character between the climate in the open and that in the soil, where conditions are even less variable. The microclimate at a given level depends on the height above the ground; there is a continuous or discontinuous vertical climatic gradient from the surface of the ground to the level of the tallest tree-tops. Since none of the tree strata is uniform in density there are also variations in microclimate from place to place at the same horizontal level.

The magnitude of these effects and the nature of the vertical climatic gradients have been investigated in some detail by Geiger (1927) in European forests with the simplest possible structure, viz. even-aged plantations in which the trees form a single layer with a closed, even canopy. The climatic gradients were found not to be uniform; their steepness varying with height above ground, the exact form of the gradient being different for different climatic factors. The nature of the gradient also depends on the intensity outside the forest of the particular factor considered; thus the gradients of wind velocity and relative humidity are different at different outside wind velocities or relative humidities respectively.

The form and steepness of the climatic gradients inside a forest will, of course, depend on its structure. In a Tropical Rain forest the gradients are more complicated than those of most temperate forests. The profile diagrams in Chapter 2 show that in a rain forest, owing to its great height and several superposed layers of trees, the horizontal barrier of crowns is of much greater depth and density than in the type of forest studied by Geiger. Such a barrier is likely to have a correspondingly greater effect in impeding air movements and interfering with radiation exchanges. It would therefore be expected that the microclimatic differences between the lower layers of a Tropical Rain forest and the open atmosphere would be greater in magnitude, if mainly in the same sense, as those between the interior and exterior of a temperate forest. The form of the gradient no doubt differs according to whether the upper (A and B) tree stories are closed or relatively open (cf. Chapter 2).

AIR MOVEMENTS

One of the most noticeable features of the lower layers of the forest is the extreme stillness of the air; this contributes much to the general feeling of oppressiveness. In the rain forest of British Guiana, which is typical in this respect, a gentle wind blows almost continuously in the tree-tops, at least during the day, so that flexible leaves are always moving; in the undergrowth the air is normally so still that smoke rises quite vertically and loose scraps of paper are never moved. Squalls such as often blow for a few seconds before rain may penetrate deeply enough into the forest to cause leaves in the undergrowth to stir. During a spell of dry sunny weather in October 1929, there was such a strong wind that on several mornings leaves near the ground were kept in motion for several hours on end. This steep gradient of wind velocity must be of considerable importance to the vegetation, because of its effect on transpiration. The mechanical effects of air movement are possibly also of some significance; delicate thin leaves such as are found in some undergrowth species would soon be torn in the relatively windy conditions of the upper layers of the forest.

In a German wood with a single layer of tree-crowns reaching a height of about 14 m. Geiger (1927) found that at low outside wind velocities the velocity inside did not diminish below the crowns, but was approximately the same at all levels below about 12 m., showing that the main braking action is exerted by the upper surface of the crowns. At higher wind velocities there was a small secondary maximum of velocity at about 6–8 m., the velocity here being slightly greater than that among the crowns or at the ground. In the Tropical Rain forest, with its more complex structure, a different relation probably holds. It might be expected that at all outside wind velocities there would be a considerable difference between the velocity in the crowns of the top-story trees and that in the undergrowth. This seems to be borne out by the figures of Allee (1926a) for the rain forest of Panama (Table 15). The period of observation was 23 days during the dry season, the time of year with the greatest amount of air movement.

For rain forest in Brazil (lat. 22° 58′ S.), perhaps sub-tropical in character but probably fairly similar in structure to a Tropical Rain forest, Freise (1936, p. 302) gives the figures shown in Table 16 for mean annual wind velocity.

During a storm in the Mambucaba locality (ibid. p. 346) a mean wind velocity of 28·8 m./sec. at 5 m. above ground outside the forest was reduced to only 1·9 m./sec. at the same height 12,000 m. within the forest.

TABLE 15. *Wind velocities in forest, Barro Colorado Island, Panama.*
(After Allee, 1926a, p. 279)

Position of anemometer	Total horizontal air movement (km.)	Average horizontal air movement (km./24 hr.)	Maximum horizontal air movement
6·5 ft. (2 m.) above ground (undergrowth)	34·8	c. 1·6	7·7 km./24 hr.
75 ft. (23 m.) above ground ('in the midst of the forest canopy')	372·2	c. 16·0	28·6 km./24 hr.
Unobstructed open place (neighbouring locality)	—	384	38·4 km./hr.

TABLE 16. *Wind velocity at two localities in the evergreen forest of southern Brazil.* (After Freise, 1936, p. 302)

	Mean wind velocity at 5 m. above ground (m./sec.)	
Position of anemometer	Mambucaba district (1908–35)	Serra dos Aymorés (1914–35)
150 m. outside forest	2·3	2·0
100 m. inside forest	0·5	0·5
1,100 m. inside forest	Too small to determine	—
2,150 m. inside forest	—	Too small to determine
8,500 m. inside forest	—	Too small to determine
12,000 m. inside forest	Too small to determine	—

INTERCEPTION OF PRECIPITATION

Trees or plants of any kind intercept a proportion of the rainfall and diminish the amount reaching the ground by a fraction depending on the number, shape and arrangement of the leaves and the branches. Since most plants can absorb water only through their roots, the fraction which does not reach the ground is lost to them and has no significance except in affecting the rate of transpiration. For temperate woods the subject of rainfall interception has been carefully studied by Hoppe (Geiger, 1927), Burger (1933) and others. For an 88-year old beechwood in Germany Hoppe found that the proportion of the rainfall passing through the crowns and that running down the trunks varied with the heaviness of the fall. With very light rain over 50 % passed through; with heavier rain over 60 % passed through and over 10 % drained down the trunks.

Freise (1936, pp. 303–4) investigated rainfall interception in evergreen forest[1] in Brazil, with the results shown in Table 17. The fraction penetrating the canopy to a rain gauge at 1·5 m. varied considerably with the character of the rainfall. In a month in which the rain fell mainly as thunder showers 26·7 % reached the rain gauge; in a month of long-continued fine rain ('land rain') as much as 35·5%. This result is at variance with the observations of Vaughan & Wiehe (1947) who found that in the upland Rain forest of Mauritius relatively more rain reached the forest floor during the season of heavy downpours than during the season when the rain is less heavy.

[1] Freise speaks of the forest as 'subtropical', but it was probably not very different in structure from Tropical Rain forest.

The rain reaching the soil is, of course, not all available for absorption by the plant, as some is lost by seepage and by surface run-off. McLean (1919, pp. 130–3) attempted to estimate what he calls the 'rainfall efficiency' of the south Brazilian Rain forest. This he defines as 'the percentage of the total rainfall descending on a given area which becomes effective in the soil, allowing not only for the stoppage of water mechanically by the leaves, but also for the effect of drainage in the soil itself'. For two stations the values obtained were 16·6 and 27·5 % respectively for a rainfall of 13·8 mm.

TABLE 17. *Approximate average balance sheet of rainfall interception in evergreen forest of southern Brazil.* (After Freise, 1936, p. 303)

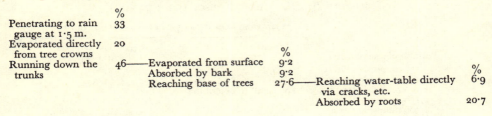

	%		%		%
Penetrating to rain gauge at 1·5 m.	33				
Evaporated directly from tree crowns	20				
Running down the trunks	46	Evaporated from surface	9·2		
		Absorbed by bark	9·2		
		Reaching base of trees	27·6	Reaching water-table directly via cracks, etc.	6·9
				Absorbed by roots	20·7

As was mentioned in the last chapter, heavy dew may occur in Rain forests. Dew forms only on the crowns of the taller trees and never inside the forest. It is unlikely that the dew is ever sufficient in quantity to run down the trunks or drip off the leaves on to the soil, but it is probably important to sun epiphytes and may be of some significance in lowering the rate of transpiration from the leaves of the trees during part of the day.

TEMPERATURE

Among the numerous published observations on temperature in the interior of Tropical Rain forest, the following are the most extensive: for Lowland Rain forest, those of McLean (1919, south Brazil), Allee (1926a, Panama), Davis & Richards (1933–4, British Guiana), Baker & Harrisson (1936, New Hebrides), Richards (1936, Sarawak), Evans (1939, south-western Nigeria), Haddow, Gillett & Highton (1947, western Uganda), Pittendrigh (1948, Trinidad). For upland (including Montane and Submontane) Rain forest there are the very full data of Brown (1919) for various altitudes on a mountain in the Philippines and Moreau's (1935a) observations on the 'Intermediate forest' of Usambara (East Africa) and Vaughan & Wiehe's (1947) on the 'Upland Climax forest' of Mauritius. Though most of these records cover only short periods, owing to the small seasonal variation of temperature they give a fairly adequate picture of the conditions.

Table 18 gives some typical figures for maximum and minimum temperatures at different heights above the ground and at different times of year. As would be expected, the undergrowth has a smaller range of temperature than the upper levels of the forest; the difference is due mainly to a lower maximum, the minimum in the undergrowth being either higher or only very slightly lower.

In the day the form and steepness of the temperature gradient are determined chiefly by the proportion of the sun's rays intercepted by the leaves and branches at different levels. A fairly steady decrease of temperature from the first story downwards would be expected, and the observations show that such a gradient

TABLE 18. *Temperatures at different levels in Tropical Rain forest (°C.)*

Locality	Barro Colorado Island, Panama (Primary Rain forest)						Near Akilla, Southern Nigeria (Primary Mixed Rain forest)			
Period of observations	9. ii.–9. iii. 24 (Dry season)			17–22. ii. 24 (Dry season)			5–11. iii. 35 (Dry season)		9–16. v. 35 (Wet season)	
Source of data	Allee (1926a)						Evans (1939)			
Height above ground (m.)	0·2	17	26	0·2	17	26	0·7	24	0·7	24
	Under-growth	Second story	'Upper part of canopy'	Under-growth	'Second story'	'Upper part of canopy'	Under-growth	B story*	Field layer	B story*
Mean maximum	—	—	—	27·1	30·8	35·7	29·7	33·9	26·8	30·9
Mean minimum	—	—	—	26·0	24·0	27·1	23·9	24·0	23·3	21·8
Mean daily range	2·9	7·3	11·8	1·1	6·8	8·6	5·8	9·9	5·5	9·2

Locality	Mt Maquiling, Philippines (Primary Dipterocarp forest, 300 m.)								
Period of observations	x. 12–i. 15			27. ii.–27. iii. 14† (Dry season)			14. viii.–11. ix. 14† (Wet season)		
Source of data	Brown (1919)								
Height above ground (m.)	—	c. 18	c. 35–40	—	c. 18	c. 35–40	—	c. 18	c. 35–40
	'Under-growth'	'Second story tree'	'Top of tall tree'	'Under-growth'	'Second story tree'	'Top of tall tree'	'Under-growth'	'Second story tree'	'Top of tall tree'
Mean maximum	26·3	27·0	32·4	27·5	27·5	32·5	26·9	27·3	31·3
Mean minimum	20·6	21·0	20·0	19·9	20·8	19·6	21·0	21·4	20·6
Mean daily range	5·7	6·0	12·4	7·6	6·7	12·9	5·9	5·9	10·7

* Cf. profile-diagram (Fig. 6) and photograph taken from this tree (Pl. III A).
† These periods have been selected by the author from Brown's data.

in fact exists, temperatures about the middle of the day being some two to three degrees lower in the undergrowth than among the crowns of the first-story trees. At night conditions are more complicated; Geiger (1927) has shown that in a European forest with a single even layer of tree-crowns there is normally at night a decrease in temperature with height from the tree-crowns to ground-level; the older view that there is a gradient in the opposite sense was based on inaccurate measurements. This night gradient is set up as follows. The ground and lower levels of the forest are sheltered and at night radiation of heat takes place chiefly from the crowns of the trees. Owing to its high density the cool air surrounding the crowns sinks; the lower levels of the forest thus become cooler than the upper. If the space between the tree-crowns and the ground

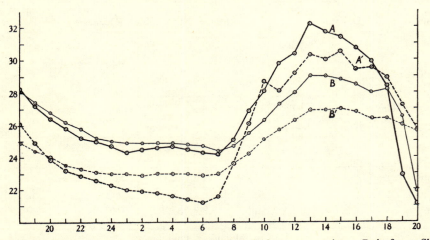

Fig. 23. Daily march of temperature in the undergrowth and tree-tops, primary Rain forest, Shasha Forest Reserve, Nigeria. From Evans (1939, fig. 7). The diagram shows the march of temperature from 18.00 hr. on one day to 20.00 hr. on the next. *A, A'* at 24 m. above ground; *B, B'* at 0·7 m. above ground; *A, B* dry season (8–9 March 1936); *A', B'* wet season (10–11 May 1936).

were sufficiently filled up to impede the downward sinking of the cooled air, a reversed gradient would be set up, i.e. the crowns would become cooler than the lower levels because they would lose heat by radiation faster than the cooled air would sink to the ground. Evans' data show that this state of affairs actually exists in some rain forests. In the forest of Southern Nigeria (cf. p. 29 and Fig. 6) the A (tallest) and B tree-stories are fairly open, but the C (lowest tree) story is very dense and has a compact surface bound together by inter-lacing lianes. We may therefore assume that vertical movements of air can take place with little hindrance as far down as the surface of the C story, but would be much impeded between the top of the C story and the ground. If this is so, it is easy to understand why the undergrowth at 0·7 m. is always warmer at night than the B story at 24 m. (cf. Evans, 1939, p. 450). The curves for the daily march of temperature on typical days (Fig. 23) show clearly the

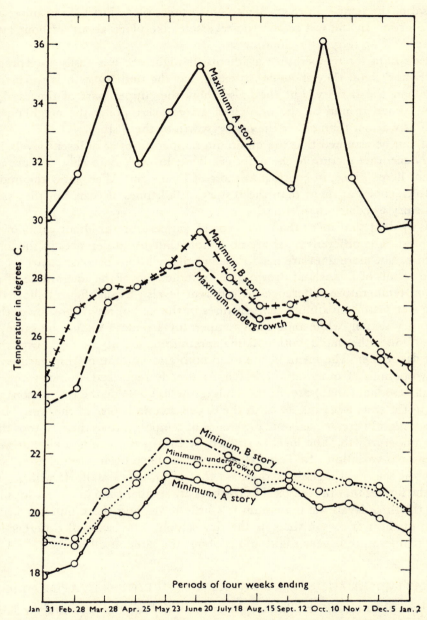

Fig. 24. Average weekly maximum and minimum temperatures in the A and B stories and in the undergrowth, primary (Dipterocarp) forest at 300 m., Mt Maquiling, Philippine Islands. After Brown (1919, fig. 6).

reversal of the temperature gradient, which takes place just before sunset and after sunrise. In the wet season the reversals take place about an hour later, probably owing to greater cloudiness.

Allee's data for temperature gradients at night are less easily interpreted. Like Evans' data, they show consistently that the undergrowth is warmer at night than the B story, but they also show the 'upper part of the canopy' (26 m. above ground on the branch of a tree overtopping the neighbouring forest) as warmer than either the undergrowth or the B story.

If it can be assumed that the minimum temperatures at different levels give a fairly accurate picture of the night gradient, Brown's data (Fig. 24) are also relevant here. These, in contrast to those of Evans and Allee, show the undergrowth as cooler at night than the B story, at all times of year, but they show the A story as cooler than either.

It is evident that more than one type of temperature gradient exists in the rain forest. The different types are no doubt related to differences in the forest structure, but more data are needed before the relations become clear.

The results of Evans and Brown give an indication of the magnitude of the seasonal temperature differences at different levels. Evans' figures show that in the wet season maximum temperatures in the undergrowth are lower than in the dry season, while minimum temperatures remain about the same; at 24 m., on the other hand, both maxima and minima are about 2–3° lower than in the dry season. The mean of the maximum and minimum temperatures for the two periods of observation (which, as already remarked, do not represent the full seasonal range) are, in the undergrowth, 26·8° in the dry season and 25·1° in the wet; at 24 m., 29·0° in the dry season and 26·4° in the wet. There is thus a slightly greater seasonal difference of temperature in the tree-tops than in the undergrowth. The locality studied by Brown represents a climate with less seasonal variation. In Table 18 two periods have been selected as typical, one in the season of highest, and one in that of lowest rainfall. It will be seen from the figures that the seasonal differences are smaller than in Evans' locality. At all the three levels the maximum is a little lower and the minimum a little higher in the wet season than in the dry season. The seasonal difference in mean temperature is very slight and is about the same at all levels.

RELATIVE HUMIDITY, SATURATION DEFICIT, EVAPORATION

The three chief measures of atmospheric humidity are the percentage relative humidity, usually calculated from wet- and dry-bulb thermometer readings or measured with a hair hygrometer; the saturation deficit, calculated from simultaneous readings of relative humidity and temperature; and the rate of evaporation measured by an evaporimeter. The chief series of data on relative humidity in the Tropical Rain forest include, for lowland forest, those of McLean (1919, S. Brazil), Davis & Richards (1933–4, British Guiana), Richards

(1936, Sarawak), Evans (1939, Southern Nigeria), Pittendrigh (1948, Trinidad), for upland forests those of Brown (1919, various altitudes on Mt Maquiling, Philippines), von Faber (Schimper, 1935, Tjibodas, Java) and Stocker (1935 *b*, Tjibodas, Java). Observations on saturation deficit in the Rain forest have been published by Davis & Richards (1933–4, British Guiana), Richards (1936, Sarawak), Baker & Harrisson (1936, New Hebrides), Evans (1939, Southern Nigeria) and Haddow, Gillett & Highton (1947, Uganda); both McLean (1919, S. Brazil) and Stocker (1935 *b*, Tjibodas, Java) have published series of simultaneous readings of temperature and relative humidity from which the saturation deficit can be calculated. Brown (1919, Philippines) and Allee (1926 *a*, Panama) have given observations on evaporation at different levels in the Rain forest.

The humidity of the atmosphere is of interest to plant ecologists chiefly because of its relation to transpiration and the most useful measure of it will therefore be that which is most likely to be correlated with the rate of transpiration. It might be supposed that direct measurements of evaporation would be of the greatest ecological value, but in practice they are of limited value owing to the difficulty or impossibility of comparing the results obtained with evaporimeters of different patterns or even with the same instrument under different conditions. We shall therefore consider here only observations on relative humidity and saturation deficit.

TABLE 19. *Relative humidity in primary Tropical Rain forest near Akilla, Southern Nigeria.* (From Evans, 1939)

	Dry season		Wet season	
Height above ground (m.)	0·7	24	0·7	24
Mean maximum (%)	94·7	97·6	96·2	94·9
Duration of maximum (hr.)	14·1	11·0	17·8	11·1
Mean minimum (%)	68·7	61·7	87·4	64·5
Mean 'half-hour minimum' * (%)	76·6	63·9	92·0	66·9
Mean daily range (%)	18·1	33·7	4·1	28·0

* See p. 168.

It will be convenient to consider mainly the observations of Evans (1939) on relative humidity and saturation deficit in Nigeria, which, though covering only short periods, are in some respects the most complete series of observations available. Thermograph and hygrograph records were made in primary rain forest on a tree at 24 m. above the ground and simultaneously in the undergrowth at 0·7 m. near the foot of the same tree. The records were very carefully checked and corrected by wet- and dry-bulb thermometer readings. Observations were made for a period near the end of the dry season and again for a period early in the wet season. As these periods did not fall in either the wettest or the driest months, the observations do not show the full seasonal range of conditions.

Table 19 summarizes the relative humidity results; Fig. 25 shows the daily march of relative humidity on two days, one typical of the dry season, the other of the wet.

It will be noticed that during the day there are numerous fluctuations imposed on the general course of the curves; for periods of a few minutes the

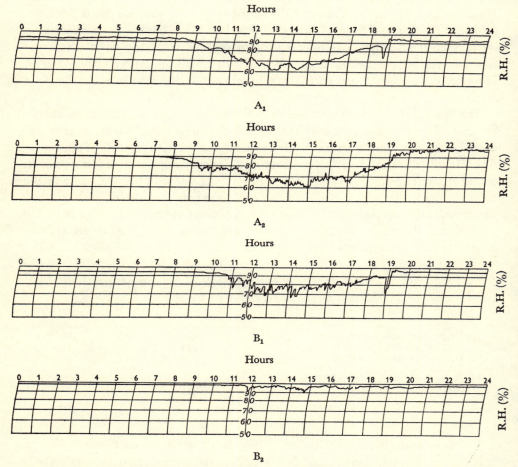

Fig. 25. Relative humidity (hygrograph records) in Tropical Rain forest near Akilla, Southern Nigeria. A_1A_2. On tree 24 m. above ground level. B_1B_2. In undergrowth 0·7 m. above ground level. A_1B_1. 9 March 1935 (dry season). A_2B_2. 11 May 1935 (wet season).

relative humidity may rise or fall as much as 10 % above or below the average for an hourly period. These fluctuations are probably caused chiefly by puffs of wind; their physiological effect is probably negligible. As the absolute daily minimum humidity is generally reached in one of these brief fluctuations, it is probable that the 'half-hour minimum', i.e. the minimum relative humidity recorded, either continuously for half an hour, or discontinuously for periods totalling half an hour, is more significant than the absolute minimum; it is this minimum which has been used to calculate the daily range of humidity.

In these data one of the most striking features is the long period at night in which the humidity does not differ significantly from saturation. This occurs both in the tree-top and in the undergrowth, but its average duration is considerably longer in the latter. In the tree-top the humidity begins to fall soon after sunrise, but in the undergrowth the fall begins later and is less rapid. Similarly in the evening the maximum humidity is established much earlier in the undergrowth than in the tree-top. During the dry season its duration is about three hours shorter in the tree-top. In the tree-top the duration of the maximum humidity is about the same in the wet season as in the dry season, but in the undergrowth it increases from about 14 to 18 hours.

In the dry season the minimum is about 7 % higher in the undergrowth than in the tree-top; here again the difference between the two levels is greater in the wet season (25 %). Thus both the duration of the maximum humidity and the minimum value show a greater seasonal variation in the undergrowth than in the tree-top, though in both seasons there is a greater daily range of humidity in the tree-top.

The daily march of saturation deficit in the undergrowth and the tree-top is shown in Fig. 26. For this purpose it is convenient to choose a typical day from the records for each season, rather than to use average hourly values for two sample periods (cf. Evans, 1939, pp. 452–3). The long night period of very small saturation deficit, corresponding to the period of maximum relative humidity, is clearly shown. A rise occurs during the day, but the curve for the undergrowth lags behind that for the tree-tops in the morning and, just as the minimum relative humidity is higher in undergrowth than in the tree-top, the maximum saturation deficit is smaller in the undergrowth. This maximum is reached at about 14 hours at both levels. In the dry season the maximum saturation deficit is over 12 mm. in the tree-top, but does not exceed 8 mm. in the undergrowth. In the wet season it reaches nearly 12 mm. in the tree-top, but never exceeds 2 mm. in the undergrowth. For comparison it may be stated that the mean saturation deficit at Kew (England) in July is 7·4 mm. (9·8 millibars) at 14.00 hr. and 1·3 mm. (1·7 mb.) at 4.00 hr. (Bilham, 1938, p. 211).

By plotting temperature against relative humidity for hourly (or other) intervals during the day, it is possible to follow the daily march of temperature, relative humidity, saturation deficit and absolute humidity (water vapour content of the air) on a single diagram. The polygons so obtained (Figs. 27–30) give much information as to changes in the humidity due to transpiration, mixing of different layers of air, etc. Families of lines can be drawn across the polygons showing constant saturation deficit and constant absolute humidity respectively.

In both polygons for the tree-top (Figs. 27 and 28) and in the dry season polygon for the undergrowth (Fig. 29) it will be noticed that the upper part of the polygon tends to run more or less parallel with the temperature axis, while the lower part more or less follows a line of constant absolute humidity, i.e. during the night, in spite of changing temperature, the relative humidity is

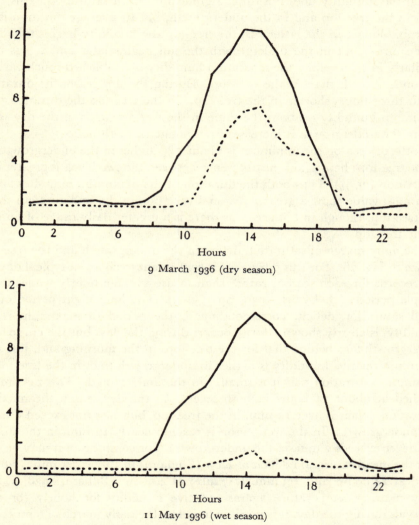

9 March 1936 (dry season)

11 May 1936 (wet season)

Fig. 26. Daily march of saturation deficit in the undergrowth and tree-tops, Tropical Rain forest, Shasha Forest Reserve, Nigeria. From Evans (1939, fig. 8). Continuous lines, 24 m. above ground on tree; broken lines, 0·7 m. above ground in undergrowth. Each point is a mean of four readings at quarter-hourly intervals.

approximately 100%, but the absolute humidity changes; during the day the absolute humidity remains approximately constant, though both temperature and relative humidity change. This behaviour is not shown by the polygon for the undergrowth in the wet season (Fig. 30), because the daily range of relative

humidity is very small compared with the random fluctuations. The full significance of these facts is not clear, but they seem to indicate that there is an analogy
between the atmosphere of the forest and a closed system of an aqueous solution
and water vapour. In such a system, starting at a relative humidity corresponding to the vapour pressure of the solution at the initial temperature and concentration, water will evaporate as the temperature rises; the solution will

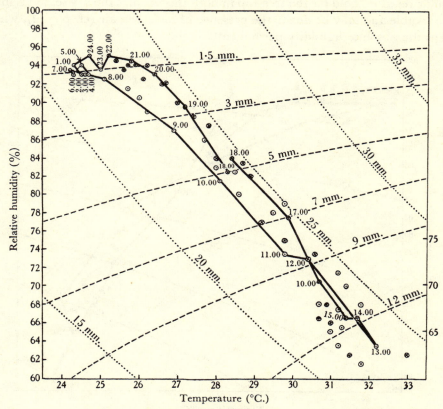

Fig. 27. Tree-top (24 m. on tree 28 m. high) in dry season (8–9 March 1936).

Figs. 27–30. Temperature and relative humidity (hythergraphs) in Tropical Rain forest, Shasha Forest
Reserve, Nigeria. From Evans (1939, figs. 3–6). The broken 'grid' lines represent constant saturation
deficit, the dotted 'grid' lines constant vapour pressure (mm. of mercury). Readings before noon ⊙;
readings after noon ⊗.

become more concentrated and the relative humidity will fall until the solution
is saturated. The curve of relative humidity will then slowly rise or fall, or
remain constant, according to the properties of the solution in question. When
all the water has evaporated, the curve will follow a line of constant absolute
humidity. In the forest a lag in both evaporation and condensation is to be
expected; the first three polygons considered seem to show evidence of this
hysteresis. The polygon for the undergrowth in the dry season shows a slight

fall of absolute humidity between 10.00 and 14.00 hr.;[1] this is probably due to the mixing of the air inside the forest with drier air outside, owing to the increased wind and convection in the course of the morning.

In the forest undergrowth air movements are so slight and the barrier between the atmosphere inside and outside the forest is so complete that it is not surprising that conditions should be comparable to a closed system. That similar relations hold for the tree-top in both the wet and the dry season is more remarkable and may be due to the presence of moist sea air (cf. p. 142), which keeps the absolute humidity permanently high.

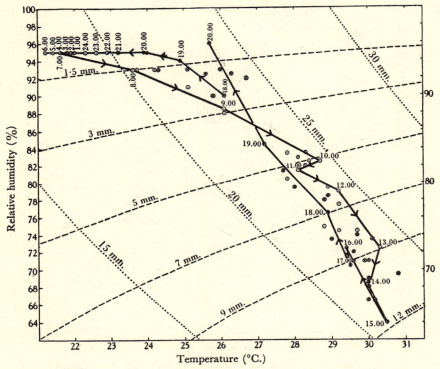

Fig. 28. Tree-top (24 m. on tree 28 m. high) in wet season (10–11 May 1936).

(See legend under Fig. 27.)

Data on relative humidity and saturation deficit in other Rain forests, especially for the upper levels, are so incomplete that no detailed comparison with Nigerian conditions is possible, but a few general statements seem justified. A long nightly period of humidity at or near saturation is characteristic of all Tropical Rain forests, in both wet and dry seasons. The duration of this period of course varies with the locality, the season and the height above ground. In the undergrowth in the wet season, or in damp climates at all times, the

[1] Nigerian standard time; in the Shasha Forest Reserve this is some ¾ hr. in advance of sun time (i.e. sunrise and sunset are at about 6.45 and 18.45 respectively with insignificant seasonal variations).

period of maximum humidity is very long and there is only a small fall of humidity during a few hours in the middle of the day. The daily minimum relative humidity, or maximum saturation deficit, occurs at about 14.00 hr. in both the undergrowth and the higher levels; its actual value does not normally exceed 10 mm. saturation deficit in the undergrowth. In the upper stories the

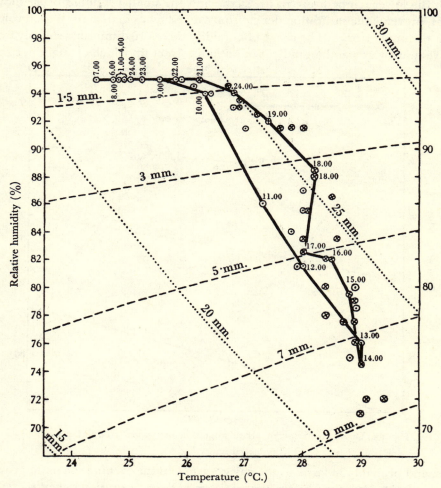

Fig. 29. Undergrowth (0·7 m. above ground) in dry season (8–9 March 1936).
(See legend under Fig. 27.)

maximum saturation deficit is, of course, greater, but even in dry conditions its average daily value probably seldom exceeds 16 mm. Both in the undergrowth and at higher levels the relative humidity and saturation deficit fluctuate throughout the day.

Thus in the Tropical Rain forest there is no humidity gradient at night, the atmosphere at all levels being near saturation. The rise of temperature in the morning causes a rise of saturation deficit (or fall of relative humidity), beginning at the top story of the forest. During the morning the warming of the lower layers of the atmosphere and the mixing of upper and lower layers by wind and convection currents cause the saturation deficit to increase even at low levels, but the evaporation from the soil and transpiration from the leaves, combined with the lower temperature in the shade and the smaller amount of air movement, prevents the saturation deficit from ever rising as high in the undergrowth as in the upper stories. The range of humidity therefore diminishes sharply from above downwards and a gradient of humidity is set up which changes in steep-

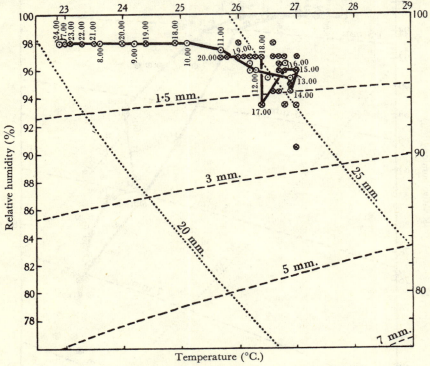

Fig. 30. Undergrowth (0·7 m. above ground) in wet season (10–11 May 1936). (See legend under Fig. 27.)

ness, and probably in form, as the day goes on. In the evening the undergrowth is in darkness some time before the tree-tops and the saturation deficit therefore falls sooner in the lower levels than in the upper.

The daily range or variation of humidity, as of temperature, is smaller in the lower than in the upper stories. The seasonal range of temperature is also greater in the upper than in the lower stories, but, if Evans' results apply generally, the seasonal range of humidity is greater in the undergrowth than in the upper stories. The seasonal range of both temperature and humidity is, however, at all levels very small compared with the daily range.

The observations of Haddow, Gillett & Highton (1947) on relative humidity, saturation deficit and temperature in Rain forest on the western border of Uganda agree closely with those of Evans for Nigeria. Measurements were made at ground-level, 16 ft. (5 m.), 31 ft. (9 m.) and 54 ft. (16·5 m.). Little difference in microclimatic conditions was found between ground-level and 16 ft., or 31 ft. and 54 ft., but a great difference between 16 ft. and 31 ft., which was, however, much reduced during wet weather. In this forest there were thus two rather sharply defined microclimatic strata or in other words the microclimatic gradient was discontinuous rather than continuous.

LIGHT

One of the characteristics of the Tropical Rain forest is the contrast between the gloom of the undergrowth and the dazzling brightness in the tree-tops and clearings. Though in the past the role of light in determining forest structure may have been over-emphasized, there can be no doubt that an accurate knowledge of the horizontal and vertical variations of the illumination is fundamental for understanding the ecology of the Rain forest as a whole, and the distribution of its individual components.

Unfortunately, the measurement of light intensity in a forest is a problem of very great difficulty. In the first place, the conditions themselves are exceedingly complex, so that measurements of many different kinds must be made before a complete picture of the illumination is obtained. Light which is not reflected directly from the upper surface of the forest may either be transmitted through the leaves, or it may pass between them forming sunflecks which vary continually in size and intensity throughout the day. In the interior of the forest, light is reflected by leaves and branches, surfaces with a great variety of optical properties. In addition, light falling on the forest will, of course, vary in intensity and spectral composition with the time of year, the time of day, and the cloudiness of the sky.

As well as the complications due to the nature of the illumination itself, there are the technical difficulties of the actual measurements. It is unnecessary to discuss here the merits and disadvantages of the various types of photometers which have been used for ecological purposes (cf. Shirley, 1935). Clearly, the ideal instrument should give readings which can easily be converted into absolute units and should have great sensitivity in the regions of the spectrum near the chief absorption bands of chlorophyll. No measurements of light intensity in forests have been made with instruments altogether meeting these requirements. Wiesner (1895), McLean (1919), von Faber (1915) and Davis & Richards (1933–4) used actinometers in which the darkening of sensitized paper is matched against a standard tint. Brown (1919) used the Livingston black-bulb actinometer. Allee (1926a) and Moreau (1935a) used illuminometers in which a visual comparison is made with the illumination produced by a standard source of light. The only type of photometer which is free from subjective errors is the photoelectric cell. This instrument was employed in the observations of Carter (1934) in British Guiana and Evans (1939) in Southern Nigeria.

The following account is merely a general picture of the conditions; it does not attempt to give a critical discussion of all these observations.

In the interior of the Rain forest there are enormous variations in the intensity of illumination. Since different types of forest often differ considerably in structure, they also often differ considerably in the average light intensity near ground-level; these variations will be considered later. Within a single homogeneous forest type there will also be unevenness of illumination due to the discontinuity of the A, and sometimes of the B, story, as well as that due to uneven spacing of the trees. The felling or natural death of trees will cause gaps and occasionally the bare crown of a deciduous tree will produce a light patch. Apart from these large-scale variations in the lighting, there is everywhere a pattern of sun-fleck and shade.

When the sun is unobscured some sun-flecks are present on the ground, even in the shadiest type of forest. Owing to their great variation in size and intensity, it is difficult to measure their total area accurately, but Carter (1934) estimated the proportion of the forest floor illuminated by sun-flecks in British Guiana at 0·5–2·5 % of the area in various types of forest. Sun-flecks are, of course, continually moving and, since the forest canopy tends to become more opaque when light falls on it obliquely, they are formed only towards the middle of the day. In Evans' (1939) observations on the daily march of illumination in the undergrowth of the Nigerian Rain forest, it was found that, if areas experiencing an illumination of more than 100 % above the mean light intensity were regarded as sun-flecks, they were present only during a period of 4 or 5 hours in the middle of the day. As was mentioned in the introductory chapter, the light intensity in the shade is greater than is often supposed. The comparatively strong diffuse lighting of the forest is possibly partly due, as Karsten (1924–6) suggested, to the large amount of light reflected by the glossy surfaces of sclerophyllous leaves.

The older measurements of light intensity in the undergrowth away from sun-flecks gave results of the order of 0·5–1 % of full daylight. McLean (1919) in south Brazil using an actinometer with a filter 'cutting out all actinic rays beyond λ 5000' obtained the following results:

'In a sun-fleck'	'In deep shade'
12 %	0·72 %

Wiesner (1895), also using an actinometer, found that in the Rain forest of Java the 'phanerogam limit' (lowest light intensity in which ground flora occurred) was in general about 0·83–0·77 %. Allee (1926a, p. 293) with an illuminometer obtained the following results at Barro Colorado Island (Panama):

Open (f.c.)	'Sun-fleck at midday' (f.c.)	'Dense shade' (f.c.)
4000–20,500	325–20,500	0·8–670
Mean: 11,000		Mean: 325 (2·95 %)

Recent observations with photoelectric cells give not very dissimilar results. In various types of forest in British Guiana Carter (1934) obtained figures varying from 10·4 to 72 % for sun-fleck intensities and 0·18–0·80 % for shade intensities.

Evans (1939) made a very detailed study of illumination in the undergrowth of primary forest in Southern Nigeria. Changes of illumination were measured during a complete period of 10 hours (Fig. 31). The march of intensity was very regular during the first and last two hours of the period of observation, but became more irregular during the middle of the day when the sun was higher and sun-flecks appeared. For most of the day the intensity remained at about 0·5–1 % of the external intensity. For short periods, totalling about 1 % of the hours of daylight, values over twice as high as this were reached, and for still

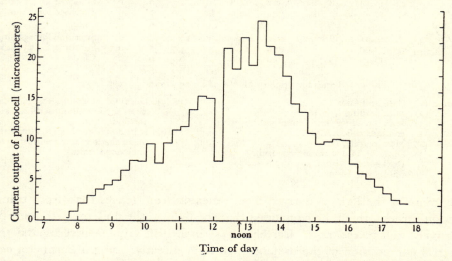

Fig. 31. Daily march of intensity of illumination in undergrowth of Tropical Rain forest, Shasha Forest Reserve, Nigeria. Measurements made at 0·4 m. above ground level on 4 March 1936, at intervals of 10 sec. from 7.40 to 14.12 hr. and of 15 sec. from 14.12 to 17.41 hr., using a Weston photoelectric cell covered by a plate of opal glass. Each horizontal line on the diagram represents the mean of all observations during a quarter of an hour. From Evans (1939, fig. 13).

shorter periods values over five times as high. Evans' results for times near midday are given in more detail in Table 20.

It may be concluded that both when the sun is clouded and when it is unobscured the shade illumination in the undergrowth is always below 1 % of the outside daylight; the sun-fleck intensity may be at least 100 times greater than that in the shade.

The diffuse light in the forest undergrowth has had a complex history of reflexion and transmission and would naturally be expected to differ considerably from daylight in spectral composition. Information on this subject is, however, very incomplete. McLean (1919) investigated the light reflected from leaves by means of a spectroscope and found that the red was reflected with less diminution of intensity than light of other wave-lengths. On the assumption

TABLE 20. *Illumination in the rain-forest undergrowth, Shasha Forest Reserve, Nigeria.* (After Evans, 1939, table VII)

Measurements with Weston photoelectric cell. Data for five stations. Cell surface 0·3 m. above ground. Stations II–V were all adjacent to the profile plot shown in Fig. 6.

Station	(1)		(2)	(3)	(4)	(5)	(6)	(7)	(8)	Weather
I	12. iv.	12.06–12.32	20·0	26·4	26·4	26	0	0	0	Cloudy
II	16. iv.	11.41–12.17	7·9	27·9	252	36	—	10·7	6·0	Sunshine for 31 min.
III	16. iv.	12.51–13.30	16·3	43·1	209	39	—	3·3	2·0	Sunshine for 20 min.
	22. iv.	10.50–11.22	12·8	33·9	33·9	32	1·7	0	0	Sunshine for 27 min.
	9. v.	10.00–10.36	3·7	31·1	44·0	34	1·5	1·0	0	Sunshine for 5 min.
						105	3·2	4·3	2·0	
IV	22. iv.	12.40–13.06	7·9	17·1	17·1	26	0	0	0	Sunshine for 23 min.
	9. v.	11.54–12.48	6·0	12·1	14·8	54	2·5	0	0	Sunshine for 48 min.
	24. v.	11.40–12.42	1·6	12·2	20·5	62	1·0	0	0	Sunshine for 3 min.
						142	3·5	0	0	
V	22. iv.	12.01–12.38	5·2	11·2	244	32	13·7	8·0	2·7	Sunshine for 25 min.
	9. v.	11.00–11.52	5·1	12·5	13·8	52	4·7	0	0	Sunshine for 35 min.
	24. v.	11.16–11.38	4·0	10·0	10·0	22	0	0	0	Cloudy
						106	18·3	8·0	2·7	
Totals for Stations II–V						**389**	**25**	**23**	**10·7**	

Column (1): Date and time.
(2): Min. reading (μA.).
(3): Max. reading excluding values in columns 6–8 (μA.).
(4): Max. reading (μA.).
(5): Duration of determinations (min.).
(6): Duration of visible sun-flecks on the cell surface (min.).

Column (7): Duration of readings > twice the mean shade reading with the sun unclouded (min.).
(8): Duration of readings > 5 times the mean shade reading with the sun unclouded (min.).

that more of the light is derived from reflexion from leaves than from transmission through them, he concluded that the light in the undergrowth must be relatively rich in red rays. Carter (1934) in British Guiana investigated the spectral composition of the undergrowth light directly, using a Bernheim cell and various filters. He found the light was relatively poor in red rays and concluded that most of it must have passed through leaves and not been reflected from their surfaces. Evans (1939), by means of a Weston cell and filters, arrived at the following estimate of the composition of the undergrowth light at one spot in Nigeria:

	Minimum percentage	
Blue (3200–5000 A.)	7·6	
Green (4700–5900 A.)	22·4	75·3
Red (beyond 6000 A.)	45·3	

The remainder (24·7 %) is not accounted for owing to the figures being calculated on the maximum transmission of the filters used; it is therefore divided between the various regions in a way impossible to calculate. When some of this remainder has been deducted for the blue and green regions so little remains compared with the large minimum percentage in the red that most of this (i.e. *c.* 40 %) must lie in the region of maximum transmission of the red filter above 7000 A. This indicates a considerable increase in the transmission of the forest canopy just beyond the red end of the visible spectrum. Evans obtained other evidence confirming this conclusion. The low proportion of red rays found by Carter was probably due to the fact that the type of cell used by him had very little sensitivity at 7000 A. and none beyond about 7300 A. The Weston cell used by Evans still retained 11 % of its maximum sensitivity at 7000 A. and 3–4 % at 8000 A. The high percentage of light near the infra-red in the undergrowth can probably not be used in photosynthesis since it is outside the range of absorption of chlorophyll.

Bünning (1947), working in the Dipterocarp forests of north Sumatra, has obtained similar results to those of Evans; he contrasts the 'grey twilight' of the tropical undergrowth with the predominantly green light of European broad-leaved forests.

In the upper levels of the forest the average light intensity is of course much higher than near the ground; there is thus a steep gradient of intensity from the full daylight of the highest tree story to the low intensities in the undergrowth. There is very little information as to the form of this gradient, but general observations suggest that in some types of forest there may be a fairly gradual diminution in the average light intensity from above downwards, while in other types there may be a level at which there is a relative abrupt fall in intensity. The latter is probably true in the Nigerian Rain forest, which has open A and B tree stories, but a very opaque C story. When climbing a tree in this forest, after passing the canopy of the C story, one seems to emerge into full daylight (cf. p. 30). The only published measurements of light intensities at various levels in the Rain forest are the following data of Allee (1926 a, p. 293), obtained at Barro Colorado Island (Panama) with a Macbeth illuminometer:

Stratum	Forest floor	'Small trees'	'Lower tree tops'	'Upper forest canopy'
Height	0	20–30 ft. (6–9 m.)	40–60 ft. (12–18 m.)	75–80 ft. (23–25 m.)
Illumination in foot-candles (relative to that on forest floor)	1	5	6	25

CARBON DIOXIDE

Though the carbon dioxide concentration of the air is not a climatic factor, it is appropriate to consider it along with the factors of the microclimate, because of its importance, together with light, as a limiting factor for photosynthesis, and therefore for the growth of plants. Further, within a stratified plant community, the carbon dioxide concentration varies and the variations, like those of the microclimatic factors, depend to a large extent on the interference by the plants with the free movement of air.

It has been known for some time that the atmosphere in the lower layers of woods, both in temperate regions and elsewhere, tends to be slightly richer in carbon dioxide than the outside air. In the lower layers of European woodlands concentrations up to about 0·04 % (by volume) have been observed (Féher, 1927, p. 316), i.e. about 1·3 times the normal concentration (which may be taken as 0·03 % by volume); higher values (up to about 0·08 % by volume) have been reported by Lundegårdh (1931, p. 271). In the upper stories of woodlands, on the other hand, there is evidence that the concentrations may be below normal, probably owing to the large amounts of carbon dioxide withdrawn from the air by the leaves when they are assimilating vigorously. Since at high light intensities the rate of photosynthesis increases or decreases almost in proportion to the carbon dioxide concentration, such deviations from the

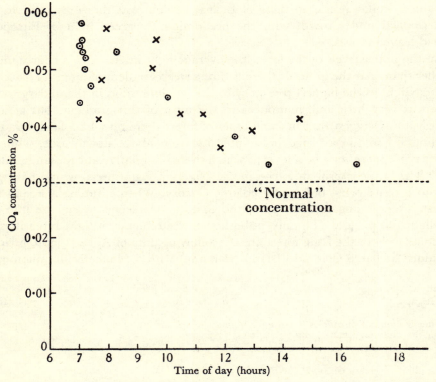

Fig. 32. Carbon dioxide concentration in atmosphere of undergrowth, Shasha Forest Reserve, Nigeria. From Evans (1939, fig. 1). × primary forest; ⊙ secondary forest.

normal concentration may have considerable ecological importance. Even at the low light intensities found in the shade of the Tropical Rain forest, a high carbon dioxide concentration would probably allow a rate of photosynthesis considerably higher than that at normal concentration.

From general indications high concentrations would be expected in the lower layers of the Rain forest. The absence of a thick litter of dead leaves and the small percentage of humus in the soil show that decomposition and carbon dioxide production by micro-organisms are very active. The many superposed stories of plants, as we have seen, offer a considerable resistance to air movements, so that mixing of the atmosphere of the upper and lower layers is much impeded. The earliest determinations of carbon dioxide concentration in the Rain forest gave, in fact, unusually high values. In the forest at Rio de Janeiro (Brazil), McLean (1919) found a concentration in still weather at a height of 1·5 m. above the ground of 0·14 vol. % in the evening and 0·34 vol. % the following morning, i.e. 4·6 and 11·3 times normal respectively. Almost equally high values were obtained by von Faber (Schimper, 1935, p. 146) at 0·5 m. in the upland rain forest at Tjibodas (Java).

Recent work by Stocker (1935 *b*), Evans (1939) and Bünning (1947) has not confirmed the existence of these very high concentrations. In Stocker's observations, made like von Faber's at Tjibodas, no values even as high as twice normal were obtained, either at 7 cm. or 1·7 m. above the ground. In fact, many of the values are below normal. Out of fifteen observations at 1·7 m., ten are subnormal and only three above normal; all the observations at 7 cm. are above normal, but only one exceeds 1·5 times normal. Stocker concludes that the earlier views were mistaken and that low rather than high concentrations are characteristic of the interior of the rain forest. Evans' observations, which were made in Southern Nigeria, at heights varying from 0·6 to 1 m. above ground, are shown in Fig. 32. The concentrations found are on the average somewhat higher than in Stocker's observations, but all of them lie between normal and twice normal (0·036–0·057 vol. % in primary forest, 0·033–0·058 in secondary). The slight differences between Evans' results and Stocker's may be due to the fact that the latter were obtained at an altitude of 1300 m., while Evans' locality was near sea-level, with a mean temperature at least 9° C. higher, and was situated in a denser type of forest. Evans points out that McLean's results are based on only two isolated determinations and that the method used was not reliable. The method by which von Faber's results were obtained is unfortunately not stated. Bünning's observations are on the whole in agreement with those of Stocker and Evans. In the lowlands and on the lower mountain slopes in a very wet region of Sumatra in forest at 0–30 cm. above ground-level he found values up to about 0·1 vol. % in the morning. In somewhat drier types of forest, similar to those studied by Evans, lower figures were obtained. Close to the surface of the ground values up to 0·2 vol. % were occasionally found.

It may be concluded that there is no good ground for believing that the carbon dioxide concentrations in the Rain forest are exceptionally high; on the contrary they do not seem to differ appreciably from those at similar levels in temperate woodlands. The view, put forward by von Faber (Schimper, 1935, p. 147), that in the rain-forest undergrowth a high carbon dioxide concentration compensates for the low light intensity may be dismissed.

The results of Stocker, Evans and Bünning show evidence of a daily drift of concentration. Stocker found the lowest values in the early morning; from 6 to 9 hr. there was a rise, followed by a gradual fall lasting till sunset. Evans, however, obtained his highest figures between 6 and 7 hr.; there was then a steady fall till evening. Bünning also found a fall in the course of the day. The fall can be attributed to carbon assimilation by the vegetation and no doubt also to the mixing of the interior and exterior air by convection currents. To explain the low values in the early morning, Stocker assumes a fall in concentration during the night due to 'binding' (*Bindung*) of the carbon dioxide. This he thinks may be due to solution in dew-drops on the leaves containing small quantities of alkalis. On quantitative grounds Evans shows that this hypothesis is untenable and suggests that Stocker's low early morning values may be due to a mass flow of air into the pores of the soil when it is cooled by radiation; this would retard the outward diffusion of carbon dioxide.

No observations have been made on carbon dioxide concentrations in the upper layers of the Rain forest, but there is no reason to expect them to be different from those at high levels in temperate forests, though possibly they might be still lower.

INFLUENCE OF FOREST TYPE ON MICROCLIMATE

In different types of rain forest differing in structure and floristic composition, it would be expected that there would be corresponding variations in the microclimate at equivalent levels. The existence of such differences was demonstrated by Davis & Richards (1933–4, table VI, etc.) in the undergrowth of four types of primary forest at Moraballi Creek, British Guiana, viz. the riverside Mora forest (*Mora excelsa* consociation), the Wallaba forest (*Eperua falcata* consociation) of bleached sand on ridges and plateaux and the Morabukea forest (*Mora gonggrijpii* consociation) and Mixed forest association of intermediate habitats. These communities differ in floristic composition (see Chapter 10), as well as in structure. From the incomplete data available it was concluded that: (i) the mean temperature in the undergrowth is lowest in the Mora forest and increases progressively through the Morabukea and Mixed forest to a maximum in the Wallaba forest; (ii) the mean saturation deficit is lowest in the Mora forest and increases to a maximum in the Wallaba forest; (iii) temperature and saturation deficit fluctuate more widely in Wallaba forest than in the other types.

In addition to these observations on temperature and saturation deficit, measurements were also made of the illumination in the undergrowth in the same four forest types and in a fifth, the Greenheart forest (*Ocotea rodiaei* consociation). The measurements were made by moving an actinometer over a constant distance in each type of forest, so as to give a rough integration of the shade and sun-fleck light intensities. By this crude method the following relative figures were obtained of the relative illumination in the undergrowth of the five forest types (expressed as a percentage of full daylight):

Mora	Morabukea	Mixed	Greenheart	Wallaba
1·33	0·61	0·67	0·81	1·43

The undergrowth of the Mora and Wallaba consociations, the types at the two ends of the series, is thus considerably better illuminated than that of the intermediate Morabukea, Mixed and Greenheart types.[1]

In the series of primary forest communities at Moraballi Creek it will be seen that the variation in temperature and saturation deficit on the one hand, and illumination in the undergrowth on the other, do not run parallel. The temperature and saturation deficit in the lower levels of the forest probably depend partly on forest structure and partly on differences in the moisture of the soil due to variations in the height of the water-table and in water-retaining properties, but the illumination must depend chiefly on the number and density of the tree stories. Mora forest has relatively open A and B stories, but is found on a flood-plain with a permanently high water-table and a relatively impermeable soil. The structure of Morabukea forest is not accurately known, but it appears to have a very dense tree canopy and, though it also grows on an impermeable soil, drainage is better. The Mixed association (from which the Greenheart type of forest differs only slightly in structure and composition) has three tree stories, the A and B stories together forming a moderately dense canopy (see Fig. 4); the soil is here both better drained and less impermeable than in the Mora and Morabukea types. The Wallaba forest has only two well-developed tree stories, which form a canopy about as open as that of the Mora forest; the soil is extremely porous and non-retentive and the situation on ridges leads to exceptionally free drainage. In the Wallaba forest both soil and structure lead to climatic conditions in the undergrowth less constant than conditions in the other types and less different from those outside the forest. The magnitude of the differences between the temperature and humidity in the undergrowth of the forest and the external climate may be regarded as largely dependent on the completeness of the barrier between the atmosphere in the forest and the atmosphere outside. In the Wallaba forest this barrier is less complete than in the Mixed and Morabukea forest. In the Mora forest it is also less complete, but the difference in structure is more than compensated for by the excess of soil moisture.

PHYSIOLOGY OF THE FLORA IN RELATION TO MICROCLIMATE

The microclimates of the rain forest will undoubtedly repay much further investigation, but some important facts have already been established. The upper strata exist under relatively variable conditions; they are exposed to the full strength of sun and wind and subjected to a wide range of temperature and humidity. In the lower layers, on the other hand, conditions are much more constant; both the daily and seasonal ranges of temperature and humidity

---

[1] Carter (1934), using a different method and working in a different part of British Guiana, determined the sun-fleck and shade light intensities in Wallaba, Mixed and three varieties of Mora forests and young secondary forest with somewhat different results.

are remarkably small. An animal ecologist (Allee, 1926*a*, p. 299) has written: 'The animals of the lower forest...would need only to avoid the sun-flecks in order to keep under environmental conditions so constant that they must excite the envy of every experimental ecologist with experience in trying to control environmental factors for land animals in the laboratory.'

Within a single plant community there are thus two very distinct superposed types of microclimate, in other words there is a very pronounced vertical environmental gradient. The shrubs and herbs (including the epiphytes) need to be adapted to only one of the two types of climate, but the taller members of the community, the A and B story trees and most of the climbers, must, at different stages of their development, be tolerant both of the poorly illuminated unvarying undergrowth climate and of the more variable conditions to which they are exposed at maturity. In forests with dense A and B stories the C story trees live under conditions similar to the shrubs and herbs of the undergrowth, but where, as in the Mixed forest of Nigeria (p. 29), the upper stories are more open, their environment more nearly resembles that of the A and B trees.

The physiological adjustment between rain-forest plants and their normal environment is probably often rather close. When isolated in clearings the taller trees usually continue to flourish, but many, perhaps most of them, require for germination and the early stages of growth shade or at least a moist atmosphere, such as that to which the seedlings are exposed in the undergrowth (Chapter 3). The shrubs and herbs of the D and E layers (and this is perhaps also true of most C story trees) seem to be intolerant to a greater or lesser extent of microclimates widely different from that which they normally experience. It is interesting that cacao, *Theobroma cacao*, in its native habitat a C story tree of the Guiana forest (Myers, 1930), grows best in cultivation in an atmospheric environment in which moisture fluctuations are least and hence requires overhead shade and suitable wind breaks (Hardy, 1935, p. 175). The existence of shade-tolerant and shade-loving species in the rain-forest undergrowth (Chapter 5) shows that even within the same stratum there may be different methods of adjustment to microclimatic conditions.

The nature and extent of the physiological differences between members of different strata are by no means well understood; what follows is merely intended to indicate that there is a range of physiological behaviour reflecting the vertical gradient of microclimate.

Water relations

The wide range in the average rate of transpiration of different rain-forest plants has already been mentioned (Chapter 6). These variations seem to be to some extent correlated with the humidity and temperature conditions in the strata to which the plants belong. All recent work on the subject points to the conclusion that, as in temperate forests, plants of the upper stories transpire more rapidly per unit leaf area than those of the lower, even when placed in an identical environment. Similarly the sun leaves of rain-forest trees transpire faster than the shade leaves. A critical study of the transpiration of tropical plants has been published by Stocker (1935 b), who worked at Buitenzorg (Java). Three tree species growing under open conditions were compared with three herbs growing in the shade. The trees—*Calophyllum inophyllum*, a common tree of Littoral woodland (p. 297), *Cassia fistula* found in (secondary) grasslands and Teak forests, and *Elaeodendron glaucum* found in Teak and Deciduous Mixed forest in east Java—unfortunately were none of them rain-forest species, but their behaviour may perhaps be not very different from that of rain-forest trees of the upper stories. The herbs—*Begonia isoptera*, *Cyrtandra pendula* and an unidentified member of the Acanthaceae—were common constituents of the shade flora in the upland (Submontane) rain forest at Tjibodas (near Buitenzorg).

In the trees transpiration was extremely low during the night, but rose to high values in the early part of the day. In the middle of sunny days the rate of transpiration (per unit area of leaf surface) showed a depression, though there was no visible wilting. In the shade herbs in their natural habitat the daily march of transpiration followed quite a different course. At night, as in the trees, transpiration fell to nil, but even during the day it remained steady and very low, rising rapidly only when a sun-fleck passed over the plant.

TABLE 21. *Suction pressure of leaf cells of plants in the Tjibodas rain forest.* (After Blum, 1933)

Suction pressure of epidermal and mesophyll cells of leaves (atmospheres)

	Mean	Minimum	Maximum
Small herbs	5·4	2·6	7·4
Large herbs	6·1	4·7	6·7
Woody climbers (lianes)	12·4	8·9	19·6
Small trees	13·0	2·6	25·5
Large trees	15·1	8·1	30·9

Two of the tree species investigated had thick, sclerophyllous leaves; the herbs also had thick leaves, but of a softer texture. Stocker shows that the sclerophyll leaves of the trees, like most leaves of this type, have a high water-content relative to their surface area (about 1·1 gm./sq.dm.), while in the thick-leaved herbs the water content is even higher (1·90 and 1·94 gm./sq.dm. in two cases). Relatively to their water content the rate of transpiration is lower in the very sclerophyllous *Calophyllum* and *Elaeodendron* than in the somewhat softer-leaved *Cassia fistula*, though relative to the surface area of the leaf the opposite is true. In all three trees, however, transpiration relative to the water-content of the leaf is much greater than in the shade herbs. The highest values in the trees were 16–20 mg./gm./min., which is equivalent to 96–120 % of the total water-content of the leaf per hour. These values are not exceptionally high compared with such temperate plants as, for example, *Helianthus tuberosus*, which with a lower ratio of leaf area to fresh weight may transpire more than 150 % of the total water-content of the leaf on a sunny day. In the shade plants the ratio transpiration/water-content never exceeded the very low figure of 2 mg./ gm./min., or 12 % of the water-content per hour.

The suction pressure of the leaf cells has been investigated by Blum (1933) in a number of plants in the upland rain forest of Tjibodas (Java). The suction pressure of the epidermal and mesophyll cells varies regularly according to the stratum to which the plant belongs; the taller the plant the higher the suction pressure. Table 21 gives a summary of some of Blum's results. Blum regards this suction pressure gradient as primarily dependent on the gradient in atmospheric humidity.

Carbon assimilation and growth

From the steep gradient in the average illumination at different levels in the Rain forest, large differences might be expected in the assimilatory behaviour of sun and shade plants, but few reliable data are available. Von Faber (Schimper, 1935, p. 106, fig. 24) has published 'ecological assimilation curves' for *Cyrtandra picta*, a typical herb in the undergrowth of the Javan rain forest and for a tropical sun plant, *Lantana camara*, which is not a rain-forest plant, but is abundant in many parts of the tropics in the early stages of succession after rain forest has been cleared. No details are given as to the methods by which the results were obtained, but for the sun plant the curve of carbon assimilation/light-intensity rises steeply up to about half full daylight (full daylight is defined as 'direct plus diffuse light at *c.* 7° S. in the mountain garden at Tjibodas, Java, in the dry season' (transl.). For the shade plant the curve rises less steeply and after about 1/20 full daylight, there is no further rise, i.e. light ceases to be a limiting factor for assimilation.

No data more definite than these seem to have been published on this subject, but the work of Guttenberg (1931) on the assimilates in the leaves of various tropical plants in the Buitenzorg garden gives evidence of considerable differences in metabolism between the members of different strata. The long series of plants studied could be arranged into four groups according to the nature and quantity of the assimilates in the leaves: (i) Those in which the leaves store much starch during the rainy season; this starch appears to act as a reserve which can be drawn upon in dry periods. More starch is found than sugar. (ii) Those which accumulate both sugar and starch, but more of the former than of the latter. (iii) Those which form starch during the day, though not in very large amounts; this disappears during the night. The sugar content of these leaves is always high, but the starch content varies periodically. (iv) Those in which the leaves never form starch. Group (i) includes typical rain-forest trees such as *Palaquium philippense* and *Pausinystalia yohimbe*, which have stiff or leathery leaves and are comparable in their behaviour with the sclerophylls of Mediterranean climates. Group (ii) includes a number of rain-forest climbers and epiphytes. Guttenberg considers that their behaviour can be associated with the fact that their assimilation is liable to interruption only for short periods. Group (iii) consists predominantly of soft-leaved species, chiefly herbs, shrubs and trees of the middle stories of the rain forest. Their carbohydrate metabolism appears to be like that of many European plants. Group (iv) includes many xerophytes, and consists of both trees and herbs, and not, as with European plants, of herbs only. None of the examples of this group seems to be a typical rain-forest species, except perhaps the epiphytic orchid *Coelogyne pulverula*.

These results may have an ecological bearing. Though there is a great variety of behaviour among the members of any one stratum or synusia, there is evidently some correlation between the carbohydrate metabolism and the texture of leaf, whether sclerophyllous or soft. Since leaf texture tends to be a character-istic of the stratum of the forest to which the plant belongs (Chapters 4 and 5), there is probably a correlation between stratum and the type of carbohydrate metabolism in the leaf. If such a correlation exists, it probably depends on the different microclimates, in particular the varying lengths of exposure to high rates of evaporation, in the different strata.

No data are available for a comparison of the compensation point (light intensity at which assimilation balances respiration at a given temperature) in plants of different strata of the Rain forest, though Stocker (1935 a) investigated the compensation point for the rain-forest tree, *Stelechocarpus burahol*, in Java and found it was not higher than in central European trees. It would be of great interest to know if among the rain-forest flora as among temperate plants the compensation point is higher in sun than in shade plants—in other words, at a light intensity at which the leaves of sun plants show a net loss of carbo-hydrates, those of shade plants can still show an excess of assimilation over respiration.

With regard to growth, though measurements have been made on many dif-ferent tropical plants, they are still insufficient to allow any general statement as to the average growth rates in different strata. Reference may be made to Brown's (1919) observations on the growth rates of tropical trees at various periods of their life, already discussed in Chapter 3.

Physiological status of the undergrowth

The physiology of the rain-forest undergrowth has always attracted much interest, and McLean (1919) has discussed the subject at some length. Because of the constantly high humidity, relatively unvarying temperature and low light intensity, the environment of the herb, shrub and lowest tree strata is very peculiar. The plant physiologists of the last century, searching everywhere for visible adaptations, expected to find in the shade flora structural features tending to increase the rate of transpiration. They assumed that the rate of intake of plant nutrients from the soil must be more or less directly dependent on the transpiration rate and therefore supposed that the rain-forest undergrowth, living in an atmosphere with a very feeble evaporating power, would necessarily tend to suffer from mineral starvation. Stahl and others drew attention to the frequency in the leaves of the undergrowth flora of hydathodes, papillose epidermal cells, 'drip-tips', etc., all of which they regarded as adaptations to promote transpiration. Some of these peculiarities have been previously discussed (Chapter 5). McLean severely criticized this point of view. In south Brazil he found that the average rate of transpiration among the shade flora was in fact very low, but nevertheless the ash content, and hence the mineral uptake, of the plants were as great or greater than in the sun flora of the same forest. Excretion of liquid water was not observed and the leaves 'showed no adaptations calculated to aid in extending their transpiration towards the limit of capacity. On the other hand, they are commonly protected by cuticle and other means against excessive water loss' (McLean, 1919, p. 53).

This conclusion has been generally accepted and it is probable that the shade flora of the Rain forest, though normally transpiring very little, suffers no disadvantage as far as its mineral nutrition is concerned. Structural features leading to increased transpiration, while perhaps not absent altogether, are certainly not conspicuous in these plants. It might at least be expected that the leaves would have a large surface area relative to their weight or bulk (it was seen in Chapter 5 that they are commonly of large dimensions), yet Stocker (1935 b) found that in the three shade herbs from the Tjibodas forest (p. 185 above) the surface area, relative to the fresh weight, was actually quite small.

In addition to having normally a low rate of transpiration, the shade flora of the Rain forest seems unable to support a high rate without setting up a permanent internal water deficit from which the plant cannot recover. This is illustrated by the behaviour of *Icacina trichantha* and *Rinorea* sp., two shrubs common in the undergrowth of the evergreen forest in Southern Nigeria (Evans, 1939, pp. 479–80). The latter is confined to the forest shade, but the former also grows on dry exposed slopes in open clearings. In a potometer placed in the forest shade cut shoots of both would stay alive and show little sign of wilting for several days. When brought into the dry air of the laboratory, however, where the transpiration rate of both species increased four to five times, the *Icacina* was still able to survive in the potometer for some time, but the *Rinorea* developed a steadily increasing water deficit and soon dried out completely. If the behaviour of this *Rinorea* is typical of the shade-loving undergrowth species (as distinct from the shade-enduring type to which *Icacina* belongs), it seems that in times of exceptional drought, or even when exposed to sun-flecks for long periods, the shade-loving plants might be liable to serious injury or death. The 'protective' features noted by McLean are thus less paradoxical than they might appear at first sight.

The ill effects of rapid transpiration on shade plants may be in part explained by a poor vascular supply to the leaves. McLean (1919, pp. 42–3) found that in three typical shade species of the south Brazilian Rain forest the ratio of the leaf area to the cross-sectional area of the xylem in the petiole was over four times as great as in three typical sun species.

Since the transpiration of the shade flora is in general very low, it is not surprising that, according to Blum (1933), the leaf cells of shade plants at Tjibodas have the lowest suction pressures yet found in any land plant.

The shade flora of the rain forest might also be expected to be peculiar in its carbon assimilation relations. The undergrowth of the Rain forest appears unexpectedly luxuriant in view of the poorness of the illumination. It might perhaps be supposed that shade plants must have special adaptations to help them in carrying on vigorous carbon assimilation at low light intensities. The theory of a 'carbon dioxide flora' has already been considered and shown to be baseless. It is clear that the leaves of tropical shade plants often have, in a highly developed form, anatomical features associated with shade leaves in other climates, e.g. thin mesophyll, lens-shaped palisade cells, etc., and it is stated by von Faber (Schimper, 1935, p. 114) that some typical tropical shade plants open their stomata at light intensities as low as 1/70–1/50 full daylight. The data are, however, quite insufficient for a general estimate of their assimilatory behaviour. The study of illumination in the Rain forest is not far enough advanced for any conclusion as to whether the shade flora does, in fact, exist in a less favourable light climate than the shade flora of other types of forest. As far as functional adaptations in the photosynthetic process itself are concerned, there is a possibility, as we have seen, that light becomes a limiting factor for assimilation at a lower light intensity in tropical shade plants than in sun plants. It would be expected, but has not yet been demonstrated, that the compensation point lies lower than in sun plants. Thus we cannot yet say whether the carbon assimilation of tropical shade plants is light-limited, either always, or for long enough for light to be the chief factor limiting their growth and distribution.

Further investigations on this subject are much needed, especially since the importance of light as the chief factor in determining the structure and composition of forest undergrowth, formerly accepted without question, is beginning to be challenged. It has usually been assumed that the occurrence of certain undergrowth species in deep shade and of others only in well-illuminated openings, for instance, was determined solely by the average light intensity; when a clearing is made there is always a great increase in the density and luxuriance of the herbaceous undergrowth and the increase of illumination would at first sight seem to be alone responsible for this. Yet recent work on temperate forest communities suggests that the scanty undergrowth of some types of temperate woodland is probably due, not to lack of light, but to the competition of the tree-roots in the soil. When the influence of the roots is removed without change of illumination, by trenching round marked quadrats, a great increase in the abundance and luxuriance of the herbaceous vegetation is observed (cf. Watt & Fraser, 1933). Coster (1935a) and Wilkinson (1939) have also found evidence that root competition may be responsible for differences in the composition and abundance of the undergrowth of tropical forests. Coster (1935 *b*) compared the abundance of undergrowth on small sample plots under trees in Java and found that the differences were correlated with the species of tree rather than with the light intensity. Under a dense stand of *Lagerstroemia speciosa* where the illumination was 1·5–2 % of full daylight (method of measurement not stated), there was a rich and varied undergrowth, but under *Swietenia macrophylla* at 2–7 % full light there was no undergrowth. Again, under a large tree of *Ficus kurzii* there was plentiful undergrowth at 1·5–2 % full light and even near the middle of the crown, where the light intensity was as low as 0·4 %, shrubs and climbing Araceae were found. Coster thought these differences were due to the nature of the root system in the tree species. Trenching experiments in teak plantations gave results similar to those obtained in European and American forests.

Assuming that the effect of the tree-roots is as important as it appears to be, the mechanism by which it operates remains uncertain. There may be competition by the tree-roots for water, salts or some other substance in short supply, or again the tree-roots may exert a favourable or unfavourable influence on the undergrowth by means of soluble exudations. Coster suggests that the critical factor is competition for oxygen in the soil, which under damp tropical conditions tends to be inadequate for root growth.

Coster's observations were made in the relatively dry lowlands of Java and it is not certain that similar results would be obtained in a typical rain forest in a wetter climate. Until experiments have been carried out, it is impossible to say whether the thin undergrowth of the rain forest and the restriction of certain species to open situations are due more to root competition or to insufficient illumination. If Coster's results are confirmed by further work, it would seem likely that the shade-loving species of the undergrowth are restricted to shade by an intolerance, not of high light intensities, but of periods of large saturation deficit; this characteristic must be combined with a tolerance of intense root competition. The shade-enduring species, which increase in abundance when openings are made, are perhaps relatively tolerant of root competition and also tolerant of large saturation deficits.

CHAPTER 8

SEASONAL CHANGES

The vegetation of temperate regions changes with the seasons and the periodicity of the vegetation coincides more or less closely with the seasonal changes of climate. The Tropical Rain forest on the other hand varies much less; as a community it has no marked seasonal 'aspects' and no 'resting' period. All the year round the foliage is the same sombre green and in every month some species are in flower; from time to time an occasional tree, bare of leaves or decked with brightly coloured young foliage, introduces some variety into the scene. Summer and winter, autumn and spring are all seen in the forest simultaneously. Only near the climatic limits of the rain-forest formation-type, where rainfall is markedly seasonal in its incidence, does periodic change become an important feature of the environment. Even there seasonal changes in the vegetation are comparatively inconspicuous and are manifested chiefly in the upper stories of the forest; in the undergrowth, where the variation in both temperature and humidity is very slight, the appearance of the vegetation remains almost the same throughout the year.

Though the vegetation as a whole may show hardly any seasonal rhythm, the component plants are more or less periodic in their activity. Most species are evergreen in the sense that they are never leafless and many flower almost continuously all the year round, but the majority have periods of 'rest' and visible activity. Periodicity is shown in the production of new leaves and shoots and to a less marked degree in flowering and other functions. Thus, in spite of the unvarying general aspect of the vegetation, the behaviour of certain species may be sufficiently regular to provide fixed points for the calendar. In the island of Espiritu Santo in the New Hebrides seasonal changes of climate are so little evident that the natives never reckon their ages in years, yet they plant their crops at regular intervals determined by the flowering or fruiting of certain trees. Yams are planted when *Erythrina indica* is in flower and sweet potatoes when *Alphitonia zizyphoides* is in fruit (Baker, J. R. & I., 1936). Similarly in Malaya the flowering of *Sandoricum koetjape* was formerly the signal for the planting of rice (Corner).

The fungi as well as the flowering plants of the rain forest are periodic in behaviour. Corner (1935) finds that in southern Malaya the forest agarics have two general 'fruiting' seasons in every normal year. A succession of species develops over a period of some 3 months, each species having its own season of about a week.

The essential difference between the periodicity of the rain-forest flora and that of other climates is that the behaviour of different species, different individuals of the same species and even different parts of the same plant is less synchronized and is often apparently unrelated to environmental conditions. Simon (1914) states that at Buitenzorg (Java) there is probably no month in the year in which some species or another is not waking from 'rest' into activity. At Peradeniya in Ceylon there is no month in which every species has its full complement of leaves (Wright, 1905). At a given moment trees of the same species and branches of the same tree may be at various stages of development, though in any single locality all the individuals of one species often behave alike. The proportion of the whole flora in a given phase of development fluctuates irregularly throughout the year; in temperate regions it undergoes a rhythmical seasonal variation.

Within the Tropical zone, the more seasonal the climate the more the periodicity of the component species of the flora tends to become synchronized and the more closely does it fit the periodicity of the climate. In the Wet Evergreen forest of Nigeria (p. 338), where there may be five consecutive months with less than 10 cm. rainfall, there is a marked reawakening of the vegetation at the beginning of the wet season, though many species are also in full activity during the dry months. In this climate the undergrowth even includes a few geophytes which die down below soil-level in the dry season (p. 98). Conditions here are transitional to those of the Tropical Deciduous or Monsoon forest in which for the great majority of the species the dry season is a resting period; the periodicity of the Nigerian Wet Evergreen forest as a whole agrees fairly closely with that of the climate.

It is generally supposed that the chief external factor controlling the periodic rhythms of tropical vegetation is seasonal drought, rather than temperature, as in temperate climates. It is likely, however, that photoperiodic responses to length of day are also an important factor affecting the periodicity of plants in both seasonal and non-seasonal tropical climates. Bünning (1947, 1948) states that in tropical plants, which are necessarily short-day plants adapted to an alternation of light and dark periods of approximately equal length, the light-adapted and dark-adapted phases pass into one another more abruptly than in temperate plants; tropical plants are consequently more sensitive to small changes in length of day than temperate plants. If this is so, the change of day-length may perhaps explain the regular seasonal behaviour of some plants in climates with no clearly marked seasonal variation of temperature or rainfall (e.g. the regular annual flowering of *Erythrina indica* in the New Hebrides, referred to above). There is some evidence, according to Bünning (1947, pp. 15–16), that the temperate species which occur native or naturalized in the mountains of the tropics are more or less insensitive to day-length. It is also significant that among ferns, which include relatively more species found in both high and low latitudes than the flowering plants, no photoperiodic responses have been detected.

It has long been realized that all the functions of plants are not necessarily quiescent at the same time. Even European deciduous trees in winter are not necessarily in a state of complete rest. This holds equally for tropical plants. The periodicity of leaf production and flowering, though often connected, are to a large extent independent and must therefore be considered separately.

LEAF CHANGE

Many observations on leaf change in tropical trees, both in botanical gardens and in their natural habitats, have been published. In addition to the data of Schimper (1903) for various parts of the tropics, there are those of Klebs (1911), Volkens (1912), Simon (1914) and Haberlandt (1926) for Java; Wright (1905), Holtermann (1907) and Smith (1909) for Ceylon; Holttum (1931, 1938 b, 1940) for the Malay Peninsula; J. R. & I. Baker (1936) for the New Hebrides; Scheffler (1901), Mildbraed (1922) and Richards (1939) for Tropical Africa; von Ihering (1923) for Brazil; Davis & Richards (1933–4) for British Guiana, and Beard (1946 b) for Trinidad. Klebs (1912, 1915, 1926), Dingler (1911) and Schweizer (1932) have studied the subject experimentally.

It is evident that there are many widely different types of behaviour among tropical trees. The average length of life of the leaves of tropical species is about 13–14 months, according to Warming & Graebner (1933), but very few, if any, renew their leaves at a constant rate throughout the year. Volkens (1912) and Klebs (1911) claimed that a few species at Buitenzorg were 'ever-growing', showing no foliar periodicity whatever, e.g. *Albizzia falcata*. Simon (1914) afterwards showed that this *Albizzia* behaves in this way only when it is young; when it is older leaf production is distinctly periodic. According to von Faber (Schimper, 1935, p. 383), some of the other supposed 'ever-growing' species are similar. It is certainly true that most rain-forest trees produce new leaves, not continuously, but in periodic flushes, so that a single shoot bears several 'generations' of leaves at the same time.

In some species all or most of the old leaves fall a considerable time before the expansion of the new, so that the tree is bare for a period of weeks or months, as in *Bombax flammeum*, *B. malabaricum*, *Cedrela toona* var. *australis*, *Hymenaea courbaril*, *Koompassia malaccensis*, *Piptadenia africana*, *Terminalia superba* and many others. Again, and this is probably much commoner in rain-forest species, the old leaves may fall almost simultaneously with the growth of the new, or only a few days before. In a few trees the old leaves fall shortly after the new leaves become fully expanded, e.g. *Polyphragmon sericeum*, *Stelechocarpus burahol* (Volkens, 1912, p. 72).

As in European deciduous trees, the old leaves are cut off by an abscission layer. They do not usually fall when fully green, though *Ficus glabella* is an exception according to Schimper (Volkens, 1912, p. 69). Generally the leaves become yellowish or flecked with yellow, sometimes turning a vivid red (Volkens, 1912, p. 69; Mildbraed, 1922, p. 110). Corner (1938, pp. 119–21) has noticed that among Malayan trees the colours assumed by the old leaves are highly characteristic of certain species and genera and of value in identification. In *Pterocymbium javanicum* and *Sterculia foetida* in Java the tree is never quite bare, but the old leaves turn yellowish, and are probably not actively photosynthetic, for some weeks before they actually fall; in others, e.g. *Ehretia javanica, Planchonia valida*, there is a long period during which the number of leaves is much reduced (Simon, 1914).

The fall of the old leaves may take place very quickly. Volkens (1912, p. 70) records that a tree of *Ficus variegata* in Java which appeared quite green and fresh became entirely leafless in 4 days; mostly, however, the process takes much longer.

The development of new leaves, when it is intermittent, may occur at regular or irregular intervals. At Singapore, Holttum (1931, 1938 *b*) found that *Parkia roxburghii* produced its new leaves regularly every 12 months with a standard deviation of only 0·24 in 10 years; the period was also a year in *Cratoxylon polyanthum, Tamarindus indica* and several other species. *Terminalia catappa, Ficus variegata* and *Peltophorum ferrugineum* produced new leaves at intervals of approximately 6 months. One tree of *Ficus variegata* had a mean period of 6·6 months for 10 years, while other individuals of the same species had a period of almost exactly 6 months. In the shrub *Breynia cernua* at Buitenzorg, Smith (1923) found that the old leaves and twigs were shed, and new leaves grown, on the average every 5½ months, the plant passing through eleven complete cycles in 5 years. A number of species with cycles not annual, nor aliquot parts of a year have been noted; a tree of *Delonix regia* at Singapore had regular periods of 9 months and one of *Heritiera macrophylla* had three successive periods of 2 years and 8 months (Holttum, 1931, 1938 *b*, 1940).

The rubber tree, *Hevea brasiliensis*, and the mango, *Mangifera indica*, are well-known examples of trees in which new leaves are produced at very irregular intervals. In the former the intervals are usually different on different individuals and also change with the age of the tree (Holttum, 1931; Schweizer, 1932), though in the relatively seasonal climate of Southern Nigeria the new leaves appear more or less simultaneously each year towards the end of the dry season on all individuals regardless of age. In the mango, different branches of the same individual produce new leaves at considerable intervals of time, so that the leaf change of a whole tree may extend over several weeks. This type of behaviour is even more marked in *Ceiba pentandra*. In West Africa it is common to see some branches quite bare while others on the same tree are in full leaf. In cultivation at Singapore the native Asiatic race or subspecies behaves irregularly in this way, but the African-American subspecies is regularly and evenly deciduous (Corner). At Singapore, *Pterocarpus indicus* also changes its leaves on one or two branches at a time, but in northern Malaya it is completely deciduous (Corner, 1940). Even in tropical species in which the whole tree has more or less the same periodicity the expansion of buds on different branches is probably seldom as nearly simultaneous as in most temperate trees. Alteration of the leaf-changing habits with age, such as is found in *Hevea*, is probably common.

The variation in behaviour among rain-forest trees is so great that no sharp line can be drawn between evergreen and deciduous species. It is convenient to regard as evergreen those in which the tree bears a substantial number of leaves at all times of the year; deciduous species are those which become bare or almost so, if only for a few days. Volkens (1912) adopted a different definition, regarding as evergreen only species in which leaves of at least two 'generations' (*Schübe*) are simultaneously present throughout the year. According to this definition trees which lose all their leaves at one time, but acquire a new set without an intervening bare period, may be classed as deciduous. This usage may have advantages from a physiological point of view, but is less useful for most purposes.

Deciduous trees, in the sense adopted here, are numerous in the Rain forest of all parts of the tropics; they are present even in climates with an almost uniformly distributed rainfall, as at Singapore and in the New Hebrides. Corner (1946) believes that there are more species of deciduous trees in the evergreen forests of the tropics than in temperate deciduous forests. In typical rain forest almost all these deciduous trees belong to the A story; the B and C stories and the undergrowth are almost completely evergreen. As the climatic limit (the drought limit, in the sense of pp. 152–3) of the Rain forest is approached, the proportion of deciduous species gradually increases, first in the upper, then in the lower, stories. In the Belgian Congo, for example, the proportion of deciduous species increases from the centre of the forest *massif* towards its edges, where the dry season is longer and more severe (Lebrun, 1936a). In the Wet Evergreen forest of Nigeria, which has a long seasonal drought, a few B-story trees, e.g. *Monodora tenuifolia* and *M. myristica*, are completely deciduous, but the majority are evergreen, though in the A story well over half the species are bare for a few weeks or months during the dry season. Beard (1942) gives the figures shown in Table 22 for the proportion of deciduous trees in four forest formations in Trinidad. The proportion of deciduous to evergreen trees in the

TABLE 22. *Proportion of deciduous (including semi-deciduous) trees in forests of Trinidad.* (After Beard, 1942, p. 14)

	Percentage of species		Percentage of individuals	
	'Upper story'	'Lower story(s)'	'Upper story'	'Lower story(s)'
Evergreen Seasonal forest*	24	Negligible	6	Negligible
Mora forest	12	0	Negligible	0
Semi-evergreen Seasonal forest	33	10	16·5	10
Deciduous Seasonal forest	50	10	66	25

* The nomenclature of the formations has been changed to agree with Beard's latest classification (1946b).

various strata is a useful diagnostic character for distinguishing tropical forest formations and it is to be regretted that more quantitative data of this kind are not available.

The relative importance of the external environment and of internal or autonomic factors in causing and regulating the rhythm of leaf change in tropical trees has been much discussed. The view that external factors are almost entirely responsible was upheld particularly by Klebs (1911, 1912, 1915, 1926), who attempted to show that even in the least seasonal tropical climates the periodicity of plants was wholly controlled by the external environment. According to him, when any one of a number of factors conditioning plant activity falls below a minimum value, the plant enters a resting phase. These external factors may be edaphic as well as climatic; when periodic leaf-fall cannot be traced to any climatic factor it may be due to the recurrent exhaustion of the nutrients immediately available in the soil.

There can be no doubt that periodicity in plants is influenced to a large extent by external factors, even in relatively non-seasonal climates. Among these external factors soil conditions probably play a part, though perhaps mainly by their effect on water supply. Thus in Brazil trees growing on calcareous soils lose their leaves more completely than those on other soils (Warming, 1892). Yet the view of Klebs is certainly extreme, for there is good evidence that little-understood internal factors, in addition to the environment, play a large part in determining periodic behaviour (Schimper, 1903; Volkens, 1912; Simon, 1914; Coster, 1923). Alternate phases of rest and activity seem to be an inherent characteristic of all living organisms. The environment can control the onset and duration of these periods; the more seasonal the climate the more closely is the inherent periodicity of the organism forced to conform to the rhythm of the climate. In the climate of the Tropical Rain forest, the internal factors have free play, since seasonal changes are so slight that there is no time of the year in which the environment compels the plant to rest if the internal factors tend towards activity and vice versa.

That the internal factors influencing periodicity are genetically determined and heritable is strongly suggested by the fact that among rain-forest trees the deciduous habit is much commoner in some families than in others. For instance, many of the Sterculiaceae and Bombacaceae are deciduous, but other families such as the Lauraceae are almost entirely evergreen. The experiments of Schweizer (1932) also show that in *Hevea* inherited factors play a very important part in periodicity. In seedlings, new leaves are produced at different times in different individuals, but cuttings from the same tree behave very much alike and produce new leaves at almost the same time.

In some species the relative importance of the internal factors as compared with the external environment is evidently greater than in others; some are thus more, and some less, plastic than others. In strongly seasonal climates teak, *Tectona grandis*, and *Bombax malabaricum* are both leafless for a long period during the dry season, but at Singapore and in West Java, where the rainfall is fairly evenly distributed, *Bombax* still has a bare period of about 3 months, while *Tectona* becomes evergreen or nearly so. Coster (1923) has found experimentally that, except towards the end of their resting period, leafless branches of *Bombax* cannot be induced to expand their leaves, but in *Tectona* the leaves can be induced to expand at any time, often by merely placing a branch in water. Beard (1946*b*) divides the trees of Trinidad into those in which the deciduous habit is obligate and those in which it is facultative. In the latter the completeness of leaf-fall varies from year to year according to the severity of the drought. Normally the leaves drop off gradually as the dry season progresses until the tree is quite bare, but in a wet year no marked leaf-fall may take place. The obligate species are most abundant in the Deciduous forest formation, the facultative in the Semi-evergreen Seasonal forest (p. 320).

The deciduous habit in temperate trees is fixed and relatively difficult to modify by the environment. Trees such as the beech, *Fagus sylvatica*, remain periodic in growth and leaf production when cultivated in Java and elsewhere in the tropics, though the date and length of the resting period changes and different branches of the same individual become out of phase with one another, which, as we have seen, often happens in species indigenous to the tropics (Schimper, 1935, p. 387; Coster, 1926*b*).

It can hardly be questioned that in trees of temperate and other strongly seasonal climates the deciduous habit is adaptive in the sense that it enables the plant to avoid excessive transpiration during periods when water lost is difficult to replace. In tropical climates, except where there is a very severe dry season, the advantage is less certain. In the evergreen forests of West Africa many tree species which are bare for a period expand their tender young leaves long before the end of the dry season, sometimes even during the very dry harmattan season when the danger of desiccation is greatest. Even in the middle Niger valley, where there is a long seasonal drought, *Acacia albida* actually loses its leaves at the beginning of the rains and becomes leafy again early in the dry season (Roberty, 1946, p. 96). In Trinidad there are no species which lose their leaves in the wet season, but some are bare only for a few days or weeks at the beginning of the dry season, becoming fully leafy when it is at its height (Beard, 1946*b*). In Ceylon and in the 'east-monsoon' region of Java also many deciduous trees unfold their tender young foliage long before the end of the dry season when injury from drought might be expected (Wright, 1905; Coster, 1923). Dingler (1911), also in Ceylon, induced certain normally deciduous species to remain leafy during the dry season, but observed no damage to the leaves. From such facts it can only be concluded that in relatively non-seasonal climates, like that of the Tropical Rain forest, trees gain no advantage from shedding their leaves at one season rather than another and could in fact retain their leaves throughout the year without risk of harm.

In Trinidad, according to Beard (1946*b*), there are no deciduous species confined to the evergreen forest; all of them reach their maximum frequency in the deciduous and semi-deciduous formations. The same is probably true of the deciduous species in evergreen forest elsewhere in the tropics, but if the deciduous habit is determined by internal as well as external factors the argument of von Ihering (1923), with reference to the flora of south Brazil, that the deciduous species must have invaded the evergreen forest from regions with a more seasonal climate, is not necessarily well founded.

GROWTH AND CAMBIAL ACTIVITY

Cambial activity in rain-forest trees has been investigated by Simon (1914). In species which lose their leaves completely, such as *Tetrameles nudiflora*, the cambium of the branches ceases to divide during the bare period and secondary growth is not resumed until the new leaves unfold. Similarly in *Terminalia catappa*, in which the foliage is for a long time discoloured and probably physiologically inactive, there is a cessation of cambial activity during this period. In *Dillenia indica* and *Ficus variegata* the leaves are fully green and apparently active most of the year, only falling at or shortly before bud-expansion; here there is also a long interruption of cambial activity ending with the expansion of the new leaves. The only strictly evergreen species studied by Simon was *Ficus annulata* var. *valida*; in this tree the periodic expansion of new leaves seemed to be always preceded by a pause in cambial activity. The conclusion was that, at least in tropical trees which expand new leaves periodically, cambial activity is also periodic. The more extensive work of Coster (1927–8) extends and confirms this result. It shows that whenever the leaves are in an active condition, the cambium is also active, both in evergreen and deciduous species.

Coster also made an extensive study of the growth zones or rings of tropical trees. Trees of non-seasonal tropical climates rarely show distinct annual rings such as are found in temperate trees. Some show no rings of any kind; others, growth zones which do not form complete rings; yet others, complete rings which are not formed annually. Coster found that in general well marked growth rings are found only in deciduous species, that is, in those in which there is a definite interruption of cambial activity. Yet the converse is not true; species in which the cambium rests periodically do not always have distinct growth rings. The anatomical structure of the rings when present varies greatly both within the same species and between different species. Age-estimation by means of growth rings is unreliable in tropical trees because even when the growth zones or rings are clearly defined they may be formed only after the tree has reached a certain age.

In tropical trees growth in length of the shoots accompanies or follows the expansion of the leaves and therefore need not be further discussed. Simon (1914) and Coster (1925) have studied the food reserves of tropical trees in relation to leaf change and periodic growth.

FLOWERING

Information on flowering seasons in the Tropical Rain forest is, for the most part, meagre and rather indefinite. The flowering of certain individual species, especially those which flower gregariously, has attracted some attention, but the incidence of flowering in formations and associations as a whole is not well known.

In evergreen tropical forest flowering generally extends throughout the year and there is no season in which a proportion of the species are not in flower, some of them blossoming almost continuously; but even in the least seasonal

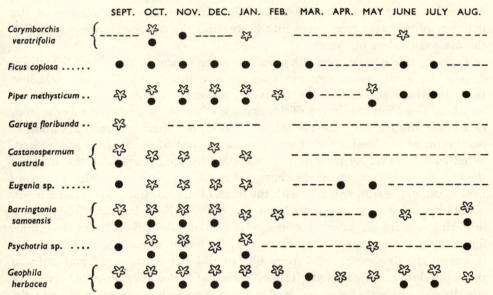

Fig. 33. Flowering and fruiting seasons of nine common plants in a non-seasonal tropical climate (New Hebrides). From J. R. & I. Baker (1936, p. 508). The selected plants range from trees to herbs. For four species no observations were made in February. Flowering is denoted by a rosette, fruiting by a black circle.

climates there are maxima of flowering at certain times of the year, which are, however, often not clearly enough marked to strike a casual observer. As might be expected, these maxima are much more pronounced where there is a distinct dry season. In the Rain forest of the New Hebrides, which has an exceptionally unvarying climate, there are two main flowering seasons, one from September to January during the wettest part of the year, coinciding with a slight rise of temperature after the end of the cooler period of the year, and another, shorter and less well marked, in May and June, at a time when rainfall is low and temperature falling (Baker, J. R. & I. (1936); see Fig. 33). Eggeling (1947, p. 64) gives the following picture of flowering in the *Cynometra alexandri* consociation of

Budongo forest, Uganda (p. 257), which has two dry seasons, one of three consecutive months with less than 10 cm. rainfall, the other shorter and less severe: 'Very few species are in flower in the period between the two rainy seasons and in the first part of the short rains. Towards the end of these rains the undershrub layer comes into flower and for almost two months the long white spikes of *Whitfieldia elongata* are everywhere conspicuous. A number of other Acanthaceae flower at the same time, but the dominant herb, *Lankesteria elegans*, does not bloom till the *Whitfieldia* is fading, its bright orange flowers brightening the undergrowth until the very end of the dry weather. In February, while *Lankesteria* is still in flower, the understory trees begin to blossom. One of the first is *Lasiodiscus mildbraedii*, followed soon afterwards by *Rinorea ardisiaeflora*, whose flowers open as the rains begin. With the coming of the wet season, the majority of the large trees burst into flower and the canopy is soon white with the massed clusters of *Cynometra*.'

Though, as we have seen, the chief flowering season in the New Hebrides is during the wetter months of the year, in most rain-forest climates, including those in which the dry season is severe, flowering occurs chiefly in the dry weather. The chief maximum of flowering is often at the end of the dry season, extending into the early part of the wet season, but there is sometimes another maximum at the beginning of the dry season, just after the rain has abated. Whether the flowering periods are controlled by the humidity of the air or of the soil, or are due to the slight change of mean temperature (which generally rises in the dry season and falls with the onset of the rainy season) is not known.

Schimper (1903, p. 254) analysed the data of Koorders and Valeton on flowering seasons in Java. For the whole island (which includes deciduous forest areas with a very long dry season) he found that 63% of the species flower in the dry season, not quite 8% in the wet season only and 29% at times unrelated to the seasons. Trimen's *Flora of Ceylon* gave a similar result; in the 'humid district', February to April and September respectively were the chief flowering seasons for woody plants, April, May and October being the wettest months. Maximum flowering of herbaceous plants was in February and September. Schomburgk's data for British Guiana show the greatest amount of flowering at the beginning and end of the short dry season (February and April) and at the beginning of the long dry season (September). In the rain-forest belt of Nigeria there appears to be a maximum flowering season at the beginning of the wet season and perhaps another at its end (Richards, 1939). In the 'Closed forest' of the Ivory Coast, which has a double rainy season, flowering takes place chiefly in the short dry season (July to August) and from August to December, the beginning of the long dry season (Aubréville, 1938, pp. 151–2). There is a marked decrease of flowering both at the height of the long dry season (January, February) and during the wettest months of the year (May and June). Trees of river banks, such as *Pterocarpus santalinoides*, are less seasonal in their flowering than those of well-drained soils.

There is some evidence that flowering is distinctly more seasonal in the upper than in the lower stories of the forest. Thus in British Guiana the A and B stories show indications of two main flowering seasons, but among the C story trees and the shrubs and herbs flowering seems to take place equally all the year round (Davis & Richards, 1933–4, p. 359). If this is so, it is doubtless to be attributed to the less variable temperature and humidity of the lower stories.

There is much variation in the flowering behaviour of individual rain-forest species, but the great majority probably flower at fairly regular intervals. In many trees individuals flower at quite different times, either in the same or in different localities and the branches of one tree may not flower simultaneously. When a species is said to flower at any time of the year this does not necessarily mean that any single individual flowers continuously, though in some species uninterrupted flowering of the same individual is normal. As with leaf change, the ill-defined flowering seasons of the community as a whole are chiefly the result of a lack of synchronization among species and individuals.

Corner (1940) classifies the trees of Malaya into 'ever-flowering' species, in which reproduction is continuous, crops of flowers and fruit being produced either simultaneously on the same individual (*Wormia suffruticosa, Adinandra dumosa*) or successively at very short intervals (*Ficus* spp.), and species which flower intermittently. In the latter flowering may occur once a year, as in many Malayan forest trees, twice a year (*Fagraea fragrans, Calophyllum inophyllum*), more than twice a year (*Eugenia grandis, Rhodamnia*) or at longer intervals than a year (various Dipterocarpaceae). The 'ever-flowering' species may start to flower when very young and continue to do so till they die; *Adinandra dumosa* begins to flower when 2–3 years old and probably lives for 100 years or more, *Wormia suffruticosa* begins at about 18 months old and flowers continuously for at least 40–50 years (Corner). Many of the smaller woody species probably behave in a similar way, but the larger trees mostly flower intermittently and do not begin until they have reached a considerable age. The intermittently flowering species which flower less often than once a year may do so at very long intervals; according to Ridley (Schimper, 1935, pp. 394–5) some Malayan species of *Hopea* and *Shorea* (Dipterocarpaceae) flower regularly every sixth year. Individuals of *Homalium grandiflorum* flower only once in 10–15 years (Corner). Some trees which flower intermittently do so at regular intervals of 7–10 months, but the flowering of different individuals is not simultaneous (*Delonix regia, Lagerstroemia speciosa*). In such instances it is evident that flowering must be determined by an internal rhythm not directly related to the seasons of the climate. In the Ivory Coast, Aubréville (1938, p. 152) also finds that some species flower continuously and others at well defined seasons, once a year, twice a year or at intervals longer than a year.

Of special interest are tropical plants which flower gregariously, that is to say, those in which all the individuals, sometimes over areas of many square kilometres, flower together at longer or shorter intervals. This is a not uncommon feature in the rain-forest flora. Spruce (1908, **1**, p. 257) says of the Amazonian Myrtaceae: 'They are remarkable for their simultaneous and ephemeral flowers. On a given day all the myrtles of a certain species, scattered throughout the forest, will be clad with snowy fragrant flowers; on the following day nothing of flowers appears save withered remnants. Hence it comes that if the botanist neglect to gather his myrtles on the very day they burst into flower, he cannot expect to number them among his "laurels".' In some cases a connexion has been traced between gregarious flowering and weather conditions. Corner (1940, p. 537) thus describes the flowering of the three cultivated species of *Coffea* in Malaya: 'On certain days of the year a shabby coffee bush is transfigured into fragrant whiteness through the sudden development of the flowers. All coffee bushes of one species will flower together in the same district on the same days, different species generally flowering gregariously on different days. The reason is that the flower buds need the stimulus of certain peculiar weather conditions to make them open. The buds develop to a certain size and then remain dormant until the necessary weather-stimulus awakens them so that in a set number of days from the stimulus they will open simultaneously. Either the nature of the stimulus or the interval between the stimulus and the opening of the flower varies in the different species and hence their different days of flowering.' Well-known examples of gregariously flowering species are various bamboos which flower at long intervals and die after ripening their fruits, and species of *Strobilanthes*. *S. cernuus* in Java flowers gregariously about every 9 years (van Steenis, 1942), *S. sexennis* in Ceylon about every 12 years (Petch, 1924). In the tree *Fagraea fragrans* gregarious flowering takes place at Singapore every year in May with great regularity; there is a second less conspicuous flowering in October and November. Holttum (1935) has shown that the chief flowering period occurs about 4 months after the break in the rainy season, which is usually in January; variation in the date of this break was closely correlated with variation in the time of flowering over a period of some 8 years. In Mauritius, Vaughan & Wiehe (1937, p. 302) find that gregarious flowering of *Homalium paniculatum* follows severe cyclones.

The best investigated examples of gregarious flowering are in the pigeon orchid, *Dendrobium crumenatum*, and certain other Malayan epiphytic orchids (Coster, 1926*a* and references given there). In these orchids flowering is gregarious over large areas; in any one species the flowers last one day or less. Flowering nearly always occurs after a thunder-shower following a dry spell. The interval between the shower and the flowering may be 8, 9, 10, or 11 days, according to the species; for some species it varies by 1 or 2 days. Experiment has shown that the effect of thunder-showers is not due to the direct action of rain or to any electrical phenomenon, but to the sudden fall of temperature. In two species the buds grow to a certain size and then enter a resting stage, as was described above for *Coffea*, from which they emerge only when stimulated by a fall of temperature. The longer they remain in the resting phase, the more easily they are aroused, and if the stimulus is delayed sufficiently long, the buds may open spontaneously.

The gregarious flowering of tropical trees and orchids seems to have much in common with the flowering in 'pulses' of temperate species of *Juncus* in which all the individuals of the same species in the same district open their flower buds together at intervals of a few days without any obvious relation to changes in the weather.

There is often an apparent relation between flowering and vegetative activity. In some tropical deciduous trees the ripening of fruit is regularly followed by the shedding of the leaves. Many species always flower when bare of leaves, and in trees in which the branches lose their foliage at different times it can sometimes be observed that flowering takes place only on the bare branches; this is often seen in *Ceiba pentandra* and occasionally in *Couratari pulchra*, a common tree of the Guiana forest. Among the deciduous trees of Trinidad, Beard (1946 b, p. 51) finds that about a half of the species flower when in leaf and fruit when bare, about a quarter flower when bare and fruit in the same or the following dry season, and another quarter flower in the dry season when bare and fruit in the following wet season. In the evergreen oaks (*Quercus* and *Lithocarpus*) and Dipterocarpaceae of Malaya and in *Mangifera indica* at Singapore flowering does not coincide with flushes of new leaves, but in other trees, such as species of *Myristica* and *Saraca*, flowers and young leaves are produced at the same time (Corner). In the 'east-monsoon' climatic region of Java, where many of the smaller trees and shrubs are 'ever-flowering', Coster (1923) finds that, on the whole, species in which shoot growth is continuous are also 'ever-flowering', while those in which it is periodic flower periodically. It has sometimes been suggested that the relation between flowering and leaf-shedding is causal and that there is an apparent antagonism between the two, so that the onset of the reproductive phase tends to coincide with a complete or partial check to vegetative development. It seems more likely that both changes depend on a common physiological cause.

Periodic flowering in tropical plants appears to be partly controlled by changes in the external environment, but the facts summarized above suggest that internal causes are also at work. As with the periodic production and shedding of leaves the evidence indicates that there are both external and internal factors. It may be that internal factors are really the basic cause of both phenomena, but, as J. R. & I. Baker (1936) point out, it is improbable that an internal physiological rhythm could be sufficiently accurate in its timing to produce the regular periodicity which is often observed. The external environment must regulate the internal rhythm and from time to time 'put the clock right'. In a temperate or strongly seasonal tropical climate environmental control is much stricter than in a uniform, relatively unvarying climate and the part of the internal or endogenous rhythms is correspondingly less evident. The internal factors in periodic phenomena in plants have been discussed in some detail by Bünning (1948), whose general conclusions are similar to those reached in this chapter.

Intermittent and gregarious flowering, as mentioned in Chapter 3, is one (though not the only) cause of good and bad seed years; it therefore deserves further study because of its bearing on natural and artificial forest regeneration. Since the length of time required to ripen the fruit varies from species to species, the season of flowering does not necessarily determine the season of seed dispersal. In seasonal climates the seeds are most often shed during dry weather which provides the most suitable conditions for their dispersal, irrespective of when flowering takes place. Germination may occur immediately, or later when conditions may be more favourable.

CHAPTER 9

SOIL CONDITIONS

The soils of the humid tropics, even if only the lowlands are considered, vary in structure, and in their physical and chemical properties, at least as much as those of any other climatic zone. The correlation between soil and natural vegetation is always close and tropical plant communities reflect differences in soil conditions no less faithfully than those of temperate climates. The common belief (for which the monotonous forest covering may be responsible) that both soil and vegetation are more uniform in the wet tropics than elsewhere has very little foundation in fact. The soil factors which affect rain-forest vegetation most obviously are perhaps for the most part not the same as those with which the European or North American ecologist is chiefly concerned, but their effects are equally evident.

Though varied in many of their properties, the majority of the soils of the rain-forest belt share certain important common characteristics. In colour they are bright red or yellow; in texture they are generally loamy or clayey, but often sandy in the superficial layers; they are usually deficient in bases, and often in plant nutrients in general; they are almost invariably acid; their humus content tends to be low and is mostly confined to the uppermost horizons; the clay fraction is relatively rich in alumina and poor in silica. To understand soil characteristics it is necessary to study soil development and so, to explain why the characteristics just mentioned are found in most rain-forest soils, the action of soil-forming factors in wet tropical climates must first be considered. Many properties of the soil of great interest to soil science or pedology in general seem at present to have little or no ecological importance; with these we need not be concerned. The main theme of this chapter is the soil as it affects the natural vegetation of the tropics; the soil properties to be discussed are chiefly those most likely to be effective ecologically as factors determining the distribution of rain-forest formations and their component associations, consociations and individual species. For a treatment of tropical soils from the wider pedological point of view reference must be made to the works of Vageler (1933), Mohr (1944) and Robinson (1949).

SOIL-FORMING PROCESSES

All soils undergo gradual development from parent materials and their properties may change fundamentally as they pass from youth to maturity. The principal factors which directly influence the course of this development are the parent rock from which the soil is formed, the vegetation which grows on it, and the climate (or rather the microclimate) to which it is exposed. Topography, especially the slope of the ground, is important because of its effects on drainage and erosion. Indirectly the soil may be influenced by man and animals which modify the vegetation. There has been much difference of opinion as to the relative importance of these various factors in soil development. Some have believed that the climate is all-powerful and that, whatever the parent material or the nature of the plant covering, the end-result, the mature soil under a given type of climate, will be the same. Others have attributed greater weight to the influence of the parent rock or to the vegetation. It is, however, preferable to regard soil, vegetation, animal life, climate and parent rock as components of a single system, the ecosystem (Tansley, 1935), and the development of the soil as influenced by all the other components of this ecosystem. At different times and under different conditions, now one, now another, component plays the predominant part. If this is so, it would be a mistake to lay exclusive emphasis on the influence of parent rock, vegetation or climate, as all three contribute to the final result. It thus follows that in an ecosystem in equilibrium, such as climax tropical forest, no one component can change without compensating changes in the others. Clearing or thinning the forest, for example, will inevitably be followed by changes in the soil profile. Again, it is not surprising that, even under the powerful influence of the tropical climate, both the mature soils and the climax vegetation may be different on parent rocks of widely different lithological character.

Weathering under wet tropical conditions

The weathering of rocks to form soil may involve both physical and chemical changes, but in the wet tropics weathering is almost entirely a chemical process. Factors, such as frost, which in other climates are chiefly responsible for physical weathering, do not operate under rain-forest conditions. Except in the initial fragmentation of sun-exposed rocks which undergo alternate heating and cooling, the main weathering process is due to water and substances dissolved in it and on rocks composed largely of silicates weathering results in hydrolysis. This goes on in a climate of which the most marked features are a high, even temperature and a permanently high humidity. Data on soil temperatures under Rain forest are scanty, but it may be assumed that everywhere in the rain forest the temperature just below the soil surface seldom rises far above 25° C.; in the Netherlands Indies at sea-level the temperature under forest at 1 m. depth is 25–27° C. (Mohr, 1944, p. 45). At such temperatures, in the presence of abundant water, hydrolysis is very rapid and on level ground weathered material may accumulate to a depth of 15 m. or more, even though there is no physical weathering. In this connexion it must also be remembered that in many parts of the tropics the process of weathering may have been going on for an extremely long period; it does not, as in most of Europe, date back only as far as the last glacial period.

Not all the weathered material overlying the parent rock can be regarded as soil, or even as immature soil. In the tropics it is more than usually difficult to draw a dividing line between the soil proper and the parent material; even in the most deeply weathered tropical soil profiles it is doubtful if the soil, in a biological sense (i.e. the layer inhabited by organisms and exploited by roots), extends deeper than 1 m. or, at most, 2 m. The underlying mass of partially weathered material (zone of alteration), which may be many times as thick as the soil itself, is best regarded as subsoil or soil-forming material.

In a wet tropical climate erosion as well as weathering is rapid and on steep slopes the soil may be very shallow even under a forest cover. For instance, in the forest on the steep slopes near Rio de Janeiro (Brazil) the average depth of soil is less than 10 cm. (McLean, 1919, p. 122). The great intensity of erosion in the tropics is witnessed by the muddiness of tropical rivers and the large amount of suspended matter they deposit when in flood. Lam (1945) quotes the figures given in Table 23 to illustrate the enormous quantities of solid matter transported in tropical rivers. Vageler (1933) has emphasized that in the tropics, owing to the great intensity of the rainfall, erosion can occur on very gentle slopes. When the rainfall during a given period exceeds what the soil can absorb, a slowly moving sheet of water spreads over the whole surface of the ground. This type of erosion is termed sheet-flow and over long periods its effects on the soil must be considerable.

TABLE 23. *Transport of solid materials by rivers.* (After Lam, 1945, p. 19)

	Rhine	Mamberamo (New Guinea)	Amazon
Water flowing (cu.m./sec.)	240	10,000	120,000
Solid matter carried (kg./year)	2000–5000	15,100	1,000,000–2,300,000
Solid matter carried (kg./year)/sq.km. of drainage area	15,000	150,000	2,000,000
Average lowering of land surface per year (cm.)	0·006	0·06	0·08

The residue from the chemical weathering of rocks which are not (like sandstones) highly siliceous consists of the products of the hydrolysis of complex silicates—bases, silica, alumina, kaolinic minerals (hydrated aluminium silicates, etc.)—and iron oxides, together with materials, such as quartz, relatively resistant to weathering (if any are present). These primary products are set free in the subsoil, immediately above the parent rock; their later history is determined chiefly by the soil water currents and the substances they carry in solution.

In a climate where precipitation on the whole exceeds evaporation, the movement of water in the soil is mainly downward, though the possibility of any considerable movement of water, of course, depends on the soil texture.[1] When precipitation is less than evaporation there is no downward, and there

[1] In some tropical soils the surface is hard and 'infilterable', allowing very little penetration of water (Chenery & Hardy, 1945).

may even be a slight upward, water movement. In the rain-forest region in general precipitation exceeds evaporation for the greater part of the year and only in months with less than 4 in. (10 cm.) rainfall may precipitation be assumed to be less than evaporation (Charter, 1941, p. 3). The downward movement of water in the soil is thus continuous almost the whole year round and even in rain-forest districts with a marked dry season it is interrupted for not more than about 4 months a year. All soils of the rain-forest region, except the most immature, are therefore strongly leached and have an eluvial horizon, or leached zone, above the illuvial horizon into which mobile substances are transported by the water currents.

It is important to realize that the composition of the soil water is not the same throughout the soil profile. In the upper layers of the soil the down-current consists of rain-water carrying little besides carbon dioxide and organic substances derived from the humus on the soil surface. Lower down, the soil water is a rock-extract and contains in solution the products of the hydrolysis of silicates.

The primary products of the hydrolysis of silicates are all soluble in the soil water under certain conditions, but their solubilities are very different. The bases are much the most soluble and, under rain-forest conditions, they are quickly carried away in the drainage water if not taken up by the roots of the trees. Mature rain-forest soils thus always have a very low base content and, since there is a rough relation between reaction and base content, are invariably acid.

Silica and the sesquioxides of iron and aluminium (Fe_2O_3, Al_2O_3) are also soluble and are carried away in the down-current of soil water. All three substances move in the form of colloidal sols and are held in solution or precipitated, depending on the content of electrolytes and other colloids in the soil water. Silica and silicates of aluminium are relatively soluble in pure rain-water, but tend to be precipitated as an irreversible gel in solutions containing humus substances or electrolytes. The oxides of both aluminium and iron, on the other hand, are relatively insoluble in pure rain-water, but can be transported by water containing humus, which acts as a protective colloid preventing their precipitation by electrolytes. The kaolinitic minerals formed by the primary weathering of rocks do not, as is often supposed, undergo further hydrolysis into free silica and alumina but are stable end-products (F. Hardy).

Now under normal rain-forest conditions the down-current of soil water contains only very small quantities of electrolytes, owing to continual leaching. For reasons which will shortly be explained, it also contains very little humus. The tendency will always be for silica and aluminium silicates to be carried downwards and the sesquioxides of aluminium and iron to remain behind. Thus, on the whole, the older a tropical soil, the poorer it becomes in silica and kaolinitic minerals as well as in bases, and the richer in sesquioxides. The ultimate end of the process would be a soil consisting of little else but alumina and iron oxides, together with a certain amount of quartz. The ratio of silica to sesquioxides (or to alumina alone) is therefore an important index to the degree of maturity of a tropical soil. Since the degree of maturity depends on the intensity of the various soil-forming factors, as well as the time during which they have acted, the silica/sesquioxide ratio is not a direct measure of the age of the soil. In the silicates of the parent rock the ratio is often about 6, but in the course of soil development under tropical conditions it may fall below 2. Since the leaching of bases and other plant nutrients is more rapid than the leaching of silica, the more mature a tropical soil, the lower, as a rule, is its fertility and agricultural productivity.

This preferential leaching of silica from a soil is called laterization. The opposite process, the removal of sesquioxides of iron and aluminium with the accumulation of silica is called podzolization. This is the typical soil-forming process of cool humid climates, though under certain conditions it can also occur in the tropical lowlands.

Tropical red earths

Space will not allow a full discussion here of the various meanings which have been given to the terms laterite and lateritic, a controversial subject on which there is still much confusion (see especially Shantz & Marbut, 1923; Martin & Doyne, 1927; Thorp & Baldwin, 1940; Pendleton & Sharasuvana, 1946). The term laterite may be applied to the end-result of laterization—a mixture of alumina and iron oxides with very little else. This end-result is never reached under a cover of rain forest, indeed in this sense laterite is rather the chief constituent of a specific horizon in certain tropical soils than a soil in itself. Though true laterite is probably not found under rain-forest conditions, many, probably most, forest soils of the damp tropics are lateritic. It is convenient to define a lateritic soil[1] as one in the clay fraction of which the silica/alumina ratio is less than 2·0 (see Martin & Doyne, 1927).

The red colour common in tropical soils formed under conditions of un-impeded drainage is due to the abundance of iron oxides. The exact shade of colour probably depends on the degree of hydration of the oxides; according to Robinson (1949, p. 408) the iron oxides found in tropical soils are less highly hydrated than those of temperate soils. The whole class of tropical soils with a red colour and showing some degree of laterization is known as the tropical red earths. Vageler (1933) draws a distinction between tropical red loams and tropical red earths. The former are less mature, and therefore less

[1] In some recent American classifications of tropical soils (Byers *et al.* 1938; Roberts, 1936; see also Hardy, 1945) the soil of large areas of tropical America is classified as podzolic. According to this American usage (Byers *et al.* 1938) podzolization is a process involving (*a*) the formation of a peaty mat of organic matter on the soil surface and (*b*) the downward removal (and often the fractionation) of iron compounds and clay. This is a vaguer conception of the process than that given above, taking no account of the Si/sesquioxide ratio. Many rain-forest soils are no doubt podzolic according to this definition which implies little more than leaching or eluviation. If the American view is accepted, a lateritic soil can also be podzolic. Mohr's (1944, p. 144) term lixivium, meaning 'drastically weathered and leached inorganic residues', has much to recommend it.

laterized than the latter. As there is a correlation between the degree of laterization and the leaching out of plant nutrients, the distinction has some agricultural importance, but it has not yet been shown to have any significance in the ecology of natural vegetation. Some rain-forest soils are yellow, rather than red in colour and are sometimes termed tropical yellow earths. The yellow colour is due to the occurrence of iron oxides more highly hydrated than those of the red earths; there does not appear to be any other important chemical difference between the two classes of soil. Mohr (1930, p. 36) believes that the yellow earths are a stage in the development of red, the yellow iron oxides being metastable and the red stable. Tropical soils subject to alternate wetting and drying are often characteristically mottled.

Lateritic red earths (in the sense used here) may be the most widespread soil type of the rain-forest region, but by no means all tropical soils are lateritic. Many types of parent material cannot undergo typical lateritic weathering. Sandstones and some acid igneous and metamorphic rocks, for example, are so highly siliceous that the de-silicification characteristic of lateritic soils is indefinitely postponed. These rocks, as shown below, may give rise to soils very different from red earths. Massive igneous rocks yield, on weathering, mainly kaolinitic materials and these, as mentioned above, are stable substances not liable to further hydrolysis. Sedimentary clays also do not give rise to lateritic soils. Lateritic weathering with the production of typical tropical red earths is best seen on fragmental volcanic rocks (F. Hardy).

Soil types determined by parent materials

On some types of parent rock, soils very different from tropical red earths may develop. In these cases the influence of the material from which the soil is formed overrides that of the climate giving rise to so-called azonal soils (in contrast to the climatic or zonal types).

Hardy & Follett-Smith (1931) obtained interesting evidence on this point in their study of soil profiles in British Guiana. Five profiles were examined, all developed in a rain-forest climate with a rainfall of over 100 in. (254 cm.). The rocks from which they were derived were igneous, varying from basic hornblende schist to acid granite. Soil development has followed a different course on the basic and intermediate rocks on the one hand and on the granite on the other. On the former rapid laterization occurs, producing a crust of 'primary gibbsitic laterite' (gibbsite = $Al_2O_3.3H_2O$) on the surface of the rock. Where drainage is good this 'primary laterite' tends to accumulate in considerable depth. Where it is less good (i.e. partially impeded), as is usually the case in lowland sites, re-silicification of the laterite takes place, apparently by means of ascending water currents in the dry season bringing up soluble silica or alkali silicates from below. This re-silicification leads to the formation of kaolinitic earths coloured red or yellow by iron oxides. On granite, however, little or no initial loss of silica takes place and no 'primary laterite' is formed. A kaolinitic red earth is produced directly from the parent rock. The end-result on basic and acid rocks is similar, but the way in which it is reached is thus very different.

An example of a tropical soil type in which not only the course of development, but the characteristics of the mature profile are completely different from those of the normal red earth type is the bleached sand now known to have a wide distribution in the rain-forest region (Richards, 1941). In British Guiana such soils are characteristically associated with Wallaba (*Eperua*) forest (p. 240), in Sarawak (Borneo) with Heath forest (p. 244). It was pointed out by Richards (1936) that the bleached sand of the Heath forest closely resembles a temperate podzol; previously it had been believed that in the tropics podzols only occurred at high altitudes under a cool climate. Shortly afterwards Hardon (1936, 1937) described in detail some bleached sand profiles from low altitudes in East Borneo and Bangka in the Malay Archipelago, close to the equator, and showed conclusively that these sands must be regarded as podzols.

The profiles examined by Hardon developed under a mean temperature of about 26° and the climate was also in other respects normal for rain forest. The annual rainfall in both localities was about 300 cm., no month having less than about 10 cm. The vegetation was the so-called padang,[1] a kind of heath-like scrub or poor forest, which, though very different from the rain forest typical of the region, is probably a stable edaphic climax (see pp. 256, 324). Ecologically and floristically padang vegetation has much in common with Heath forest. In structure and physiognomy it is closely analogous to the muri of British Guiana, a kind of scrub dominated by *Humiria floribunda* var. *guianensis*. Muri also appears to be an edaphic climax and is related to the Wallaba (*Eperua*) forest of its own region in much the same way as padang to Heath forest.[2]

The description and source of the analytical data for two of Hardon's padang profiles are reproduced in Table 24. The data show that these sands have all the normal features of a podzol. There is a much bleached A (eluvial) horizon, the upper part of which is stained blackish with humus. Below there is an illuvial B horizon in which precipitation of humus and sesquioxides occurs. In one of the profiles this B horizon is consolidated into a hard pan. The texture is coarse and porous throughout the profile: the clay fraction is higher in the B horizon than in the A, but even in the former it is under 10 %. The silica/sesquioxide ratio of the clay fraction is high in the A horizon (3·62–8·64) and low in the B (0·31–0·42), showing that sesquioxides, not silica, are being leached from the A horizon: deposition of sesquioxides takes place in the B horizon. This is quite unlike what occurs in a lateritic red earth profile. As would be expected from the

---

[1] In Malay *padang* means merely meadow or open land not covered by trees. The term is used here for want of a better; 'Tropical heath' is a possible alternative.
[2] Muri, unlike padang, is often associated with grassy savannas, as well as with high forest (p. 324).

high rainfall and the porous texture of the soil, the content of bases and plant nutrients generally is exceedingly low, even compared with other tropical soils.

The surface of the soil is covered by very acid humus. This has a very high carbon/nitrogen ratio (57), indicating incomplete decomposition of the organic materials. Evidently this humus is of a *mor* rather than a *mull* character (see p. 218). Deeper in the soil the percentage of humus is no higher than in other lowland tropical soils, but the depth of penetration is greater and some accumulation of humus occurs in the B horizon.

TABLE 24. *Profiles of lowland tropical podzols.* (After Hardon, 1937)

Locality	Horizon	Depth (cm.)	Description	Percentage of sand (0.05-2 mm.)*	Percentage of clay (0.0005- 0.005 mm.)*	Organic matter (%)	Molecular ratio $\dfrac{SiO_2}{Al_2O_3+Fe_2O_3}$ (for clay fraction)	pH	CaO
Padang Loewai, E. Borneo	A$_0$	0–20	Brownish black half-decomposed organic material	—	—	—	—	2.8	—
	A$_2$	10–120	Loose white quartz sand	97.3	0.5	0.1	—	6.1	0.008
	B	120–170	Yellowish brown quartz sand	93.9	2.8	1.2	—	5.4	0.002
Bangka	A$_0$	0–10	Half-decomposed black organic material mixed with coarse sand	—	—	—	3.62	2.7	—
	A$_1$	10–25	Loose greyish black humic quartz sand	95.6	1.6	—	7.17	3.9	0.022
	A$_2$	25–40	Loose greyish white quartz sand	94.0	0.6	—	8.64	6.1	0.032
	B$_1$	40–70	Hard pan. Dark brown very compact quartz sand	86.9	7.2	5.2	0.42	3.9	0.029
	B$_2$	70–100	Loose light brown quartz sand	92.1	4.6	—	0.31	4.6	0.035

* The percentage of gravel (> 2 mm.) and of silt (0.005–0.05 mm.) have been omitted.

The padang soils of East Borneo and Bangka are developed from alluvial quartz sands, the Heath forest podzols of Sarawak from Miocene sandstones. According to Hardon the development on the padangs of a podzol rather than a tropical red earth depends chiefly on the porous nature of the soil and its initial deficiency in bases. The coarse soil texture allows rapid movement of water through the soil, and bases set free from the humus are leached even more quickly than in most tropical soils. It appears that under these conditions acid humus can accumulate to some extent even in a hot, wet climate.

Tropical podzols seem to have a wide distribution in the eastern tropics, as padang vegetation is said by Hardon to occur in Siam (Thailand), as well as in many parts of the Malay Archipelago. Podzols bearing Heath forest are widely distributed in Borneo.

Bleached quartz sands, which are most probably tropical podzols, are also widespread in the American rain-forest region. In British Guiana there are large areas of 'white sands' bearing savanna, muri, Wallaba (*Eperua*) or other peculiar types of rain forest (Chapter 11). The 'Valencia sand' of Trinidad (Hardy, 1940) is probably a similar soil. The profiles of these sands closely resemble those of Hardon's padang soils. The white sands of Guiana, which are found on ridges and plateaux in considerable depth, were formerly believed to have arisen by the decomposition of granite *in situ*. Recent evidence, however, shows that they are more probably tertiary alluvial deposits formed in a shallow sea, though ultimately derived from granite. In their past history they must have lost most of their finer fractions, together with most of any bases they may originally have contained.

On some bleached sands in the lowlands of British Guiana a type of forest exists dominated by *Dimorphandra conjugata* (dakama); the surface of the soil in this type of forest is covered with a layer of peaty humus about 1 ft. (30 cm.) thick (T. A. W. Davis). The profiles of these very interesting soils have not been examined, but it seems likely that they are podzols with an unusually thick A (raw humus) horizon. As these soils are not flooded or permanently water-logged, it is hard to understand why such large quantities of humus should accumulate (p. 256). The reason is perhaps connected with the porous texture of the soil which, by allowing the surface to dry very readily, prevents the decomposition of the humus. The chemical or physical nature of the *Dimorphandra* leaves may also be partly responsible.

Bleached sands like those of Guiana are widely distributed in the Amazon basin, especially in the Rio Negro region (Spruce, 1908, pp. 206–7, 303–4).

A 'white' sand, similar in character to the bleached sands of South America and the eastern tropics is found in the island of Mafia, off the coast of Tanganyika (East Africa). The vegetation is heath-like and is dominated mainly by the ericaceous shrub *Philippia mafiensis* (Greenway, 1938).

A striking feature of these bleached sands in all parts of the tropics is that the streams draining them are always 'blackwater' streams, that is they have clear water, nearly black by reflected light, orange-brown by transmitted. 'Blackwater' rivers are a well-known feature especially in tropical America; the Rio Negro derives its name from its black water. Streams draining red earth soils have clear or turbid, but always colourless water. The brown colouring is evidently due to humus compounds in a finely dispersed colloidal condition: where 'blackwater' drains into the sea, the brown matter is precipitated by the electrolytes in the sea-water. Why the humus is carried out from these sands in solution in the drainage water is uncertain. It has been suggested (Richards, 1936, p. 26) that it remains in colloidal solution because the bleached sands lack electrolytes which might precipitate it, as well as clay and iron oxides which would form adsorption complexes with it. Hardon (1937) attributes the brown colour of the streams on padangs to iron humates which fail to precipitate in the B horizon of the soil owing to its high acidity. 'Blackwater' streams are also found flowing from Tropical Moor forest (peat swamps, see p. 292), but where no extensive swamps are known to exist 'blackwater' is a trustworthy guide to the presence of bleached sands.

There is thus abundant evidence that podzols are of common occurrence in the tropical lowlands and are not confined, as was formerly thought, to cool climates north or south of latitude 40°, or, nearer the equator, to high altitudes. Their presence in the tropics, where climatically they seem out of place, evidently depends more on the nature of the soil-forming material than on the soil-forming processes themselves.

Whether other soil types determined chiefly by the nature of the parent rock exist in the rain-forest region is not yet known. According to Hardy (1935, p. 157) limestones give rise in the wet tropics to soils very similar to the red and yellow earths formed from rocks with a less high base content. Leaching soon removes nearly all the calcium carbonate and other bases and the soil becomes acid. The properties of the mature soil are then determined largely by the impurities in the original limestone. Bare limestone rocks in the tropics, however, have a peculiar flora (p. 223) and soils overlying limestone, because of excessive drainage, often support a 'drier' type of vegetation than soils overlying other rocks in the same climate.

Swamp forest soils

Since the movement of water in the soil plays such an important part in soil development, tropical soils formed under conditions of impeded drainage, or where drainage is chiefly lateral, differ fundamentally from those formed where there is free downward drainage. The effect of partially impeded drainage in bringing about the re-silicification of 'primary laterite' has already been referred to; in this section we shall consider the extreme case of swampy rain-forest soils in which there is a water-table which rarely falls more than a few inches below the surface.

These swamp soils have been very little investigated, but they seem to fall into two rather sharply separated groups, swamp soils with little humus accumulation, and swamp soils with a surface layer of peat or humus. The latter develop under oligotrophic (extremely mineral-deficient) conditions, the former where the water supply is eutrophic (relatively rich in bases, etc.).

The non-peaty swamp soils are mostly silts or clays of heavy texture, differing from the red earths of well-drained sites in their greyish, whitish or pale yellow colour. Though the general colour is not red, there may be reddish brown, orange or yellow mottlings. According to Mohr (1930, p. 23), such soils consist chiefly of 'kaolin' and silicic acid and the vegetation depends chiefly on the incoming drainage water for its supply of nutrients. The humus layer is very thin and the penetration of organic matter very shallow. These non-peaty swamp soils of the tropics would seem to have much in common with the gley soils of other climates. From the ecological point of view they differ from better drained soils in their superabundance of water, poor aeration, and possibly in their chemical properties.

The peaty type of swamp soil is much less widely distributed than the non-peaty, but has been more studied. Till Potonié (Potonié & Koorders, 1909) drew attention to the occurrence of swamp peats of considerable thickness in Sumatra, peat formation was generally supposed to be impossible at the temperatures found in the tropical lowlands (p. 218).[1] After the publication of this paper extensive peat swamps were discovered in many parts of the Malay Archipelago and Peninsula, in tropical America and in tropical Africa. In many respects these peat swamps are comparable with the peat bogs or moors of temperate regions and as their plant covering usually consists chiefly of trees they may be termed Tropical Moor forests (*Waldmoore*).

An excellent account of the peat deposits and Moor forests of the East Indies has been given by E. Polak (1933). Two chief types of moor are recognized in this paper, ombrogenous and topogenous. The former occur in the coastal plains of Sumatra, Java, and probably New Guinea; they require a high rainfall for their development—probably at least 2000 mm. per annum. The latter are less extensive; they are determined chiefly by the topography and are less restricted to regions of high rainfall. They occur on lake shores, etc., both in the lowlands and on the mountains. The vegetation of the ombrogenous moors is tall ever-green forest; it is in fact a specialized type of Tropical Rain forest (Chapter 13); that of the topogenous moors is often herbaceous. *Sphagnum* seldom occurs except on upland topogenous moors.[2]

Ombrogenous moors are often found immediately behind the coastal man-grove swamps, but their ground water is neither salt nor brackish. They consist of masses of peat of considerable extent (no figures are given, but the largest of them would appear to be a few kilometres in diameter). The peat becomes thicker from the edges towards the nearly level centre, so that the shape of the whole mass is that of a biconvex lens, the lower surface being below the level of the surrounding land, the upper surface above. In the thickest part the peat may attain a thickness of over 7 m. It is a reddish or greyish brown, porridge-like mass; only the tree-roots give a firm foothold. From the central portion of the moor 'blackwater' streams with very acid water (pH *c.* 3·0) flow out radially; thus the water supply comes chiefly from the rainfall. These moors are extremely oligo-trophic, the peat giving only a very small residue of mineral matter when burnt.

The profile of some of these moors has been examined and seems to indicate that they have borne forest vegetation throughout their development; remains of trees form the main bulk of the peat at all horizons. Underlying the peat there is a bleached clay.

[1] Potonié believed that the European coal of the Carboniferous period might have been formed at tropical temperatures, under conditions similar to those in present-day tropical peat swamps.

[2] In South America *Sphagnum* probably occurs in the lowlands, in Moor forests as well as in certain types of savanna (p. 323). In Malaya and Borneo *Sphagnum* is sometimes found in open places in lowland peat swamps, but not in great abundance.

As Polak points out, the form, hydrography and chemical characteristics of these ombrogenous moors show them to be analogous to the raised bogs (*Hochmoore*) of the Temperate zone. The topogenous moors, which develop in hollows and have a depressed, not lens-shaped, profile, are eutrophic and are the tropical equivalent of fen.

The accumulation of peat at tropical temperatures is clearly due partly to waterlogged, and therefore anaerobic, conditions. According to Mohr (1944, p. 108), tropical soils formed under water conditions are unfavourable for the growth of most fungi and for all bacteria other than facultative or obligate anaerobes. The rate of breakdown of plant debris thus becomes slow relative to its rate of formation (cf. p. 218). Because in many tropical swamps no humus accumulation takes place, it is evident that some edaphic factor in addition to waterlogging is necessary for peat formation. Mohr suggests that this is an oligotrophic water supply extremely deficient in bases. Though Polak does not accept this view, the fact that in Borneo Moor forests and bleached sands (podzols) are often found in close proximity suggests that it is near the truth.

Peat swamps, probably similar to the oligotrophic ombrogenous moors of the East Indies, are known to occur in various parts of tropical America. Ule (1908 *a*) and Bouillienne (1930) describe them from Amazonia. The 'pegass' swamps of British Guiana are evidently very similar. They occur among the coastal 'swamp savannas' (p. 322) over fairly extensive areas and inland as small swamps near the heads of streams in the white sand area. The 'pegass' consists of soft wet peat several feet deep; the water table is sometimes within a foot of the surface even in dry weather.

SOILS AT HIGH ALTITUDES

With increasing altitude both soil and vegetation change in response to the change of climate. Tropical Rain forest gives place to rain forest of a different type (Submontane Rain forest) and eventually to Montane or 'Temperate' Rain forest (Chapter 16). The factor determining the change in the case of both vegetation and soil is primarily the fall in the average temperature. Owing to the steep slopes, soils on mountains are usually subjected to intense erosion and are often very immature. Of the change in the soil-forming processes and the mature soil types little is known. In Java, according to Mohr (1930, p. 124), on ascending the mountains the red earths give way to yellow, then to bleached soils. The increase in the humus content of the soil with increasing altitude is very marked and there is evidence that laterization tends to give way to podzolization as the chief process in soil development. Several other authors have recorded podzolization at high altitudes in the tropics. Senstius (1930–1) found that in Java and the Philippines podzolization took place above 2000 m., though true podzols were not formed, but in the mountains of New Guinea Hardon (1936) records typical podzols at about 2000 m.

The soils of Montane Rain forest are very little known. In the Montane ('Warm Temperate') Rain forest of the Usambara Mountains (East Africa) Pitt-Schenkel (1938, p. 55) found both 'tropical forest podzols' and soils resembling temperate brown forest soils.

BIOTIC FACTORS IN SOIL DEVELOPMENT

Vegetation affects soil development in the first place by modifying the soil climate. The conditions for soil development under primary Rain forest will differ considerably from those on cultivated land or under open secondary vegetation in the same climate. The effect of the forest cover in intercepting rainfall and modifying the temperature and humidity at the surface of the ground was demonstrated in Chapter 7. Later in this book, when dealing with the destruction of rain forest and the secondary succession, the subject will be further considered.

The second way in which vegetation affects the soil is by the addition of humus. The importance of humus in soil development depends partly on its colloidal properties and partly on the bases and other inorganic substances released during its mineralization.

The humus of the soil is formed from the dead leaf litter on its surface. The common idea that a thick layer of litter is characteristic of tropical forests is completely mistaken. In the Rain forest the average thickness of the leaf litter is only 1–3 cm. Usually it is insufficient to cover the whole surface so that here and there patches of bare soil show through. Only in hollows does a thicker litter accumulate. The conversion of litter into humus is so rapid that it keeps pace with the continual rain of dead leaves from above.

The decomposition of wood, bark and dead leaves into humus, and the eventual breakdown (mineralization) of humus so formed into carbon dioxide, water and mineral matter, is mainly the work of the soil flora and fauna, but it is possible that there is also a purely chemical oxidation of humus in which micro-organisms are not involved (see Mohr, 1944, pp. 103–5). The nature of the biological and biochemical processes taking place under rain-forest conditions is very little understood. The biological aspect of the problem has been discussed by Corbet (1935), and Vageler (1933, pp. 131–9) deals briefly with soil organic matter from a biochemical point of view.

The flora of tropical soils consists of bacteria and fungi, and sometimes also of algae. The fauna includes protozoa, earthworms and insects of all kinds, especially ants and termites, as well as larger creatures, such as burrowing mammals and reptiles. All that it is possible to say here about the activities of the soil flora is that both bacteria and fungi play a large part in the breakdown of organic matter. Very little is known about the soil fungi of the rain forest, but it is believed that at high temperatures they are generally of less importance than the bacteria (Mohr, 1944, p. 107). Animals, especially termites and ants, probably play a large part in the decomposition of plant materials. Termites are very abundant and are active on the trunks and branches of the trees as well as in the soil itself. They are largely responsible for·the destruction of dead timber and so effective is their work that the dead logs, stumps and boughs lying on the forest floor usually last only a very short time. Earthworms occur in rain-forest soils, but not constantly or abundantly, and their activities do not seem to be of much importance.

The details of the action of the soil flora and fauna are thus still very obscure; yet certain general relationships between the activity of the flora and fauna as a whole and environmental conditions, especially temperature, are becoming clear. With increasing temperature the rate of formation of organic matter by plant communities increases up to an optimum and then decreases. The same is true of the conversion of this organic matter into humus (humification) and of humus into inorganic matter (mineralization). Thus both for humus formation and humus breakdown there are maximum, minimum and optimum temperatures, but they are not the same for the two processes. According to Mohr[1] (1930, pp. 17–20) the addition of organic matter to the soil starts at 0° C. and reaches its optimum at about 25° and its maximum at about 40°. The destruction of soil organic matter has a higher temperature range, the soil micro-organisms not becoming active till between 10° (or somewhat lower) and 15°. Their optimum is not reached till 35° or higher and their maximum is above temperatures normally met with in the soil. From 0° to about 20° there is an increasing surplus of organic matter, above 20° it diminishes and at 25° the curves for formation and breakdown cross, so above this temperature no humus will accumulate. These temperature relations only apply if the soil is adequately drained and aerated; under anaerobic conditions, as in the swamp soils considered above, the relations are different, as has already been mentioned.

In lowland rain forest, where the temperature near the surface of the soil is about 25° and moisture is abundant all the year round, no humus should accumulate, provided the soil is sufficiently aerated. In practice humus is rather more abundant than would be expected theoretically but, except in the top 10–20 cm., the content of organic matter does not often exceed 1–2 %. According to Vageler (1933, p. 7), tropical soils often contain more humus than their colour would suggest: this is because the humus is often much lighter in colour than in temperate soils and easily masked by the bright tints of the iron oxides. By far the greater part of the humus in tropical rain-forest soils is usually confined to the top 10 cm., but the depth of penetration varies with the texture of the soil. In a porous sand the humus tends to be washed much deeper than in more impervious soils.

Perhaps because of its rapid decomposition, the humus of rain-forest soils of the red or yellow earth type is normally not very acid in reaction. It would appear to resemble the 'mull' rather than the 'mor' of temperate forest soils.

[1] See also Mohr (1944, pp. 96–109).

Though under tropical conditions the quantity of humus formed is usually small, it has considerable effects on soil development. It has already been stated that the presence of humus tends to retard the leaching of silica. Humus thus has a conservative action on soil development and under a rain-forest cover laterization is prevented from proceeding to its furthest extent. In other ways also humus tends to maintain the fertility of the soil. It contains considerable quantities of bases and plant nutrients which as it becomes mineralized are added to the upper layers of the soil. These substances are then washed down into the deeper layers where they are absorbed by the roots of the vegetation, carried up to the leaves and stems, and returned again to the soil in the humus. In this way the vegetation tends to counteract the downward leaching of soluble substances and bring them back to the superficial layers. The adsorptive properties of humus also tend to retard leaching by holding soluble bases in the upper part of the soil.

Because of the action of humus in retaining and setting free bases the reaction of the upper layers of the soil is often less acid than that of the deeper layers of the profile, though an increase of acidity with depth in tropical soils may also be due to other causes (Doyne, 1935).

THE CYCLE OF PLANT NUTRIENTS

In the rain-forest climate, as in all climates in which the movement of water in the soil is predominantly downwards, the trend of soil development is always towards impoverishment. Soluble substances are continually being washed down into the deeper layers of the soil and removed in the drainage water. The most important common characteristic of all rain-forest soils, whether of the red earth or the podzol type (except perhaps some very immature soils) is thus their low content of plant nutrients. This being so, it seems paradoxical that rain-forest vegetation should be so luxuriant. The leached and impoverished soils of the wet tropics bear magnificent forest, while the much richer soils of the drier tropical zones bear savanna or much less luxuriant forest. This problem has been considered by Walter (1936) and Milne (1937) for African forests, and by Hardy (1936a) for those of the West Indies, and all these authors reach a similar conclusion. In the Rain forest the vegetation itself sets up processes tending to counteract soil impoverishment and under undisturbed conditions there is a closed cycle of plant nutrients. The soil beneath its natural cover thus reaches a state of equilibrium in which its impoverishment, if not actually arrested, proceeds extremely slowly.

Fresh plant nutrients are continually being set free by the decomposition of the parent rock. Provided the horizon in which they are set free is not too deep for the tree-roots to reach, a part of these nutrients is taken up by the vegetation in dilute solution. Some of these substances are fixed in the skeletal material of the plant—the cell walls—others remain dissolved in the cell sap. Eventually all of them are returned to the soil by the death and subsequent decomposition of the plant or its parts. The top layers of the soil are thus being continually enriched in plant nutrients derived ultimately from the deeper layers. The majority of the roots, including nearly all the 'feeding' roots, are in the upper layers of the soil. Most of the nutrients set free from the humus can be taken up again by the vegetation almost immediately and used for further growth. The loss, if any, must be very slight; Milne (loc. cit. p. 10) has shown that in the Usambara Rain forest the electrolyte content of the streams is very low. It can thus be seen that in a mature soil the capital of plant nutrients is mainly locked up in the living vegetation and the humus layer, between which a very nearly closed cycle is set up. The resources of the parent rock are only necessary in order to make good the small losses due to drainage.

The existence of this closed cycle makes it easy to understand why a soil bearing magnificent Rain forest may prove to be far from fertile when the land is cleared and cultivated. When the forest is felled the capital of nutrients is removed or set free in the soil and the humus layer is often destroyed at the same time by burning and exposure to the sun. As Milne (loc. cit. p. 10) says: 'The entire mobile stocks are put into liquidation and, as is usual at a forced sale, they go at give-away prices and the advantage reaped is nothing like commensurate with their value.' Crops planted where rain forest has been cleared may thus do very well for a few seasons, benefiting from the temporary enrichment of the soil, but before long, unless special measures are taken, a sterile, uncultivable soil may develop. On the ordinary system of native cultivation practised in rain-forest areas it is rare for more than two or three harvests to be obtained in succession without a long intervening period of 'bush fallow'. Even in British Honduras, where the annual rainfall is not more than about 180 cm., the yield of maize on forest clearings falls from about 800–1000 lb. (350–450 kg.) in the first year to about 400–600 lb. (180–270 kg.) in the third (Charter, 1941, p. 16). On very poor porous rain-forest soils such as the Wallaba podzol in British Guiana (p. 242) it may be impossible to obtain even one crop. The changes in the soil due to the clearing or thinning of the forest emphasize the delicate equilibrium of soil and vegetation in a natural rain forest.

Since the poverty of rain-forest soils in plant nutrients has been realized, some authors have been inclined to see great significance in the fact that a large proportion of rain-forest plants, like those of many other types of vegetation, contain mycorrhizal fungi. Janse (1897) examined seventy-five species from primary forest in West Java and found that sixty-nine of them possessed mycorrhiza. Data are not available for other regions, but there is little reason to doubt that a careful examination of the root systems of rain-forest plants would show that a very large percentage of them are associated with mycorrhizal fungi. The role of mycorrhiza is so little understood and the subject as a whole so controversial that there can be little use in discussing it here. The mycorrhiza of rain-forest plants may be of great ecological significance, but at present little can be said with certainty as to what that significance may be.

In this connexion reference should also be made to the abundance of leguminous trees in many Rain forests. Since the soils are so poor in nutrients the possession of root nodules containing nitrogen-fixing organisms is probably of considerable advantage to rain-forest plants. In five forest types at Moraballi Creek, British Guiana (p. 236), Davis & Richards (1933–4) found great differences in the proportion of Leguminosae. Among trees 4 in. in diameter and over the sample plot figures were as follows:

Forest type	Mora	Morabukea	Mixed	Greenheart	Wallaba
Percentage of Leguminosae (individuals)	59	33	15	14	53

Leguminosae are thus much more abundant in the Mora and Wallaba forest types than elsewhere. The Mora community grows on a swamp soil subject to frequent flooding, the Wallaba community on a highly leached podzol; the other forest types are found on red or brown soils of the red earth group (see Chapter 10), in which nitrification is possibly more active and leaching less intense. There is perhaps some significance in these facts, and in the abundance of Leguminosae in rain forest in general. In rain-forest consociations the single dominant tree species is frequently leguminous (Chapter 11).

THE SOIL AS A DIFFERENTIAL ECOLOGICAL FACTOR

It has already been stated or implied in several places that in the moist tropics, as elsewhere, differences of soil manifest themselves in differences in the composition of the plant communities. Edaphic factors are not only responsible for variations in floristic composition within primary Rain forest itself, but there is evidence that soil conditions may even result in the development of scrub, e.g. the padang and muri communities referred to above (p. 211), or some types of savanna (Chapter 15), as edaphic climaxes in a rain-forest climate. This is contrary to the view which was for long generally accepted—that in the tropics soil differences are of little ecological importance, certainly of less importance than in the Temperate zone. Only recently has enough knowledge been obtained of the composition of tropical plant communities to dispel this misconception.

The subject will be further discussed in later chapters; meanwhile two examples may suffice to show how close is the correlation between the soil and the climax vegetation.

Among the five primary forest communities at Moraballi Creek, British Guiana, two show a particularly striking correlation with soil conditions. The Wallaba forest (consociation of *Eperua falcata*) is confined to bleached ('white') sand capping plateaux and ridges; the Greenheart forest (consociation of *Ocotea rodiaei*) occurs on brown sand, which is usually found in belts along the contour of the ridges. The two communities are sometimes found in proximity and when this is so the boundary between them is extraordinarily sharp, coinciding exactly with the change from the bleached to the brown sand. In one transect the transition was complete within a distance of less than 150 m. (see Davis & Richards, 1933–4, fig. 1, p. 121). Elsewhere in British Guiana an even more abrupt boundary between these types of Rain forest has been observed; Wood (1926, pp. 4–5) says of the Bartica-Kaburi district:

'In most cases it was possible to step in one stride from the white sand where the greenheart never occurred to the brown sand where it did, generally at some point on the slopes down from the flat ridge to the creeks. The most striking instance was seen...where a nest of *akushi* [leaf-cutting] ants had thrown up soil over an area about 30 ft. [9 m.] square. On the upper half of the nest the soil was white, on the lower half brown, and the dividing line was sharp enough to lay a hand across. Two trees were growing out of the nest so close together that a man would have to go sideways to pass between and they occurred one just on the white soil and the other just on the brown soil below. The upper tree was the last wallaba and the lower the first greenheart on that slope.' Though the boundaries between the other forest types at Moraballi Creek were generally not as sharp as that between the Wallaba and Greenheart consociations, a close relation between forest type and soil was evident in all of them.

The other example is the distribution of the Ironwood forest of Borneo and Sumatra, a peculiar consociation dominated by *Eusideroxylon zwageri* to the almost complete exclusion of other species of trees (p. 258). Gresser (1919) observed that this type of forest is confined to pure or loamy sand. The later investigations of Witkamp (1925) showed that *Eusideroxylon* may actually be used as a geological indicator. In the Koetai district of Borneo the rocks consist of Tertiary sands, sandstones, clays and clay-shales, with occasional limestone and marl. The ironwood disappears or becomes rare on passing from sand or sandstone to clay, shale, marl or limestone, and the boundary is sharp, except where the rocks themselves merge gradually. Most of the district is an anticline, and sandstone, clay, etc. occur as parallel bands. Exactly correlated with this, the ironwood communities occur in long narrow bands, marking the outcrops of sand and sandstone.

These instances are by no means isolated and clearly illustrate the decisive part often played by soil in determining the distribution of rain-forest communities. The effect of soil differences is often overlooked because tropical vegetation types may be very similar superficially and detailed floristic analysis is needed to demonstrate their distinctness.

Though the effectiveness of soil differences in determining the composition of tropical plant communities can no longer be questioned, the edaphic factors concerned are not necessarily the same as those which chiefly operate in the ecology of temperate vegetation. The edaphic factors usually regarded as of most importance in temperate regions (apart from maritime vegetation) are soil acidity (pH), content of calcium carbonate and other bases, soil texture, water-supplying capacity and aeration, while recently the redox potential of the soil has been claimed as ecologically significant. In the wet tropics it is doubtful if acidity or base status are often of great importance, since nearly all of the soils are acid and more or less markedly deficient in bases. The differential edaphic factors, as far as our present very imperfect knowledge extends, seem to be most often water-supplying capacity, aeration and soil depth; they thus seem to be mainly physical rather than chemical, though Hardon's (1937) demonstration, that plants growing on the extremely leached lowland tropical podzols of Malaysian padangs (p. 211) have a much lower base content than the same species growing on a more fertile soil, suggests that different degrees of base deficiency may sometimes be an important ecological difference. The exceptional deficiency of tropical podzols in plant nutrients of all kinds (which is much greater even than that of normal rain-forest soils) may be one of the factors responsible for their highly characteristic vegetation. The presence in many, but not in all, tropical soils of large quantities of aluminium (see below) is possibly another chemical edaphic factor of differential value.

Whether calcicole and calcifuge species exist in the flora of wet tropical regions is a question which has been variously answered. Van Steenis (1935a) has discussed it briefly. Climax rain-forest communities characteristic of calcareous soils are unknown because in a humid tropical climate mature or even moderately mature soils never have a high lime content. In some parts of the tropics, e.g. Trinidad and British Honduras, the forest on limestone ridges differs considerably from that on the non-calcareous rocks of the surrounding lowlands, but it is clear that the difference is mainly due to excessive drainage and the thin covering of soil over the porous limestone rather than to any chemical factor. Thus in the moister parts of Trinidad the limestone ridges bear Semi-evergreen Seasonal forest, unlike the Evergreen Seasonal forest of the sand and clay soils in the same climate, but where the rainfall is lower similar Semi-evergreen forest occurs on non-calcareous clay not overlying limestone (Beard, 1946b). Here the water-supplying capacity of the soil is certainly the crucial factor. Charter (1941) has also pointed out that species almost confined to calcareous soils in British Honduras occur in Trinidad in a drier climate on non-calcareous sands and clays.

In the Malay Peninsula, Java and Borneo there are steep-sided limestone crags with a highly characteristic flora. These hills are almost bare of soil, except in hollows, and do not support high forest. Klein (1914) drew attention to certain species which appeared to be peculiar to such hills and regarded these as true calcicoles, i.e. as dependent on the chemical rather than the physical properties of the limestone. Van Steenis (loc. cit.) is however of the opinion that most of these plants are cremnophytes (rock-crevice plants) rather than calcicoles and, according to him, some of them also occur in similar rocky habitats on non-calcareous rocks. The present author's own observations on the flora of limestone rocks and of non-calcareous sandstone cliffs in Sarawak (Borneo) tend to confirm the views of Klein rather than those of van Steenis; the limestone rocks at Bidi and Bau have a highly characteristic flora quite unlike that of the sandstone, including the remarkable *Moultonia singularis* and the fern *Phanerosorus* (*Matonia*) *sarmentosus* which occurs on limestone rocks in several widely separated localities and seems never to occur on non-calcareous rocks. The problem needs a more critical investigation than it has yet received.

If the existence of calcicole species in the wet tropics is still doubtful, it is fairly well established that some rain-forest plants are calcifuge. The avoidance of limestone by *Eusideroxylon zwageri* has already been noted. Though it is sometimes stated that the Dipterocarpaceae as a family avoid limestone (van Steenis, 1935, p. 59), Symington (1943) states that Dipterocarpaceae are in fact absent on the limestone hills in the south of the Malay Peninsula, but probably only because these hills are inaccessible to species of dipterocarps suited to xerophilous conditions; in the north of Malaya, nearer the semi-deciduous and deciduous forests of Burma and Thailand, several dipterocarps are found on the limestone hills and at least one of these (*Hopea ferrea*) seems to show a distinct preference for limestone.

The possibility that the abundance of aluminium in tropical soils acts as a differential ecological factor is raised by Chenery's work (1948 and unpubl.) on aluminium-accumulating plants. Many tropical species, both woody and herbaceous, belonging chiefly to large cosmopolitan or pan-tropical families (Rubiaceae, Melastomaceae, several families of pteridophytes) have been shown to accumulate large quantities of aluminium in their tissues. These plants are abundant in both primary and secondary rain-forest communities and Chenery believes they play a fundamental part in succession and soil-forming processes in the tropics. It is possible that some of the differences between the vegetation of tropical soils rich in aluminium, such as lateritic red earths, and siliceous soils relatively poor in aluminium (e.g. podzols) may be partly due to the differences in the abundance of aluminium.

In spite of some instances of the apparent importance of chemical edaphic factors it seems probable, as has already been said, that physical edaphic factors, especially those affecting the water supply of the plant and the supply of oxygen to its roots, play a larger part in the ecology of tropical vegetation. The occurrence of characteristic plant associations on river margins and land liable to flooding, of others on sites with exceptionally free drainage, the close correlation observed in British Guiana between soil texture and the climax forest communities (Chapter 10), all these are ecological phenomena for which physical edaphic factors must be mainly, though not necessarily exclusively, responsible. The precise nature of these factors awaits further investigation.

A similar view of the relative importance of chemical and physical edaphic factors in the tropics has been put forward by Hardy (1936 b) with reference to the climax forest communities of Trinidad. In the south-central district of the island Hardy studied the distribution of two varieties of Semi-evergreen Seasonal forest and three of Evergreen Seasonal forest. When a map of the vegetation was superimposed on a soil map, a certain coincidence became obvious. A number of individual species were also noted as occurring only on sand, clay, silt, marl or mud-flow soils. Hardy considers that the main factor controlling the distribution of the species and forest communities in this region is the water-supplying power of the soil, which depends on its physical properties as well as on topography and climate. Soil, climate and topography interact in a complex manner, sometimes tending to counteract one another, sometimes acting in the same direction.

'Soil type appears to be significantly important mainly with respect to those physical features that decide its moisture relations. Its chemical properties and attributes appear to exert little or no influence, except in so far as they affect the behaviour of soil water. Among the chemical features, lime and magnesia contents and humus content alone appear to be important in this connection. Other chemical factors (nitrogen supply, phosphate content, potash content, etc.) seem to be quite subsidiary in deciding the broad distribution of vegetation types, although they assume important roles when forest lands are utilized for the growing of commercial crops.' (Hardy, loc. cit. p. 28.)

In the more recent work of Beard (1946 b), the prime importance of the physical edaphic factors is equally strongly emphasized, and the author concludes that it is the moisture relations of the habitat, which are the resultant of the three factors, soil, topography and climate, which chiefly govern the distribution of vegetation in tropical countries such as Trinidad.

In British Honduras there are two main groups of forest communities: those on soils derived from limestones and those on soils derived from non-calcareous rocks such as granodiorites, slates, schists, quartzites and sandstones. Though the two groups of communities are found on soils differing in base content and other chemical properties, Charter (1941) believes their distribution is largely controlled by moisture relationships.

There is some evidence that aeration as well as the water-supplying power of tropical soils is often a potent ecological factor. This is especially likely to be true in the wettest climates and under conditions of topography or soil texture leading to impeded drainage. Chenery & Hardy (1945), in their interesting description of the 'moisture profiles' of certain Trinidad soils, showed that of the soils they examined, only three free-draining sandy soils were adequately aerated down to the full depth of the profile at all seasons of the year. The others were saturated with water at least during the wet season and inadequately aerated except in the surface layers. Different plant species undoubtedly vary very greatly in their tolerance for poorly aerated soils. Coster (1935 b) has suggested that the trees of the Java lowlands have lower soil oxygen requirements than those of the more porous and better drained upland soils. His view that competition for oxygen by the tree-roots is of great ecological importance in tropical forest has been referred to above (p. 190).

Another factor closely bound up with aeration is soil depth. Tropical trees, though broadly speaking shallow rooted, vary in the depth of penetration of their root systems. In Chapter 4 it was shown that there is a close relation between the buttressed habit and the depth of the root system, buttresses developing on superficial lateral roots and being seldom present in trees with a well developed tap-root; it was also shown that the frequency of buttressing in different types of Rain forest shows a definite relation to the effective depth of the soil. Where the soil is actually shallow or, owing to a high water-table, 'effectively' shallow, the root systems tend to be superficial and buttressing is common. Sykes (1930) has described a *Gossweilerodendron-Cylicodiscus* type of Mixed Rain forest in Nigeria which is confined to deep sandy soils; he suggests that the reason for this is that the dominant species need room for the development of their long tap-roots. Though it is possible that soil depth may sometimes act in this purely mechanical way, it must clearly interact with aeration in limiting the thickness of the soil layer which the tree-roots can exploit.

Part III

FLORISTIC COMPOSITION OF CLIMAX COMMUNITIES

Die uebergrosse Mannichfaltigkeit der bluehtenreichen Waldflora verbietet die Frage woraus diese Wälder bestehen.

The excessive diversity of the flora and its richness in flowering plants forbid one to ask 'What is the composition of these forests?'

HUMBOLDT (quoted by Kurz, 1878).

CHAPTER 10

COMPOSITION OF PRIMARY
RAIN FOREST (I)

The most important single characteristic of the Tropical Rain forest is its astonishing wealth of species. Whether we consider a large area such as the Amazon basin or the island of Borneo, or merely a small sample plot one or two hectares in extent, the number of species, especially of trees, is far greater than in any other type of forest community. This great floristic richness is due partly to conditions likely to favour a high rate of speciation, especially the climate, which is favourable to plant growth and reproduction at all seasons, but it is no doubt also largely due to the great age of the large tropical land masses, which has made possible the persistence of vegetation more or less like that of the present day from an unknown but certainly distant geological period.

The exact number of species in the rain-forest flora is not known for any whole country or region, but the following figures give some idea of the magnitudes to be expected. In the 'Outer Provinces' (territories excluding Java) of the Netherlands East Indies, where the forests are predominantly Rain forests, there are, according to Endert (1933), about 3000 species of trees with trunks over 40 cm. in diameter, belonging to some 450 genera. In the Malay Peninsula, a much smaller area, about 2500 species of trees are known (Foxworthy, 1927). There is no recent estimate for the great *hylaea* of the Amazon, but Huber (quoted by Mildbraed, 1930a) supposed that it contained at least 2500 species of 'large trees'; in the State of Pará alone there are more than 1000. These figures are probably underestimates. In the '*forêt dense*' (a term including the 'Mixed Deciduous forest' as well as true evergreen forest) of the Ivory Coast there are some 596 described species of trees reaching 10 cm. diameter or more, belonging to 276 genera and 61 families (Aubréville, 1938). The number of shrubs, herbs, lianes and epiphytes in these areas is unknown, but, though considerable, is doubtless smaller than the number of trees. Van Steenis (1938) estimates the total phanerogamic flora of Malaysia (including the Malay Peninsula and New Guinea) at about 20,000 species. Even the Malay Peninsula alone, most of which was until recent times uniformly covered by climax Rain forest, has not less than 9000 species (Corner). Trinidad, an island 1754 sq. miles (4542 sq.km.) in extent, has 1289 species of flowering plants of which 364 are trees, including palms (Beard, 1946b).

It is not easy to find comparable figures for temperate floras to set beside these, but it is certain that no plant communities, except possibly the South African and Australian sclerophyll scrub, surpass the Tropical Rain forest in wealth of species.

The three great tropical regions are by no means equally rich. The Indo-Malayan Rain forest, smaller in extent than that of tropical America or Africa, is probably richer in species than either. The African forest is floristically relatively poor and uniform. The total number of phanerogams known from the Belgian Congo and Ruanda-Urundi, a vast area which includes much deciduous forest, savanna and mountain vegetation as well as Tropical Rain forest, is only 9705 species, 1631 genera and 170 families (Robyns, 1946). The flora of Borneo, on the other hand, a much smaller and less diversified area, was estimated by Merrill (1915) at not less than 10,000 species. The comparative poverty of Africa is especially marked in certain families usually regarded as particularly characteristic of rain-forest regions. In the Belgian Congo, according to Robyns, there are only eight Lauraceae and thirty species of palms; the Orchidaceae, represented in other parts of the tropics in bewildering variety, here reach only the relatively modest total of 377 species, as compared with an estimated 5000 in Malaysia (van Steenis, 1938). These regional differences in floristic richness are not easy to explain. Mildbraed (1922) ascribes the poverty of the African Rain

TABLE 25. *Number of species of trees in small sample plots of primary Tropical Rain forest*

Locality	Type of forest	4 in. (10 cm.)	8 in. (20 cm.)	12 in. (c. 30 cm.)	16 in. (41 cm.)	Size of sample plot (hectares)	Source of data
		Number of species of trees					
		diameter and over					
SOUTH AMERICA							
Moraballi Creek, Brit. Guiana	Mixed Rain forest association	91	55	39	32	1·5	Davis and Richards (1933–4)
,, ,,	Morabukea forest (*Mora gonggrijpii* consociation)	71	45	29	21	1·5	,,
,. ,,	Wallaba forest (*Eperua falcata* consociation)	74	49	30	16	1·5	,,
Massa Mé Forest Reserve, Ivory Coast	Mixed Rain forest association	74	58	35	c. 23	c. 1·4	Aubréville (1938)
AFRICA							
Okomu Forest Reserve, Nigeria	Mixed Rain forest association	c. 43	—	21	—	0·5	Richards (1939)
		c. 60	—	32	—	1·0	
		70	51	40	31	1·5	
Fernando Po	Mixed Rain forest association	—	—	15	—	1·0	Mildbraed (1933 *b*)
Cameroons	Mixed Rain forest association	—	—	34	—	0·5	Jentsch (1911)
Mauritius	'Upland Climax forest'	52	33	18	11	1·0*	Vaughan & Wiehe (1941)
ASIA							
Mt Dulit, Sarawak	Mixed Dipterocarp association	—	98	—	32	1·5	Richards (1936)
Marudi Forest Reserve, Sarawak	Heath forest	—	56	—	12	1·5	Richards (1936)
Mt Maquiling, Philippines	Mixed Dipterocarp association (*Parashorea-Diplodiscus* association)	43	27	15	11	0·25	Brown (1919)

* Sum of ten small plots.

forest to past climatic changes; whatever the true explanation, it can hardly be due to any ecological factor operating at the present day. The Rain forest of remote islands, such as Mauritius and the Fiji group, doubtless owing to long isolation and barriers to immigration, is also floristically poor.

Some figures for the approximate number of species on small areas of Rain forest (sample plots) are given in Table 25. These figures will be discussed in the next chapter.

Plant formations are normally built up of distinct sociological units, consociations, characterized by single dominant species, and more homogeneous mixtures, associations, in which no one dominant forms a clear majority of the whole population. It has already been stated that the Rain forest also consists of consociations and associations, but, owing to the large number of species and the difficulty of separating many of them in the field, these sociological units are seldom very readily distinguishable. Until quite recently the Tropical Rain forest was regarded as a vast mixture in which little or no order could be discerned. Since no distinct associations or consociations were recognized, it was tacitly supposed that the Rain forest in, for instance, South America, formed a single mixed association fluctuating in composition from place to place. The progress of taxonomy and the application of quantitative plant sociological methods in the last 20 years has now placed the recognition of rain-forest associations and consociations on a firm basis.

METHODS OF STUDY

The methods of studying the floristic composition of tropical forests are necessarily very different from those used for temperate plant communities. They must usually be less exact, but it is of great importance that they should as far as possible be quantitative and should involve the counting of individuals on sampling units (plots or belts) of known area. The great majority of the species in a rain-forest community are tall trees, the flowers and fruits of which are overhead and far out of sight. A pair of strong binoculars is always useful for examining the trees, but often a sweet scent, the buzzing of insects or fallen corollas on the ground are the only indications that a tree is in flower. The only reliable methods of obtaining specimens for identification are felling and climbing, though shooting down branches with a gun and even specially trained monkeys have sometimes been used. Owing to the relatively non-seasonal climate, flowering is spread over a large part of the year. As was seen in Chapter 8 some trees flower less than once a year and a few species not more than once in 10 years. To collect herbarium material of all the species even in a small tract of forest is thus at best a time-consuming, and often an impossible, task. Usually, therefore, we have to be content with exact identifications for only a proportion of the species present, though a rough estimate can generally be made of the number and proportions of the remaining unidentified species.

For the day-to-day recognition of the species it is necessary to use non-floral characters extensively, especially those of the bark and trunk which are visible from the ground. Those who know only the meagre tree flora of the temperate zone may find it surprising that these non-floral characters are sufficiently varied to be of much value in dealing with a tree flora running into thousands of species. Yet the trunk and bark of tropical trees provide a great variety of useful diagnostic characters. The form of the trunk, whether straight or curved, smooth or fluted, the buttresses when present, the colour and texture of the bark, the arrangement of the lenticels, the colour, odour and appearance of the blaze, the macroscopic appearance of the wood—all these provide characters which tend to occur in highly specific combinations and together they enable the practised eye to determine a large majority of the tree species, even in a rich flora. With their help keys can be constructed to the tree flora of whole regions as has been done, for example, by Foxworthy (1927) in his book on the timber trees of the Malay Peninsula.[1] It should be noted that while trunk and bark characters usually run parallel with those normally used by taxonomists, they do not always do so; thus occasionally species of the same genus are very different in trunk, bark and blaze, while unrelated species may be much alike. The appreciation of these inconspicuous and often subtle characters requires a sharp eye and considerable experience. It cannot be learned in a few weeks or months and in practice few Europeans ever become as expert as natives who spend their whole lives in the forest. For them distinguishing between the hundreds of species of trees is a part of daily experience and most of the species have vernacular names.

Among the Arawak Indians of British Guiana, for example, several hundred vernacular names of trees are in common use and, for the most part, each name represents only a single taxonomic species. Some vernacular names include several species, e.g. 'baromalli' = two related species of *Catostemma*, 'warakusa' = *Inga* spp. (excluding a few outstandingly distinct species). Only a few names include members of different genera or unrelated families, e.g. 'kokeritiballi' = several genera of Sapotaceae, 'kulashiri' = various Meliaceae, Flacourtiaceae and Sapindaceae. In some countries certain trees have no native names or are grouped together under vague collective terms, e.g. 'medang' in Malaya. Most trees have names because they are used as timbers, medicines, poisons, etc., or because they have some superstitious interest.

It should be emphasized that while any forest-dwelling native may know the names of some trees, only a few experts will know the names of a large number. In tropical countries with well-organized forestry services some of these native experts have been trained to work with European foresters. By constant practice they acquire a knowledge of the tree species of their district which is astonishing in its extent and accuracy. In naming a tree they seldom hesitate unless they

[1] Reference may also be made to Endert's work on the trees of the Netherlands East Indies (1928) in which extensive use is made of non-floral characters.

meet a species new to them. Expert native 'tree-finders' exist in British Guiana, Nigeria, Malaya, in the Ivory Coast and doubtless in most other tropical countries.

When the trees of a sample area have been enumerated by means of their vernacular names it will be necessary to translate as many of these names as possible into their taxonomic equivalents. This is a laborious process involving much careful checking and comparison. The use of each name must be repeatedly checked in the field to find out to what species or group of species it corresponds. As many collections as possible of herbarium material of the same 'vernacular species' should be made for subsequent study by specialists. It should be remembered that the same native name may be used for different species in different districts, so that an equivalent which has been carefully checked in one area cannot be used without further investigation in another. When a native name is applied to a group of similar species, it may happen that in a given district, or in a particular sample plot, only one of these species is present; within the plot or limited district the name can therefore be safely treated as the equivalent of one species. Approximate information should not be despised; it is often useful to know that a certain genus is present, even if it is impossible to decide which species or how many.

When listing the species on a sample plot or transect both the vernacular names and their botanical equivalents should usually be given, even when the equivalent of every name has been worked out. The evidence for the correctness of the equivalents will probably vary in completeness. Thus if herbarium material of a certain 'vernacular species' has been collected only once, the evidence that the name in question is the equivalent of a particular taxonomic species is less complete than if ten collections have been made. An equivalent based on material collected actually on the plot or transect under consideration is more reliable than when it is based on material collected in another locality in the same district or in another district. A scheme of conventions to indicate the 'degree of reliability' of the names used may be adopted (e.g. Richards, 1939, p. 53 and tables 6–9). Whenever an identification is based on a collected specimen a serial number should be quoted and the herbarium where the specimen is deposited should be indicated.

Systematists accustomed only to working with temperate floras will perhaps doubt the value of methods of studying floristic composition which involve the use of native 'tree-finders' and vernacular names; yet such methods can give results of great scientific value provided sufficient precautions are taken. In any ecological investigation involving the listing of species which cannot be immediately identified in the field there are two stages at which errors are liable to arise; one is in the determination of the specimens collected and the other in deciding which individuals are in fact specifically identical with the specimens collected. The possible error at the second stage depends on the taxonomic ability and experience of the field worker and can never be entirely eliminated.

When a native 'tree-finder' is employed the possibilities of error are necessarily somewhat increased but, if care is taken, the lists of species, though inevitably less complete and accurate than for a well investigated temperate flora, may have a high degree of reliability.

The problems involved in choosing the shape and size of the sampling unit are not essentially different in tropical forests from those in other plant communities and need only be mentioned briefly here. In forestry surveys in the tropics a method of sampling by means of belt transects has usually been adopted. Since, however, most primary Rain forests are interspersed with patches of secondary forest (on sites of old native fields, wind-falls, etc.) and intersected with numerous stream valleys, in which the floristic composition is different from that on the intervening high ground, compact sample plots are often preferable for ecological investigations. In the Rain forest, as in other plant communities, the number of species at first increases with increasing area almost linearly and then falls off, the curve of species/area having an approximately logarithmic form (Fig. 34; see also curves of Eggeling (1947) for forest communities in Uganda and of Vaughan & Wiehe (1941) for the 'Upland Climax forest' of Mauritius). Theoretically the ideal sample plot should not be smaller than the area at which the number of species first attains its maximum value; in practice the size chosen must be a compromise between the theoretical requirements and the time and facilities available. In the author's experience square plots 400 × 400 ft. (area c. 1·5 ha.) give satisfactory results in most types of Rain forest. For rapid reconnaissance a convenient method is to enumerate all the trees on both sides of a path for a distance of 0·5–1 km. (Aubréville, 1938, p. 208).

A study of the floristic composition of a forest community should embrace not only all the tree strata (which should if possible be listed separately), but the shrub and herb layers and the lianes and epiphytes as well. For the study of the shrub and herb layers no special methods are necessary. In both layers the majority of the individuals are young trees and there may be considerable difficulty in distinguishing between young trees and shrubs not in flower. For many purposes it is sufficient to make lists of the species in the lower layers, collecting specimens of those which cannot be identified in the field and estimating the frequency by eye, using the usual symbols (a., l.f., v.r., etc.). More precise results can be obtained by the use of small quadrats, but because both the shrub and herb layers are usually sparse the quadrats must generally be larger than 1 sq.m. On the quadrats the individuals may be counted or the frequency, percentage cover, etc. of each species may be estimated according to the methods of Continental ecologists. If the lower layers are merely listed, the list can be made over the whole of the plot used for the trees; if quadrats are used they should be either spaced at random within the large plot or transect or placed at arbitrarily chosen regular intervals. Sometimes the lower layers of a forest community are less uniform in composition than the tree layers (light and shady patches, swampy and drier areas, etc.); in this case the various types of undergrowth should be studied separately.

The synusiae of lianes and epiphytes offer special difficulties. Lianes can be studied satisfactorily only if a plot is clear-felled and even then they often form a tangled mass in which it is very hard to determine the number of individuals

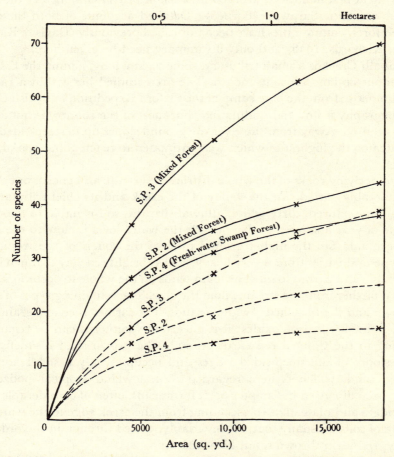

Fig. 34. Species/area curves for rain-forest associations in Nigeria. From Richards (1939, fig. 7). S.P. 2–4 refer to three sample plots. Continuous lines for trees 4 in. (10 cm.) diameter and over, broken lines for trees 12 in. (30 cm.) diameter and over.

of each species. On a clear-felled plot the epiphytes can also be listed and, if necessary, counted; at any one time only a small proportion of the species are usually in flower. In some open types of forest it is sometimes possible to recognize the epiphytes and estimate their frequency from the ground, using binoculars (see p. 119).

PRIMARY FOREST COMMUNITIES IN BRITISH GUIANA

Before discussing the composition of rain-forest associations and consociations in general terms (Chapter 11) it will be useful to describe the primary forest communities in two limited areas which the author has had the opportunity of studying at first hand, viz. Moraballi Creek in British Guiana and the Baram basin in northern Sarawak (Borneo). Detailed accounts of both these areas and their forest communities have been published previously (Davis & Richards, 1933–4; Richards, 1936) and only a summary need be given here.

Moraballi Creek is a small tributary, some 24 km. long, joining the Essequibo River about 80 km. above its mouth. The area studied lies within a radius of a few kilometres from the base camp of the Oxford Expedition to British Guiana. The topography is low and undulating; but most of the country is not actually flat or swampy. Away from the river the ground slopes up to steep-sided ridges and plateaux, the highest of which were estimated to reach about 100 m. above sea-level.

The underlying rocks of the whole district are granite and gneiss, but the soils are remarkably varied. In the valley of the creek and its chief tributaries the soil is a fine-textured, dirty white, alluvial silt; this soil is liable to flooding at any time of year and even in dry periods the water-level is close to the surface of the ground. On the higher ground most of the soil is of the tropical red earth type and varies from a very heavy, sticky, bright red clay to much lighter, more friable and porous loam. The loam tends on the whole to occur at a higher level and farther from the streams than the clay. The remaining types of soil are the brown and the bleached ('white') sands. The former is coarse-grained and porous, stained with iron oxides; it is uncertain whether it can be regarded as belonging to the tropical red earth group.[1] The brown sand is chiefly found on the slopes of the steep-sided ridges and occupies only a small area. The bleached sand, which is here regarded as a lowland tropical podzol, has already been discussed in Chapter 9. It forms caps, often of considerable depth, to the ridges and plateaux. In ascending from the creek edge to the watersheds a catena of soils is usually, but not invariably, passed through in the order: silt, red clay, red loam, brown sand, bleached sand.

The climate of the whole district is, as far as is known, uniform and typically equatorial. The annual rainfall is over 250 cm. There are two maxima of rainfall in the year and two dry seasons, but no month has an average of less than 10 cm.[2].

Rain forest, much of it primary, or only slightly affected by felling, covers the whole district and extends without interruption north-west to join that of the

---

[1] Dr E. M. Chenery suggests that the brown sand may be derived from bleached sand which has become secondarily infiltrated with water carrying iron compounds washed out of the sand above, i.e. that it represents a kind of truncated B horizon of the podzol capping the ridges.

[2] Meteorological data for Mazaruni Station (27 km. distant); see Tables 10–12 and Fig. 22.

Orinoco basin and eastwards through Surinam and French Guiana to join the vast forest of the Amazon. The northern edge of the Guiana forest lies some 50 km. north of Moraballi Creek.

In the primary forest five main communities can be distinguished. Though some of them have minor variants or facies, they are on the whole constant in composition and well characterized. The boundaries between the different communities are sometimes very abrupt (see p. 221), but more often gradual. The five communities, in the order they are usually met with from the creek to the ridges, are the Mora consociation (dominant *Mora excelsa*), the Morabukea consociation (dominant *Mora gonggrijpii*), the Mixed forest association (many dominants), the Greenheart consociation (dominant *Ocotea rodiaei*) and the Wallaba consociation (dominant *Eperua falcata*). Some of these communities are known to have a wide distribution in tropical America, others are more local. The Mora community has a wide distribution in the coastal region of British Guiana; it extends inland to the Pakaraima Mountains (where it occurs as fringing forest in the savannas (Myers, 1936)), eastwards into Surinam and north-west to the Orinoco delta, reappearing in a somewhat modified form in Trinidad (Beard, 1946a). Morabukea forest is known in various parts of British Guiana and in Surinam. Greenheart as a species occurs only in British Guiana and the neighbouring parts of Surinam and Venezuela; it is likely that it forms the dominant of consociations only near the centre of its range. Wallaba forest is probably widely distributed in north-eastern South America. The caatinga forests on bleached sand in the Rio Negro region of the Amazon (Spruce, 1908, **2**, pp. 206–7 and 303–4) appear to be very similar to this community. (The caatingas of the Rio Negro are not to be confused with the south Brazilian caatingas which are deciduous forests.)

The quantitative data in the following descriptions are based on single typical sample plots of each community, each 400 × 400 ft. (*c*. 1·5 ha.).

Mora forest. This occupies the lowest ground, forming broad strips on the silt of the flood plain on either side of the creek and its main tributaries. It is also found, but less commonly, on steep rocky slopes with an impermeable and very shallow clay soil. The common feature of the two habitats seems to be shallowness of soil, in one case due to the high water-table, in the other to erosion which prevents a deep soil from accumulating.

The structure of the Mora forest was not examined in detail, but it is probably similar to that of the Trinidad Mora forest described by Beard (see Fig. 7*b* and p. 34), except that the upper tree stories are more open and the under-growth consequently lighter (average illumination at breast-height 1·33 %; see p. 182).

The A tree story is over 30 m. high and is composed chiefly of the dominant species; below this there are mixed B and C stories, a shrub (D) story and a herb layer which is vaguely separable into strata of tall and dwarf herbs respectively. The luxuriance of the tall herbs, such as *Carludovica* sp. and various Scitamineae is characteristic. *Mora excelsa* forms over half the total stand in the larger diameter classes and a considerable percentage in the smaller classes, as the following figures show:

Diameter class	Percentage of *Mora excelsa*	Diameter class	Percentage of *Mora excelsa*
4–7 in. (10–19 cm.)	13·6	All trees 4 in. (10 cm.) and over	23·4
8–11 in. (20–29 cm.)	13·5		
12–15 in. (30–40 cm.)	35·4		
16–23 in. (41–60 cm.)	54·6	All trees 16 in. (41 cm.) and over	67·2
24 in. (61 cm.) and over	91·3		

Mora excelsa is in fact the most abundant species in every diameter class except the 8–11 in.; seedling moras are also abundant and on the plot there were 2·7 seedlings under 3 ft. high, and 2·0 3–6 ft. high per square yard, on the average in the drier parts, though in the wetter parts there are fewer. The commonest trees after mora are *Pterocarpus officinalis* and *Pentaclethra macroloba*, both B story species. Each formed 10·2% of the stand 4 in. diameter and over. Altogether there were about sixty species of trees reaching 4 in. diameter on the plot. The greater number of these, including mora itself and *Pentaclethra*, are found in most of the other four primary forest communities. *Pterocarpus* and a few other species appear to be peculiar to the Mora consociation.

Morabukea forest. This community, dominated by a related species of *Mora*, has a less well-defined habitat. Characteristically it occurs on the slopes of the lower flat-topped hills, sometimes extending over their summits; it avoids the flood-plains and prefers heavy red clay soils, though in some situations, for instance on low-lying ground at the foot of slopes, it grows on loamy or almost sandy soil.

Little is known of the structure of this type of forest, but there are probably three tree stories. Owing to the density of the A and B stories, illumination in the undergrowth is very poor (average 0·61 %); this is the darkest of the five types of primary forest. Characteristic features are the dense thickets of seedlings of the dominant species, the comparatively thick litter of dead leaves, the abundance of small saprophytes and the scarcity of all other ground herbs.

Mora gonggrijpii is the most abundant species in almost every diameter class, but in the larger classes its dominance is slightly less complete than that of *M. excelsa* in the Mora consociation. The percentages in the various classes are as follows:

Diameter class	Percentage of *Mora gonggrijpii*	Diameter class	Percentage of *Mora gonggrijpii*
4–7 in. (10–19 cm.)	17·6	All trees 4 in. (10 cm.) and over	25·9
8–11 in. (20–29 cm.)	16·5		
12–15 in. (30–40 cm.)	19·6		
16–23 in. (41–60 cm.)	61·3	All trees 16 in. (41 cm.) and over	60·7
24 in. (61 cm.) and over	59·3		

After the dominant the most abundant species are *Eschweilera sagotiana* (7·4% of trees 4 in. and over) and *Catostemma commune* (5·0%). Both these belong to the A·story. No other species reaches 5%. The total number of species on the plot is seventy-one (4 in. and over). Few, if any, of these are peculiar to Morabukea forest; the list of species as a whole closely resembles that for the Mixed forest, the main differences being the smaller number of species and the high percentage of *Mora gonggrijpii*.

Mixed forest. The Mixed forest association, which has no single dominant, is characteristic (at Moraballi Creek) of rather sticky loamy soils, intermediate between the heavy clays on which the Morabukea consociation is found and the brown sands of the Greenheart forest. It covers much of the lower hilly land of the district and occurs on hilltops when they are not capped with bleached sand.

The structure has already been described in detail (p. 25, Fig. 4). A noteworthy feature is that the number of trees per unit area is greater than in either of the two preceding communities (*Mora* consociation 310 (4 in. and over) per hectare, Morabukea consociation 309, Mixed association 432). The illumination in the undergrowth is poor (0·67%), but somewhat better than in the Morabukea forest; because of this, perhaps, herbaceous undergrowth is much more abundant.

The most characteristic feature of the floristic composition is the large number of dominants. In the class of trees 16 in. (41 cm.) diameter and over the most abundant species is *Eschweilera sagotiana*, which forms 15·6% of the stand. In the 4 in. and over class however this species forms only 6·1%, while *Pentaclethra macroloba* (B story) forms 11·3%. Five species each form over 5% in this class. The composition of the Mixed association probably fluctuates in different parts of the Moraballi Creek district and *Eschweilera* and *Pentaclethra* are probably not always the most abundant species. The total number of species of trees on the sample plot was ninety-one, the highest in any of the communities except the Greenheart consociation. Included in this number is the majority of the species in the other types of forest except some Mora and Wallaba forest species and possibly a very few confined to the Morabukea and Greenheart forest.

Greenheart forest. As has been said, this community is limited in the Moraballi Creek district to a single type of soil, a light reddish-brown sand, occurring chiefly on the sides of ridges the tops of which are covered with bleached sand bearing Wallaba forest. Where Greenheart and Wallaba forest adjoin, the boundary may be very sharp (see p. 221).

The structure of Greenheart forest has not been specially studied, but is probably very like that of the Mixed forest, though trees of large dimensions are somewhat more frequent. The number of trees 4 in. diameter and over per hectare is 519, i.e. considerably higher than in the Mixed forest, but much lower than in the Mora and Wallaba forest. The herb layer is well developed and has a rather distinctive composition.

Ocotea rodiaei is easily the most abundant species in the larger diameter classes, but in the smaller classes it forms only a small proportion of the stand, as the figures show:

Diameter class	Percentage of *Ocotea rodiaei*	Diameter class	Percentage of *Ocotea rodiaei*
4–7 in. (10–19 cm.)	0·5	All trees 4 in. (10 cm.) and over	9·4
8–11 in. (20–29 cm.)	3·3		
12–15 in. (30–40 cm.)	1·1		
16–23 in. (41–60 cm.)	38·4	All trees 16 in. (41 cm.) and over	43·4
24 in. (61 cm.) and over	60·0		

The most abundant subordinate species are *Pentaclethra macroloba* (9·1 % of trees 4 in. and over), *Eschweilera sagotiana* (7·6%) and *Licania venosa* 5·6%. All three of these are among the most abundant species of the Mixed forest. Greenheart forest in fact resembles the Mixed association so closely in composition that it could be regarded as a special facies or 'lociation' of it rather than a distinct consociation. The total number of tree species on the plot is ninety-five, the highest number of all the communities, though perhaps not significantly greater than in the Mixed forest. An extensive investigation might show that a few of these are peculiar to the Greenheart forest.

Wallaba forest. The last of the five forest types is the consociation of *Eperua falcata*, a tree which, like the two moras, belongs to the Leguminosae. In structure, floristic composition and habitat Wallaba forest is very distinctive and differs more from the other four communities than they differ from one another. It is strictly limited to the bleached sand soils (podzols) and stops sharply at the boundary of this soil type. At Moraballi Creek Wallaba forest was the only type of primary forest occurring on bleached sand, but elsewhere in the Guiana Rain forest other communities occur on similar soils (see pp. 255–6).

The structure of Wallaba forest, in which the B story is almost absent, has been already described in Chapter 2 (Fig. 8 b). A peculiar feature is the extremely large number of trees per unit area (617 per hectare 4 in. diameter and over), nearly twice as many as in the Mora forest. The shrub layer is remarkably dense. Herbaceous plants are few in species and individuals, but highly characteristic, a proportion of the species being confined to this type of forest. The poor development of the herb layer is not due to lack of light; illumination in the undergrowth is greater than in any of the other forest types (1·43 %). Other characteristic features of the Wallaba forest are the scarcity of buttressed trees (Fig. 11 and p. 66) and of large lianes. Some species of the latter are probably characteristic of this consociation.

Diameter class	Percentage of *Eperua falcata*	Diameter class	Percentage of *Eperua falcata*
4–7 in. (10–19 cm.)	10·0	All trees 4 in. (10 cm.) and over	21·1
8–11 in. (20–29 cm.)	17·3		
12–15 in. (30–40 cm.)	32·7		
16–23 in. (41–60 cm.)	66·7	All trees 16 in. (41 cm.) and over	67·0
24 in. (61 cm.) and over	75·0		

Eperua falcata is the most abundant species in every diameter class, except the smallest, and in the larger classes it is overwhelmingly dominant. *Eperua falcata* regenerates abundantly; six random counts on the sample plot gave an average of 1·8 seedlings under 2 m. high per sq.m.

After the dominant the next most abundant tree species are *Catostemma fragrans* and *Licania buxifolia*. Both form less than 5 % of trees 16 in. diameter and over, but 15·1 and 12·2 % respectively of trees 4 in. and over; they thus form a considerable proportion of the whole stand. In this type of forest there is a tendency to 'gregariousness' (or a high ratio of individuals to species) which expresses itself in the great abundance of certain subsidiary species, as well as in the presence of a single dominant (Table 29). There were seventy-four species of trees 4 in. and over on the plot, a much smaller number than in the Mixed and Greenheart forest, but somewhat more than in the Mora and Morabukea forest. About three-quarters of the species, among them *Eperua falcata* itself, are found in the other forest types. The remainder, including *Catostemma fragrans* and *Licania buxifolia*, are apparently confined to Wallaba forest, at least in the Moraballi Creek district. The *Eperua* consociation thus has a very distinctive floristic composition. Not only are certain species nearly or quite confined to it, but those species which occur in other types are found here in very different relative proportions. Further, it is of interest that there are shrubs, herbs and probably lianes peculiar to the Wallaba forest.

The chief characteristics of the five primary forest communities at Moraballi Creek are summarized in Table 26. From a comparison of their floristic composition a number of facts emerge. A large number of species are common to all five communities (see Davis & Richards, 1933–4, tables II–VI), but their relative proportions are different in each. Some species are probably confined to each of the five communities, but since only one plot of each was listed it would be misleading to attempt to enumerate these species. The differences in composition between the different communities are not equally great; the Morabukea, Mixed and Greenheart forests are much more similar to each other than to the Mora and Wallaba forests.

The climate of all five types is alike and all are primary communities which have probably been little affected by human activities. It is evident that the composition of each type is determined by the characteristics of the soil, or by a combination of soil and topography. The correspondence between the boundaries between the communities and the soil boundaries is often strikingly exact. The degree of similarity or difference between the soils reflects very closely the degree of difference in floristic composition; the Morabukea, Mixed and Greenheart soils have much in common, but the Mora and Wallaba soils are very different both from each other and from all the rest.

It is also a striking fact that the five types form a regular series, so that when arranged in the order in which they normally occur—Mora, Morabukea, Mixed, Greenheart, Wallaba—their characteristics vary according to a definite plan.

Some increase or decrease more or less steadily through the series, e.g. the number of trees per unit area increases steadily from a minimum in Mora to a maximum in Wallaba. Other characteristics reach their maximum or minimum towards the middle of the series and fall or rise respectively towards the two ends. Thus the percentage of the most abundant tree species is greatest at the two ends of the series; it falls regularly towards the middle and reaches a minimum in the Mixed type with no single dominant. The number of species per plot, on the other hand, is greatest in the middle and falls towards the two ends. The degree of dominance of single species and the total number of species are related

TABLE 26. *Forest types at Moraballi Creek, British Guiana*

Forest Type	Mora	Morabukea	Mixed		Greenheart	Wallaba
Habitat:						
Texture of soil	Fine silt	Heavy silt	Light loam		Sand	Light sand
Index of texture (Hardy) (lower sample)	40	44	18		5	0
Illumination at breast height (%)	1·33	0·61	0·67		0·81	1·43
Floristic composition:						
Dominant or most abundant species	*Mora excelsa*	*Mora gonggrijpii*	*Eschweilera sagotiana*	*Pentaclethra macroloba*	*Ocotea rodiaei*	*Eperua falcata*
As percentage of trees 4 in. (10 cm.) diam. and over	23·4	26·6	6·1	11·3	9·4	21·1
As percentage of trees 16 in. (41 cm.) diam. and over	67·2	60·7	15·6	6·7	43·4	67·0
No. of tree species 4 in. (10 cm.) diam. and over (approx.)	60	71	91		95	74
No. of tree species 16 in. (41 cm.) diam. and over (approx.)		21	32		33	16
Other features:						
No. of trees 4 in (10 cm.) diam. and over (per hectare)	310	309	432		519	617
No. of trees 16 in. (41 cm.) diam. and over (per hectare)	45	60	60		87	67

and vary inversely (see p. 260). Both characteristics seem to depend ultimately on the nature of the soil. The Mora and Wallaba soils, it seems reasonable to assume, are relatively unfavourable to plant growth, the one because of shallowness and poor aeration, the other because it is excessively porous and probably, like other podzols, strongly leached and deficient in plant nutrients. The Mixed forest soil, on the other hand, is probably the optimum soil of the series; that this is so is suggested by the fact that the Indians when choosing a site for shifting cultivation prefer sites in Mixed forest. Attempts at agriculture on Wallaba soils have completely failed (Milne, 1940). The Morabukea and Greenheart soils are probably intermediate. It thus appears that the most 'mixed' type of forest is found on the optimum soil and that single-species dominance depends on un-favourable soil characteristics of one kind or another. This principle appears to be of wide application and extends to the composition of Rain forest generally. It seems as if on the optimum soil a great variety of species, some tolerant and some less tolerant of varying soil conditions, flourish together, but on less favourable soils a selective effect excludes the more exigent species and gives a competitive advantage to those which are relatively tolerant.

PRIMARY FOREST COMMUNITIES IN SARAWAK

The two types of primary lowland Rain forest studied by the author at Mt Dulit and Marudi in the Baram basin of northern Sarawak are instructive when compared with the five primary types in British Guiana. The climate here is similar in essentials to the climate of Moraballi Creek, but the flora itself is of course entirely different and there are no species common to the two areas.

The two forest types, known as the Mixed forest and the Heath forest respectively, were studied on the slopes and foothills of Mt Dulit, an escarpment some 1100–1400 m. high, lying to the south-west of the Tinjar River, a tributary of the Baram. Heath forest was also examined at Marudi (Claudetown), situated on the Baram about 80 km. from Mt Dulit in low-lying country.

Except where the natural vegetation has been destroyed by cultivation, almost all the lowlands of Borneo are covered by Rain forest. In the Mt Dulit district shifting cultivation has destroyed most of the primary forest on level ground, but on the steeper hill slopes there is still virgin forest. Four types of primary forest were recognized: (i) lowland Mixed Rain forest, (ii) Heath forest, (iii) Submontane Rain forest, and (iv) Montane or 'Mossy' forest. (iii) and (iv) occur only above about 450 m. and are dealt with in Chapter 16. The Mixed forest and Heath forest are both lowland types and occur under similar climatic conditions, the differentiating factor apparently being soil. Mixed forest is found on sticky red or yellow loams, soils of the tropical red or yellow earth type, and is the prevailing type of forest at low altitudes while Heath forest occurs on coarse bleached sands, similar in their characteristics to the Wallaba soil of British Guiana, and also belonging to the lowland tropical podzol type. Where the two types of soil are divided by a sharp boundary, there is also a sudden transition between the two types of forest. Heath forest was met with only locally on the dip-slope of Mt Dulit from about 750–1100 m. and at Marudi at about sea-level.

Both the sticky yellow loam and the bleached sand are alike derived from Miocene sandstone and it is not clear what has directed the course of soil development, in some places towards laterization, in others towards podzolization. Possibly different strata of the sandstone differ slightly in texture and give the soil, as it were, a bias in one direction or the other.

The Mixed forest and the Heath forest differ markedly in physiognomy and structure and also in floristic composition, though, owing to the richness of the flora and the little previous work on it, the composition could be compared only in very general terms.

Mixed forest. The prevailing forest type of the Mt Dulit region is typical of the primary forest of the Malayan region. The structure has been fully described in Chapter 2. The outstanding features of its floristic composition are the extremely large number of tree species present and the small proportion of the stand formed by any one of them. It is thus an extreme example of a mixed forest or rain-forest association. On the sample plot the number of species 8 in. (20 cm.) diameter and over[1] was estimated at over ninety-eight and there are at least thirty-two species 16 in. (41 cm.) diameter and over. The larger diameter-classes are therefore about as rich floristically as the Guiana Mixed forest, but the smaller classes are much richer. Out of 261 trees 8 in. diameter and over not more than twelve (4·5 %) belonged to the same species. The most abundant tree on the plot belonged to the B story, all the individuals being under 16 in. diameter. In the class 16 in. and over the two most abundant species (both unidentified) were each represented by six individuals out of a total of sixty-two (9·7%), but these formed as little as 3·4 and 2·7% respectively of the whole stand 8 in. diameter and over. The Mt Dulit Mixed forest is thus even further from the dominance of a single species than the Mixed forest of Guiana. It is noteworthy that, though no single species formed more than a very small proportion of the whole stand, one family, the Dipterocarpaceae, forms at least 17% of trees 8 in. diameter and over and at least 44% of those 16 in. and over. There is thus a 'family dominance' of the Dipterocarpaceae in the upper stories and this forest, like most of the lowland forests of Indo-Malaya, may be described as a mixed Dipterocarp forest. 'Family dominance' seems to be a rather common feature of tropical forests, both with and without single dominant species (p. 254).

Heath forest. Heath forest (*Heidewald*) was the name given by Winkler (1914) to a characteristic type of 'subxerophilous primary forest' met with on sandy soils in south-east Borneo. The climax community found on bleached sand in Sarawak agrees with Winkler's description both in the species of which it is composed and in many physiognomic features and the same name may be adopted for it. Winkler suggested the term Heath forest because the community, like European heaths, was associated with a sandy soil and because many of its components had a xeromorphic facies (small sclerophyllous foliage, etc.). From the nature of the climate it is unlikely that the xeromorphy of Heath forest (which is entirely evergreen) is related to seasonal drought; it seems rather to be of the same nature as the imperfectly understood 'physiological xeromorphy' of temperate heath and bog plants growing on a mineral-deficient, very acid soil.

---

[1] Owing to the difficulties of investigation it was not possible to include trees down to 4 in. diameter in the enumeration as at Moraballi Creek.

The Heath forest of Mt Dulit and Marudi differs strikingly from the Mixed forest in its denser, small, woody undergrowth (D–E strata), better illumination in the lower levels and in the smaller amount of buttressing of the trees. The number of trees per unit area is greater. The stratification was not investigated, but it is possible that there are only two well-developed tree stories instead of three as in the Mixed forest. Lianes are few and small. Ground herbs are very few and both the shrub and herb layers include species which appear to be rare or absent in the Mixed forest.

A sample plot was listed in the Koyan valley (Mt Dulit) and at Marudi. The two samples were somewhat different in composition, as might be expected from the difference of altitude and locality, but agreed in their main features. Both plots were less rich in species than the Mixed forest; on the Koyan plot there were about fifty-five tree species of 8 in. diameter and over, on the Marudi plot fifty-six, as against about ninety-eight on the Mixed plot. More important than this is the fact that both plots show a *tendency* to single-species dominance, one species forming a much larger proportion of the stand than any species in the Mixed forest, though a smaller proportion than the single dominants of the Guiana consociations. On the Koyan plot the conifer *Agathis borneensis* formed 15·2 % of trees 8 in. and over and 35·2 % of those 16 in. and over. On the Marudi plot *Agathis* was also plentiful, but the most abundant species was an unidentified dipterocarp which formed 12·0 % of trees 8 in. and over and 35·9 % of those 16 in. and over. Both plots are therefore samples of mixed associations rather than consociations, but ones in which a single species forms a substantial proportion of the stand. As in the Guiana consociations the tendency to 'gregariousness' is shared by more than one species, thus the second most abundant species on the Koyan plot forms 12·4 % of trees 8 in. and over and 30·8 % of trees 16 in. and over (Table 29). Similarly at Marudi the second most abundant species (a dipterocarp) formed 11·4 % of trees 8 in. and over. There is a marked family dominance of the dipterocarps on the Marudi plot (40·5 % of trees 8 in. and over, 51·3 % of those 16 in. and over); on the Koyan plot dipterocarps are much less plentiful, no doubt because of the higher altitude (the Dipterocarpaceae are characteristically a lowland family). Enough species were identified in the two Heath forest plots to indicate that the flora in general is very different from that of the Mixed forest and probably includes a large proportion of peculiar species, among the lianes as well as the trees, shrubs and herbs. The occurrence of the pitcher plants *Nepenthes*, for example, is characteristic of Heath forest and they were never seen in the Mt Dulit Mixed forest. *Agathis* and several other species typical of the Sarawak Heath forest are also abundant in the Heath forest of South and South-east Borneo (Winkler, 1914; Diels & Hackenberg, 1926), distant some 700 km. or more.

The comparison of the Heath forest and Mixed forest shows that in Borneo, as in Guiana, the forest on bleached sand (podzol) is poorer in species and less mixed in composition than the corresponding mixed type on tropical red earth. In both Borneo and Guiana the forest on the relatively unfavourable soil shows a tendency to single-species dominance, but in Borneo, probably owing to the greater floristic richness of the region as a whole, this tendency is much less pronounced. In Borneo, as in Guiana, the bleached sand can be regarded as having a selective action, the more tolerant Mixed forest species occurring on it together with other species which are never found in Mixed forest.

TABLE 27. *Comparison to show parallel edaphic differentiation of forest types, British Guiana and Sarawak (Borneo)*

| | British Guiana | | Sarawak | | |
| | Mixed | Wallaba | Mixed | Koyan Heath | Marudi Heath |
Forest Type					
Soil (lower samples):					
Texture	Light loam	Light sand	Loam	Sand	
Index of texture (Hardy)	18	0	18–27	5–13	0
Illumination (breast height)	0·67 %	1·43 %	Moderate	Good	
Structure and physiognomy (tree strata):					
No. of trees 8 in. (20 cm.) diam. and over per hectare	232	325	184	232	246
No. of trees 24 in. (61 cm.) diam. and over per hectare	6	2	29	29	7
Woody undergrowth	Thin	Dense	Thin	Dense	
Herbaceous undergrowth	Few species, many individuals	Very few species, few individuals	Many species and individuals	Few species	
Lianes	Large and numerous	Few and scarce	Large and numerous	Few and scarce	
Buttressing of trees	Many strongly buttressed trees	Few strongly buttressed trees	Many strongly buttressed trees	Few strongly buttressed trees	
Floristic composition:					
No. of tree species 8 in. (20 cm.) diam. and over (approx.)	55	49	98	55	56
No. of tree species 16 in. (41 cm.) diam. and over (approx.)	32	16	32	18	12
Percentage of most abundant species among trees 8 in. (20 cm.) diam. and over	*Pentaclethra macroloba* 13	*Eperua falcata* 32	'Medang lit.' 5	*Agathis borneensis* 15	'Tekam' 12
Percentage of most abundant species among trees 16 in. (41 cm.) diam. and over	*Eschweilera sagotiana* 16	*Eperua falcata* 67	'Meranti daging' and 'Marakah batu' each 10	*Agathis borneensis* 35	'Tekam' 36

It is very significant that in other respects besides the tendency to single-species dominance the Bornean Heath forest shows far-reaching resemblances to the Wallaba forest of British Guiana. Some of these resemblances are brought out in Table 27. The two communities are so similar in their general aspect that in the Heath forest it is as if every individual plant of the Wallaba forest had been replaced by one of similar habit and general appearance, but with different systematic affinities. Not only do the Heath and Wallaba forest types resemble one another closely, but both differ from the corresponding mixed types in precisely the same respects. Heath forest and Wallaba forest are in different hemispheres and have no species and very few genera in common so the resemblances must be entirely due to the similarity of climate and soil. The resemblance between the vegetation of distant regions with a different flora but similar climate is well known and is well illustrated by the similarity of structure and physiognomy between Mediterranean sclerophyll vegetation in different parts of the world or between the mixed rain forest associations of tropical America, Africa and Malaysia. Here there is another phenomenon, equally significant but less familiar, a parallel edaphic differentiation of plant communities under similar conditions of climate.

CHAPTER 11

COMPOSITION OF PRIMARY
RAIN FOREST (II)

Most Tropical Rain forests, as has already been stated, are of mixed composition, though the Rain forest does not always consist, as earlier observers supposed, of associations without single dominants. Though primary communities with single dominants occur, as we shall see, in all the main geographical divisions of the rain forest, British Guiana appears to be exceptional both in the number of single-dominant communities which occur and in the large area which they cover. Even at Moraballi Creek the Mixed forest is the most widely distributed forest type and probably occupies a larger proportion of the ground than any other community. In some parts of the tropics, for instance throughout West Africa, single-dominant climax primary communities seem to be absent, or nearly so, and all the primary forest on well-drained sites appears to form a single, very extensive mixed community, which fluctuates in composition from place to place, but is almost impossible to separate into more than one distinct association. The Mixed forests of British Guiana and Borneo are thus typical Rain forests in that they show scarcely any tendency towards single-species dominance.

COMPOSITION OF MIXED RAIN FOREST

Composition in small samples

The floristic composition of two sample plots of typical Mixed forest was described in some detail in the last chapter. In Table 28 the data for these plots are compared with those for some other small samples of mixed communities and in Table 25 (Chapter 10) figures were given for the number of species in sample areas of Mixed forest in various localities. From both tables several interesting facts emerge, but the significance of all of them is not at present easy to estimate. It is evident from Table 28 that even in widely separated regions Mixed Rain forest shows considerable uniformity in the general features of its composition. The facts also suggest that the Mixed forest of different geographical regions shows small local peculiarities in its composition, though from a small number of samples it would be unwise to draw very definite conclusions.

The number of species per unit area shows a considerable range, especially among trees of the smaller diameter classes. In any one district the number is fairly constant, the differences between different districts probably depending mainly on the floristic richness of the region as a whole. This, as was suggested earlier, seems to depend on historical factors rather than on the environment at the present day. Only for the West African forest is a considerable number of figures for species per unit area available; in addition to the data given in the table, other figures have been published by Jentsch (1911), Mildbraed (1930 a, 1930 b, 1933 b) and Richards (1939). It is uncertain whether the table covers the whole range of variation in floristic richness. Jentsch's plots in the Cameroons, which are not directly comparable with the author's figures, are possibly even richer than the Mt Dulit Mixed forest plot; on the other hand, some small areas of Rain forest may have fewer species per unit area than the relatively poor Mauritius plot (the poverty of which may be due to its situation on an oceanic island).

TABLE 28. *Composition of Mixed Tropical Rain forest*

	Asia Mt Dulit, Sarawak	S. America Moraballi Creek, Brit. Guiana	Africa		
			Okomu Forest Reserve, Nigeria	Massa Mé Forest Reserve, Ivory Coast	Crown land Macabé, Mauritius (alt. 550 m.)
Locality					
Source of data	Richards (1936)	Davis & Richards (1933–4)	Richards (1939)	Aubréville (1938)	Vaughan & Wiehe (1941)
Size of plot in hectares (approx.)	1·5	1·5	1·5	c. 1·4	1·0
No. of trees per hectare:					
4 in. (10 cm.) and over	—	432	390	530	1710
8 in. (20 cm.) and over	184	232	223	214	331
16 in. (41 cm.) and over	44	60	47	39*	60
No. of species:					
4 in. (10 cm.) and over	—	91	70	74	52
8 in. (20 cm.) and over	98	55	51	58	33
16 in. (41 cm.) and over	32	32	31	c. 23	11
Ratio individual/species:					
Species 4 in. (10 cm.) and over	—	7·1	8·3	10·0	32·9
Species 8 in. (20 cm.) and over	2·7	6·0	6·2	5·2	10·0
Species 16 in. (41 cm.) and over	1·9	2·8	2·3	2·4	5·5
Percentage of most abundant species:					
4 in. (10 cm.) and over	—	*Pentaclethra macroloba* 11	*Strombosia retivenia* 30	*Pachylobus deliciosus* 15	*Eugenia glomerata* 18
8 in. (20 cm.) and over	'Medang lit' 5	*Pentaclethra macroloba* 13	*Strombosia retivenia* 35	*Pachylobus deliciosus* 17	*Eugenia glomerata* 19
16 in. (41 cm.) and over	'Marakah batu' 'Meranti daging' each {10, 10}	*Eschweilera sagotiana* 16	*Pausinystalia* sp. (or spp.) 14 (or less)	*Piptadenia africana* 13	*Mimusops maxima* 25

* Figure for trees 40 cm. diam. and over.

The number of species will of course depend partly on the number of individuals per unit area. This may vary within wide limits and, as the Moraballi Creek communities show, is related to soil and other conditions of the environment. In the mixed Evergreen Seasonal forest of Trinidad the number of trees per acre in the wetter districts is nearly double that in the drier (Beard, 1946 b, p. 63).

The ratio of individuals/species has been termed the *Mischungsquotient* by Mildbraed (1930a), who considers it a useful means of characterizing different types of rain-forest community. Mildbraed, however, overlooks the obvious fact that while the number of individuals varies linearly with the size of sample plot, the number of species does not (see species/area curves, Fig. 34). The *Mischungsquotient* can therefore only be used for comparing plots of the same size. In the examples in Table 28, if we omit the Mauritius plot which has an abnormally large number of individuals per unit area and a small number of species, its range of variation is not great. On all the plots the average number of individuals per species is greater for trees of the smaller diameter-classes than for the larger.

TABLE 29. *Number of species represented by 1, 2, 3, ... individuals in Mixed and Single-dominant Rain forests*

Number of individuals

	I	1–5	1–10	11–20	21–30	31–40	41–50	51–60	61–70	71–80	81–90	91–100	Over 100
Mixed forests:													
1. Moraballi Creek, British Guiana	21	68	79	4	4	1	2	—	1	1	—	—	—
2. Okomu Forest Reserve, Nigeria	23	48	58	7	2	1	—	—	1	—	—	—	1
3. Mt Dulit, Borneo*	41	85	97	1	—	—	—	—	—	—	—	—	—
Single-dominant forests (or associations with tendency to single-species dominance):													
4. Mora (*Mora excelsa*) forest, Moraballi Creek, British Guiana	22	41	47	9	1	2	—	—	—	—	—	—	1
5. Morabukea (*Mora gonggrijpii*) forest, same loc.	25	49	61	3	3	1	—	—	—	—	—	—	1
6. Greenheart (*Ocotea rodiaei*) forest, same loc.	31	63	78	6	—	3	2	2	—	2	—	—	—
7. Wallaba (*Eperua falcata*) forest, same loc.	14	45	55	11	4	—	1	—	—	—	—	—	3
8. Freshwater Swamp forest, Shasha Forest Reserve, Nigeria	11	19	24	6	2	3	—	—	—	1	—	—	1
9. Heath forest, Koyan Valley, Borneo*	18	39	47	7	—	—	2	—	—	—	—	—	—
10. Heath forest, Marudi, Borneo*	13	42	49	3	1	2	1	—	—	—	—	—	—
11. Ironwood (*Cynometra alexandri*) forest, Budongo, Uganda	3	3	5	2	—	—	—	1	1	—	—	—	2

* For trees 8 in. (20 cm.) diameter and over (other data for trees 4 in. (10 cm.) and over). 1–10 from Davis & Richards (1933–4), Richards (1936, 1939); 11 from Eggeling (1947).

The percentage of the most abundant species shows a number of features which are possibly significant. On all the plots the most abundant species in the diameter class '4 in. and over' is different from that in the class '16 in. and over'. Thus the species which approaches nearest to dominance among the larger trees does not form the largest proportion among the smaller. Each stratum probably has a different 'most abundant' species. Among the larger trees (16 in. and over) no one species forms more than about one-sixth of the stand, except in the Mauritius forest where one species forms about one-quarter. On three of the plots in the table the most abundant species in the '4 in. and over' and '8 in. and over' classes forms a *smaller* proportion of its class than the most abundant species in the '16 in. and over' class; in other words, there is even less tendency towards single-species dominance among the lower strata of the forest than among the upper. In the two West African plots, of which the opposite is true, the upper strata may have been artificially depleted by felling.

The 'mixed' nature of these forests may also be expressed in another way, by stating the number of species in each plot represented by 1, 2, 3, ... individuals. The results for several Mixed forest plots (and for comparison some plots of single-dominant communities) are given in Table 29. It will be observed that in Rain forests of all types the great majority of the species are represented by very few individuals. In both Mixed and Single-dominant forests at least half the species present are represented by five individuals or fewer on a 400 × 400 ft. plot (equivalent to from less than 1 % to about 2 % of the stand, depending on the number of individuals per unit area) and considerably more than half by ten individuals or less. In the Mt Dulit Mixed forest as many as eighty-five species, out of a total of ninety-eight (8 in. diam. and over) are represented by five individuals or fewer. In Mixed forest it is rare for any species to be represented by more than 100 individuals (the Mixed plot at Okomu Forest Reserve, Nigeria, is exceptional in this respect), but in Single-dominant forests several species (three in the case of the Guiana Wallaba forest) reach this level of frequency.

Variations in Mixed forest over large areas

So far only small uniform samples of Mixed forest have been discussed; it is also necessary to consider how the composition varies over much larger areas. Does the Mixed forest over an extensive area or a whole geographical region vary in composition in a regular manner or does the composition merely fluctuate fortuitously from place to place? If regular changes in composition exist, can different mixed associations dependent on variations of soil or climate be recognized within any one region? Are the variations in the composition of the tree-layers correlated with variations in the composition of the undergrowth (shrub- and herb-layers)? These questions have been answered in various ways, depending on different authors' conceptions of the nature of an association.

The evidence is at present insufficient for a definite answer, but some of the data deserve consideration.

Meijer Drees (1938) compared statistically the composition of small sample plots of 'high forest' (Mixed Rain forest) in the Netherlands Indies and concluded that only one association was present. In the lowland rain forest of Trinidad (Evergreen Seasonal forest of Beard) Marshall (1934) recognizes three 'associations': the *Carapa guianensis-Eschweilera subglandulosa* association, the *Mora excelsa* association and the *Carapa guianensis-Licania biglandulosa* association. In the first of these he distinguishes four 'subdivisions' of different composition, each occurring under slightly different conditions of soil and topography. Marshall's *Mora* 'association' is a single-dominant type and is here regarded as a consociation. The forest communities of Trinidad have been reclassified by Beard (1946*b*), who treats all the Evergreen Seasonal forest as a single *Carapa-Eschweilera* association with four 'faciations'. All Marshall's 'associations' and 'subdivisions', including the single-dominant *Mora* community, are regarded by Beard as faciations.

The 'closed forest' of the Gold Coast is considered by Chipp (1927) to form several communities: a *Cynometra-Lophira* association, and *Lophira-Entandrophragma*, *Entandrophragma-Khaya* and *Triplochiton-Piptadenia* 'pre-climaxes'. The last is equivalent to what is usually termed Mixed Deciduous forest (see Chapter 15) and the application to it of the term 'pre-climax' is open to criticism. The other communities correspond to the Mixed Rain forest of Nigeria and the Ivory Coast, but it should be noted that the species after which they are named are not the only or even the chief, dominants; they are merely abundant A story species selected more or less arbitrarily.

Chipp's concept of the association has been criticized by Aubréville (1938), who holds that no true associations are distinguishable in the Mixed Rain forest of West Africa. Before considering Aubréville's criticisms, it will be useful to summarize his own description of the composition of the *forêt dense* of the Ivory Coast, which is contiguous to that of the Gold Coast. This is the fullest account available of the variations in composition of Mixed forest over a large area.

Aubréville shows, first of all, that the individual tree species composing the rain forest have several different types of distribution. Many, e.g. *Lophira procera*, are found more or less throughout the rain-forest area; others are much more local. Some are limited to a single compact area, which may be comparatively large, as in *Chidlovia sanguinea*, or only a few square kilometres in extent, as in *Stemonocoleus micranthus*. Others again have discontinuous areas, e.g. *Entandrophragma utile*. The causes of these puzzling differences of distribution in what was probably until recently a continuous and apparently uniform tract of forest are obscure and probably historical. (A theory to account for the rather similar problems of distribution in the Guiana rain forest has been put forward by Davis (1941).)

Because of these differences in the distribution of its species, the forest must necessarily differ in composition from place to place. Though there are estimated to be some 596 species of large trees in the forest of the whole of the Ivory Coast, the number in any one locality is much smaller. On a single hectare there are about forty (diameter limits not stated) and on a slightly larger homogeneous area about seventy.[1] Each locality has a characteristic assemblage of abundant species, which do not exceed twenty in number. Often the population consists of one to four 'dominant' species ('dominant' is used by Aubréville in a sense different from that adopted here) and one to six 'abundant' species; the numerous other species are represented by single, or very few, individuals. The 'dominant' and 'abundant' species vary from place to place; a species which is 'abundant' in one locality may be almost absent in a neighbouring one. Over a wide extent of country almost every possible combination of species is met with, but the 'dominants' are always a selection from a group of not more than perhaps twenty species. Thus the forest is a collection of species of similar ecological requirements which occur in combinations fluctuating in composition from place to place.

These combinations vary not only in space, but, according to Aubréville, they also vary in time, the forest at any one spot continually changing in composition. Aubréville's theory of forest regeneration, which we have called the Mosaic theory, has already been discussed in Chapter 3.

If this is an accurate picture of a large tract of Mixed forest, it follows that no true associations can be recognized in it. Chipp's 'association' and 'pre-climaxes' would thus represent merely a few arbitrarily chosen combinations of species, constant neither in time nor in space. The whole Mixed forest in this case must be regarded as a single association of fluctuating composition.

Aubréville admits that where there are peculiar local soil conditions, true associations may exist. In these the population is more homogeneous than in the Mixed forest and at times one may meet with almost pure stands of one species. An example of these 'edaphic associations' is the combination of species characteristic of river banks in the Ivory Coast, including *Cynometra megalo-phylla*, *Hymenostegia emarginata*, etc. This combination of species is probably to be regarded as a fragment of a riparian community similar to the creekside Mora forest of Guiana and the West African Fresh-water Swamp forest described in Chapter 13), rather than as an association of the Mixed forest.[2]

This account of the composition of the Ivory Coast forest agrees well with Mildbraed's (1930a, 1930b) description of the Likomba forest in the Cameroons,

[1] The number of species on the author's sample plot (area 1·5 ha.) of Mixed forest in the Okomu Forest Reserve, Nigeria, was seventy (4 in. and over).

[2] Groups of species characteristic along the banks of rivers and streams are of course found in all types of Rain forest, just as *Alnus*, *Salix*, etc., by temperate rivers. Examples are the 'Saraca streams' fringed by *Saraca* spp. and the 'Neram rivers' fringed by *Dipterocarpus oblongifolius* in the Mixed forest of Malaya (Corner, 1940, p. 42). Some of these species are probably confined to the actual river margin and are not found in wider belts of riparian forest.

a much smaller area. There he found a marked preponderance of certain species in certain areas, but concluded that the forest was a mixture of species in varying proportions rather than a series of true associations.

For a Mixed forest community to qualify for recognition as a distinct association it should have a characteristic composition remaining constant over a wide area, the most abundant species must retain their places permanently and not give way to others, and we should also expect that a characteristic floristic composition would be shown in all the strata, including the shrubs and herbs. The evidence is inadequate, but, it must be admitted, the presence of several associations in the Mixed Rain forest within a single geographical region has not yet been clearly established.

Family dominance in Mixed forest

Reference has already been made to 'family dominance' in Mixed forests, that is to say, the numerical preponderance of species of the same family or of a group of related genera in some types of Mixed forest. The most striking example is the family dominance of the Dipterocarpaceae in the lowland tropical forests of the Indo-Malayan region west of Wallace's line. In the greater part of the Rain forest (mainly mixed in composition) from Ceylon to Indo-China and Borneo the dominance of this family is well marked and a number of species of one or several genera are usually among the most abundant species in any one area. It would also be possible to speak of the family dominance of Leguminosae in some South American forests, or of Meliaceae in some West African forests. The significance of the phenomenon, which is certainly widespread, is not understood. It is usually believed that competition between plants or animals is most severe between the most nearly related species. On this view it is surprising to find several species of the same genus or family co-dominant in the same habitat. Family dominance probably deserves further study because of the light it may throw on competition in mixed forest communities, still a very obscure subject.

COMPOSITION OF RAIN FOREST WITH SINGLE DOMINANT SPECIES

Among the primary forest communities in British Guiana and Borneo described in the last chapter there were four, the Mora, Morabukea, Greenheart and Wallaba types of Moraballi Creek, in which a single species of tree forms a substantial proportion of the whole stand and may therefore be regarded as the only dominant species; in another of these communities, the Heath forest of Borneo, no one species could be said to be actually dominant, but there was a 'tendency' towards single-species dominance.

All five of these forest types appear to be climax communities in stable equilibrium with their environment. There is nothing to suggest that any of them are seral stages in either a primary or a secondary succession. The *Mora excelsa* community of the flood-plain (though not that of stony hillsides) is a tropical equivalent of European *Auenwald* or 'mature fen woodland' and is the final stage of a hydrosere (see Chapter 13). By American ecologists it would be regarded as a 'subclimax', but as it shows no tendency to develop into the Mixed forest association (or into anything else) it seems better to regard it as a 'physiographic' or 'edaphic' climax (see p. 291). Analogous communities occur on flood-plains in other parts of tropical America, but whether they have single dominants or a tendency towards single-species dominance is not known.

Besides the four single-dominant communities at Moraballi Creek a number of types of Rain forest dominated by one, or in a few cases two, species have been reported, both from British Guiana and from other parts of the tropics, but very little is known about the status, habitat requirements, extent and floristic composition of most of them. Some are probably stages in secondary successions. In the successions which follow the clearing or burning of Rain forest (Chapter 17) the temporary dominance of one species is a not uncommon feature. In the early stages of such successions a single species may at times form an almost pure stand, e.g. the consocies formed by *Musanga cecropioides*, *Trema guineense*, etc., in tropical Africa. Some types of single-dominant forest, which at first sight might appear to be primary types, are certainly seral, thus the remarkable *Eucalyptus deglupta* (*E. naudiniana*) Rain forest of New Britain seems to occur mainly on areas which have been burnt, while the nearly pure, often even-aged, stands of *Octomeles sumatranus* which are a feature of some parts of New Guinea occur on recently deposited alluvium (Lane-Poole, 1925 a, b). There remain, however, some single-dominant types for which there is at least no positive evidence suggesting a seral nature. The scanty information about these may now be summarized.

Among the single-dominant types recorded for British Guiana, other than those already described, the best known is that formed by the leguminous tree *Dicymbe corymbosa* (clump wallaba). This species, which has a remarkable habit of self-coppicing so that a large proportion of the mature trees consist of clumps of up to a dozen large trunks springing from a common base, was mentioned in Chapter 4. *Dicymbe* is dominant over very large areas in British Guiana; it does not occur as near the coast as Moraballi Creek, but is abundant in the Bartica-Potaro district and Myers (1936) also met with it near the Pakaraima Mountains in the far interior. The species also occurs, and may be presumed to be locally dominant, in the valley of the River Uaupés in the Amazon basin. The degree of dominance is very high; T. G. Tutin found in British Guiana that *Dicymbe* formed 57 and 52[1] % respectively of all trunks 2 in. (5 cm.) diam. and over on two sample plots each 40 × 20 m. On both these plots *Eperua falcata* (wallaba) was subdominant and the associated species seemed to be like those of the Wallaba consociation. The ecological requirements of this community are obscure; in some places it occurs on bleached sand side by side with the *Eperua*

[1] 37 and 29 % respectively if each 'clump' is reckoned as one tree.

consociation (which it is apparently invading); elsewhere it occurs on tropical red earth similar to that occupied by consociations of *Mora gonggrijpii,* indeed *Dicymbe* and *M. gonggrijpii* sometimes occur mixed.

Near the coastal savannas of British Guiana from the Corentyne River in the east to the Essequibo in the west (p. 324) large areas are covered with consociations of *Dimorphandra conjugata,* another member of the Leguminosae. The same type of forest reappears on the Pomeroon River and inland in the Essequibo-Kaburi region, but apparently not on the Rupununi River or elsewhere in the far interior. The dominant is a tree about 38 m. high, which reproduces freely, chiefly vegetatively from root suckers and epicormic shoots on fallen trunks. It forms almost pure stands or may be mixed with the related *Dimorphandra hohenkerkii*; other associates are bania (*Swartzia* sp.?), kakarua (Sapotaceae) and *Catostemma fragrans,* a subdominant of Wallaba forest. Sometimes *Dimorphandra conjugata* forms merely a poor scrub 4–5 m. high. The consociation is confined to bleached sands and is remarkable for the thick layer of raw humus or peat which accumulates chiefly as mounds round the tree bases. In both the *Dimorphandra* and the Wallaba consociations patches of scrub ('muri bush', pp. 265 and 324) are often found (T. A. W. Davis).

On the lower Corentyne River and in the Mahaicony district forests dominated by *Aspidosperma excelsum* (Apocynaceae) occur locally on the lower slopes of the hills (T. A. W. Davis). Near the savannas of the Guiana plateau Myers (1936) found, as well as Mixed forest and the *Mora excelsa* and *Dicymbe corymbosa* consociations, forest types dominated respectively by *Peltogyne* sp. and by an unidentified tree with the native name, asheroa. The former occurs as almost pure stands on bare sandy flood-plains and is probably a seral stage, but the latter, which covers extensive flat areas at an altitude of 1300–3000 ft. (400–900 m.), avoiding slopes, may be a climax type. It should be noted that all the dominants of pure consociations in British Guiana, except *Ocotea* (greenheart), *Aspidosperma excelsum,* and possibly asheroa, belong to the Leguminosae.

There seem to be no definite records of single-dominant forests on the mainland of tropical America outside the Guianas, but in the island of Tobago there is a peculiar type of Rain forest, probably a climax, found on shallow igneous soils on exposed sites above 800 ft. (240 m.), to which Beard (1944*b*) has given the name, 'xerophytic rain forest'; in this *Manilkara bidentata* forms nearly half the stand of trees of 4 ft. (120 cm.) girth and over. The *Dacryodes-Sloanea* association of the Lesser Antilles also tends to become a single-dominant community towards the edge of its range (see p. 261).

In West Africa and the Belgian Congo there are large areas of forest, comparable with the Mora forest of the Moraballi Creek flood-plain, liable to flooding and also probably the final (climax) stage of a hydrosere. In these forests, which are further referred to in Chapter 13, the floristic composition is very different from that of the Mixed forests on well-drained sites; it is of interest that in some of them there is evidence of a tendency towards single-species dominance, or even towards the formation of nearly pure stands of one species. In a sample plot of forest liable to flooding in the Shasha Forest Reserve, Nigeria, the total number of species was low (thirty-eight '4 in. and over', eleven '16 in. and over') compared with the Mixed forest plots in the same district (except one which was probably not a climax community).

On well-drained land in the African Rain forest consociations with a single dominant show an interesting localization in distribution. In the belt of ever-green forest fringing the Gulf of Guinea practically no single-dominant com-munities of climax status have been observed, but in parts of the great equatorial forest *massif* of Central Africa there are very extensive areas of apparently climax forest dominated by a single species. Lebrun (1936a) divides the equatorial forest of the Congo basin into three 'subdistricts': (a) the Central basin, (b) the North-eastern basin, (c) the Eastern basin (transitional to the montane forests of East Africa). In (a), the richest of these regions floristically, the forest is invariably mixed and single-dominant forests are practically unknown, but in (b) at least two-thirds of the area is occupied by forests dominated by either *Macrolobium dewevrei* or *Cynometra alexandri*.

The consociations of *Macrolobium dewevrei* were first described by Mildbraed (1914, 1922), who reported their occurrence west of Ruwenzori, in the Ituri forest and in several localities in the southern Cameroons. Further data are given by Vermoesen (1931) and Lebrun (1936a). *M. dewevrei* is often very strongly dominant; in a sample hectare in the Uele-Itimbiri district Lebrun found that it formed 94 % of the trees over 20 cm. diameter. *M. dewevrei* is a tall tree branching quite low down, thus giving the community a peculiar physiognomy. Herbaceous undergrowth is well developed and seedlings of the dominant are abundant. The consociation is found on deep sandy soils (Vermoesen, Lebrun), or according to Mildbraed, on yellow or yellowish grey loamy clay; undoubtedly it is edaphically determined. According to Lebrun's map it covers an area of many hundreds of square kilometres, running westwards in a broad band along the eastern and northern margins of the Congo forest. The dominant sometimes grows scattered or in small groups in Mixed forest, as well as in pure consociations; as a constituent of Mixed forest it extends as far west as Nigeria where, however, it is restricted to river-banks and swampy situations.

Similar consociations, but somewhat less extensive and less pure, are formed by *Cynometra alexandri*. These are found in the extreme north-east of the Congo forest to the north and north-west of Ruwenzori at an altitude of 900–1000 m. above sea-level (Lebrun, 1936a) and also extend into western Uganda. Eggeling (1947) has given a detailed account of the composition and structure of the *Cynometra* (ironwood) consociation in the Budongo forest (Uganda). Here the dominant forms 20–35 % of all trees 4 in. (20 cm.) diam. and over, and as much as 75–90 % of those 16 in. and over. In an area in the Belgian Congo examined by Lebrun it formed about 50 % of trees over 20 cm. The number of other tree species in the Budongo *Cynometra* forest is small; on one sample plot (400 × 400 ft.)

Eggeling found only eleven species and on another twenty-five, as compared with fifty-eight and fifty-three respectively on two similar Mixed forest plots. The community has three tree stories (p. 35), but the A story consists merely of exceptionally large *Cynometra* trees emerging from the B story. *Strychnos* sp. and three species of *Celtis* are the most abundant of the other species in the B story; *Lasiodiscus mildbraedii* is usually dominant in the C story, but sometimes it is absent and replaced by *Lepidoturus laxiflorus*. This type of forest has a characteristic appearance. Lianes are very scarce and the scanty C tree story and shrub layer make visibility possible for long distances. The herb layer is not well developed and the ground herbs become parched and half-wilted in the dry season, as in a much drier type of forest. The soil is of the tropical red earth type, but down to the upper limit of the B 2 horizon (30 cm.) it is loamy and of a powdery texture.

Cynometra alexandri also occurs in Budongo forest as a constituent of Mixed forest, where, it is interesting to note, it reaches larger dimensions than in the consociation.[1] It is also a significant fact that Eggeling has found convincing evidence that in Budongo, where the forest as a whole is extending its area at the expense of grassland, Mixed forest is a seral community and the *Cynometra* consociation the climax. The latter is slowly invading and replacing the former.

A gregarious habit, similar to that of *Macrolobium dewevrei* and *Cynometra alexandri*, is also shown by the African rain-forest species *Berlinia ledermannii*, *B. polyphylla*, *Brachystegia* sp. and *Tessmannia parvifolia*. From Mildbraed's remarks (1914, 1922) it would appear that none of these forms extensive communities comparable with the two just described. In Uganda a rain-forest community occurs between 4500 and 5000 ft. (1400–1500 m.) in which *Parinari excelsa* (Rosaceae) forms 'about 80 % of the canopy' (Eggeling). It is uncertain whether this community should be rightly classified as Tropical Rain forest; it is perhaps a local variant of Lebrun's Transition forest (p. 358).

It is interesting, and possibly significant, that all the species reported as occurring as single dominants on well-drained sites in the African rain forest (excepting *Parinari*) belong, like the majority of the Guiana single dominants, to the Leguminosae.

In the Rain forest of the Indo-Malayan region single-dominant types other than seral stages are rare and occur only over relatively small areas. Forest communities distinct from the prevailing Mixed Dipterocarp association of the region are found in habitats similar to that of flood-plain Mora forest in Guiana, but there is little definite information about their composition. The Moor forests of tropical raised bogs are referred to below (p. 292).

Among Malayan forest types on well-drained sites those dominated by the Borneo ironwood, *Eusideroxylon zwageri*, are of particular interest, as the

---

[1] At Moraballi Creek, British Guiana, *Mora excelsa* also grows to a larger size when growing as a solitary individual in Mixed forest than as a member of its own consociation (Davis & Richards, 1933–4, p. 111).

dominant is a timber tree of considerable economic value. This species is widespread in Sumatra and Borneo as a scattered B story tree in Mixed Dipterocarp forest. Locally, however, it becomes strongly dominant and may form almost pure forests (Gresser, 1919; Witkamp, 1925; Koopman & Verhoef, 1938). The peculiar structure of the Ironwood forests was described in Chapter 2. Only here and there does a solitary *Koompassia*, *Shorea* or *Intsia* raise its crown far above the general level. According to van der Laan (1925, 1926*a*, 1926*b*, 1927), who has published some data on the composition of these forests, there may be as many as thirty-three trees of the dominant over 20 cm. diam. per ha., forming up to 30 % of the timber volume. Gresser says that in the Djambi district of Sumatra the *Eusideroxylon* patches were mostly under 100 ha. in extent, but one of 373 ha. and another of as much as 1068 ha. were seen. Koopman & Verhoef state that *Eusideroxylon* 'complexes' up to 10,000 ha. are found in the Djambi and Palembang districts. The dominant regenerates abundantly and there is little other undergrowth.

The Ironwood forests are usually found on flat ground near rivers and on the immediately adjoining slopes. Gresser attributes the apparent inability of the tree to colonize slopes to the heavy egg-shaped fruits, which roll downhill (p. 94). As was mentioned in Chapter 9 (p. 222), *Eusideroxylon* consociations are limited to sandy soils and show a very striking correlation with the distribution of sandstones and sand.

Another Malayan forest type with a single dominant is formed by *Dryobalanops aromatica*, the kapur or Borneo camphor, a lofty tree belonging to the Dipterocarpaceae. Like *Eusideroxylon*, the dominant also occurs as solitary individuals scattered through Mixed forest, but as a single dominant it is found only in small localized patches. Such patches of *Dryobalanops* forest have been recorded in various parts of the Malay Peninsula, chiefly on the eastern side, and in Sumatra. The best known of these, now much altered by sylvicultural treatment, is at Kanching, near Kuala Lumpur. Van Zon (1915) has described the *Dryobalanops* forests at Bengkalis in Sumatra. The largest patch was not more than 20 ha. in extent and seven others less than 5 ha. were seen. All were situated in the midst of normal Mixed Dipterocarp forest. On a sample plot of 0·25 ha., out of thirty-one trees over 20 cm. in diameter, ten were *Dryobalanops*. Regeneration was abundant. These patches did not appear to be associated with special soil conditions. A similar, but less marked, gregarious habit is shown by other rain-forest dipterocarps, for instance by *Shorea curtisii* (Pl. IX), which dominates small societies on steep slopes in the hill rain forests of the Malay Peninsula; these are conspicuous because of the pale bluish-green foliage of the tree (Foxworthy, 1927, p. 39).

In the Poso district of Celebes, Steup (1930) has described a forest type with a single dominant which is apparently a climax, though it is possible that it would be better classified as a unit of a deciduous or semi-deciduous formation than as a type of true Rain forest. Patches of forest were found dominated either by *Diospyros* sp., or by *Diospyros* in association with the palm *Livistona rotundifolia*, to the almost entire exclusion of other species. They occurred on steep stony slopes and ridges and some had an area of several hectares. Steup (1932) has also described a forest type (climax?) in Celebes with a 'tendency' towards single-species dominance, similar to that shown by the Heath forest of Sarawak (Chapter 10). At Gunong Klabat on a plain of loose volcanic sand there is a forest in which *Elmerrillia ovalis* is the most abundant species over about 1570 ha. On sample transects, totalling 25·3 ha., 11 % of the trees over 20 cm. diam. belonged to this species.

In Papua, at the foot of the Hydrographer Range, there is a type of forest in which *Intsia bijuga* and *Anisoptera polyandra* are co-dominant (Lane-Poole, 1925 *a*, *b*). The former constituted 51 % of the stand (diameter limits not stated), the latter 43 %, other species only 6 %.

More detailed knowledge is clearly needed about the composition of single-dominant forests; adequate information is available only for the four Moraballi Creek consociations and the *Cynometra alexandri* forest of East Africa. From these it would be unsafe to draw very far-reaching conclusions.

In the Moraballi Creek single-dominant types the proportion of the dominant ranges from 43 to 67 % of the larger trees (16 in. diam. and over), and from 9 to 27 % of the whole stand (4 in. and over); in the *Cynometra* community, according to Eggeling's data, the percentage of the dominant is sometimes greater than in any of these Moraballi Creek communities and this may well be also true of the *Macrolobium dewevrei* consociation of tropical Africa and the *Eusideroxylon* consociation of Borneo and Sumatra. There is every intermediate between typical Mixed forests and communities like the Sarawak Heath forest, in which there is only a 'tendency' to single-species dominance on the one hand, and communities approaching a pure stand of one species on the other. The ratio of species to individuals (*Mischungsquotient*) of the Guiana single-dominant types varies from 3·9 to 6·3 for trees 16 in. and over on a 400 × 400 ft. plot; for the *Cynometra* forest the ratio on two plots is 18·2 and 7·5 respectively. These are of course higher ratios than in the Mixed forest communities (Table 28), but among trees 4 in. diam. and over the ratio varies from 6·5 to 12·4 in the Moraballi Creek single-dominant communities, which is not very different from some examples of Mixed forest. In one of Eggeling's *Cynometra* plots the ratio is as high as 53·9.

The number of tree species per unit area is smaller in the *Cynometra* consociation, and in all the Moraballi Creek consociations, than in the corresponding types of Mixed forest. It is also smaller in the Sarawak Heath forest than in the Mixed forest of the same region. This reduction in the number of species is to be expected, since the larger the proportion of the stand formed by any one species the less room there must be for others. This point is strikingly illustrated by the *Dacryodes-Sloanea* association, the climax community of the lowland rain forest in the Lesser Antilles. Table 30 shows the relation between the number of species and the proportion of the whole stand formed by *Dacryodes excelsa*, which is the most abundant species throughout the association. The number of species in the community is thus roughly inversely proportional to the percentage formed by the most abundant species. It is significant that the *Dacryodes-Sloanea* association is floristically richest in Dominica, which is the centre of its area of distribution; with increasing distance from the centre the dominant occupies an increasingly large place in the community and the number of subordinate species diminishes.

TABLE 30. *Variation in the composition of the Dacryodes-Sloanea rain-forest association.* (Data from J. S. Beard)

Island	Percentage of *Dacryodes excelsa*	Number of species per 10 acres (4·05 ha.)
Dominica	20	60
St Lucia	11	41
St Vincent	25	39
Grenada	40	23
St Kitts	40	18

There is a tendency in some single-dominant types, but not in all, for several subdominant species in addition to the dominant to form a considerable proportion of the stand. This is a marked feature of the British Guiana Mora and Wallaba forest types and of the Sarawak Heath forest, but is not shown by the Morabukea forest or by the African *Cynometra* consociation; in these latter single-dominant types there is, as it were, no near competitor with the dominant. 'Family dominance', mentioned earlier as a common feature of the composition of Mixed forests, is also shown by some single-dominant forests. In the Guiana Wallaba forest, for example, in which *Eperua falcata* is the dominant, another species of *Eperua*, *E. jenmani*, is a plentiful subordinate species and several other of the more abundant (though not subdominant) species belong to the same family as the dominant, Leguminosae forming 53 % of the whole stand. In the Mora forest the dominant, the three subdominants and several subordinate species, forming 59 % of the stand in all, belong to the Leguminosae.

Single-dominant communities, as might be expected, differ in structure as well as in floristic composition from Mixed forests; these differences have already been discussed (Chapter 2).

RELATION OF MIXED AND SINGLE-DOMINANT PRIMARY COMMUNITIES

From the data presented in this chapter and the last we may attempt, though very tentatively, to draw a general picture of the floristic composition of the Tropical Rain forest as a whole. In each of its major geographical divisions (or formations)—the tropical American, the African and the Indo-Malayan-Australasian—by far the greater part of the primary forest consists of Mixed associations composed of a very large number of species no one of which is dominant by itself, the great majority of the species being represented (in a small sample) by very few individuals. The failure of any one species to gain the upper hand in the mixture may be due, as Aubréville suggests, to all the species having very similar ecological requirements and responding in a very similar way to slight variations in the environment. As would be expected on grounds of probability, under these conditions the composition of these mixed communities will fluctuate in space and probably in time as well. These fluctuations are of no ecological significance and merely indicate that the mixture is in a state of dynamic equilibrium. Within any one geographical area the Mixed forest forms a single association.

These very extensive Mixed associations are diversified in river valleys, swamps and other areas with a permanently high water-table by forest types of different composition, some of which are consociations. There are also smaller areas of land with good or normal drainage which are occupied by types of forest with a single dominant or a tendency to single-species dominance, varying greatly in extent, individuality and 'degree of dominance'. On the one hand there are some, like the Wallaba community of British Guiana, the *Cynometra alexandri* community of Africa and the Heath forest of Sarawak, which have a highly distinctive floristic composition and extend over large areas of country. Such communities seem fully comparable with the consociations of temperate forest formations. At the other extreme there are types like the *Dryobalanops* forest of Sumatra and Malaya which amount to little more than local societies dominated by a single gregarious species.

The data would seem to indicate that single-dominant communities are usually limited to special soils or combinations of soil and topography; further investigation may show that this is true of all of them. In the last chapter the hypothesis was put forward that the dominance of single species in rain-forest communities is associated with unfavourable, i.e. non-optimal, soil conditions. The evidence in favour of this view is not yet conclusive. As far as the single-dominant communities of British Guiana are concerned it is very strong, but for those of other parts of the tropics further data are needed. If the idea is accepted at least as a working hypothesis, we can regard the habitat as having a sifting or selective effect on the very numerous potential dominants of rain-forest formations. In Mixed forest environmental conditions are almost equally favourable or unfavourable to most of these dominants, but under the special conditions which lead to single-species dominance some species are wholly eliminated while others gain a relative advantage.

If the picture which has been drawn is correct, there would be a close analogy between the Tropical Rain forest, with its extensive mixed associations and relatively localized consociations, each formed by selection from the related type of Mixed forest, and the 'Mixed Mesophytic Forest' of North America, as Braun (1935) has described it. This forest in places approaches the Tropical

Rain forest in floristic richness. A large part of it consists of a mixed association in which a considerable number of species shares dominance. In southern Ohio, for example, some six species are more or less equally co-dominant. Towards the limits of the geographical range of the mixed mesophytic forest climax (or formation) variants of the association begin to appear; these are due, not to any pronounced regrouping of the dominants, but to the dropping out of some of them altogether. On the other hand, in the area of maximum development of the mesophytic forest 'variety in expression, due to segregation and regrouping of the dominants results in many, at first sight entirely unlike, communities with different dominants' (Braun, loc. cit. p. 514). There is a striking resemblance between the 'dropping out of dominants' near the limits of distribution of the mesophytic forest and the impoverishment, accompanied by the increasing dominance of one species, at the periphery of the *Dacryodes-Sloanea* rain forest association of the Lesser Antilles (p. 261). In the 'variety of expression' in the area of maximum development of the North American mesophytic forest there is an analogy to the parallel development of a mixed association and several consociations under different soil conditions in the British Guiana Rain forest. The more mesophytic of the variants of the North American forest Braun regards as climatic climax communities; others with smaller moisture requirements are restricted to dry slopes and ridges and are regarded as 'subclimaxes' or 'physiographic climaxes'.

These variants Braun terms 'association segregates'; she considers them to have become gradually differentiated since the Tertiary period from the originally uniform mixed and undifferentiated climax. In Europe the conception of an undifferentiated mixed forest climax is unfamiliar because Pleistocene glaciations have destroyed the whole of the Tertiary mixed forest, except perhaps in certain relict areas such as the western Caucasus.

In the Tropical Rain forest a similar process of differentiation may well have gone on and the single-dominant types, including those with only a slight tendency to single-species dominance such as the Bornean Heath forest, may be regarded as 'association segregates'. The Mixed associations of course represent the undifferentiated climax. The process of segregation has not taken place to the same extent in all geographical regions. In the Guianas segregation is far advanced and there are several highly differentiated association segregates. The reason for this can only be guessed; it may be the great age of the region as a land mass or it may depend on local conditions leading to strongly marked differentiation of soil types. In tropical Africa and Asia segregation is much less in evidence. Forest types such as the *Dryobalanops* forest of Malaya are perhaps association segregates in embryo.

THE THEORY OF THE CLIMAX

The views tentatively put forward in the last section raise the question: which types of primary Rain forest represent the climatic climax? What is the relation to the climatic climax of types like the Guiana Wallaba forest with a single dominant and those like the Bornean Heath forest with a tendency to single-species dominance? How should they be classified and how are they to be regarded from the point of view of the climax theory?

The climatic climax is generally regarded as a position of stability in the development of vegetation. The stability is relative to the human time-scale; in geological time all plant communities undergo gradual change. Two views are held as to the development of climaxes in a given climate. According to one, the Monoclimax theory, all vegetation in a given climatic region tends to develop towards one and the same climatic climax, except where special conditions of soil, e.g. strongly leached sandy soil, or topography, e.g. steep slopes, permanently swampy conditions, hinder normal development and lead to the establishment of subclimaxes. The alternative view, the Polyclimax theory, regards the development of vegetation as leading not to a single climatic climax, but to a series of climax types, all equally stable, but adapted to different conditions of soil, slope, drainage, etc.

The theory of the climatic climax, indeed the whole structure of ecological concepts and theories, has been built up on the study of temperate vegetation, mainly that of North America and Eurasia. It can hardly be doubted that the study of tropical plant communities will greatly modify these concepts and theories, but our knowledge of tropical vegetation is not yet sufficient to justify a full discussion of the theory of the climax as applied to rain-forest formations.

It is already certain, however, that what is known about the floristic composition of rain-forest communities cannot be harmonized with the monoclimax view of the climatic climax. The primary forest types of Moraballi Creek, for instance, are all found under the same climate and there is no reason to consider any one of the five as less stable than the others. No process of development and no change of soil conditions is known or can be imagined which would, for example, convert the Wallaba forest on bleached sand into Mixed forest on red loam or vice versa. The Mora forest, liable to flooding, shows no tendency to develop into Mixed forest and the Mora soil could not conceivably develop into a Mixed forest soil. Much the same can be said of the relation of the Bornean Heath forest and Mixed forest. It seems that all the Moraballi Creek primary types must be regarded as of equivalent status, as the Polyclimax theory demands. Each depends on a different, but equally permanent, combination of soil and topography developing under the same set of climatic conditions. We have at Moraballi Creek what may be termed a catena of climaxes.

There is a close analogy between soil development and the development of the climax. Formerly it was supposed that under one climate one, and only one, climatic soil type could develop; this view is the counterpart of the Monoclimax theory of climax development. Now it is generally admitted that factors other than climate, e.g. parent material, topography, etc., may play a predominant part in soil development, so that not one, but many different, mature soil types can be found side by side within the same climatic region. The tropical red earths and the lowland tropical podzols, described in Chapter 9, are an example of this. The English school of plant ecologists have compromised between the Monoclimax and Polyclimax theories of vegetational development by regarding one climax formation in each climatic region (that found on the most widespread or 'normal' combination of soil and topography) as the climatic climax and others in the same region as edaphic or physiographic climaxes. If this view is adopted the Mixed forest of Moraballi Creek (and other Mixed rain-forest associations) can be regarded as a climatic climax and the Mora and Wallaba consociations as edaphic climaxes. The Morabukea, and Greenheart communities can also be considered as edaphic climaxes, or, since they differ comparatively little from the Mixed forest, they can be termed 'lociations' (Clements, 1936) of the climatic climax. The distinction now commonly adopted between 'zonal' and 'azonal' soils may be recalled here. Zonal soils are the 'normal' climatic soil types of each climatic region, in the development of which climate and vegetation play the principal part; azonal soils, such as lowland tropical podzols, show the predominant effect of parent material, age or topography. In a similar way we might regard Mixed rain-forest associations as the zonal climax formation and other stable types of primary Rain forest, dependent on less frequent soil-topography combinations, as azonal formations.

It will be shown later (Chapters 13, 15) that other types of vegetation besides the primary forest types considered in these last two chapters, e.g. certain types of savanna and the muri, padang and Moor forest already referred to in Chapter 9, are apparently stable types of vegetation, occurring in the same climate as the Rain forest, but under special conditions of soil and topography. If this is so they deserve the rank of climax as much as the forest itself and, if the suggestion made above is accepted, can be regarded as azonal climaxes.

Part IV

PRIMARY SUCCESSIONS

CHAPTER 12

THE PRIMARY XEROSERE AND THE RECOLONIZATION OF KRAKATAU

The successions or seres leading to the establishment of stable climax Rain forest are classified, somewhat arbitrarily perhaps, into primary successions or priseres starting on soil not previously occupied by plants, for instance new soil formed from bare rock or arising by the silting up of lakes or seas, and secondary successions or subseres starting where previously existing vegetation has been destroyed or damaged by felling, burning, flooding, etc. The subseres, which in the Rain forest are mainly due to human interference, will be dealt with in Chapter 17; in this and the two following chapters the rather fragmentary information on the priseres will be considered.

THE RECOLONIZATION OF KRAKATAU

The most spectacular example of a primary plant succession of which there is any record is the recolonization of the volcanic island of Krakatau after its original vegetation had been destroyed by the great catastrophe of 1883.[1] Though many of the facts are already well known, a short sketch of the development of the plant communities after the eruption will form an appropriate introduction to the priseres leading to the development of Tropical Rain forest.

The history of the island after the eruption

The recolonization of Krakatau has given rise to a voluminous literature, partly because the subject has unfortunately become involved in an embittered controversy (Backer, 1929; Ernst, 1934; Docters van Leeuwen, 1936; cf. also Turrill, 1935). The question at issue is whether any of the original vegetation of the island survived the eruption. All the earlier visitors to the island believed that it was totally destroyed, but Backer in 1929 brought forward reasons for doubting the correctness of this conclusion; he suggested that part at least of the original flora may have remained alive as seeds, spores, underground parts or fragments, probably in the ravines on the higher part of the island which were not explored until many years after the eruption. Doubts as to the complete destruction of the fauna had previously been expressed quite independently by R. F. Scharff. The arguments for survival, especially those of Backer, cannot be

---

[1] Though the island was previously covered with vegetation the recolonization is classified as a prisere because after the eruption the whole surface of the island became covered with new soil-forming materials.

lightly dismissed. From the nature of the evidence a rigid proof of the 'sterilization hypothesis' is unattainable; as in most questions of plant geography, we
must be content with high probability, without complete certainty. In the light
of the counter-arguments put forward by Ernst (1934) and Docters van Leeuwen
(1936) one is forced to admit that the probability of complete sterilization is very
high indeed. Those who wish to follow the arguments on the two sides must
refer to the works named; all that will be attempted here is to give a brief and
straightforward account of the development of vegetation on the island. Though
for the study of plant and animal dispersal in relation to sea-barriers the controversy is crucial, from the present point of view it is of secondary importance
whether a few survivors from the old flora took part in the recolonization or not.
Whatever the outcome of the controversy, the student of plant succession cannot
agree with Backer (1929, p. 286) that '...the Krakatao-problem...is of no
importance at all for botanical science', though he will certainly regret that the
earlier explorations of the island were not more frequent and more thorough.

Krakatau is one of a group of small volcanic islands situated in the Sunda
Straits between Java and Sumatra and about 40 km. from both. Early in 1883
it was about 9 km. long and 5 km. broad, rising to a peak 2728 ft. (822 m.)
above sea-level. At this date the whole island was covered with luxuriant
vegetation. About the nature and composition of this vegetation next to nothing
is known, but there is every reason for supposing that it was mostly Tropical
Rain forest similar to that now existing in the neighbouring parts of Sumatra.
Near the shore various littoral plant communities were no doubt present and
towards the summit of the peak there were probably different types of rain
forest, presumably belonging to what is here termed the Submontane and
Montane Rain forest formations (see Chapter 16). In May 1883 the volcano,
which had long been regarded as extinct, began to be active and the activity
gradually increased till it reached a climax on 26 and 27 August. On those
two days occurred the famous eruption, the sound of which was heard as far
away as Ceylon and Australia. More than half the island sank beneath the
sea, the peak being split in two, though its highest point still remained. The
surviving parts of Krakatau were covered with pumice-stone and ash to an
average depth of about 30 m. and a new marginal belt 4·6 sq. km. in area was
added to the southern coast. During the period of volcanic activity the bulk of
the vegetation was certainly destroyed. As has been stated, until recently all
investigators were unanimous in thinking that the island was completely sterilized
and that all organisms found there after the eruption must have migrated across
the sea. The vegetation of two neighbouring islets, Verlaten Island and Lang
Island, was almost certainly destroyed at the same time. That of Sebesi Island,
19 km. from Krakatau, was much damaged, but not altogether destroyed.

After the great eruption of 1883 there was no further volcanic activity and the
heavy rainfall (over 2600 cm. per annum) soon cut deep gullies in the new
covering of ash and pumice. For a while the island remained without any

vegetation and E. Cotteau, who visited Krakatau in May 1884, found it still a desert; the only living thing he saw was one spider.

When the botanist Treub arrived in June 1886 there was, however, already a considerable amount of vegetation. In his account of what he found Treub (1888) emphasizes the sharp distinction which already existed between the flora of the beach and that of the interior of the island. On the beach there were the following nine species of flowering plants: *Calophyllum inophyllum*,[1] *Cerbera manghas*, *Erythrina* sp., *Hernandia peltata*, *Ipomoea pes-caprae*, *Scaevola frutescens*, the grass *Pennisetum macrostachyum* and two unidentified Cyperaceae. In addition he found seeds or fruits of several other species. Except for the *Pennisetum*, all these are species common on recently emerged coral islands, as Treub noted; they are widely distributed pan-tropical or Indo-Malayan sea-shore species. Inland the most striking feature of the vegetation was the abundance of ferns, of which there were eleven species including *Acrostichum aureum*, *Pteridium aquilinum*, *Pityrogramma calomelanos*, *Dryopteris* and *Pteris* spp.; the following flowering plants were also found: *Blumea humifusa*, two species of *Conyza*, *Scaevola frutescens*, *Tournefortia argentea*, *Wedelia biflora* and the grasses *Neyraudia madagascariensis* var. *zollingeri* and *Pennisetum macrostachyum*. These flowering plants grew scattered and as individuals were far less numerous than the ferns. Some of both the flowering plants and the ferns were species usually found near the sea. Docters van Leeuwen suggests that their occurrence inland may have been partly due to the absence of competition, but ecologically the conditions must have been similar to those on a sandy shore. Besides these vascular plants Treub found two unidentified mosses and noted that the surface of the ash and pumice was everywhere covered with a thin crust of blue-green algae (Cyanophyceae); six species were collected, belonging to the genera *Anabaena*, *Hypleothrix*, *Lyngbya*, *Symploca* and *Tolypothrix*. The gelatinous and hygroscopic layer formed by these algae was probably important in providing a favourable substratum for the germination of seeds and spores. At this date, then, the inland vegetation consisted of a two-layered associes, the upper layer formed chiefly of ferns, the lower of Cyanophyceae. The total number of species of vascular plants recorded was twenty-six. Owing to the roughness of the ground Treub was unable to ascend far into the interior of the island.

The vegetation of Krakatau was not again examined until 1897, 14 years after the eruption. The party which then visited the island again included Treub and the results are described in a brief paper by Penzig (1902). Since 1886 the vegetation had made great progress and closed plant communities had begun to appear.

On the beaches there was a well-developed '*Pes-caprae* formation' (associes dominated by *Ipomoea pes-caprae*), a characteristic community of sandy tropical shores (see p. 296). In this certain species belonging to the '*Barringtonia* formation' (Littoral woodland community, see p. 297) were found, including

[1] These and following identifications are as emended by Docters van Leeuwen (1936).

Terminalia catappa and *Barringtonia asiatica*. *Calophyllum inophyllum*, a characteristic tree of this community, was not recorded on this occasion, though Treub had found young specimens on the beach during his previous visit in 1886. *Casuarina equisetifolia*, another very characteristic tree of the Littoral woodland, had already established itself; on Verlaten Island at this date it was beginning to form a closed stand.

The interior of the island now looked very different from what Treub had seen 10½ years before. Instead of vegetation consisting chiefly of ferns, there was now a dense growth of grasses, in places taller than a man. The species forming this savanna or 'grass-steppe' were, on the lower parts of the hills, chiefly *Saccharum spontaneum*, *Neyraudia madagascariensis* var. *zollingeri* and *Pennisetum macrostachyum* and higher up *Imperata cylindrica* var. *major* (the well-known alang-alang grass) and *Pogonatherum paniceum*. Associated with the grasses in the lower region, there were various dicotyledons, including several climbers. The dicotyledons included several beach plants, for instance *Scaevola frutescens* and *Vigna marina*. Higher up in the hills isolated shrubs, various dicotyledonous herbs and ferns were intermixed with the grasses. On steep tufa walls the vegetation remained much as it was before, consisting of ferns and a crust of algae. In the latter a number of Cyanophyceae not collected by Treub were found, also five species of diatoms; though the latter were not recorded in 1886, their existence was probably merely overlooked. In addition to algae and some sixty-four species of vascular plants, the 1897 expedition found two mosses, one liverwort and one agaric. As in 1886, it was found impossible to penetrate very far into the interior of the island.

The next thorough examination of the vegetation was made in 1906, when Krakatau was visited by a party of botanists. The results of this expedition are given in the well-known book by Ernst (1908). Many further changes had by then taken place in both the shore and the inland vegetation.

The plant communities on the shore now covered a wider belt, probably because the beach itself had widened by accretion. Instead of consisting only of the '*Pes-caprae* formation', the vegetation of the beach consisted of two distinct zones. The outer one was the *Pes-caprae* associes, better developed and richer in species than in 1897. Behind this was the '*Barringtonia* formation' forming a discontinuous but well-developed belt of woodland, not everywhere closed, but similar in essentials to the corresponding community on the coast of Java. Much of the closed portion of this woodland apparently consisted of a more or less pure stand of *Casuarina equisetifolia*; elsewhere the chief trees were *Barringtonia asiatica*, *Calophyllum inophyllum*, *Hibiscus tiliaceus* and *Terminalia catappa*. As well as trees, the woodland included climbers, shrubs, grasses and other herbs. An interesting feature was the occurrence of coco-nut palms, but they may have been planted.

Ernst's description of the interior of the island is much less satisfactory than that of the beach, but it is clear that up to a considerable altitude the vegetation was still a savanna consisting mainly of the same species of grasses as those noted by Penzig. The associated plants included Cyperaceae, climbers such as *Vigna*, *Canavalia* and *Cassytha*, shrubs of *Tournefortia* and *Scaevola* and various ferns. 'Beyond the strand-forest the whole gently sloping surface of the south-east side of the island was covered with the steppe-like vegetation which we have described and this extended in a dense mass into the wild ravines and on to the steep sides far up on the cone. The uniformity of the jungle of fresh and decaying stems of grasses and reeds is only occasionally broken by the occurrence of a fallen tree or shrub.' (Ernst, 1908, p. 31.) A deep ravine about half-way up the slopes of the peak was particularly striking because of its luxuriant growth of trees and shrubs, but it proved impossible to reach it and, like all the previous expeditions, Ernst's party could not approach at all near to the peak.

Further details about the inland vegetation at this period were given by Backer (1909, 1929), who visited Krakatau again (he had been a member of Ernst's expedition) two years later. On the south-east of the island, behind the beach vegetation, he found here and there narrow strips of mixed woodland consisting of *Ficus fistulosa*, *F. fulva*, *Macaranga tanarius*, *Melochia umbellata* and *Pipturus incanus*, growing from 5 to 15 m. high. All these trees are species characteristic of the secondary Rain forest of Java and the Malayan region generally (p. 392). Above this mixed forest stretched the grass savanna dominated by *Saccharum spontaneum* where isolated specimens of all five of the mixed forest trees were seen.

Backer succeeded in exploring the wooded ravine which Ernst had failed to reach and ascended to a height of about 400 m. The forest in ravines was best developed between 300 and 400 m. above sea-level; it was formed by a very few species of trees with herbs, etc. beneath. Among the ten species not previously recorded from the island was the gesneriaceous shrub *Cyrtandra sulcata*, which, as we shall see, later became extremely abundant. According to a member of the party which accompanied Backer, the vegetation *above 400 m.* consisted of trees and ferns. It is thus likely that the fern community, which in 1886 had covered most of the island, had persisted at the higher altitudes longer than at the lower, where it had been ousted by the grass savanna; it had, in fact, receded upwards.

From 1908 to 1919 no botanical explorations of Krakatau were made and the vegetation of the island suffered slight interference. In 1916 a German obtained a concession to work building stone and settled in the island where he remained for several years. He made a garden and was probably responsible for introducing a few new plants; otherwise his occupation seems to have left little trace. In 1919 a fire was started which burnt much dry grass, but according to Docters van Leeuwen the trees and shrubs were merely singed and the damage was negligible. It does not appear that the natural course of the plant successions on Krakatau has ever been seriously affected by man.

In 1919 a party including Docters van Leeuwen made a more thorough investigation of the vegetation than any that had been made previously. All regions of the island were examined, including the upper slopes, and the summit of the peak itself was reached. Since 1919 there have been frequent visits and the development of the vegetation has been carefully recorded. This is fortunate, for during this period very important changes have taken place. A good account of conditions in 1919 and of subsequent changes up to 1932 has been published by Docters van Leeuwen (1936).

The littoral plant communities were essentially the same in 1919 as they had been in 1906, but in the meanwhile there had been much erosion of the beach and both the *Pes-caprae* and the *Barringtonia* formations were reduced in area; the *Pes-caprae* formation formed only a very narrow strip in 1919 and the *Barringtonia* formation had also been washed away in many places. The composition of both these communities was fairly typical by 1919, though even by 1932 they did not include all the species which are found in similar vegetation in Java and Sumatra. Docters van Leeuwen regards the *Pes-caprae* and *Barringtonia* communities as permanent, stable and not liable to change into anything else (see below, p. 298). The *Casuarina* consocies is, however, certainly seral, for though it still remained in 1919 it was clearly being replaced in many places by the mixed *Macaranga-Ficus* woodland which had first been noted by Backer in 1908. The tall old *Casuarina* trees were being overgrown by lianes; some had already fallen down and others were obviously dying. Beneath them, probably owing to the shade, no young casuarinas were coming up; instead there were young trees of *Macaranga*, *Ficus* spp., *Pipturus incanus*, *Premna integrifolia*, etc. Since 1919 the replacement of the *Casuarina* consocies has continued, though in some places new *Casuarina* woods have developed. According to Docters van Leeuwen (loc. cit. p. 254), the *Casuarina* consocies is sometimes succeeded, not by the *Macaranga-Ficus* associes, but by the *Barringtonia* community.

Up to 1919 the vegetation of the lower slopes behind the beach had looked much as it had done since 1898, but by 1932 it was completely transformed. The strip of mixed *Macaranga-Ficus* woodland first noted in 1908 had developed and extended. It was now so shady that there were few ground herbs, though young trees were abundant. The greater part of the middle region in 1919 was, however, still a grass savanna dominated by *Saccharum spontaneum*, associated with *Imperata* and various dicotyledonous herbs, but everywhere among the grass there were scattered trees growing singly or in groups. These trees were *Macaranga*, *Ficus*, etc. and foreshadowed the development of mixed woodland from the savanna. In some places the shade under the trees was sufficient to suppress the grasses and allow shade species, such as the ground orchid, *Nervilia aragoana* to grow. This tendency towards the development of woodland was much more marked in the ravines where a luxuriant growth of trees and shrubs replaced the grasses and there was even some accumulation of humus. At low altitudes in the ravines *Macaranga* was the commonest tree; higher up the composition was more mixed. On the steep tufa walls, on the other hand, the succession lagged behind; where the surface was sufficiently stable the grass *Pogonatherum* was dominant.

After 1919 the groups of trees steadily increased in size and number until nearly the whole savanna was converted into a mixed *Macaranga-Ficus* wood. By 1928 the whole south-eastern side of the island was wooded up to about 400 m. above sea-level. This woodland had spread both upwards from below and outwards from the ravines. During the transition from savanna to woodland the fern *Nephrolepis* was sometimes locally dominant for a time. The *Saccharum* consocies still survived on ridges here and there, but by 1931 it had almost gone, though whenever the soil is laid bare by a landslip it reappears as a stage in the subsere. The mixed woodland seems to be similar in most respects to certain types of the secondary forest (Chapter 17) which appears everywhere in the Malayan region after the primary Rain forest has been cleared. Though lianes are very abundant, the mixed wood is not impenetrable. In many places the chief plant in the undergrowth is *Nephrolepis*, either alone or mixed with *Selaginella plana*. Here and there *Costus sericeus* is seen. This mixed wood is richer in species than the grass savanna, but still far less rich than primary Rain forest.

One of the most valuable results of Docters van Leeuwen's 1919 expedition was the exploration of the highest part of the island. Almost all of this was found to be covered with a dense growth of the shrub *Cyrtandra sulcata*, with very few other species intermixed. In the ravines above 400 m. this plant became increasingly common and at still higher altitudes it was dominant everywhere except on the ridges, where *Saccharum*, ferns and a few other herbs were found. The branches of the *Cyrtandra* were loaded with epiphytic ferns and bryophytes, giving the community the aspect of a 'mossy' Montane forest (Chapter 16). Since 1919 the *Cyrtandra* consocies has undergone as striking a change as the grass community at lower altitudes. The tree *Nauclea purpurascens*, which in 1919 occurred scattered through the *Cyrtandra* scrub, especially in ravines, has become more and more common. By 1929 it had become dominant in the *Cyrtandra* zone up to 750 m. and old trees had reached a height of 25–30 m. In 1932 *Nauclea* had spread still more and it will clearly not be long before most of the summit region consists of a *Nauclea* forest with *Cyrtandra* as an undershrub.

General view of the successions

From the account which has been given, it will be seen that it is possible to construct a fairly good picture of the main outlines of succession on Krakatau since the eruption, though there are many gaps which cannot be filled. For instance we do not know when or how the *Cyrtandra* consocies established itself in place of the fern community which may be presumed to have preceded it. Some further information may be gained by studying the colonization of Anak Krakatau, a volcanic crater which has emerged from the sea near Krakatau during a period of renewed volcanic activity which began in 1927. The general course of events on Krakatau can be summarized in a diagram (Fig. 35).

The successions on the neighbouring Verlaten and Lang Islands, have been, as far as is known, similar to those on the main island.

The development of vegetation on Krakatau has not yet reached a stable climax stage, but the general course of future changes can be predicted with some confidence, at least for the middle and upper regions of the island. In the former it may be expected that the *Macaranga-Ficus* woodland will develop by a series of changes into a stable climax rain forest to some extent similar to the mixed primary rain forest of the neighbouring parts of Sumatra and Java. How long this development will take it is difficult to guess, but the study of secondary

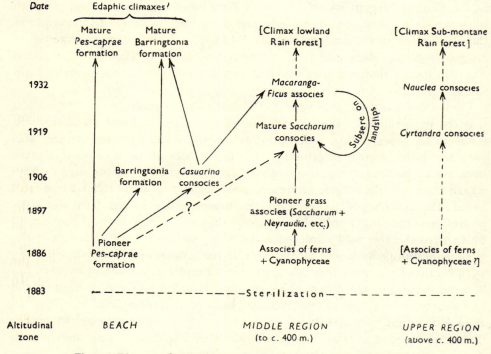

Fig. 35. Diagram of successions on Krakatau since the eruption of 1883.

successions (Chapter 17) suggests that it will be much longer than from the great eruption to the present day. Owing to the isolation of Krakatau it is unlikely that the number of species in this forest will ever approach that in the Rain forest of the larger islands, which in the distant past were connected to the mainland of Asia by land bridges. The zone above 400 m. is evidently developing towards a different climax and there can be little doubt that this will resemble the Indo-Malayan Montane Rain forest (see Chapter 16); on large mountain masses this climax does not appear till comparatively high altitudes, but on isolated peaks on small islands, such as Krakatau, it may begin to occur within a few hundred metres of sea-level (see p. 347).

The succession in the middle region of Krakatau is an excellent example of a xerosere, or primary succession on bare rock, but, except in its earliest stages, it differs little, as will appear in Chapter 17, from the secondary succession on abandoned cultivated land, etc. Its outstanding feature is the successive dominance, first of cryptogams—here chiefly algae and ferns—later of herbaceous flowering plants and finally of trees. This is typical of nearly all xeroseres in regions where the natural climax is forest; but in a secondary succession on land already covered with soil the cryptogamic stage is omitted. On bare rock lichens and bryophytes generally act with algae and assist in soil formation as pioneers; bryophytes were observed on Krakatau as early as 1886, but neither they nor lichens seem to have been of much importance during the earlier phases of the succession. The part of algae, however, was a significant one and though Treub in 1886 found ferns and flowering plants as well as algae, the algae with bacteria may have been the earliest colonists. All these algae, it should be noted, were blue-greens or diatoms; no green algae were recorded. This is not surprising; in the damp tropics Cyanophyceae are usually the dominant algae on bare rock or soil (Fritsch, 1907, 1936).

The temporary dominance of ferns is a less usual feature. It is characteristic of some, but not of all, secondary successions in the rain-forest region. At Krakatau it may have been due, partly to the early arrival of spores, which are of course easily dispersed by the wind, and partly to the tolerance of ferns for soils deficient in mineral nutrients (see p. 398).

TABLE 31. *Life-forms of flowering plants on Krakatau.* (After Docters van Leeuwen, 1936, p. 191)

Date	Number of species (flowering plants)	Phanerophytes (%)	Chamaephytes (%)	Hemicryptophytes (%)	Cryptophytes (%)	Therophytes (%)
1886	15	59	7	20	7	7
1897	49	70	6	14	6	4
1906	73	78	6	8	3	5
1934	219	65	6	12	2	15

Though the middle region of Krakatau shows clearly phases in which cryptogams, herbs and trees are dominant in turn, it is an interesting fact that the Raunkiaer life-form spectrum (for all three regions taken together) has remained approximately constant throughout these phases. Docters van Leeuwen gives figures (Table 31) for the proportion of different life-forms. Phanerophytes have thus formed the majority of the species in the flora of Krakatau from the earliest stages of the successions. The statistics given in Chapter 1 clearly showed the overwhelming dominance of phanerophytes in climax Rain forest; it is significant that this dominance is also shown, though not to such an overwhelming extent, throughout the primary xerosere. Subsequent chapters will show that a preponderance of phanerophytes, during most of their stages, is a feature of all successions, primary and secondary, which culminate in the development of Rain forest. This is in agreement with the conclusion of Raunkiaer (1934, see especially pp. 197–8), who maintained that the biological spectrum of primary and secondary seral vegetation (because it is determined primarily by the climate) is similar to that for the whole flora of the same region. It should be noted, however, that hemicryptophytes and therophytes, which are absent from the flora of climax Rain forest, play a considerable part in the seral vegetation of Krakatau.

In the Krakatau successions, as in all successions on bare ground in fairly moist climates, there has been a change from open to closed vegetation. Already at the time of Treub's second visit (14 years after the eruption) there were closed plant communities in many places. Afterwards more and more of the vegetation became closed, though even at the present time open communities persist in places where the succession lags behind, as on unstable soil and steep rock walls.

In one point the Krakatau successions differ from most, perhaps from all, primary successions in temperate regions, that is, in the continual increase in the number of species. There are no exact figures for the number of species in the various communities separately, but Docters van Leeuwen (loc. cit. p. 209) gives the following totals (angiosperms and gymnosperms) for the whole of Krakatau and the two neighbouring islets, Lang Island and Verlaten Island:

1883	1886	1897	1908	1920	1928	1934
0	26	64	115	184	214	271

In the colonization of bare ground in temperate regions, the number of species is at first small, rises to a maximum and later declines as the climax is approached (Warming, 1909, p. 356); the decline is usually contemporaneous with the change from an open to a closed community. On Krakatau—and this appears to be true for the beach and summit, as well as for the middle region—there are no signs of a decline in the number of species, only of a steady increase. The *Saccharum* consocies is richer than the fern associes which preceded it, and the *Macaranga-Ficus* woodland is in turn richer than the *Saccharum* consocies. If, as we expect, the *Macaranga-Ficus* woodland is at length replaced by a mixed Rain forest, the number of species will certainly increase still further. A continual increase in species up to the climax stage, without a subsequent decline, is not peculiar to Krakatau, but is characteristic of tropical successions in general.

The change on Krakatau from open to closed vegetation is accompanied by a change from a mixed community to one dominated by a single species; thus the fern community seems to have been a mixed associes, while in the grass community which succeeded it the single species, *Saccharum spontaneum*, soon became the sole dominant. Later on, however, mixed communities reappear; the *Macaranga-Ficus* woodland, like the climax Rain forest which may ultimately be expected to succeed it, is a community with several dominants. The decline in the number of species in the later stages of most temperate successions is generally attributed to the increasing intensity of competition, but since competition also increases in intensity during tropical successions without any corresponding decline in the number of species, this cannot be the only cause. The smaller number of species in the final, as compared with the middle, stages of temperate successions appears to be really due to the fact that temperate climaxes are usually dominated by one, or only a small number of dominants, which is a consequence of the poverty of the flora in woody species.

In analysing the causes behind any plant succession three groups of factors have to be taken into account: first the supply of seeds, fruits and other dispersal units, secondly the competitive and co-operative effects of plant on plant and finally the changes in the habitat brought about by the climate and the reactions of the vegetation itself. The interactions of these three groups of factors are of course exceedingly complex. Species appear and disappear and in any given case it is generally impossible to trace the appearance or disappearance to a single factor.

TABLE 32. *Means of dispersal of seed-plants (angiosperms and gymnosperms) on Krakatau, Lang Island and Verlaten Island.* (Data from Docters van Leeuwen, 1936, p. 240)

	Percentage of species					
Means of dispersal:	1886	1897	1908	1920	1928	1934
Wind	62	44	28	35	42	41
Sea	38	47	52	36	33	28
Animal	0	9	20	19	23	25
Man	0	0	0	9	2	6
Total no. of species	26	64	115	184	214	271

In the recolonization of Krakatau the supply of dispersal units must obviously have been of great importance. Even if we assume that none of the original flora survived, we can only deduce the most probable method of arrival of each species from whether its seeds, fruits or spores appear to be adapted to wind, water, animal or human dispersal. The figures in Table 32 have been calculated on this basis. These figures show that the proportion of wind-borne species was slightly, but probably significantly, higher in 1886 than later on. Animal-carried species did not appear at first, while human-dispersed species were absent till after the arrival of the German settler in 1916. We are certainly justified in concluding from these data that difficulties of dispersal have acted as a limiting factor to the rate of succession, in other words that some species were late in appearing not because the habitat was not yet suitable for them, but because their dispersal units did not happen to get carried to the island. That even now there must be many species absent on Krakatau only because they have been unable to surmount the sea-barrier is shown by the absence of the Loranthaceae, a family of parasites which is abundant in both Sumatra and Java. Many species for which the island provides suitable habitats will doubtless never succeed in reaching it.

Changes in the environment

While the means of dispersal of the species has certainly greatly influenced the plant successions on Krakatau, we must not underestimate the part played by changes in the environment. Immediately after the eruption in 1883 Krakatau was not a very attractive habitat for plant life. The material thrown out by the volcano covered the whole surface of the island to a depth of many metres and was loose, shifting and extremely porous; it consisted of pumice blocks of all sizes and of volcanic ash or pumice sand. The unattractiveness of this substratum was increased by the complete lack of shelter from sun or wind. Chemically it was not so unfavourable as might be expected, for analysis (Ernst, 1908, p. 49) showed that it contained most of the ordinary plant nutrients excepting combined nitrogen and phosphorus, many of them in a soluble form. In course of time the pumice sand, which was not a soil in the strict sense, but merely a soil-forming material, developed into a true soil. When the profile was examined in 1928 (van Baren, 1931) analysis showed that a remarkable amount of leaching of sodium and potassium had taken place. The silica/alumina ratio (p. 208) of the soil was smaller than in the parent material, indicating that the lateritic weathering characteristic of volcanic rocks in wet tropical climates had already made some progress.

By 1886 erosion had already produced four different types of substratum from the original uniform covering, viz. pumice sand, which still covered the greater part of the area, sea sand mixed with pumice on the shore, the soil of the ravines consisting partly of old unweathered volcanic soil and partly of eroded pumice sand, and finally bare unweathered lava. By the action of the climate, bacteria and the newly arrived plant life itself, the pumice sand gradually became a comparatively favourable medium for plant growth. Nitrogen compounds were added to it, first by rain alone, later also by the activities of nitrogen-fixing organisms. When soil bacteria first began to appear is not known, but Ernst (1908, p. 51) examined soil samples in 1906 and found bacteria almost as numerous as in the soils of Java; organisms of several different biological types, including nitrogen-fixers, were found. The abundance of Cyanophyceae in the early stages of recolonization is perhaps significant, since certain genera have been claimed to have the power of fixing atmospheric nitrogen. How the phosphorus, which must by now be present in the soil, has arrived is not clear; presumably it must have come either in blown sea-spray or in atmospheric dust.

Once a fairly continuous plant cover had been built up, the environment must have become rapidly more favourable to plants. The shading of the soil would completely alter the microclimate at its surface, raising the average humidity and decreasing the range of temperature. The roots of plants would help to bind the sand, making it less loose and unstable. The dead remains of the vegetation would also add humus to the soil, modifying both its physical and chemical properties. As the vegetation became taller and herbaceous plants gave place to trees all these effects would become more pronounced.

Thus, as in all primary successions, the plants have modified the environment so as to make it more favourable for themselves. Less exigent species have given way to more exigent, with which they cannot compete under the conditions they have themselves helped to create. The general trend is from a xerophytic to a more mesophytic environment, in other words, from more to less extreme conditions.

OTHER XEROSERES

The greater part of this chapter has been devoted to the successions on Krakatau because they are the most striking, and best documented, examples of successions from bare rock in a climate where the Tropical Rain forest is the climax type of vegetation. Opportunities for observing xeroseres in the tropics are, however, not infrequent, particularly in regions of volcanic activity, though few records of such successions have been kept. A promising field for such investigations is the volcanic islands of the Samoa group in the Pacific, where there are lava flows of various ages, some of them of known date. There is a typical humid rain-forest climate and various stages in the development of forest on the lava flows can be seen. According to Vaupel (1910), the lava flows on the north side of the island of Savaii, formed during an eruption in 1905, were entirely bare of vegetation when he examined them (1906–7). Other lava flows on the same island, about 100–150 years old, were already covered in favourable places with a shade-giving growth of trees, rooted in the detritus in the cracks of the lava, on soil accumulated on its surface or sometimes perhaps penetrating right through it to the soil beneath. Older residents could remember that 50 years earlier there had been nothing but ferns and small bushes on these areas. The young lava fields of Samoa are also mentioned by Rechinger (1910, pp. 225–7), who says they are termed 'mu' (burning) by the natives, owing to the high temperatures reached by the bare black lava in the sun. He describes the vegetation (clearly a fairly early stage in the succession) as xerophytic in character and consisting chiefly of low bushes with hard leathery leaves, such as *Morinda citrifolia*, species of *Gardenia*, *Fagraea* and *Loranthus*, growing with xerophilous pteridophytes such as *Nephrolepis rufescens*, and *Lycopodium cernuum*, as well as a few grasses and the parasitic *Cassytha filiformis*. The other lava flows carry a well developed closed vegetation consisting of a dense undergrowth from which giant trees of *Eugenia* spp., *Rhus simarubifolia*, *Ficus* spp., *Calophyllum speciosum*, etc. rise up. In these forests mosses, orchids and grasses are common on the ground and ferns and species of *Freycinetia* on the tree-trunks.

Beard (1945) has described the successions on the Soufriere of St Vincent, a volcano in the West Indies which was devastated by an eruption in 1902. The successions at various altitudes on the mountain show much similarity to the successions at corresponding altitudes on Krakatau. On the Soufriere, in the altitudinal zone within which Tropical Rain forest is the climax, the primary succession has closely resembled the secondary succession in the same region. The earliest tree colonists are light-demanding species with efficient means of seed dispersal, e.g. *Ochroma* and *Cecropia* spp.; the same species frequently occur as dominants in young secondary forest on abandoned cultivated land (p. 397). In the higher zones on the mountain the seral communities tend to resemble in physiognomy formations which occur elsewhere as climax vegetation at much higher altitudes; thus there are associes similar to Elfin woodland (Chapter 16) paramo or alpine tropical grassland, and alpine 'tundra'—all these are like types of vegetation normally found much higher. Beard ascribes this to the less favourable, more xerophytic, conditions during the earlier stages of succession. Hardy (1939) examined the soils of the Soufriere in 1933 and found that the content of nitrogen and organic matter was comparable with that in the cultivated soils of St Vincent. Thus within some 30 years a soil had developed which was probably capable of supporting climax vegetation.

Other information on xeroseres in the wet tropics is very scanty. Docters van Leeuwen (1936) reviews a number of instances of colonization of land devastated by volcanoes but, since in most of these cases the original vegetation was only partly destroyed, the subsequent successions are secondary or mixed rather than truly primary. Vaughan & Wiehe (1937, 1941) have described various plant communities in Mauritius which are believed to be stages in the xerosere leading to the Upland Climax forest (a type of Tropical Rain forest, see Chapter 11). They emphasize that this is a succession dominated by woody plants; herbs and grasses play little part in it. Chevalier (1909) has briefly described an interesting succession on bare granite rocks in the Fouta Djalon mountains in West Africa. The climax here is probably Mixed Deciduous forest (see Chapter 13) rather than true evergreen Rain forest. The first colonist is a sedge, *Trilepis pilosa*. The succession makes little progress so long as this plant remains alive, but when it is killed by drought or grass fires, seeds of various other plants afterwards germinate on the soil it has formed. A community of herbaceous plants grows up, which is later invaded by woody plants. A similar succession can be observed in Southern Nigeria, e.g. on the Idanre Mountains.

CHAPTER 13

HYDROSERES

In fresh-water lakes, ponds and the slower reaches of rivers the land constantly tends to encroach on the water. In this process of *Verlandung*, or formation of new land from water, vegetation plays an essential part. Aquatic plants offer resistance to water movements, slowing them down and so increasing the rate of deposition of suspended matter. Later amphibious plants help to consolidate the sediments and by adding to them their dead remains, change their texture and increase their bulk. This leads to a general rise in the level of the ground relative to that of the water and accompanying the rise there is a hydrosere or plant succession starting with various types of aquatic vegetation and usually ending, in a forest climate, with a type of forest similar to, though perhaps not quite identical with, the climatic climax. During the course of the hydrosere there is a gradual change from open water to relatively dry conditions and as with the xerosere the vegetation itself tends to produce a mesophytic environment.

In the tropics examples of hydroseres, often on a grandiose scale, are plentiful. By the smaller rivers, and in the swamps and lagoons which occupy large areas in many tropical lowlands, every stage in the succession from open water to closed forest can often be seen. The banks of the large tropical rivers also sometimes show well developed hydroseres, but, as they are generally densely populated, their natural fringe of vegetation is often lost or much modified. Circumstances do not as a rule allow continuous observation of the changes in the vegetation in any one spot, but the successive stages of the hydrosere can be inferred with reasonable certainty from the zonation of the plant communities.

The complete story of the hydrosere has scarcely been worked out anywhere within the area of the Tropical Rain forest, but there are many more or less fragmentary observations which can be pieced together. The well-known fact that aquatic and marsh plants are usually far more widely distributed than those of dry land makes this relatively easy. In the New World, for instance, many of the species taking part in the hydroseres range through the whole Tropical zone and much the same is true of tropical Africa and the Indo-Malayan region. Some species of water plants are even common to the Temperate and Tropical zones and some important genera of the tropical hydrosere are represented by closely allied species in the Old and New Worlds, e.g. *Raphia, Carapa, Symphonia*. The hydrosere therefore tends to be similar in widely separated regions, not only in the general course of events, but in the structure and physiognomy of its successive phases. Its earlier stages are to a considerable degree unaffected by differences in climate.

The local variations which are met with probably depend on differences in the water-level régime (season and height of flooding), differences in the nature of the sediments deposited and chemical differences in the water itself, but the influence of these factors has been little studied.

HYDROSERES IN TROPICAL AMERICA

A straightforward hydrosere has been described by Kenoyer (1929) from Gatun Lake in the Panama Canal. Five stages are distinguished:

(1) Floating aquatic associes. *Salvinia auriculata, Jussieua natans, Utricularia mixta, Pistia stratiotes, Eichornia azurea.*

(2) Water-lily associes. *Nymphaea ampla*, together with the preceding species.

(3) Emergent aquatic associes. Most abundant species: *Typha angustifolia* and the fern *Acrostichum danaeifolium*, associated with *Crinum erubescens, Hibiscus sororius* and *Sagittaria lancifolia.*

(4) Reed swamp associes. This consists of *Cyperus giganteus, Scirpus cubensis* and other Cyperaceae, together with large grasses such as *Phragmites communis* and *Gynerium sagittatum*, the dicotyledonous herb *Jussieua suffruticosa* and ferns.

(5) 'Marsh scrub' (carr) associes. *Dalbergia ecastophylla* and the tall aroid *Montrichardia arborescens.*

The 'marsh scrub' community would probably in time give place to swamp forest, but such a community has not yet developed.

This very simple succession is very like many temperate hydroseres and several temperate genera and even species take part in it. It is also very similar to hydroseres leading to a now completely destroyed 'mesophytic forest' climax in Porto Rico (Gleason & Cook, 1926). Successions from herbaceous swamp to Fresh-water Swamp forest also occur in Trinidad and have been briefly described by Marshall (1934). Beard (1946*b*) also describes the swamp and 'marsh'(seasonal swamp) communities of Trinidad, but does not discuss their successional relationships. The 'swamp savannas' or 'water savannas' which occupy large areas in the coastal region of Surinam (Lanjouw, 1936), British Guiana and Venezuela are treeless swamps and are stages in hydroseres, but here also the successions have not yet been studied.

Hydroseres on the banks of large rivers are somewhat different from those in lakes and swamps as the conditions are complicated by the scouring effect of the current, seasonal changes of water-level and, near the sea, by tides. Perhaps the most impressive example anywhere in the world of this kind of succession is that seen in the *furos* or channels of the delta region between the mouth of the Amazon proper and the Rio Pará. The water here is tidal, but more or less fresh. New islands of alluvium are continually being built up and consolidated by the growth of vegetation. The first plants to colonize the islands are supplanted by others which can only establish themselves in their shelter. As the succession proceeds the island grows in area and the earlier colonists are pushed outwards to form a marginal belt. Round the larger islands there is thus a zonation representing successive stages in the hydrosere. A similar succession and zonation can be observed on the convex banks of meanders in the channels. These successions have been excellently described by Huber (1902) and further details are given by Bouillienne (1930).

The first pioneers on the mud are *Montrichardia arborescens*, a robust aroid growing to 2–3 m., and *Drepanocarpus* (*Machaerium*) *lunatus*, a leguminous shrub. These two species form practically pure communities, one seeming to exclude the other; often the two consocies alternate every 10 m. or so. Both species have seeds which float and are abundant in the river drift. Germination is exceptionally rapid and the seedlings can establish themselves between two successive tides; in this respect these plants resemble mangroves. In the shelter provided by the Montrichardietum and Drepanocarpetum a quite different kind of vegetation develops, a 'sudd' or floating sward consisting of masses of the grass *Hymenachne amplexicaulis*, together with *Eichornia azurea* and *E. crassipes*. This last community is much more extensively developed higher up the Amazon above the influence of the tides. There it may become rooted and play a part in the succession, but in the *furos* region its role is not important.

The second stage in the succession is the invasion of the *Montrichardia* and *Drepanocarpus* communities by the mangrove *Rhizophora racemosa*. This forms a consocies extending as far upstream as the influence of the tides; it is not, like most mangroves, confined to salt and brackish water (see p. 304). In places *Rhizophora* is replaced by other mangroves, *Avicennia tomentosa*, *A. nitida* and *Laguncularia*; these too grow here in water which is said to be never salt or even brackish. The mangroves soon shade out *Montrichardia* and *Drepanocarpus* and beneath them there is only a very scanty undergrowth.

After a time seedlings of the palms *Euterpe oleracea* and *Mauritia flexuosa* and of the dicotyledonous trees *Cecropia palmata* and *C. paraensis* establish themselves beneath the mangroves. The genus *Cecropia* is one which is also very important in the secondary succession throughout the tropical American Rain forest (p. 397). The palms and cecropias soon overtop and suppress the mangroves; they may lead directly to the establishment of a *varzea* or Fresh-water Swamp forest liable to flooding. This consists of gigantic trees and is similar in general aspect to the climax Rain forest. At other times (the accounts are not quite clear about this) the succession seems to follow a different course and the *Rhizophora* community is followed by one of the palm *Raphia vinifera* var. *taedigera*. Behind the *Rhizophora* or *Raphia* belt there may be either one dominated by the tall palm *Mauritia flexuosa* growing as a pure community or intermixed with other palms such as *Euterpe oleracea*, *Manicaria saccifera* and *Maximiliana regia*, or a belt of Cecropietum dominated by the two species mentioned above. The trees *Virola surinamensis* and *Carapa guianensis* are characteristic of the palm and *Cecropia* belts. The reasons for these variations in the succession are not known, but whatever the course of events in the intermediate stages, *varzea* forest is eventually developed.

The *varzea* varies much in character in different places. Its average height is from 15 to 30 m. Very characteristic is its irregular profile in which the huge dome-shaped crowns of *Ceiba pentandra* are conspicuous. Other conspicuous trees in this swamp forest are *Carapa guianensis*, *Hevea brasiliensis* and *H. guyanensis*, *Symphonia globulifera*, *Virola surinamensis* and many Leguminosae. Palms are frequent and where the forest reaches the water's edge there is often a fringe of *Pachira aquatica*.

This *varzea* forest of the *furos* region of the Amazon, as described by Huber and Bouillienne, is very similar in structure and floristic composition to the Mora forests of the British Guiana flood-plains, dealt with in Chapter 10. In the *varzea*, however, *Mora* itself is entirely absent and it is not known whether there is any tendency to single-species dominance. *Varzea* and Mora forest may both be regarded as edaphic climaxes.

Much less is known about the hydroseres on the lower Amazon above the delta region, though some information can be gleaned from papers by Bouillienne (1930) and Ule (1908 a). Here the tides are smaller, but seasonal floods cause large variations in the water-level. The climate is more continental than in the delta. Bordering the river there is a *restinga* or 'dam' formed by the deposition of the coarsest sediments; beyond it the ground is lower and more swampy. The following seral communities have been described: (i) various types of floating aquatic vegetation, including a community dominated by the giant water-lily, *Victoria amazonica* and a 'sudd' of *Echinochloa polystachya* and *Eichornia*; (ii) *campos de varzea*, grassy swamps, which, like the sudd, are chiefly dominated by *Echinochloa polystachya*, *Hymenachne amplexicaulis* and *Paspalum repens*; (iii) a zone of the shrubs *Salix humboldtiana* and *Alchornea castaneifolia* which lines the river banks. Round the edge of the *campos de varzea*, forest starts abruptly. The margin of this forest is, (iv) a zone of Cecropietum formed of several species, including *C. palmata* and *C. paraensis*; it consists of groups of trees rather than a continuous belt. Each group is usually even-aged and the Cecropietum lasts for only one generation (see p. 399). The succeeding community is (v) *varzea* forest which grows mainly on the *restingas* or 'dams' and so consists mainly of narrow strips. Two successional stages of *varzea* are recognized, young *varzea*, in which the chief trees are quick-growing species such as *Calycophyllum spruceanum*, *Inga* spp. and *Triplaris surinamensis*, and mature *varzea* forest, which is similar in composition to that of the delta region, except that the most abundant palm is *Astrocaryum jauari*, *Mauritia* being almost absent. In the eastern part of the lower Amazon these *varzea* forests are strikingly less luxuriant than those of the delta, probably owing to the drier climate, but further up the river, where the climate becomes moister, they again become taller.

Ule (1908 a) has briefly described the succession on sandbanks in the Juruá, a tributary of the upper Amazon. Here the first stage is a herbaceous marsh consisting of grasses and various other herbs. When the bank becomes higher *Salix humboldtiana* and *Alchornea castaneifolia* appear, followed by *Cecropia* spp.

On the 'blackwater' tributaries of the Amazon (see p. 213), such as the Rio Negro, the riverside vegetation, according to Ule (loc. cit.) is quite different from that of the main river and 'whitewater' tributaries, the *Cecropia* communities and some of the commonest species of the other rivers being quite absent.

Some data on the succession on river banks in British Guiana are given by Rodway (1894), Davis (1929) and Myers (1935). On the slower-flowing parts of the Guiana rivers the early stages in the colonization of alluvium are similar to those of the Amazon delta. *Montrichardia arborescens* is the first species to arrive and it is followed by a scrub consisting of *Drepanocarpus lunatus* and one or two other leguminous species. As in the Amazon there is a floating sudd which anchors itself to the *Montrichardia* or to the bushes. The chief grass in the sudd is *Panicum elephantipes*. The later stages of the succession have not been worked out, but Davis has described a number of communities from the north-west district of British Guiana which are probably late stages in the hydrosere. Where the water is brackish the river banks have a narrow fringe of the mangrove *Rhizophora mangle*. In the more or less brackish swamps behind this there is 'truli bush', a type of forest consisting of dicotyledonous trees with an understory consisting largely of the palm *Manicaria saccifera*. Where the swamps are extensive and the water-level rather lower 'truli bush' is replaced by 'kirikowa bush' dominated by *Iryanthera paraensis*. Other types of swamp forest in this district are 'ité savannah', dominated by the palm *Mauritia flexuosa* with an understory of *Clusia cuneata*, etc., and Mora forest dominated by *Mora excelsa*; this is like the Mora forest at Moraballi Creek, described in Chapter 10, except that palms are more abundant.

Myers (1935) describes the types of river-bank vegetation met with successively in ascending the Guiana rivers. Near the estuaries there is Rhizophoretum, which is sometimes associated with sudd. Next comes a fringe of *Pachira aquatica* and *Pterocarpus officinalis*, associated with stands of *Montrichardia* and *Drepanocarpus lunatus*. Above this is a belt of mixed swamp forest, the edge of which is hidden by a thick curtain of creepers. Next there is swamp forest with no differentiated fringe and finally, at a considerable distance from the sea, tall Rain forest comes right down to the river banks. These various kinds of marginal vegetation do not, of course, represent a straightforward sequence of stages in a hydrosere but, as Myers points out, depend partly on factors such as the diminishing width of the stream, decreasing salinity and the change in the character of the sediments. The change in the bank vegetation along the Guiana rivers is exactly analogous to the change, alluded to by Corner (1940, p. 42) and Symington (1943, p. xviii), seen in passing up Malayan rivers from mangrove to riparian forest of *Dipterocarpus oblongifolius* (neram), etc. and then to a community in which species of *Saraca* are abundant ('*Saraca* streams').

HYDROSERES IN TROPICAL AFRICA

By the lakes and slow rivers of Africa stages of the hydrosere can often be seen to great advantage, especially on the west coast, where there are chains of shallow fresh-water lagoons hundreds of miles long, into which the rivers discharge. The successions, however, have been little studied.

In south-western Nigeria (Richards, 1939) the succession in the lagoons and waterways begins with submerged and free-floating aquatic communities. The former consist of such plants as *Ceratophyllum demersum* and *Utricularia* spp., the latter of a well developed sudd of grasses, *Pistia*, etc. The sudd drifts about in large masses and may become a serious obstacle to navigation. When the water becomes shallow enough, a rooted floating-leaf community appears, usually dominated by *Nymphaea lotus*. These purely aquatic communities are succeeded by a herbaceous swamp consisting of grasses or *Cyperus papyrus*, or by a community of *Pandanus* sp. Sometimes this may be followed by a stage dominated by shrubs, such as *Alchornea cordifolia*, but generally the next stage seems to be a palm swamp dominated by one (or more?) species of *Raphia*. This leads to the development of tall Fresh-water Swamp forest consisting chiefly of dicotyledonous trees.

This Swamp forest has a general similarity to climax Rain forest, but is usually more open and irregular in structure; floristically it is usually poorer than the primary forest of dry ground, but many species are common to both. The Fresh-water Swamp forest of West Africa probably varies much in character according to local conditions, but a type which is probably widespread has been described by Richards (1939, p. 42) from the Shasha Forest Reserve in Nigeria. In this forest tall trees are abundant, but not evenly distributed, so that patches of deep shade with little undergrowth are mingled with gaps filled only with thickets of bushes and lianes. The dominant plant in these thickets is *Leea guineensis*, a straggling shrub supported by prop roots. The commonest tree is *Mitragyna* sp. (probably *M. ciliata*), which, though scarcely dominant, forms as much as 36 % of trees 16 in. diam. and over. The abundance of this or a closely related species is characteristic of fresh-water swamps over a wide area in tropical Africa; on islands and low-lying flooded river banks in the Bas-Kasaï and Lake Leopold II districts of the Belgian Congo *Mitragyna stipulosa* (or more likely its recently distinguished segregate, *M. ciliata*) forms pure stands 'like veritable plantations' (Vermoesen, 1931, p. 181, quoting E. & M. Laurent). Some of the subordinate species of the Shasha Swamp forest are confined to swamps, others occur also on dry ground. In the latter group are a number of species which on dry ground are commoner in secondary than in mature forest. The frequency of these in the Swamp forest is probably accounted for by the relatively good illumination.

The succession from open water to tall Fresh-water Swamp forest can be seen in a telescoped form as a zonation on the banks of the larger Nigerian rivers. On the Osse, for instance, the successive zones are: (i) submerged aquatics and sudd of *Pistia*, etc.; (ii) Pandanetum; (iii) Raphietum; (iv) Fresh-water Swamp forest.

Chipp (1927), in his account of the vegetation of the Gold Coast, describes a number of seral stages very similar to those seen in south-western Nigeria. Among the later seral units which he recognizes are: (i) 'Lagoon Marsh associes', consisting chiefly of grasses, a species of *Anadelphia* being one of the dominants; (ii) a *Calamus-Ancistrocladus-Raphia* associes or pure consocies of *Raphia hookeri*, which appears to be very similar to the palm swamps of Nigeria; (iii) a *Tarrietia-Anopyxis* associes, corresponding to the Fresh-water Swamp forest of Nigeria.

From Angola, Gossweiler & Mendonça (1939) have described a number of communities which are evidently similar to stages in the hydrosere of Nigeria and the Gold Coast. Eggeling (1935) describes in some detail a hydrosere in the Namanve swamp on Lake Victoria in East Africa. Though the altitude is 3730 ft. (1140 m.), the general course of the primary succession is remarkably like that of the West African hydroseres. The stages recognized are:

(1) Floating-leaf and submerged aquatic vegetation. Species of *Nymphaea*, *Ceratophyllum*, *Trapa*, etc.

(2) 'Fern and sedge community.' A mixture of various ferns, sedges, grasses and other herbaceous plants.

(3) *Limnophyton* swamp. An associes dominated by *Cyperus papyrus* and the grass *Miscanthidium violaceum*, with *Limnophyton obtusifolium* as subdominant.

(4) Papyrus swamp. *Cyperus papyrus* alone dominant, but with many subsidiary species.

(5) *Phoenix* swamp. Trees 20–30 ft. (6–9 m.) high, including dicotyledons such as *Mitragyna stipulosa* as well as the palm *Phoenix reclinata*.

(6) Rain forest.

The Papyrus swamp may pass directly into *Phoenix* swamp or there may be an intervening stage dominated by *Miscanthidium*. The *Phoenix* swamp forms a narrow belt between the Papyrus and *Miscanthidium* communities on one side and the surrounding forest on the other. Towards the forest margin another palm, *Raphia monbuttorum*, is found.

In the great rain-forest massif of the Congo basin hydroseral stages must be extensively developed, but very little information about them is available. Lebrun (1936a), however, distinguishes two main types of swamp forest in the Congo: (a) liable to seasonal flooding, soil firm and dry at other times, (b) permanently swampy (*forêt marécageuse*). The former, which seems to resemble the Nigerian Swamp forest described above, has a less dense tree cover than the forest of terra firma; it is relatively poor in species and the chief genera of trees include *Chrysophyllum*, *Cynometra*, *Maba*, *Oubanguia* and *Uapaca*. A herb layer is lacking except for a few species with bulbs or rhizomes which develop aboveground shoots during the season of low water. The second type, richer in species, has more trees per unit area than the dry-land forest, though none of them becomes very large. Most of the trunks are buttressed or stilt-rooted. The 'dominant' species include *Chrysophyllum laurentii, Uvariastrum pynaertii* and species of *Parinari, Mitragyna, Copaifera* and *Uapaca*. Species of these last three genera tend to form more or less homogeneous stands. The undergrowth is rich in species, but the herb layer is not continuous. A number of aquatic and riverside communities are also described in the same paper.

Trochain (1940) has given a full account of the hydrosere and aquatic plant communities in Senegal. Though the climax is here 'climatic savanna', the earlier seral stages are similar to those in forest regions.

HYDROSERES IN THE EASTERN TROPICS

In the eastern tropics ecological and phytogeographical studies have been much hampered by the extreme richness of the flora. It is not surprising therefore that information about hydroseres in the rain-forest region of Malaysia and tropical Asia is exceedingly meagre. No connected account of a hydrosere in even one locality seems to have been published, though a large number of marsh and swamp plant communities have been described.

Various submerged and floating communities in the lakes of Java and Sumatra are dealt with by van Steenis & Ruttner (1932), but most of the lakes in question are at fairly high altitudes and it is uncertain whether their vegetation is similar to lowland waters where the succession might lead to some kind of Rain forest. In the slow rivers of Borneo a sudd is developed and this, together with various types of river and swamp vegetation, is mentioned by van Steenis (1935a) in his fully documented enumeration of Malayan vegetation types, but no indication is given of their successional relations.

Various types of swamp forest, most of which must be regarded as late stages of hydroseres, are also described by Endert (1920) in an account of the forest types of Palembang (Sumatra) and by Champion (1936) in his classification of the forests of India and Burma. Burmese swamp forests have been described by Kurz (1875) and by Stamp & Lord (1923). Corner (1940, p. 42) gives notes on some interesting swamp forest types in the Malay Peninsula. As almost nothing is known of the successional and ecological relations of these swamp forest types, it would be unprofitable to discuss them here in detail. It may be noted, however, that in many of these swamp forests the number of tree species is relatively restricted and there is a tendency for one or a small number of species to be 'gregarious' (dominant). In many parts of the Malayan region a very characteristic consocies of swampy ground is formed by *Melaleuca leucadendron*; this community is often a stage in a subsere or deflected succession after burning (Endert, 1932) and does not seem to belong to the primary hydrosere.

In the lowlands of New Guinea there are vast areas of swamp vegetation, including forests of the sago palm (*Metroxylon*), which provide the staple food for many of the native tribes. Beyond the notes of Lam (1945, chapter 4) and Lane-Poole (1925 b) there are almost no data about these interesting types of vegetation.

SWAMP FOREST CLIMAXES

In Chapter 9 it was pointed out that the majority of tropical swamp soils are not peaty and contain little if any more humus than soils with normal drainage; but where the drainage water is very oligotrophic (poor in dissolved mineral matter) peat formation can occur, leading to the development of ombrogenous moor forests, the tropical equivalent of the raised bogs (*Hochmoore*) of the Temperate zone, on lens-shaped masses of peat many feet in thickness. Corresponding to the two types of swamp soil—the normal (non-peaty) swamp soil with a relatively eutrophic water-supply and the peat soil with an extremely oligotrophic water-supply—there appear to be two distinct types of hydrosere leading to different types of climax forest.

In the course of any hydrosere the level of the soil is gradually raised until it is at, or slightly above, the normal level of the water. In eutrophic waters the raising of the soil level is due mainly to the accumulation of inorganic sediments; in oligotrophic waters it is chiefly the result of the accumulation of plant remains. In either case the soil level rises no further once it has reached the height of the highest water level, since the conditions for silt accretion or peat formation then cease to exist. The hydrosere thus ends with the formation of 'dry' ground, but ground in which the water-table is still comparatively near the surface at least during part of the year. In the tropics such ground is capable of bearing high forest more or less similar to climax forest in structure and physiognomy, but different in floristic composition, because the majority of the species must be specialized and suited to conditions in which the soil is water-logged or imperfectly drained to within a short distance below the surface. Such forest must be regarded as an edaphic rather than a climatic climax. The final stage of a hydrosere thus cannot be typical (climatic climax) Rain forest unless there is some topographical change (e.g. one leading to a general lowering of the water level).

Non-peaty swamp forest

Examples of non-peaty climax swamp forests have been already described, e.g. the Mora forest of British Guiana (Chapter 10), the *varzea* of the Amazon, the Fresh-water Swamp forest of Nigeria, etc. All these communities appear to be the final (edaphic climax) stage of hydroseres in relatively eutrophic waters (conditions unsuitable for peat formation).

Moor forest (peaty swamp forest)

Moor forests, developing in oligotrophic waters, as stated in Chapter 9, are widespread in the rain-forest region, especially in the Malay Peninsula, the Malay islands and New Guinea. Owing to the difficulty of exploring them, they have been little studied. E. Polak (1933) has given a detailed account of the stratigraphy and hydrography of the Moor forests of Sumatra and Borneo, but describes the vegetation itself only very briefly; some information about its floristic composition can be gleaned from Endert (1920), Sewandono (1937) and Symington (1943) (further references in van Steenis, 1935*a*, pp. 89–90).

Moor forests are evergreen and are dominated by dicotyledonous trees. The height of the trees may be as much as 30 m. at the edge of the moor and diminishes gradually towards the centre, where the vegetation may consist of dwarf forest (*Krüppelholz*) interspersed with pools of open water (Polak, loc. cit. p. 16). Though the trees of the Moor forest are smaller in diameter and less tall than those of typical Rain forest, the total volume of timber is considerable and the Moor forests of the Rio islands and Sumatra's east coast have been extensively exploited, mainly by Chinese wood-cutters on the so-called *panglong* system. Seedlings of the chief tree species are abundant and for this reason, together with the absence of grasses such as *Imperata* which invade forest clearings on dry land (Chapter 17), the forest regenerates very readily (van der Koppel, 1945). The numerous knee-roots and pneumatophores (p. 74) of the trees make the surface of the ground an impenetrable tangle.

The flora includes palms, species of *Pandanus* and *Podocarpus*, as well as representatives of most of the chief families normally present in the Indo-Malayan Rain forest, including the Dipterocarpaceae, but many (probably most) of the species are peculiar to this type of vegetation; there is an affinity to the flora of Heath forest (p. 244). The total number of species is restricted; Sewandono estimates the number of tree species in the Sumatra Moor forests at under a hundred. There is a very strong tendency for the species to be gregarious and a single species may form an almost pure community over a large area. The floristic composition changes gradually from the edge of the moor towards the centre. On the moor on the Paneh Peninsula in Sumatra the following zones are met with from the edge inwards: 'floating forest'; forest on peat less than 0·5 m. thick; forest on thick peat, consisting of the following subzones: (i) forest with thick undergrowth of the palms *Licuala* and *Zalacca*, (ii) dense forest, (iii) high forest of thin-stemmed trees mixed with dwarf trees, (iv) dwarf forest dominated by *Tristania* (Polak, *loc. cit.* p. 17). Endert speaks of *Campnosperma macrophyllum* as forming nearly pure forests in the Moesi delta in Sumatra; in the centre of the moors, where the peat is thickest, *Tristania obovata* and *Ploiarium alternifolium* are dominant. The liane *Nepenthes ampullaria* climbs to the tops of the trees.

The concentric zones of vegetation presumably represent different stages in the hydrosere, but the details of the succession have not been worked out. Polak's studies of the peat stratigraphy have shown that woody plants are dominant from the early stages in the development of these moors.

Though extensive peat swamps are known to exist in tropical America, very little is known about them. Bouillienne (1930) describes an *igapo* (peat swamp) forest in the Amazon delta. The trees are up to 30 m. high, with little branched crowns, and pneumatophores are a conspicuous feature. Palms, epiphytes and common herbaceous swamp plants such as *Montrichardia arborescens* are abundant.

The pegass swamp vegetation of British Guiana (p. 216) and the peculiar 'carapite forest' of Dominica in the West Indies (Beard, 1944c) are perhaps similar to Moor forest vegetation. 'Carapite forest' is found in water-logged situations over an iron pan; it has 'an extremely thick surface mat of raw humus'. The trees grow to a height of some 90 ft. (27 m.) and the most abundant species are *Amanoa caribaea* (carapite) and *Tapura antillana*.

GENERAL FEATURES OF TROPICAL HYDROSERES

Since data are lacking for hydroseres in oligotrophic waters, it is possible to speak generally only about successions in relatively eutrophic waters. From the very incomplete information available, it is clear that under these conditions the hydrosere takes a very similar course everywhere in the tropics and in essentials follows the same lines as in temperate regions.

The first stage consists of communities of free-floating and submerged aquatic plants. These form communities mostly very similar in structure and floristic composition to the analogous types of vegetation in the Temperate zone; many of the genera, and even the species, are the same or closely related. A peculiar feature of the tropical hydroseres is the development of a sudd or free-floating community generally dominated by grasses. A full discussion of this type of vegetation, in its *locus classicus* the Nile, is given by Deuerling (1909); the only similar communities in temperate waters are floating grass swards such as the 'hover' of the English Broads and the *plav* of the Danube delta (Pallis, 1916).

When the water becomes shallower owing to silting, rooted floating-leaf vegetation appears, consisting of water-lilies etc., and this prepares the way for communities of emergent aquatics, plants rooted below the water surface, but with most of their shoot system in the air. Where conditions are unsuited to free-floating, submerged and floating-leaf vegetation, as in the Amazon delta, the succession may be begun by these emergent aquatics.

The emergent stage is succeeded by scrub or low forest communities which may be compared with the 'carr' of temperate countries. This leads on to the final stages of the hydrosere which are forest.

Thus in the earlier phases of the hydrosere herbaceous plants are generally dominant; later they are replaced by woody plants. Both in temperate and tropical hydroseres monocotyledons play an important part in the intermediate stages. In temperate regions they dominate the reed swamp stage; in the tropics, though there is often a reed swamp stage, one in which palms are dominant is even more constantly present. Locally also, species of *Pandanus* are important in these middle stages. This temporary dominance of monocotyledons may be due to the exceptional tolerance of a badly aerated root environment which many of them seem to possess, combined with their limited height-growth, which excludes them from the later stages of the succession.

The last phase of the succession is tall forest dominated by dicotyledonous trees. In this the water level is still at, or quite near, the soil surface, even during the dry season. Mature Swamp forest is clearly the tropical equivalent of the *Auenwald* and 'mature carr', dominated by species of *Salix, Alnus*, etc., occupying similar habitats in Europe. At first sight it often appears very similar, both in structure and general aspect, to primary Mixed Rain forest on sites with unimpeded drainage; many species also are common to both. Closer examination, however, discloses a number of important differences between the two types of forest. The number of trees per unit area is larger in Swamp forest, but in spite of this the canopy is more open and less regular than in dry-land forest. In consequence of the good illumination the undergrowth tends to be dense and to include light-loving species, among them some which are otherwise characteristic of young secondary forest on sites with normal drainage. Swamp forests differ floristically from mixed primary forest on dry land not only in the presence of numerous characteristic swamp species in all strata, but also in a tendency to single-species dominance. The total number of species per unit area is less than in dry-land forests, but greater than in the earlier stages of the hydrosere, for in the tropical hydrosere, as in the xerosere, there is a progressive increase in the number of species, not an initial rise to a maximum, followed by a fall.

This final stage of the hydrosere, dominated by tall woody species, in which the floristic richness is greatest, is, as we have seen, a stable edaphic climax. In the absence of external change leading to a permanent lowering of the water level, further development to the climatic climax is impossible because the high water-table and consequent poor aeration of the soil exclude most of the characteristic dominants of the Mixed Rain forest.

CHAPTER 14

COASTAL SUCCESSIONS

In coastal successions the starting-point is a marine deposit such as mud, sand or coral, either below or above high-tide mark. As in other primary successions there is a progression from extreme conditions, which only a small number of specialized species can tolerate, to less extreme conditions suitable for a much greater number of species. In the earlier stages, environmental factors, such as the effects of tides, wave action, sea winds and salt water, as well as the nature of the substratum, play a much larger part in determining the nature of the vegetation than climatic factors. During the course of the succession the importance of these non-climatic factors, relative to the climatic factors, becomes less and the vegetation gradually becomes similar to the climatic climax of the region. It is therefore easy to understand why the earlier stages of these coastal successions tend to be alike, not only throughout the region in which tropical rain forest is the climatic climax, but far beyond its limits.

Thus, the so-called '*Pes-caprae* formation', a plant community characteristic of sandy tropical shores, is found from the equator to well outside the tropics, while mangrove vegetation reaches its northern limit, much impoverished in species it is true, in Bermuda, the Gulf of Akaba in the Red Sea and in southern Japan; its southern boundary is in Natal and northern New Zealand (see Schimper, 1935, pp. 585–7). These coastal communities thus extend through a wide range of climates in which the climax vegetation varies from tropical rain forest to desert, and the later stages of the successions naturally differ in the different climatic regions.

The foundations of our knowledge of tropical shore vegetation were laid by Schimper (1891) in his classical memoir on the Indo-Malayan strand flora. It will be convenient to follow his classification of the seral communities of the shore into the *Pes-caprae* and *Barringtonia* formations, characteristic mainly of exposed sandy shores, and the Mangrove and *Nipa* formations found chiefly on sheltered muddy shores. The first two of these communities can establish themselves only above the level of ordinary high tides; the other two may begin their development as far down as low-water mark. In intermediate types of habitat mixed or intermediate communities may develop and there are perhaps some types of tropical shore vegetation which will not easily fit into this classification. Successions on rocky sea coasts are more akin to xeroseres than to other coastal successions and will not be dealt with here.

VEGETATION OF SANDY SHORES

On sandy shores in the Indo-Malayan region the *Pes-caprae* formation forms a zone of low-growing herbaceous plants, among which the most constant (and often the dominant) species is *Ipomoea pes-caprae*. Behind it, where the natural vegetation has not been destroyed, there is often a narrow belt of Littoral woodland in which *Barringtonia speciosa* and *B. racemosa* are characteristic trees. Both these communities have a wide distribution on the shores of the tropical Indian and western Pacific oceans. On the coasts of the West Indies the *Pes-caprae* formation reappears, accompanied by a zone of woody vegetation closely corresponding to the *Barringtonia* formation. In reality both the herbaceous and woody zones of tropical shores can be subdivided and the transition from one to the other may be gradual or abrupt. The zonation is best displayed on wide gently sloping beaches and in such places a system of sand dunes may develop. Where the beach is narrow the zones may be telescoped or only one may be present. Since Schimper's time the plant communities of sandy shores in the tropics have often been described (Ule, 1901; Cheeseman, 1903; Schenck, 1904; Tansley & Fritsch, 1905; Guppy, 1906; Diels, 1906; Rechinger, 1908 and 1910, pp. 207–10; Børgesen, 1909; Gleason & Cook, 1926; Christophersen, 1927; Chipp, 1927; Däniker, 1929; Raunkiaer, 1934; Stehlé, 1935; Chapman, 1944, etc.; further references on Indo-Malayan beach vegetation in van Steenis, 1935 a).

The *Pes-caprae* 'formation' would be more correctly termed the *Ipomoea pes-caprae–Canavalia* associes, since in both the eastern and western tropics species of the leguminous genus *Canavalia* are almost as constantly present as the *Ipomoea*. This community is the tropical analogue of the strand vegetation of European and North American shores to which species such as *Agropyron junceum* and *Arenaria peploides* belong; at least one species (*Salsola kali*) is common to both types of vegetation. The *Pes-caprae* formation is a mixed community of plants, many of which, including *Ipomoea* itself and *Canavalia*, have a trailing habit, sending long runners over the surface of the sand. Some of the species, including the *Ipomoea*, occasionally occur inland, while others are restricted to sea beaches. The majority of the species are probably halophytes and are unharmed by occasional submergence in sea-water and by the high salt-content of the soil. Like other halophytes some of them are succulent. The seeds or fruits of most of them float in sea-water and are dispersed by this means; they germinate chiefly among the drift, which provides them with humus and a store of moisture.

In addition to several species of *Ipomoea* and *Canavalia*, various other dicotyledons such as *Sesuvium portulacastrum* and grasses including *Sporobolus virginicus* are generally found in this zone; in the eastern tropics *Spinifex* spp., *Thuarea involuta*, *Zoisia matrella* and others are common. Shrubs, for instance *Suriana maritima*, *Scaevola frutescens*, and *Tournefortia gnaphalodes* (West Indies), may occur scattered or in groups. Round the taller members of the community sand may accumulate as embryo dunes and these may fuse to form dune ridges.

Notes on the physiognomy and biology of these shore plants will be found in several of the papers mentioned above and in Schimper (1903, 1935). Their life-forms are dealt with at length by Raunkiaer (1934, ch. v).

The *Pes-caprae* formation may form a complete cover to the ground, but towards its seaward edge it generally consists of a more or less open community of plants, of which the pioneers first established themselves in the drift. On sandy shores in Ceylon (Tansley & Fritsch, 1905) an outer and an inner *Pes-caprae* zone may be recognized, the former open and consisting entirely of species confined to the seashore, the latter closed and including various inland weeds as well as true maritime species. On the coastal sand dunes of Barbados (Gooding, 1947) there are two zones comparable with those of Ceylon, viz. (*a*) an outer pioneer zone, without *Ipomoea*, of scattered low-growing plants, the chief of which are *Sporobolus virginicus*, *Philoxerus vermicularis* and *Euphorbia buxifolia*, and (*b*) an inner zone in which *Ipomoea* is dominant, accompanied by various subordinate species, including in some places *Canavalia maritima*[1] and often a belt of *Tournefortia gnaphalodes* on the seaward edge; the vegetation here is closed or becomes so as the distance from the sea increases. Further inland there is a scrub zone dominated by *Coccoloba uvifera* (see below p. 298). Raunkiaer (1934, ch. 5) distinguishes several 'facies' of the *Pes-caprae* formation.

In dry climates, for instance on many of the drier coral islands of the Pacific (Christophersen, 1927), communities of low-growing halophytes often extend far inland and cover extensive areas. Most of the species of these communities also occur in the typical narrow *Pes-caprae* belt.

On many tropical shores shrubs and trees become more and more frequent with increasing distance from the shore and there may thus be a quite gradual transition from the mainly herbaceous *Pes-caprae* community to a mainly woody type of vegetation; in other places, however, the transition is sudden so that the Littoral woodland or *Barringtonia* formation forms a solid wall behind the low-growing vegetation nearer the foreshore.

The typical *Barringtonia* formation is found on sandy shores in the damper tropical regions of the Indian and Pacific oceans. It is well developed only in thinly populated regions, as elsewhere the trees are generally cut down for firewood or cleared to make room for coconut plantations. It may form a dense belt of woodland or the trees may form scattered groups with open spaces between, somewhat as in a savanna. At its best, for example in the Solomon Islands and on the coast of New Guinea, the *Barringtonia* woodland may form a narrow zone of trees reaching a height approaching that of inland Rain forest. Some of the most constant and characteristic trees of this community are *Barringtonia speciosa* and *B. racemosa*, *Calophyllum inophyllum*, *Terminalia catappa*, *Thespesia populnea*, *Hibiscus tiliaceus*, *Tournefortia argentea*, *Pandanus tectorius* and *Casuarina equisetifolia*. The last-named often grows in extensive pure stands, especially on rapidly accreting shores, such as sand spits at river mouths. Most

---

[1] In Barbados this species is not common, but in other West Indian islands it is dominant in this zone.

of these species rarely or never occur inland, but some at least of them are not tolerant of immersion in sea-water. Burkill (1928) noted in Pahang (Malay Peninsula) that after being flooded by an exceptionally high tide the leaves of many of the trees in the *Barringtonia* zone turned yellow and dropped off. Many of the trunks are gnarled and twisted or lean seawards. The abundance of lianes and the density of the undergrowth make the woodland difficult to penetrate and in areas of high rainfall epiphytes are common on the branches. The biology of the species in the *Barringtonia* formation has been fully discussed by Schimper (1891), Tansley & Fritsch (1905) and others.

Like the *Pes-caprae* formation the *Barringtonia* formation may be subdivided into zones, the smaller trees and shrubs such as *Scaevola frutescens*, *Tournefortia* and *Pandanus* spp. generally growing on the seaward fringe and the taller species further inland. In some places a gradual transition can be observed from the *Barringtonia* woodland to the inland forest characteristic of the region.

On the shores of the West Indies there is a zone of woody vegetation corresponding very closely to the *Barringtonia* formation of the eastern tropics. Sometimes it forms a belt of woodland with trees up to 20 m. high and interlaced with lianes, but often, especially on very exposed shores, it is dwarfed to a mere scrub or thicket. The small tree or shrub *Coccoloba uvifera* is generally dominant, at least on the seaward side of this zone, though other species such as *Hippomane mancinella* and *Chrysobalanus icaco* are almost equally characteristic.

Towards the seaward margin, adjoining the *Pes-caprae* vegetation, *Coccoloba* often forms a nearly pure community with very few other species intermixed, but further inland its dominance is less complete and other trees, shrubs and herbs appear. Some of the subsidiary species, e.g. *Terminalia catappa* and *Thespesia populnea* are pan-tropical species which also occur in the *Barringtonia* formation of the eastern tropics and several genera are represented by related species in the *Coccoloba* and *Barringtonia* communities respectively. In Trinidad, Beard (1946b) distinguishes two littoral communities on sandy shores: the *Coccoloba-Hippomane* association of scrub or thicket form, and the taller association of *Manilkara bidentata* with the palm *Roystonea oleracea*. The latter usually occurs further from the shore than the former and merges into the inland forest. The width of the *Coccoloba* zone varies with the exposure and is probably controlled by sea winds.

Though much has been written about the vegetation of sandy shores in the tropics little attention has been given to its successional aspects. It is clear that the zones of vegetation between the bare foreshore and the forest or other inland community behind are, at least potentially, stages in a succession. Since the characteristics of the various shore communities are probably determined by the dominant factors of the environment—the salt content of the soil, the liability to occasional flooding by sea-water and the prevalence of strong winds carrying salt spray— it seems unlikely that the succession will become actual and cease to be potential as long as the shore line is stable and not advancing seawards. When new sand is accumulating, by the action of winds, waves and currents, and by the dune-forming activities of the vegetation itself, the plant communities may undergo rapid change and the zones of vegetation will then be in fact stages in a succession.

Interesting observations bearing on this have been made on the Barbados sand dunes by Gooding (1947). Here, as on other coastal sand dunes, the salt content of the soil diminishes with increasing distance from the sea; at the same time, owing to the addition of plant remains to the soil, its water content increases, as can be seen from the following figures (for one locality):

	Percentage of salt in soil water (mean)	Percentage of water in sand (mean)
Pioneer zone (50 m. from sea)	5·3	0·26
Ipomoea pes-caprae zone (100 m. from sea)	1·8	0·48
Coccoloba zone (150 m. from sea)	0·4	1·65

The nature of the soil, together with the effects of the wind (shown in the dwarfed and wind-pruned appearance of the *Coccoloba* bushes), probably prevent the respective zones from advancing towards the foreshore, but in small localized areas an actual succession could sometimes be seen to take place, demonstrating that the three zones are in fact stages in a psammosere or succession similar to that found on sandy shores in Europe and North America. In this succession the plants themselves play an important part, by collecting sand and by adding their dead remains to the soil. Where no actual succession is taking place, each zone is in equilibrium with its environment; each can then be regarded as an edaphic climax not encroaching on the neighbouring zones.

Similar results would no doubt be obtained in other areas. On Krakatau, as we have seen, active successions have been observed on the seashore. In general we can probably regard the zones on sandy tropical shores as stages in what under certain conditions may become an active succession.

MANGROVE VEGETATION AND THE '*NIPA* FORMATION'

The word mangrove is used both for an ecological group of species inhabiting tidal land in the tropics and for the plant communities composed of these species. In the first sense mangroves are evergreen trees and shrubs belonging to several unrelated families. They share similar habitat preferences and a similar physiognomy; they are also similar in their physiological characteristics and in their structural adaptations, most of them having pneumatophores or 'breathing roots' and many a more or less marked tendency to vivipary. Mangroves are perhaps the most remarkable of all examples among plants of epharmonic convergence or resemblance between unrelated species living in a similar habitat. Associated with the true mangroves there are species with similar but less strongly marked characteristics; for these Tansley & Fritsch (1905) suggested the convenient name semi-mangrove.

Mangrove vegetation, the 'Mangrove formation' of Schimper, is a complex of plant communities covering large areas fringing sheltered tropical shores. It varies in character from forest 30 m. high or more to a poor scrub barely 2 m. high. The monotonous dark green shiny foliage of the mangrove belt and its apparently impenetrable tangle of aerial roots, is one of the most characteristic features of the landscape on many tropical coasts. The term 'tidal forest' which has sometimes been applied to mangrove vegetation is not entirely accurate as mangroves may grow from below the level of the lowest to above the level of the highest tides and are sometimes found on coasts where there are no tides at all. Along estuaries mangrove vegetation penetrates far inland. Typically mangroves are found on sheltered muddy shores where the land is encroaching on the sea, often at a rapid rate, but they also grow on coral reefs and on sandy shores where there is little accretion (Chapman, 1944). Where, however, accretion is going on, the mangroves play an active part in it. By obstructing the currents they speed up the rate of deposition and by binding the soil with their roots and adding humus they help to consolidate it; a kind of marine peat may even be formed. As the ground level rises the mangroves spread further and further seawards and thus play a similar part in salt and brackish water to reed swamp and other hydroseral communities in fresh water.

In brackish estuarine areas semi-mangroves such as the palm *Nipa fruticans* form extensive communities often adjoining those of the true mangroves. These semi-mangrove communities constitute the '*Nipa* formation' of Schimper, though the actual *Nipa* consocies is only one of a series of communities of this kind in the eastern tropics. In America and Africa there are analogous communities, for instance the *Manicaria saccifera* swamps of north-eastern South America and a *Pandanus* community on the west coast of Africa (p. 306).

Mangrove vegetation extends, as has been mentioned, to about 32° N., and in the southern hemisphere even further from the equator, but it reaches its greatest luxuriance and floristic richness in the wet tropics where the Tropical Rain forest is the climatic climax, above all on the coasts of the Malay Peninsula and the neighbouring islands. Two mangrove formations can be recognized, an eastern one on the coasts of the Indian and western Pacific oceans, and a western on the coasts of America, the West Indies and West Africa. The Australian mangrove differs in some respects from the rest of the eastern formation and perhaps deserves to be considered a distinct 'sub-formation'. The eastern and western formations are essentially similar in their physiognomy and ecological relationships, but the eastern mangrove is much the richer in species. Though all the genera of the western mangrove are found in the eastern, all the species are different. The Fiji and Tonga islands in the Pacific are unique in possessing both an eastern species of *Rhizophora* (*R. mucronata*) and a western one (*R. mangle*).

Mangroves, because of their many structural and physiological peculiarities, are of outstanding scientific interest; they are also of considerable economic importance, not only indirectly, because of their part in reclaiming land from the sea, but also directly as a source of firewood, timber and tan-bark. Con-

PLATE I

Tropical Rain forest, 800 m. above sea-level,
Arfak Mountains, New Guinea

PLATE II

B. Undergrowth of Mora forest, Moraballi
Creek, British Guiana

The buttressed tree is *Mora excelsa*, belonging to the
A story. The small tree on the right is *Duguetia* sp. and
shows the habit characteristic of trees in the C story.

A. Canopy of Rain forest, Moraballi Creek, British Guiana, seen
from about 34 m. above ground

On the horizon at the extreme right is an 'outstanding' tree with its whole
crown clear of the surrounding forest.

PLATE III

A. Canopy of Rain forest, Shasha Forest Reserve, Nigeria, seen from 24 m. above ground

In foreground, C story trees bound together into a dense mass by lianes; in background 'emergent' trees of the A and B stories.

B. Undergrowth of Rain forest, Okomu Forest Reserve, Nigeria

The large tree is *Entandrophragma angolense* var. *macrophyllum* showing its characteristic surface roots.

PLATE IV

B. Stilt-rooted tree (*Tovomita* sp.), Moraballi

A. Buttressed tree (*Mora excelsa*), Moraballi

PLATE V

A. A cauliflorous tree (*Napoleona* sp.),
Idanre, Nigeria

B. An epiphytic fern (*Asplenium africanum*), Shasha
Forest Reserve, Nigeria

PLATE VI

A. Forest with numerous lianes, Shasha Forest
Reserve, Nigeria

The roots of a strangling fig (*Ficus* sp.) are seen attached to
the trunk of the large tree.

B. A cauliflorous liane (*Pararistolochia flos-avis*), Shasha
Forest Reserve, Nigeria

PLATE VII

A. A saprophyte (*Leiphaimos aphylla*) on a tree
stump, Moraballi Creek, British Guiana

B. *Thonningia sanguinea*, a parasite on tree-roots,
Shasha Forest Reserve, Nigeria

PLATE VIII

A. Morabukea forest (consociation of *Mora gonggrijpii*), Moraballi
Creek, British Guiana

The undergrowth consists largely of seedlings and saplings of the dominant.

B. Wallaba forest (consociation of *Eperua falcata*),
Moraballi Creek, British Guiana

Two trees of *Eperua* sp. in middle of picture; note absence of buttresses.
The grass-like plants to left of nearer *Eperua* are *Tillandsia* sp., an
epiphyte which has fallen from the tree tops and taken root in the
ground. Another *Tillandsia* can be seen growing a few feet from the
ground on the small tree to the left.

PLATE IX

Tropical Rain forest, 350–550 m., Malay Peninsula

Hill Dipterocarp forest, with society of *Shorea curtisii*
(light coloured crowns) in centre.

PLATE X

B. Fresh-water Swamp forest, Shasha Forest
Reserve, Nigeria
The stilt-rooted tree is *Uapaca staudtii.*

A. Fresh-water Swamp forest, Akilla, Nigeria
In the opening in the foreground the dominant plant is
Cyrtosperma · senegalense. The broad-leaved tree is *Mitragyna ciliata*
and the palm is *Raphia* sp.

PLATE XI

B. Mature mangrove (*Rhizophora mucronata* consociation) at low tide, Aroe Bay, East Sumatra

The trees are c. 40 m. high.

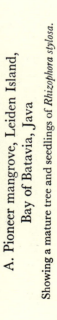

A. Pioneer mangrove, Leiden Island, Bay of Batavia, Java

Showing a mature tree and seedlings of *Rhizophora stylosa*.

PLATE XII

A. Aerial view showing Gallery forest and treeless savanna,
Siba area, Uganda

The savanna is dominated by *Pennisetum purpureum*.

B. South American savanna. South of R. Orinoco, road from
Upata to Euasipati, Venezuela

The trees are *Curatella americana*, with *Byrsonima crassifolia* and *Bowdichia virgilioides* in smaller numbers. Ground flora of short 'bunch grasses' (*Trachypogon* etc.). A white sandy top-soil overlies impermeable red clay. Rainfall *c.* 130 cm. per annum.

PLATE XIII

Montane Rain forest with conifers, alt. *c*. 2000 m.,
Arfak Mountains, New Guinea

PLATE XIV

A. Shifting cultivation. Native 'farm' in Rain forest.
Near Akilla, Nigeria

Maize crop in foreground.

B. Secondary succession. Vegetation after 5 years on abandoned
native 'farm'. Near Akilla, Nigeria

In right foreground, *Veronina conferta* (large leaves), young *Fagara* sp. (pinnate leaves).

PLATE XV

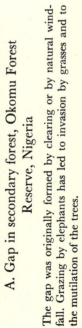

A. Gap in secondary forest, Okomu Forest Reserve, Nigeria

B. Old secondary forest, Okomu Forest Reserve, Nigeria

The gap was originally formed by clearing or by natural windfall. Grazing by elephants has led to invasion by grasses and to the mutilation of the trees.

sequently few types of tropical vegetation have been more studied than mangrove communities. No attempt can be made here to deal fully with the subject, which involves many problems with little bearing on the main theme of this book. Some of the most valuable modern contributions to the extensive literature on mangroves are those of Watson (1928), Davis (1940) and Chapman (1944). The summary by van Steenis (1935 a) on the mangroves of the Netherlands East Indies and the general review by von Faber (Schimper, 1935) are also useful. Guppy (1906) deals at length with the problems of the geographical distribution of mangroves and von Faber (1923), Troll & Dragendorf (1931), Walter & Steiner (1936–7) and Chapman (1944) have dealt with various aspects of their physiology and autecology. Here we are interested in mangrove communities chiefly from the successional point of view.

One of the most striking features of mangrove swamps is the zonation of the dominant species more or less parallel with the shore line. At the inland margin there is generally a gradual transition to Fresh-water Swamp forest or some other type of non-maritime vegetation. Like that on sandy shores, this zonation suggests the existence of at least a potential succession and where rapid accretion is going on actual successions can readily be observed. In the course of these successions there are changes in a great many environmental factors, especially the relative level of land and water. These factors interact in a very complex manner and the mangrove successions are therefore by no means simple. An insight into the nature of these successions will most easily be obtained if we focus attention chiefly on two areas which have been intensively studied, the mangrove forests of the Malay Peninsula (Watson, 1928) and those of Florida (Davis, 1940). It will be convenient to consider the Florida mangrove first, since there the small number of species makes the problems simpler.

In Florida mangrove vegetation is extensively developed as far north as the southern limit of 'killing' frosts. All the species of the western mangrove formation are present and, as elsewhere, these species are arranged in zones which can be correlated with the frequency of tidal immersion, the nature of the substratum, the activity of accretion and erosion, and the salinity of the ground water. The growth of the vegetation in one zone prepares the way for that in the next. One community thus succeeds another till eventually an inland type of vegetation not tolerant of immersion in sea-water is established. The following main types of mangrove communities can be distinguished:

(1) The 'pioneer *Rhizophora* family'. This consists mainly of young plants of *Rhizophora mangle* growing on almost continually submerged soil. With the *Rhizophora* are associated the submerged marine angiosperms *Thalassia testudinum* and *Cymodocea manatorum* and in some places the grass *Spartina alterniflora*. Seedlings of the mangroves *Avicennia* and *Laguncularia* are frequent. The viviparous seedlings of *Rhizophora* float in the sea and become established on shoals, sandbanks, etc., as well as at the outer fringe of the mature mangrove forest, if the water is calm enough and not too deep for them to take root. The young plants send out

tiers of prop-roots and soon anchor themselves firmly in the loose substratum. There are usually about twenty to thirty young *Rhizophora* plants to 100 sq. ft.

Sedimentation and the accumulation of plant and animal debris raise the level of the soil and help to consolidate it, so that once the plants are established they can withstand strong wave-action.

(2) The 'mature *Rhizophora* consocies'. In course of time the 'pioneer *Rhizophora* family' develops into a mature community, a forest up to 30 ft. (9 m.) high. The soil level is now considerably higher and the stilt roots of the trees are so entangled that they are very efficient in catching debris of all kinds. The trees are so strongly anchored that only a violent hurricane can dislodge them.

The marine angiosperms are less abundant here. The *Rhizophora* community may persist for a long time as a nearly pure consocies or it may change rapidly by the invasion of other species which are able to grow on the firm soil and in the shelter provided by the *Rhizophora*. If the community changes, an inner zone of the mangrove *Avicennia*, generally associated with low-growing salt-marsh plants, may develop.

In many parts of the coast there is no pioneer off-shore stage outside the mature *Rhizophora* consocies. In such places, though the mangrove community does not spread by means of seedlings, the trees may send out roots into the deeper water, sediments may accumulate among these roots and the swamp may thus extend slowly seawards.

(3) The '*Avicennia*–salt-marsh associes'. Behind the outer *Rhizophora* consocies, on land which is regularly or only occasionally submerged, there is typically a zone dominated by *Avicennia nitida*. This forms an open forest, in striking contrast to the thickly tangled *Rhizophora* consocies, with an undergrowth of succulent shrubs, such as *Batis maritima* and *Salicornia perennis* and salt-marsh grasses. Where the ground level is low there may be little undergrowth, but typically the dense stand of the salt-marsh plants among the pneumatophores of the *Avicennia* is characteristic. The '*Avicennia*–salt-marsh associes' is best developed on land which is not regularly flooded by the tide.

The *Avicennia* trees are not rapidly replaced by natural regeneration and as they die the forest becomes more and more open, with an increase in the salt-marsh vegetation or an invasion of species from the *Conocarpus* associes which often adjoins the inner edge of the *Avicennia* zone. If the trees disappear entirely an open salt-marsh or meadow may take the place of the mangrove swamp.

(4) The '*Conocarpus* transition associes'. The semi-mangrove *Conocarpus erectus*, though not always abundant, is the most characteristic species in this community; it occupies a zone seldom reached by the tides, immediately inland from the *Avicennia* associes. Like the *Avicennia* zone, the *Conocarpus* associes is an open stand of trees and shrubs with an undergrowth of low-growing salt-marsh plants. A large number of species of trees, shrubs, grasses and other ground herbs, as well as ferns and epiphytes, grow in this community. Some of these

plants also grow in the 'hammock' (inland evergreen) forests; many of the trees are poorly developed and are merely strays from inland communities.

In many places, though not everywhere, there is evidence that the soil of the *Conocarpus* associes has been built up by mangroves and it therefore certainly forms part of the mangrove succession. On its inland margin the *Conocarpus* zone passes into non-halophytic communities: 'hammock forests' (p. 368), pine forests, a palm and hardwood association or a dune associes. In some of the 'hammock' forests' there is evidence that the forest soil is underlain by mangrove soils, but elsewhere there is no evident successional relation between the mangrove and the inland vegetation. Occasionally the true mangrove vegetation adjoins the inland climax forest with no intervening *Conocarpus* zone.

(5) The 'mature mangrove forest association'. On parts of the Florida coast there is a very tall luxuriant mangrove forest in which about 60 % of the tall trees are *Rhizophora* and about 30 % *Avicennia*. The mangrove *Laguncularia racemosa*, *Conocarpus* in small numbers and various palms and other inland plants are also found in it. There is no recognizable zoning within the community. The ground is regularly and deeply flooded by high spring tides and, during the wet summer season, by fresh water, but ordinary tides do not reach it and the flood water is seldom strongly saline. A characteristic feature is the peaty soil, 10–14 ft. (3–4 m.) deep, formed of mangrove remains. This community appears to be very stable and there is little evidence that it will develop further.

In addition to these five main communities of the Florida mangrove there are others which are less important. For instance, locally under a not very well-defined set of conditions there are societies or pure stands of *Laguncularia*. There are also communities of dwarf mangroves, but these are not clearly distinguishable from the normal tall mangrove communities.

The successional relations between the various mangrove communities are complex and have not been fully elucidated, but several hypothetical seres have been recognized. Davis' views on the relations of the main mangrove and associated communities are summarized in Fig. 36. The main line of the succession appears to be from the 'pioneer *Rhizophora* family' to the 'mature *Rhizophora* consocies' and from this through the '*Avicennia*–salt-marsh associes' and the '*Conocarpus* Transition Associes' to 'hammock forest', the climatic climax of the region (p. 368). The position of the 'mature mangrove forest association' is not clear. Davis regards it as a stable 'subclimax' type not tending to change into any other type of vegetation and therefore not part of the main succession from open sea water to inland forest. We may perhaps regard it as an edaphic or physiographic climax analogous to the Moor forest or tropical raised bog dealt with in the last chapter, but developing in saline or brackish water.

If we assume that the four communities—pioneer *Rhizophora*, *Rhizophora* consocies, *Avicennia* associes and *Conocarpus* associes—represent the successive stages, it will be evident that great changes in the environment take place during the course of the succession. Owing to the interaction of the vegetation and its habitat, the ground level rises. In the pioneer stage the vegetation is almost continually under water, but in the later stages the frequency of submergence diminishes till in the *Avicennia* and *Conocarpus* stages tidal flooding becomes quite infrequent. Accompanying these changes in the relative level of land and water there are changes in salinity, as can be seen from Table 33. The average salinity

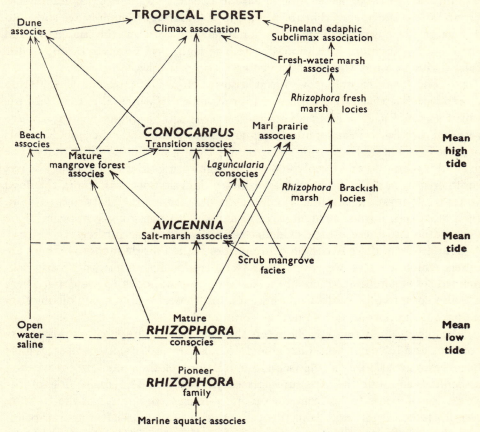

Fig. 36. Successional relations of mangrove and of some of the associated plant communities in Florida. After J. H. Davis (1940).

reaches a maximum in the *Avicennia* stage and afterwards decreases until it approximates to the value for inland soils. In none of the mangrove communities is salinity constant. It varies most in the *Avicennia* consocies where the ground and surface water is only occasionally renewed by the tide. Here the salinity may rise to very high values in dry weather, while after heavy rain it may fall very low. Apart from the remarkable ability of the seedlings to establish themselves under tidal conditions, and the plant's tolerance of prolonged flooding at all stages of its development, the faculty of growing in media of high and often variable salinity is the chief physiological characteristic of mangroves. Yet since *Rhizophora* and even *Avicennia* grow naturally where the ground water is apparently perfectly fresh (see p. 285), it may be concluded that at least the American mangroves are salt-tolerant and not salt-demanding, that is to say they are facultative rather than obligate halophytes. This view is supported by a certain amount of experimental evidence.[1]

TABLE 33. *Habitat conditions of mangrove communities in Florida.*
(After J. H. Davis, 1940, table 6, p. 349)

| Community | No. of samples | Salinity (%) | |
		Soil solution	Surface water
Pioneer *Rhizophora* family	7	2·91	3·185
Mature *Rhizophora* consocies	14	3·56	3·488
Avicennia–salt-marsh associes	13	5·11	3·684
Conocarpus transition associes	8	1·80	0·144
'Hammock' forest association	8	0·35	(No surface water)

Outside Florida the western mangrove communities have nowhere been studied in great detail but, as far as can be judged from the available data, the zonation and the course of the succession are everywhere essentially similar. The low-growing salt-marsh vegetation associated particularly with the *Avicennia* zone in Florida is absent nearer the equator. In Puerto Rico, Gleason & Cook (1926) describe a *Rhizophora mangle* zone on the seaward edge of the mangrove, followed successively by an *Avicennia* and a *Conocarpus* zone further inland. The fern *Acrostichum aureum* which is present in the *Avicennia* and *Conocarpus* zones in Florida, but not particularly abundant, is here dominant in extensive patches or belts on the landward margin of the mangrove. Though at first sight the fern zone might appear to be a stage in the primary succession following the *Conocarpus* community, Gleason & Cook found evidence that it was more probably a secondary type following the destruction of the mangroves by felling. A similar *Acrostichum* zone on the inner edge of the mangrove belt has been recorded in other parts of the world. When there is no interference the *Conocarpus* community in Puerto Rico is believed to be succeeded by a brackish-water swamp forest, dominated by *Pterocarpus officinalis*, and eventually by 'mesophytic forest', the climatic climax. In the relatively dry climate of Guadeloupe (Stehlé, 1935) the mangrove communities are succeeded by *Acrostichum* and this in turn by a moist meadow community of *Paspalum* spp., *Kyllinga* etc., but whether the meadow represents a stage in the primary succession is not clear.

In Trinidad and British Guiana, and doubtless also on many other low-lying tropical American coasts, there is often a gradual transition inland from mangrove to fresh-water swamp (herbaceous or forest) vegetation. In such places the transition most probably represents an actual succession, as in Puerto Rico. In

---

[1] Von Faber (1923; also in Schimper, 1935) disagrees with this view, but the eastern mangroves to which he refers are possibly different in this respect.

Jamaica, according to Chapman (1944), the mangrove communities may be eventually succeeded by 'dune-strand vegetation', thorn scrub or a fresh-water reed-swamp of *Typha domingensis*.

The extensive mangrove forests of West Africa consist of the same species as the tropical American mangrove and therefore, as has been said, it can be considered as part of the same formation. Information about the zonation and ecology af the West African mangroves is scanty: observations on the mangroves of the Cameroons and some notes of Schimper on those at the mouth of the Congo have been published by Walter & Steiner (1936–7). A short account of the Congo mangrove has also been published by Pynaert (1933). In the Cameroons *Rhizophora mangle* forms the pioneer community on the seaward fringe and *Avicennia* grows further inland. Still further from the shore the mangrove passes into a brackish-water community dominated by a species of *Pandanus* in which palms (*Raphia* sp., *Calamus* sp., *Phoenix reclinata*) are common. This community resembles some which we have already noted (p. 289) as forming part of the fresh-water hydrosere in West Africa. At the Congo mouth the communities are similar and it is noteworthy that here, as in America, *Rhizophora* sometimes grows in quite fresh water.

We may now turn to the more complex eastern mangrove formation and consider in detail the mangrove communities of the Malay Peninsula, as described by Watson (1928).

On the Malayan coasts the mangrove forests are exceptionally tall and luxuriant and consist of many more species than in Florida; Watson lists some seventeen 'principal' and twenty-three 'subsidiary' species. The communities generally form well-marked zones which are illustrated in Fig. 37, though other distributions are sometimes met with. The following five main types of community may be recognized:

(1) The *Avicennia-Sonneratia griffithii* type. In Malaya the pioneers are not species of *Rhizophora*, but *Avicennia alba* and *A. intermedia*, or sometimes, on deep mud rich in organic matter, *Sonneratia griffithii*. These pioneer forests establish themselves on shoals or sandbanks out at sea which are exposed at neap tides, or along the seaward edge of existing forests. *Avicennia intermedia* grows on a comparatively firm clayey substratum which is easy to walk on, *A. alba* and *Sonneratia* on softer and blacker mud. On the clay soils the *Avicennia* is normally succeeded by *Bruguiera caryophylloides*, but where *Sonneratia* is the pioneer, *Rhizophora mucronata* usually follows on.

(2) The *Bruguiera caryophylloides* type. This occurs at a higher level than the preceding and forms a nearly continuous pure belt behind the *Avicennia* forest along the west coast of the Malay Peninsula, interrupted only by small stretches where *Avicennia* forest merges directly into *Rhizophora* forest. The soil is a firm stiff clay above the reach of the ordinary tides and only flooded during the day or two before and after the spring tides. This type is found chiefly on the sea-face and is usually absent both on shoals and in river forests.

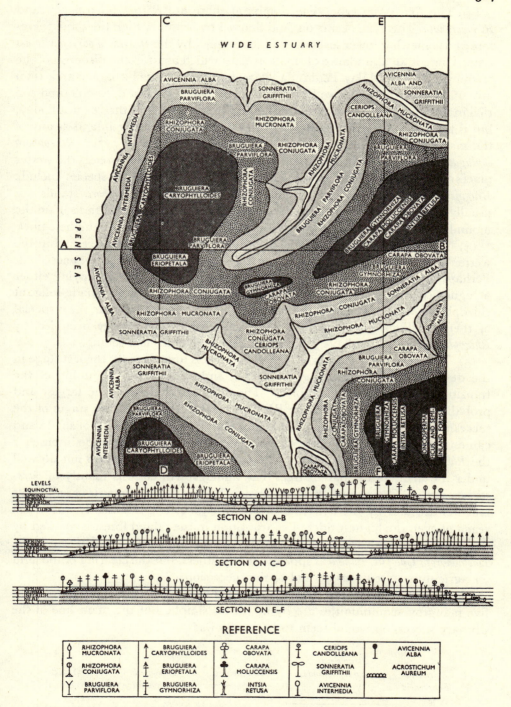

Fig. 37. Diagram of typical distribution of more important species in the mangrove forests of the Malay Peninsula. After Watson (1928).

(3) The *Rhizophora* type. The dominants here are *Rhizophora conjugata* and *R. mucronata*. This type occurs on land flooded by ordinary high tides and therefore at a somewhat lower level than that occupied by the *Bruguiera caryophylloides* type. There is an abundance of small streams which assist in the dispersal of the seedlings. In the Malay Peninsula *Rhizophora* forest covers a larger area than any other type of mangrove forest. Typically it grows on a dark-coloured soil rich in humus and with an admixture of fine sand; it will not thrive on stiff clay, but is tolerant of quite dry soils and will maintain itself as far inland as any of the mangrove communities except the *Bruguiera gymnorrhiza* type. *Rhizophora conjugata* probably covers larger areas then *R. mucronata* which prefers the wetter places such as the banks of streams and creeks. Associated species include *Bruguiera parviflora* and *Xylocarpus granatum*. The older *Rhizophora* stands are usually invaded by the fern *Acrostichum aureum*; this tends to establish itself on the mounds made by burrowing prawns, out of reach of any but the highest tides.

(4) The *Bruguiera parviflora* type. This type may grow as a pure stand in the wetter areas. Some believe that it precedes type (3), others that it succeeds it. Perhaps it may be regarded as an 'opportunist' community occupying either a higher or a lower position than the longer-lived and more shade-tolerant rhizophoras which are liable to invade it where conditions favour the dispersal of their seedlings. It certainly often grows on areas formerly occupied by *Rhizophora*, but it may have colonized such areas after clear-felling.

(5) The *Bruguiera gymnorrhiza* type. This type is undoubtedly the last stage in the development of the mangrove forests and heralds the beginning of the transition to the inland Rain forest. The dominant species is the largest and probably the longest-lived of the Rhizophoraceae; in the earlier stages of the succession its occurrence is markedly sporadic. It is very tolerant of shade and can establish itself in *Rhizophora* communities where the shade is too intense for the *Rhizophora* itself to regenerate. Its own seedlings, however, do not do well under the shade of the parent tree. The undergrowth consists mainly of *Acrostichum aureum*. The ground-level has by now been raised by deposits of sediments and humus, and by the activities of prawns, till it is permanently above the reach of even the highest tides. The transition to the inland forest is marked by the invasion of such species as *Xylocarpus moluccensis, Intsia bijuga, Ficus retusa, Daemonorops leptopus, Pandanus* spp. and many others, but the clearing of the land for agriculture often puts an end to the succession.

From what has been said it will appear that the successional relations of these five mangrove communities are by no means simple, but the main line of the primary succession would seem to be as follows:

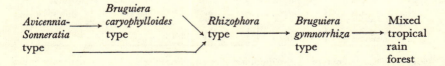

| Avicennia-Sonneratia type | → | *Bruguiera caryophylloides* type | ↘ | *Rhizophora* type | → | *Bruguiera gymnorrhiza* type | → | Mixed tropical rain forest |

The position of the *Bruguiera parviflora* type is, as already explained, somewhat uncertain. There are many local variations in the succession and it must not be assumed that all mangrove swamps pass through the entire sequence of development or that all the types are necessarily present, even in extensive forests.

The most important environmental change taking place during the succession is the raising of the ground level and the consequent decrease in the frequency of tidal flooding. With respect to tidal flooding the species may be classified as follows: (i) Species growing on land flooded by all high tides; no species normally exists under these conditions, but *Rhizophora mucronata* will do so exceptionally. (ii) Species growing on land flooded by 'medium high tides'—*Avicennia alba*, *A. intermedia*, *Sonneratia griffithii* and, on river banks, *Rhizophora mucronata*. (iii) Species growing where they are flooded by 'normal high tides'; the majority of the species thrives under these conditions, but the rhizophoras tend to become dominant. *Sonneratia griffithii* continues to thrive, but cannot reproduce itself. (iv) Species growing on land flooded by spring tides only. Such areas are a little too dry for the rhizophoras and suit *Bruguiera gymnorrhiza* and *B. caryophylloides* better. (v) Species on land which is flooded by equinoctial or other exceptional tides only. Here *Bruguiera gymnorrhiza* is the chief mangrove, but *Rhizophora conjugata* and *Xylocarpus moluccensis* still survive.

It will be noticed that though the stage in the succession in which *Bruguiera caryophylloides* is dominant has been considered (and can be proved) to be earlier in time than that of the rhizophoras, it grows at a higher, less frequently flooded level. The explanation of this is not entirely clear, but Watson suggests that the structure of the root system of *B. caryophylloides* leads to a great accumulation of raw humus and hence to a raising of the soil level. A system of streams subsequently develops draining *away* from the sea and these, by erosion, carry away the silt and lead to a fall in the soil level. When this subsidence has taken place the soil is rich in humus and more suitable for the establishment of *Rhizophora* than before the colonization by *Bruguiera*.

On the coasts of Sumatra and Borneo there are very extensive mangrove swamps in which both the communities present and their successional relations are much as in the Malay Peninsula. The east coast of Sumatra, according to Troll & Dragendorf (1931), is fringed with an almost unbroken mangrove belt which in many places is several kilometres wide. The dominant species in the pioneer zone is *Avicennia alba*, often associated with *Sonneratia alba*. Further inland these give place to species of *Rhizophora*, *Bruguiera*, *Xylocarpus*, etc. and at the landward edge of the swamps there is generally a zone of the palm *Nipa fruticans* with which *Sonneratia acida* is often associated. The ground water in the *Nipa* zone is brackish and not salt. The upper limit reached by the tides and the limits of salt and brackish water by no means coincide and, as Troll & Dragendorf emphasize, some mangroves, notably *Bruguiera gymnorrhiza*, *Sonneratia acida* and the semi-mangrove *Nipa*, like the western mangroves, can grow in perfectly fresh water. Still further inland, where the natural vegetation has not been destroyed, the *Nipa* zone is followed by a wide belt of Moor forest (p. 292). In Amboina in the Moluccas (Troll & Dragendorf, 1931) (Fig. 38) the mangrove communities are essentially similar to those of Sumatra, the pioneer species being *Sonneratia alba*.

In the Malay Peninsula and the Greater Sunda Islands the eastern mangrove formation is seen at its highest and most complex development. Elsewhere on the coasts of the Indian and Pacific oceans, especially outside the area of the

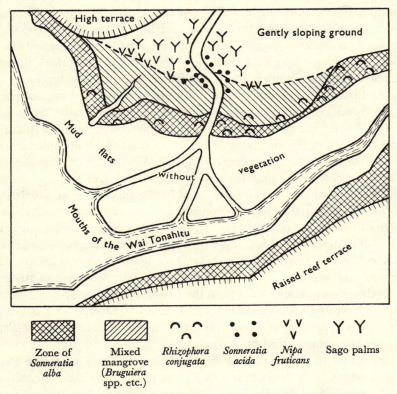

Fig. 38. Distribution of mangrove communities in Amboina (East Indies). After Troll & Dragendorff (1931).

Tropical Rain forest, environmental conditions are often less favourable and the fewer species of mangrove are present. From a maximum in the Malayan region, the floristic richness of the mangrove diminishes westwards towards Africa, and eastwards, northwards and southwards in the Pacific, till at the furthest limits of the eastern mangrove, as in the Red Sea and in southern Japan, only a single species may be present. In the central Pacific the mangrove formation as such disappears entirely and in Tahiti, for instance, habitats suitable for mangroves are occupied by thickets of *Hibiscus tiliaceus*, a tree normally characteristic of littoral woodland and semi-mangrove communities.

With the impoverishment of the flora the zonation and succession of the mangrove communities also necessarily become simpler.

Frequently in the eastern mangrove formation *Avicennia alba,* and species of *Sonneratia* are the pioneers, but it should be noted that in some places, e.g. Fiji (Mead, 1928) and New Caledonia (Däniker, 1929), this part, as in the western mangrove, is played by a *Rhizophora* (*R. mucronata*). Schimper's view (1891), based mainly on his own observations in the land-locked lagoons of Tjilatjap and Negombo in Java, was that *Rhizophora* is always the pioneer in the eastern mangrove, while Troll & Dragendorf (1931) held that it always succeeds some other species; neither view is justified.

On the east coast of Africa the climate at sea-level is nowhere moist enough to support typical Rain forest and the mangrove vegetation is correspondingly poorly developed. The greatest luxuriance and floristic richness is attained on the stretch of coast between Tanga and Mozambique, but even here only five mangrove species are present. Near Tanga, on the coast away from large river mouths, the mangrove vegetation forms a belt from somewhat above the level of the lowest ebb tides to above the limit reached by ordinary spring tides (Walter & Steiner, 1936–7). Four zones can be distinguished: (i) a pure stand of *Sonneratia alba* on the seaward face, forming the pioneer zone; (ii) a zone of *Rhizophora mucronata*; (iii) a narrow zone of *Ceriops candolleana*; (iv) a zone of *Avicennia marina. Bruguiera gymnorrhiza* occurs as isolated trees, mainly in the *Rhizophora* and *Avicennia* zones. The shrub *Lumnitzera racemosa* and other plants are found in the *Avicennia* zone.

Further inland the *Avicennia* zone is followed by a belt entirely bare of vegetation, probably because the variations in the salt-content of the soil are greater than any plant can tolerate. Where the natural vegetation has not been destroyed a non-halophilous evergreen scrub, in which *Adansonia digitata* is a characteristic tree, follows the bare zone. Near the mouths of rivers the zonation is somewhat different; *Rhizophora* is the pioneer and on the landward edge of the mangrove zone there is a belt of *Acrostichum aureum.* The mangrove vegetation here passes insensibly into fresh-water swamp communities. It seems clear that in the dry climate of the East African coast mangrove vegetation is only succeeded by non-halophilous types of vegetation under the locally moist edaphic conditions at the river mouths; elsewhere the *Avicennia* community is the final stage of the succession and must be regarded as an edaphic (or physiographic) climax.

Enough has now been said about mangrove vegetation in the western and eastern tropics to show the general course of the succession. Under some conditions—which will generally obtain on sheltered muddy coasts in regions where the Tropical Rain forest is the climatic climax—mangrove vegetation is succeeded by Fresh-water Swamp forest or perhaps sometimes directly by climax Rain forest, under other conditions mangrove communities themselves form a stable climax, depending on local conditions of relief and topography. In the former case the later stages of both the mangrove succession and the primary hydrosere may be identical. The mangrove succession shows interesting resemblances to the fresh-water hydrosere in eutrophic waters and also important differences from it. In both types of succession there is a progression from extreme to less extreme conditions and in both the early dominance of one or two pioneer species ultimately gives place to a more mixed community of woody species. An important difference from the hydrosere is that in the mangrove succession herbaceous plants and palms play very little part. The pioneers as well as the dominant species which follow them are dicotyledonous trees, a further illustration of the overwhelming preponderance of woody plants and the phanerophyte life-form in the vegetation of the tropics.

Part V

TROPICAL RAIN FOREST UNDER LIMITING CONDITIONS

CHAPTER 15

RAIN FOREST, DECIDUOUS FOREST AND SAVANNA

Schimper (1903, p. 260; 1935) distinguished four lowland tropical 'climatic formations', which he defined as follows:

'*Rain forest* is evergreen, hygrophilous in character, at least 30 m. high, but usually taller, rich in thick-stemmed lianes and in woody as well as herbaceous epiphytes.

'*Monsoon forest* is more or less leafless during the dry season, especially towards its termination, is tropophilous in character, usually less lofty than the rain forest, rich in woody lianes, rich in herbaceous, but poor in woody epiphytes.

'*Savanna forest* is more or less leafless during the dry season, rarely evergreen, is xerophilous in character, usually, often much, less than 20 m. high, park-like, very poor in underwood, lianes and epiphytes, rich in terrestrial herbs, especially grasses.

'*Thorn forest*, as regards foliage and average height, resembles savanna forest, but is more xerophilous, is very rich in underwood and in slender-stemmed lianes, poor in terrestrial herbs, especially in grasses, and usually has no epiphytes. Thorn-plants are always plentiful.'

In addition to these four lowland formations with a structure more or less that of forest or woodland, Schimper recognizes two others not dominated by trees, Tropical grassland and Tropical desert; the former occurs either as 'Savanna' (grassland with evenly or unevenly scattered trees) or as 'Steppe' (grassland without trees).

These six formations occur as analogous, but floristically different, communities in the tropics of America, Africa and Asia, and Schimper regarded each as well-defined and readily recognizable by its physiognomy and structure. 'With at least 180 cm. of rainfall, the high forest alone predominates. In regard to rainfalls of 150–180 cm. no data are available. With 90–150 cm. of rainfall there is a struggle between xerophilous woodland and grassland. Xerophilous woodland gains the victory when greater heat and more prolonged rainless periods prevail during the vegetative season; grassland succeeds when a milder temperature, a more even distribution of rainfall during the vegetative season, and windy, dry or frosty seasons prevail' (Schimper, 1903, p. 282). The six formations are each determined, in Schimper's view, by a distinct type of climate; within the tropical zone we can thus distinguish a rain-forest climate, a monsoon-forest climate, a grassland climate, etc. Subsequent work has made it clear that edaphic factors can locally overcome or modify the effects of climate.

Belts of evergreen forest (Fringing forest, Gallery forest of Schweinfurth), for example, often penetrate far into savanna regions along the valleys of rivers and watercourses; these are dependent on the permanent moisture in the soil, which compensates for seasonal dryness of the climate.

Schimper's conceptions have the merit of simplicity and have enabled much progress to be made in the description of tropical plant communities; but with the gradual increase of knowledge they have inevitably become inadequate. From the ecological point of view their most serious defect is that one of the six 'climatic formations' is probably not in fact climatic. Tropical grassland (here, following Schimper, both savanna and treeless 'steppe' are included in the term) can occur in close proximity to luxuriant Rain forest as well as to forest of a much 'drier' type and it is found under a wide variety of climatic conditions. In fact, as Beard (1946b) has recently pointed out, there is no such thing as a 'tropical grassland climate'. Evidence is accumulating in favour of the view that all lowland tropical grasslands and 'open' savannas are biotic climaxes usually due to fire, or edaphic climaxes due to soil conditions unfavourable to trees, or stages in a hydrosere. It is extremely doubtful whether any tropical grassland is in fact a true climatic climax.

Several attempts have been made recently to meet the need for a new system of classification and nomenclature for tropical vegetation to replace that of Schimper; among the most noteworthy are those of Champion (1936) and Burtt Davy (1938). None of these schemes has yet been generally accepted.

The various formations constituting the transition between rain forest on the one hand and desert on the other form a sequence which is closely parallel throughout the tropics, but, as was emphasized by Richards, Tansley & Watt (1940), much more knowledge of these plant communities is required before a classification can be constructed applicable to both the Old and New Worlds. What are chiefly needed for this purpose are data on structure (which can be accurately recorded by means of profile diagrams) and on the physiognomy of the component species (which is also capable of exact measurement or description). Eventually it may be possible to give precise definitions of tropical plant formations in terms of the number and height of their component strata, their life-form spectrum, the percentage of evergreen and deciduous, mesophyllous and microphyllous, etc. species found in them.

Until this detailed knowledge is available it would be unprofitable to discuss at length these difficult problems of classification and nomenclature. The aim in this chapter is to give an accurate picture, for a few selected areas, of the relations of rain forest to the 'drier' forest types beyond its climatic limits, and to savanna.

RAIN FOREST, DECIDUOUS FOREST AND SAVANNA IN NORTH-EASTERN SOUTH AMERICA

Forest types of Trinidad

The island of Trinidad, separated from the mainland of South America by straits 11 km. wide, is a particularly favourable area for studying Rain forest at its climatic limits. The island, though only 4543 sq. km. (1754 sq. miles) in extent, is remarkably varied in soil and topography and there is also a wide range of climate. The rainfall varies from over 250 to less than 120 cm., and (what is even more significant for the vegetation) there are large variations in its seasonal distribution, some parts of the island having over 4 in. (10 cm.) in every month of the year and others a long annual drought. Some 23 % of the area is reserved forest and, owing to the population having been sparse until comparatively recently, much of this forest is primary and comparatively little modified by human activities.

The forest types of Trinidad were first described by Marshall (1934) and have been more recently reclassified by Beard (1944a, 1946b), who has added a wealth of precise data about their structure, physiognomy and floristic composition, of which a short summary will be given here. In the lowlands and 'Lower Montane zone' (foothills up to about 2500 ft. (760 m.)) Beard recognizes four climatic formations, each of which appears to be a separate climax and is represented by one or more associations or consociations, viz.: (i) Lower Montane Rain forest, (ii) Evergreen Seasonal forest, (iii) Semi-evergreen Seasonal forest, and (iv) Deciduous Seasonal forest. These four types form a series, the first occupying the areas of optimum climate, the last the areas with the longest and most severe seasonal drought, and the other two occupying areas with intermediate conditions. On the neighbouring Gulf Islands and on the mainland of Venezuela, though not in Trinidad itself, there are two still more xerophilous formations of the same series, namely Thorn woodland and Cactus scrub. In addition to the main climatic formations just enumerated, there is another, Littoral woodland, determined by the local climatic conditions of the sea-coast, and a number of edaphic formations with which we are not here concerned.

Full information about the climatic limits of the formations is not available, but their requirements may be stated in general terms as follows:

	Lower Montane Rain forest	Evergreen Seasonal forest	Semi-evergreen Seasonal forest	Deciduous Seasonal forest
Total annual Rainfall (cm.)	Over 180	Over 180	130–80	80–130
Duration of dry season	None	3 months each with under 10 but over 5 cm.	5 months each with under 10 but over 2·5 cm.	5 months each with under 10 cm., two under 2·5 cm.

Little is known of the climatic conditions in Thorn woodland and Cactus scrub. The relation of these formations to rainfall is considerably modified locally by

318

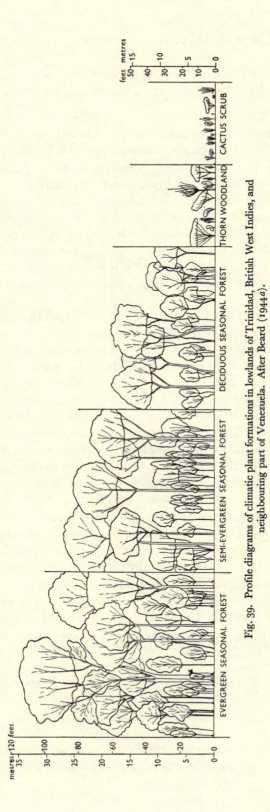

Fig. 39. Profile diagrams of climatic plant formations in lowlands of Trinidad, British West Indies, and neighbouring part of Venezuela. After Beard (1944a).

soil conditions (see p. 223), differences in the water-supplying capacity of the soil compensating to some extent for differences in rainfall.

(i) *Lower Montane Rain forest.* This is a fairly typical Rain forest, though dwarfed and somewhat modified by exposure to wind. Almost all the species are evergreen. Tree species are very numerous, but lianes and epiphytes poorly developed. In Trinidad this formation is represented by only one association, the *Byrsonima spicata-Licania* association.

(ii) *Evergreen Seasonal forest.* This formation closely resembles typical rain forest and like it consists of three stories of trees (Fig. 7 *a*, p. 32); these are at 30 m. and upwards, 12–27 m. and 3–9 m. respectively, so that the community is somewhat lower and less luxuriant than the best developed Rain forest. Apart from the lower height of all the stories, the chief difference from 'true' Rain forest is that the highest tree story is still more discontinuous. Individual trees may reach large sizes, but trees 3 m. or more in diameter are scarce. 'The general impression is of an occasional huge tree in the midst of smaller growth and the closely ranked columnar effect of rain forest is lacking' (Beard, 1944*b*, p. 138).

The larger trees branch relatively lower down than in Rain forest and clean boles over 20 m. in length are rare. This formation is mainly evergreen, but some species of large trees have a short leafless period lasting from a few days to a few weeks and may be described as semi-deciduous. In the highest story about 17 % of the species are deciduous and 10 % semi-deciduous, but of the individuals only 3 and 0·4 % are deciduous and semi-deciduous respectively. The lower stories are even more completely evergreen.

As in typical Rain forest buttressing is a prominent feature of many of the first-story trees and the leaves of an overwhelming majority of the trees fall into the 'mesophyll' size-class of Raunkiaer (p. 84). Lianes and epiphytes are fairly abundant. Four 'faciations' of Evergreen Seasonal forest are recognized; more or less the same species occur in all of them, but in different proportions. In one 'faciation' *Mora excelsa*[1] is dominant, but this community is regarded here as a distinct consociation (see pp. 34, 237).

From the above description it will be seen that the structural and physiognomic differences between Evergreen Seasonal forest and the typical lowland Tropical Rain forest are slight. Much of what has usually been reckoned as Rain forest, both in South America and elsewhere (including probably the Nigerian Rain forest dealt with on pp. 29 etc.), is similar to the Evergreen Seasonal forest of Trinidad and probably belongs to an analogous formation. Though it is certainly an advantage to recognize the differences between Evergreen Seasonal forest and the more luxuriant Rain forest found in climates with a less marked dry season, it is doubtful whether it is desirable to narrow the term Tropical Rain forest so as to exclude communities of this kind. On grounds of convenience it would be preferable to preserve the well-established broad

[1] In Beard's 1944 paper the '*Mora* faciation' of Evergreen Seasonal forest was regarded as a consociation of 'true rain forest'.

conception of Tropical Rain forest and to regard the Evergreen Seasonal forest as a 'subformation' (or perhaps in Clements' terminology as a faciation) of the Tropical Rain forest formation-type.

(iii) *Semi-evergreen Seasonal forest*. This formation and the next appear to be the approximate equivalents of Schimper's Monsoon forest; it seems desirable, however, as Beard points out, not to use the term Monsoon forest for forest communities outside south-eastern Asia, the area in which it was originally applied.

In Semi-evergreen Seasonal forest there are only two tree stories, a layer at 20–26 m. forming a more or less closed canopy and a lower layer at 6–14 m. Trees of large girth are rare, most of the mature trees averaging about $\frac{1}{2}$ m. in diam. Branching begins at a low level and the crowns tend to be umbrella-shaped. The upper story consists of both evergreen and deciduous species; from 26 to 42 % of the species and from 21 to 30 % of the individuals, depending on the association, are regularly deciduous. Of the remaining trees in the first story some are truly evergreen, but most are facultatively deciduous, the amount of leaf-fall varying from year to year, so that in a wet year most of the trees lose very few leaves, while in a severe dry season the whole canopy may appear leaf-less. The lower story is mostly evergreen, but also includes a considerable number of deciduous species. The leaves of the upper story are predominantly 'meso-phylls', but in the lower story there is a notable proportion of 'microphylls'.

There are few strongly buttressed trees. Thorns are not frequent, but some species in both tree stories have thorny trunks. Lianes are better developed than in any other formation in Trinidad; all large trees are heavily loaded with them. Epiphytes are relatively scarce.

There are six associations belonging to the Semi-evergreen Seasonal forma-tion. All of them probably exist in habitats liable to more severe seasonal drought than the Evergreen Seasonal forest, but the drought is not wholly climatic and is due to a combination of topographic, edaphic and climatic factors.

(iv) *Deciduous Seasonal forest*. In Trinidad itself the Deciduous Seasonal forest formation is represented by two communities, both of which appear to be secondary associes considerably affected by human interference. On the neighbouring island of Little Tobago, however, it is represented by what seems to be a true climax formation (Beard, 1944 b) and Beard's description is partly based on the latter area.

The Deciduous Seasonal forest is low and consists of two tree stories, an upper open one of scattered trees up to 20 m. high, and a lower layer at 3–10 m. Trunks fork or branch low down and are frequently crooked; many of the lower story trees tend to grow in clumps. There are few very large trees; about $\frac{1}{2}$ m. is the maximum diameter.

Over two-thirds of the trees in the upper story are deciduous and their deciduousness is obligate, taking place with unfailing regularity every dry season. The lower story is almost entirely evergreen. About half the leaves in the upper story are 'mesophylls' and about half 'microphylls'; in the lower story the majority of the leaves are 'mesophylls', but there is a large proportion of 'microphylls'.

Buttressed trees are absent, but a few species have spiny or thorny trunks. Lianes are somewhat rare and epiphytes absent or scarce.

(v) *Thorn woodland*. This formation, which as already mentioned does not exist in Trinidad, is found in the neighbouring parts of Venezuela as a fairly open scrubby community consisting largely of hard-leaved, microphyllous, evergreen trees 3–10 m. high, belonging chiefly to the Leguminosae (Mimosoideae and Caesalpineae); most of these trees are thorny. There is no grass on the ground (as there often is in similar types of vegetation in other parts of the world) and the only low-growing vegetation consists of a few Bromeliaceae and succulents.

(vi) *Cactus scrub*. This formation is met with on the island of Patos, between Trinidad and Venezuela, and on the mainland. It is an open type of vegetation dominated by columnar cacti and prickly pears (*Opuntia*) mingled with scattered gnarled micro- or leptophyllous bushes and terrestrial Bromeliaceae. There is no grass and the ground is often bare.

The six climatic formations described by Beard obviously form a continuous series, each type grading into the next. Though the dividing lines are more or less arbitrary, the divisions proposed by Beard seem on the whole both practically and theoretically convenient.

Taken as a whole the series shows a striking relation between the structure and physiognomy of the vegetation on the one hand and the climate on the other (Fig. 39). As the seasonal drought (whether due solely to climate or to a combination of climate with other environmental factors) becomes more severe, the stature of the vegetation decreases and its structure becomes simpler, the number of layers decreasing. The deciduous habit becomes commoner, first in the upper stories, later in the lower. That the lower stories should remain evergreen when the top story has become mainly deciduous is of course to be expected from their moister microclimate. There is a tendency towards a general reduction in leaf size; 'microphylls', which are rare in Rain forest, become increasingly common towards the dry end of the series. Buttressing decreases (but this may be merely a function of the decreasing average size of the trees) and thorniness becomes an increasingly common feature.

The plant formations of Trinidad are doubtless outliers of formations more extensively developed on the mainland of South America. Seasonal forests more or less similar to those of Trinidad are widely distributed in the interior of British Guiana (Davis, 1933; Myers, 1936), the Amazon basin (Ducke, 1938, quoted by Beard, 1944a), Venezuela (Myers, 1933; Pittier, 1939, quoted by Beard, 1944a), Colombia (Dugand, 1934), southern Brazil (Warming, 1892) and in Central America, as well as in Cuba (Bennett & Allison, 1928) and elsewhere in the Antilles. Thorn woodland and Cactus scrub exist in various parts of the West Indies, especially the Greater Antilles (e.g. Jamaica, Brockmann-Jerosch, 1925), and in Venezuela and elsewhere in the mainland of South America. The well-known caatingas of southern Brazil (Lützelburg, 1925–26; Ule, 1908*b*, 1908*c*) are similar communities (to be distinguished from the caatinga forests of the Rio Negro region, see p. 237). It is thus probable that everywhere round the margin of the tropical American Rain forest, where the natural climax vegetation has not been destroyed or much modified by human interference, a series of formations closely parallel to those of Trinidad will be found. How far the scheme of classification proposed by Beard is of general application throughout the New World tropics, time alone can show.

Savanna in the West Indies and north-eastern South America

So far nothing has been said about the relation of savanna to Rain forest in Trinidad and north-eastern South America. This is a complex problem which cannot be exhaustively dealt with here.

It should be noted that the term 'savanna', like heath, moor, etc., is a vernacular word (probably of Carib origin), to which it is difficult, if not impossible, to give a precise scientific connotation. In the West Indies, 'savanna' means practically any kind of low-growing vegetation dominated by grasses or other herbaceous plants; it is even applied to parks and grass plots in towns. The term is often used for communities dominated by grasses or Cyperaceae growing in permanently swampy situations. Of this type are the 'water savannas' of the Orinoco delta and the coastal region of British Guiana (Myers, 1933) and the swamp savannas on the coast of Surinam (Lanjouw, 1936), as well as the *savanes basses* and *savanes tremblantes* of French Guiana (Benoist, 1925). Similar, too, are probably the *campos de varzea* of the lower Amazon (Bouillienne, 1926, 1930). Such savannas are hydroseral stages and have little in common with the savannas we are concerned with here.

Typical savannas are dominated by grasses, sometimes by sedges (*Rhynchospora* spp., etc.), generally with a considerable admixture of herbaceous and small shrubby dicotyledons. Trees and bushes, usually more or less xeromorphic in appearance, are seldom absent over extensive areas. Characteristically these grow in scattered groups, as well as in strips along rivers and watercourses (Fringing or Gallery forest), giving the landscape a park-like appearance (Pl. XII B). In 'orchard savannas' low-growing, often crooked, trees are evenly scattered through the grass, so that there is a resemblance to a derelict apple orchard. Among the trees of the South American savanna, palms such as species of *Mauritia* and *Copernicia*, are often conspicuous. The definition of a savanna proposed by Lanjouw (1936)—'savannas are plains in the West Indian islands and northern South America covered with more or less xeromorphic herbs and small shrubs and with few trees and larger shrubs'—is serviceable, though it is hardly practicable to limit the application of the term to a single geographical region.

Savannas, even so defined, are very heterogeneous in physiognomy, floristic composition and ecological status. Though in recent years some excellent descriptions of the South American savannas have been published, far too little is yet known about them for any adequate classification to be put forward.

In Trinidad there are two kinds of true savanna (Beard, 1946*b*; see also Myers, 1933 and Marshall, 1934), lowland and mountain. In the lowland savannas of Piarco, Erin, etc., the vegetation consists of grasses (species of *Paspalum*, *Axonopus*, *Trachypogon ligularis*, etc.) and other herbs, among which small gnarled trees of characteristic savanna species (chiefly *Curatella americana* and *Byrsonima crassifolia*) are scattered. In the well-known Aripo savannas the ground is covered with Cyperaceae and grasses, intermixed with species of *Polygala*, *Utricularia*, *Drosera*, patches of *Sphagnum* and many other plants. There are occasional clumps of bushes and scattered groves of tall *Mauritia* palms, but *Curatella* is absent. The mountain savanna at St Joseph is a grassland of *Trachypogon ligularis* with scattered bushes and trees; over part of the area these are chiefly *Curatella* and *Byrsonima* as at Piarco etc., but elsewhere they are mainly *Myrcia stenocarpa* and *Roupala montana* which do not occur in the lowland savannas.

The Trinidad savannas are of small extent and occur as pockets in large areas of forest (mostly Evergreen Seasonal); it is very difficult to attribute their presence to climatic factors. There appears to be no tendency for the forest to invade the savannas and though some of them are frequently burnt, not all of them are subject to fire; Beard states that the Aripo savannas have never been known to burn. Grasslands are found in some parts of Trinidad which are known to have arisen by the destruction of forest for shifting cultivation followed by repeated burning. In these grasslands, which are a stage in a deflected succession (see Chapter 17) and may be termed secondary savannas, the dominant grass is *Imperata brasiliensis* which is never abundant in the true savannas (a closely related species plays a very large part in secondary and deflected successions in the Old World tropics, see Chapter 17); both the herbaceous and the woody flora is quite different from that of the primary savannas now being discussed. The rich flora of the primary savannas, which includes a number of species which are endemic or of highly discontinuous distribution, is strong evidence for their natural origin and great antiquity. In the lowland savannas drainage is impeded and the soil is waterlogged for part of the year (though not as swampy as in the 'water savannas' mentioned above) and very dry for the rest; the 'hog-wallowed' (hummocky) surface is characteristic. In the mountain savanna of St Joseph drainage is freer, but there is a very shallow gritty soil overlying impermeable quartzite which also becomes very desiccated in the dry season. It is perhaps the alternation of waterlogged and very dry soil conditions which excludes the forest vegetation from the savanna areas.[1] Beard regards the

---

[1] A similar explanation has been suggested for the absence of trees from the 'dambos' or valley grasslands of Northern Rhodesia (Michelmore, 1939, pp. 283–4).

Trinidad savannas as occupying senile soils on relics of an ancient land surface which has not been forest-covered since Pleistocene times.[1]

In British Guiana there are two main savanna areas (apart from the 'water savannas'), both of which cover very large tracts of country: (*a*) the Berbice savannas lying west of the Berbice River about 60–90 miles inland, (*b*) the savannas of the interior; these are in the basin of the Rupununi and in the neighbourhood of the Pakaraima Mountains and Roraima; they extend over the Brazilian frontier into the basin of the Rio Branco. The Berbice savannas have been described by Martyn (1931) and Follett-Smith (1930); they consist of rolling plains at an altitude of 80–90 ft. (24–28 m.) above sea-level, dominated by grasses and Cyperaceae. Here and there are 'islands' of bushes and trees and strips of fringing forest are found in the stream valleys. A slightly different vegetation is found in moister depressions, termed 'pans', as well as on large ant-hills which become colonized by bushes and trees. The soil is mainly a brown sand; where it changes to a coarse bleached sand (similar to the bleached sand of the Wallaba forest, see p. 240) the typical savanna flora is replaced by 'muri bush', a poor scrub of bushes and small trees (chiefly *Humiria floribunda* var. *guianensis, Clusia nemorosa*) interspersed with patches of very scanty vegetation including grasses, the fern *Schizaea incurvata* and lichens. This 'muri bush', as mentioned in Chapter 9, is physiognomically very like the so-called padang vegetation of Malaysia, which occurs under closely similar conditions of soil and climate. The savanna is surrounded by evergreen Rain forest and, as in many savanna areas, the transition from savanna (or 'muri bush') to forest is often abrupt. Patches of savanna or muri, often only a few hectares in extent are sometimes found as pockets in otherwise unbroken Rain forest, far from the main savanna areas. According to T. A. W. Davis (MS. notes), the Rain forest in the immediate neighbourhood of the Berbice savannas is usually Dakama forest, dominated by *Dimorphandra conjugata* (p. 256); further away the *Eperua* (Wallaba) consociation is usually met with. There is in fact a zonation of communities and a typical transect from forest to savanna might show the following sequence of zones: (i) *Eperua* consociation; (ii) *Eperua-Dimorphandra* association; (iii) *Dimorphandra* consociation; (iv) Transition zone in which 'bania' (*Swartzia* sp.), *Catostemma fragrans, Licania* sp. and 'kakarua' (Sapotaceae) are prominent; (v) 'muri bush', (vi) savanna. None of the forest types bordering the Berbice savannas is deciduous or even semi-deciduous; there is thus no evidence here for a transition: evergreen forest → deciduous forest → savanna.

The interior savannas are much more extensive. In the Rupununi-Rio Branco region (Myers, 1936) they consist largely of blue-green expanses generally dominated by the grass *Trachypogon plumosus*, with which other grasses and

---

[1] Charter (1941) has expressed views very similar to those of Beard as to the origin of 'pine ridge', a type of savanna in British Honduras (Central America) in which *Pinus caribaea* is the chief tree; he regards the pine ridge soils as over-mature (senile) and supposes that at an earlier stage in their development they carried evergreen forest.

many species of dicotyledonous herbs are associated. Trees, among the commonest being *Curatella americana* and *Byrsonima* spp., grow thinly scattered. The grassland is interrupted by strips of fringing forest and by 'islands' of low woodland varying from 20–30 m. to several kilometres across. In the moister depressions (*baixas*) the chief association is one consisting chiefly of Cyperaceae, in which groves of the palm *Mauritia flexuosa* take the place of the dicotyledonous tree 'islands' of higher ground. The fringing forests and other forest areas adjacent to the Rupunini-Rio Branco savannas are not Rain forest, but include a large proportion of deciduous species (Myers, 1933; Davis, 1933) and should probably be classified as Semi-evergreen Seasonal forest; the flora of the tree 'islands' is somewhat different, with fewer deciduous species. The upland savannas of the Pakaraima Mountains are like those of the Rupununi, but their vegetation is more varied.

The savannas of Surinam (Lanjouw, 1936, see below, p. 326) and of French Guiana (Benoist, 1925) closely resemble those of Berbice, consisting partly of open grassland and partly of 'muri bush'. In the Surinam savannas Cyperaceae are more often dominant than grasses; the adjacent forest is evergreen. Similar patches of savanna reappear in the lower Amazon region (Bouillienne, 1926).

The largest savanna areas of northern South America are the Venezuelan *llanos*, which are familiar through the classical description of Humboldt (1852). A modern account of their vegetation has been given by Myers (1933), who recognizes two types: (*a*) Bunch grass savannas. The chief dominants here are the grasses *Hyparrhenia rufa* and *Andropogon condensatus*; trees, especially *Curatella americana*, *Bowdichia virgilioides* and the palm *Copernicia tectorum*, occur in fringing forests and scattered elsewhere in very varying density. (*b*) High grass savannas. Here the tall grass *Paspalum fasciculatum* is overwhelmingly dominant. The forests bordering the *llanos* are mainly of a dry deciduous type (Schimper, 1935).

Ecological status of savanna in north-eastern South America

Savannas have been variously regarded as a climatic climax formation, as a biotic climax determined mainly by grazing and periodic burning or as an edaphic climax dependent on special soil conditions. Schimper (1935, p. 424) defines the climatic requirements of savanna as an annual rainfall of 90–150 cm., an effective dry season of 4–5 months and a range in the monthly mean temperatures of up to 14°. Under these conditions, according to Schimper, there is a struggle between savanna and woody vegetation, the issue of which is decided partly by climatic factors such as the amount of rainfall during the growing season, and partly by edaphic factors. According to this view then, the most important factors determining the occurrence of savanna are climatic. In the South American savanna areas we are now considering it is possible, or even likely, that the chief determining factors are not always the same, but there is remarkably little evidence that climate is ever decisive. The existence of savanna under a wide range of climatic conditions, its frequent occurrence in close proximity to typical Rain forest, the abrupt boundary between forest and savanna and the frequency of small pockets of savanna in extensive areas of high forest are all facts which make a climatic theory very difficult to uphold.

The Trinidad savannas, as we have seen, are difficult to reconcile with Schimper's view and Lanjouw's (1936) very clear discussion of the Surinam savannas leads to a similar conclusion. The annual rainfall in the latter (data for three stations) varies from 2056 to 2219 mm. and in the dry season there are from two to three consecutive months with less than 100 mm. If these figures are compared with those in Table 12 (p. 141) it is evident that the savanna is here well within the climatic limits of evergreen Rain forest, which is frequently found where the seasonal drought is considerably more prolonged. On the other hand there is much evidence that both edaphic factors and fire have played a large part in determining the vegetation. These Surinam savannas are situated on low plateaux with well-drained, more or less porous soils. In some places the soil is a coarse bleached sand, in others a sandy clay. Owing to the heavy rainfall, both types of soil are excessively leached and probably very deficient in plant nutrients. They appear to be of podzol type and the soil conditions would seem to be like those of the British Guiana Wallaba forest and the Bornean Heath forest and padang (Chapter 9) but, perhaps owing to the flat topography, the savannas are liable to seasonal waterlogging. The vegetation is frequently burned and the clear-cut boundary between the savanna and the surrounding forest can hardly be due to anything but fire. Lanjouw supposes that the savanna areas were originally covered by Rain forest similar to that still existing in the neighbourhood. Leaching of the soil has caused the replacement of the Rain forest by 'savanna forest' (we should perhaps understand by this a type of vegetation similar to the 'muri bush' or Dakama forest mentioned above (p. 324), rather than the climatic, seasonally deciduous Savanna forest of Schimper's definition). The savanna forest, being more liable than Rain forest to burning, has been converted by repeated fires into savanna, which is here a fire-climax, though its distribution is primarily determined by edaphic causes.

The status of the Berbice savannas in British Guiana, which are very similar to those of Surinam, can probably be interpreted in the same way. The situation of the Berbice savannas is similar and the rainfall regime is also similar, though the seasonal drought is somewhat more severe. The soils have been studied in some detail by Follett-Smith (1930), who recognizes three main types: acid brown sand, grey sand and coarse white sand (the last type, as mentioned above, being associated with the occurrence of 'muri bush'). These savannas are regularly grazed by cattle and frequently burned, both intentionally, to provide young grass for the stock, and accidentally. Here also the savanna vegetation does not seem to be determined primarily by the climate, but partly by the nature of the soil and partly by burning and grazing.

Both the Rupununi-Rio Branco savannas of Guiana and the Venezuelan *llanos* occur in regions where the annual rainfall is less (probably about 1150–1500 mm.) than in the Surinam and Berbice savannas, and the seasonal drought longer. The Rupununi-Rio Branco savannas are not heavily grazed, but they have been frequently and regularly burned over a very long period. Myers (1936) is doubtless correct in regarding their present vegetation as a fire-climax, but soil conditions (of which little is known) may also be important. The *llanos*, according to Myers (1933), are strongly affected by periodic fires and are also grazed—in some parts heavily overgrazed. The 'high grass' savanna type is seasonally flooded and perhaps all the *llanos* are liable, like the lowland savannas of Trinidad, to waterlogging during part of the year. On the little evidence available it seems probable that the vegetation, like the other savannas we have considered, may be primarily determined by edaphic factors, though much modified by fire. In both the Rupununi-Rio Branco region and the Venezuelan *llanos* the climatic climax appears to be deciduous or semi-deciduous forest.

For all the savanna vegetation of north-eastern South America and the neighbouring islands information on the climate, soil, frequency of burning and intensity of grazing is still extremely meagre, but from what has been said it seems fairly clear that none of the savannas of this region can be regarded as a climatic climax. The majority of them have been influenced by fire. Myers (1936, p. 176) says: 'I have never seen in South America a savannah however small or isolated or distant from settlement which did not show signs of more or less frequent burning.' This statement may be an exaggeration, as some savannas, like Aripo savanna in Trinidad, seem not to be subject to fire and the existence of such areas makes it improbable that burning alone is responsible for the inability of high forest to invade savanna. Frequent burning has given the savannas their sharp boundaries and has perhaps favoured the grasses and herbaceous vegetation at the expense of the woody plants, but the north-eastern South American savannas in general seem to be primarily an edaphic climax modified into a fire-climax by repeated burning. Those, like Aripo, which are not burned, may indicate the nature of the original edaphic climax from which the others are derived.

RAIN FOREST, DECIDUOUS FOREST AND SAVANNA IN SOUTH-EASTERN ASIA

In Asia the most typical development of the Tropical Rain forest is found in the Malay Peninsula and in the neighbouring islands of Sumatra and Borneo. Throughout this large area the lowland climate is equatorial and is characterized by a heavy and evenly distributed rainfall. To the north and west of it, in Siam, Burma and in tropical and subtropical India, though locally there are considerable tracts of evergreen forest, the climate is on the whole more seasonal and there is usually an annual dry period of several consecutive months each with less than 10 cm. (4 in.) of rain. The vegetation in consequence is mostly more or less tropophilous (in Schimper's sense) and the climax types range from semi-deciduous and deciduous Monsoon forest to Thorn woodland; in places, for instance in the centre of the 'Dry zone' of Burma and in the driest parts of India, highly xerophytic semi-desert vegetation prevails.

In the Malay Peninsula signs of the transition from the evergreen Rain forest to the deciduous Monsoon forest are first apparent in the northern Malay States (Perlis, Kedah, northern Kelantan), where typically 'Burmese' trees such as *Albizzia lebbek* and *Lagerstroemia floribunda* begin to be frequent; a few of these species penetrate even further southwards. At the southern fringe of the Malayan Rain forest, in the southern and eastern islands of the Malay Archipelago, there is also a transition from an equatorial climate to one with a dry season and deciduous forest replaces evergreen forest as the climax. In the Sunda Islands the boundary between the non-seasonal Rain forest and the seasonal Monsoon forest crosses Java, the western part of which has a rain forest climate (though the lowland Rain forest has been almost entirely destroyed), contrasting with the monsoon climate of the eastern part.

The great evergreen Rain forest *massif* centring on Singapore thus divides the seasonal Monsoon forest of south-eastern Asia into two widely separated areas; it is therefore a striking fact that there is a considerable floristic resemblance between the deciduous forests of Siam and Burma on the one hand, and those of the eastern Sunda islands on the other (see van Steenis, 1935, pp. 83–5). Such a typical Monsoon forest tree as the teak (*Tectona grandis*), for instance, occurs abundantly in both areas, while in the intervening rain-forest region it is not only not native, but cannot be successfully cultivated. Teak and other species with a like distribution have been regarded, though perhaps on insufficient grounds, as introduced in the south-eastern area (see van Steenis, 1935, p. 85 and references given there).

The seasonal (or Monsoon) forests of the eastern Sunda Islands are inadequately known and have for the most part been so much modified by fire and other biotic influences that their natural structure is hard to reconstruct. For this reason the transition from Rain forest to Monsoon forest will be considered here in detail only as it is seen in Burma. Even in Burma the effects of fire and shifting cultivation are so widespread that the recognition of the climatic climax types offers considerable difficulties.

The forest types of Burma are fairly well known through the work of Kurz (1875), Stamp (1925) and the detailed study by Stamp & Lord (1923) of part of the 'riverine tract'. Champion (1936) has also published a comprehensive and elaborate classification in which a uniform system of nomenclature for the forest communities of the whole of India and Burma is proposed.

In Burma there is a gradation of climate from the wet region of the Arakan Coast in the west, and of Tenasserim and the Irrawaddy delta in the south, to the 'Dry belt' which occupies the central part of the Irrawaddy basin. The west and south have a heavy rainfall and a short dry season, while the 'Dry belt' has a rainfall in places as low as 20 in. (51 cm.), mostly falling as heavy downpours on a few days in the wet season, the rest of the year being almost completely rainless. According to Stamp (1925), Burma can be divided into three parts, separated by the isohyets of 40 in. (102 cm.) and 80 in. (203 cm.):

(i) Areas with over 80 in. rainfall. Here the climax vegetation is evergreen forest and is usually described as Tropical Rain forest. The climate is unsuitable for typical monsoon-forest trees—'in plantations teak will indeed grow rapidly, but the trees in many cases become hollow and fluted early' (Stamp, 1925, p. 7).

(ii) Areas with 40–80 in. rainfall. Here the climax types fit Schimper's definition of Monsoon forest (p. 315), the dominant tree species losing their leaves in the dry season (the old leaves are shed in January or February and the new leaves appear about a month before the rains break in June). The chief teak forests are found within this area.

(iii) The 'Dry belt' with less than 40 in. rainfall. In this area there are no tall forests; the climax vegetation varies from low open woodland, falling into Schimper's category of Savanna forest, to Thorn forest and semi-desert scrub.

Within the three zones of vegetation thus delimited a number of climatic climaxes or formations can be recognized. The chief of these, according to the scheme of Stamp (1925), are set out in Table 34. It is noteworthy that in constructing this scheme Stamp has adhered to Schimper's four main divisions, Rain forest, Monsoon forest, Savanna forest and Thorn forest (see Stamp & Lord, 1923) as a basis. All the climax formations recognized are dominated by woody plants. Though savanna, or savanna-like communities, are found in Burma, e.g. the kaing grasslands or elephant-grass jungles, all of these appear to be seral, forming part of biotic or deflected successions (in the case of kaing grasslands perhaps of a primary succession). The same may perhaps be said, though with less certainty, of the various types of bamboo and cane brakes (communities dominated by bamboos and *Calamus* spp.). The chief floristic and physiognomic features of the formations in Table 34, as described by Kurz (1875) and Stamp (1925), may be summarized as follows:

1. Evergreen Dipterocarp forest. Tall forest dominated by a mixture of species in which *Dipterocarpus* spp. and other Dipterocarpaceae play a leading part. Floristically this formation is much richer than any of the others; Kurz mentions finding 300–350 species of trees in an area of barely 8–9 sq. miles (20–30 sq. km.) and says that a similar area in even the richest deciduous forest would contain barely 70–80 species. There are three stories of trees in addition to shrub and herb layers. Trees of the top story are seldom less than 150–200 ft. (45–60 m.) high and consist of a mixture of evergreen and deciduous species; some of the species, e.g. *Dipterocarpus* spp., shed their leaves annually, but expand the new leaves before the old ones have dropped off. The middle and lowest stories

TABLE 34. *Chief lowland climatic climaxes of Burma.*
(After Stamp, 1925)

The names according to the terminology of Champion (1936) are given in square brackets [ ].
The principal (dominant or most abundant) species are given in round brackets ( ).

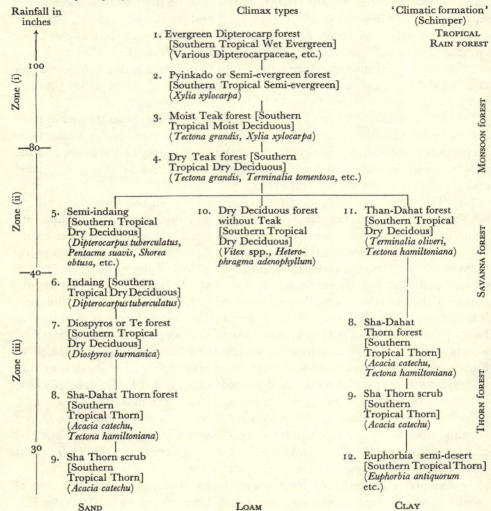

Rainfall in inches	Climax types			'Climatic formation' (Schimper)
	1. Evergreen Dipterocarp forest [Southern Tropical Wet Evergreen] (Various Dipterocarpaceae, etc.)			TROPICAL RAIN FOREST
Zone (i) 100	2. Pyinkado or Semi-evergreen forest [Southern Tropical Semi-evergreen] (*Xylia xylocarpa*)			MONSOON forest
	3. Moist Teak forest [Southern Tropical Moist Deciduous] (*Tectona grandis, Xylia xylocarpa*)			
—80—	4. Dry Teak forest [Southern Tropical Dry Deciduous] (*Tectona grandis, Terminalia tomentosa*, etc.)			
Zone (ii)	5. Semi-indaing [Southern Tropical Dry Deciduous] (*Dipterocarpus tuberculatus, Pentacme suavis, Shorea obtusa*, etc.)	10. Dry Deciduous forest without Teak [Southern Tropical Dry Deciduous] (*Vitex* spp., *Heterophragma adenophyllum*)	11. Than-Dahat forest [Southern Tropical Dry Decidous] (*Terminalia oliveri, Tectona hamiltoniana*)	SAVANNA forest
—40—	6. Indaing [Southern Tropical Dry Deciduous] (*Dipterocarpus tuberculatus*)			
Zone (iii)	7. Diospyros or Te forest [Southern Tropical Dry Deciduous] (*Diospyros burmanica*)		8. Sha-Dahat Thorn forest [Southern Tropical Thorn] (*Acacia catechu, Tectona hamiltoniana*)	THORN forest
	8. Sha-Dahat Thorn forest [Southern Tropical Thorn] (*Acacia catechu, Tectona hamiltoniana*)		9. Sha Thorn scrub [Southern Tropical Thorn] (*Acacia catechu*)	
30	9. Sha Thorn scrub [Southern Tropical Thorn] (*Acacia catechu*)		12. Euphorbia semi-desert [Southern Tropical Thorn] (*Euphorbia antiquorum* etc.)	
	SAND	LOAM	CLAY	

consist almost entirely of evergreens, 'the lofty trees forming, as it were, a leaf-shedding open forest above the lower stratum of evergreen trees' (Kurz, 1875, p. 31). Palms, including *Calamus* spp., and bamboos form part of the undergrowth. The herb layer is sparse and consists chiefly of ferns and dicotyledonous herbs, never of grasses. Epiphytes and lianes are abundant.

From this description it is evident that the Evergreen Dipterocarp formation of Burma closely resembles typical Tropical Rain forest, but differs in the larger proportion of deciduous trees; it would appear to be closely analogous to the Evergreen Seasonal forest of Trinidad and tropical America.

This formation occurs in parts of southern Burma both in the Pegu Yomas with an annual rainfall of 80 in. (203 cm.) or more, and more typically in Tenasserim, where the rainfall is 150–200 in. (380–500 cm.).

2. Semi-evergreen or Pyinkado forest. The characteristic tree in this formation is *Xylia xylocarpa*, a deciduous leguminous tree which has a similar distribution in Burma to teak (*Tectona grandis*), but tolerates a moister climate. This species is often dominant; other prominent species are *Lagerstroemia* sp., *Dipterocarpus alatus* and *D. turbinatus*. Bamboos occur and climbers and epiphytes are usually abundant. Details of structure are lacking, but in most respects this type of forest seems to be like the last, except in having a still greater proportion of deciduous trees.

Usually Semi-evergreen forest is found under a rainfall of 80–95 in. (203–242 cm.), but on shallow porous soils it may be found under considerably higher rainfall.

3. Moist Teak forest. The deciduous *Tectona grandis* is the most characteristic species in this formation and the next, but it is very rarely dominant and seldom exceeds 10 % of the stand. In Moist Teak forest *Xylia xylocarpa* is the most constantly associated species and is numerically more important than teak. The trees grow to 120 ft. (37 m.) on sandstone in the Pegu Yomas, but only to 80–90 ft. (24–27 m.) under less favourable conditions. The forest as a whole may not lose its leaves till the end of March or it may be already bare in February, but the leafless period does not exceed three or four months. The shrub layer is not well developed, but bamboos play an important part. A grass covering on the ground is exceptional and terrestrial ferns are few. Climbers are conspicuous, but few in individuals, and do not interfere with the open appearance of the forest. Orchids and other epiphytes are not conspicuous.

Moist Teak forest is perhaps comparable in structure and physiognomy with the Semi-evergreen Seasonal forest of Trinidad. It is widely distributed in Burma where the rainfall is between 60 and 90 in. (152–229 cm.) (up to 100 in., 254 cm., in northern Tenasserim).

4. Dry Teak forest. Though *Tectona grandis* is again a characteristic species in this formation, most of the associated species are different from those in Moist Teak Forest. In the latter most of the species are those common in Evergreen Dipterocarp and Semi-evergreen forest; here they are mostly species of the Indaing (formation 6) and other dry types. The bamboo *Dendrocalamus strictus* is constant and characteristic in Dry Teak forest, while *Bambusa polymorpha* and *Cephalostachyum pergracile* are the typical bamboos of Moist Teak forest. This formation is more influenced by soil than the three moister formations and there is a striking difference between the Dry Teak forests on Tertiary sandstones and those on ancient gneisses, schists and limestones.

Dry Teak forest is widely distributed and its lower limit of rainfall is about

40 in. (102 cm.); the upper limit is very dependent on the nature of the soil; on poor sandy soils Dry Teak forest may be found with as much as 75 in. (191 cm.), but on loamy and more fertile soils it may be replaced by Moist Teak forest at 50–55 in. (127–142 cm.).

5–6. Semi-Indaing and Indaing. The first of these types is transitional to formation 4 and can scarcely be regarded as an independent formation. In formations 5 and 6 the most abundant tree is generally 'in' (*Dipterocarpus tuberculatus*; Indaing = 'in' forest), which often forms consociations, sometimes nearly pure. The characteristic species in addition to *D. tuberculatus* are in Semi-Indaing, *Pentacme suavis*, *Shorea obtusa* and *Terminalia tomentosa*, in Indaing, *Pentacme suavis*, *Melanorrhoea usitata* and *Terminalia tomentosa*. Indaing is open and sunny; the trees are widely spaced and climbers and large undergrowth are not abundant enough to obstruct visibility. The average height of the trees varies from 30 to 80 ft. (8–24 m.); except for 'in' itself and a few other species the trunks are more crooked and stunted and the branches disproportionately thick. Indaing becomes quite leafless in the dry season, though many of the trees produce their young leaves long before the rains begin; most of the trees flower when bare of leaves. The undergrowth consists often of a tangle of grasses and also includes a cycad and sometimes the palm *Phoenix paludosa*. Bamboos are frequently absent over wide areas. The ground flora is scanty and there is much bare soil. Epiphytes, especially orchids, are usually abundant.

Semi-Indaing and Indaing occur in areas with a rainfall varying from about 25 in. (64 cm.) in parts of the 'Dry belt' to as much as 60–80 in. (152–203 cm.) in the Pegu Yomas, but the distribution is determined as much by soil as by climate. The most usual soil is light sand, often containing a hard iron pan, but the same formation is also found on laterite, bare limestone and on dry ridges of gneiss and hardened sandstone with little soil.

7. *Diospyros* or Te forest. A formation of mixed composition in which te, *Diospyros burmanica*, is especially typical; *Pentacme suavis* and *Terminalia tomentosa* are also abundant. In structure and physiognomy *Diospyros* forest would appear to be a typical Savanna forest (according to Schimper's definition); the trees are low-growing and the undergrowth consists almost entirely of grasses (mainly *Andropogon* spp.). An erect *Calamus* and a stemless *Phoenix* are the only palms. Bamboos and various climbers occur and there are a few epiphytes. Ferns, terrestrial and epiphytic, are widely distributed but few in individuals.

The rainfall limits of this formation are about 34–38 in. (86–97 cm.); the soil is sandy, similar to that typical for formations 5 and 6.

8. Sha-Dahat Thorn forest. This has the characteristic structure of Thorn forest, though sometimes it is as open as savanna. The chief trees are *Acacia catechu* (sha) (which is also common in formations 7 and 11 and occurs here in both tree and bush form) and *Tectona hamiltoniana* (dahat) which is here a small tree. *Acacia leucophloea* is locally abundant on the eastern side of the 'Dry belt'. Nearly all the trees except *Tectona hamiltoniana* are thorny; none is more than

about 30 ft. (9 m.) high. Woody climbers are abundant and the chief undergrowth is grass.

This type of vegetation is abundant in the 'Dry belt' on clays and loams at a lower rainfall than that required by formation 11 and also occurs on sand at a still lower rainfall (less than that required by formation 7).

9. Sha Thorn scrub. A low scrub in which the most characteristic plant is *Acacia catechu*; this grows to a height of only 1·5–2 m., and is still lower and sparser in the centre of the 'Dry belt'. In the eastern part of the 'Dry belt' *A. catechu* is replaced by *A. leucophloea*. *Tectona hamiltoniana* is found as a bush less than 1 m. high. Grass is almost the only undergrowth and there is much bare soil. Some rosette plants occur.

The upper rainfall limit of this formation is about 40 in. (102 cm.). It occurs at a rainfall of less than 35 in. (89 cm.) on 'kyatti-mye' (alkali) soils formed from the Peguan Clays under conditions of low rainfall. On sandy soil the rainfall limit is higher and a rainfall which on clay would support only Sha Thorn scrub supports Indaing or *Diospyros* forest on sand.

10. Dry Deciduous forest without teak. This formation is similar in structure and physiognomy to Dry Teak forest; it occurs under similar climatic conditions to Semi-Indaing and Indaing, but on hill slopes with poor stony loam soils. *Vitex* spp. and *Heterophragma adenophyllum* are common trees, but it is difficult to name any species as especially characteristic. Bamboos are often so abundant as to exclude other grasses; the characteristic species is *Dendrocalamus strictus*.

11. Than-Dahat forest. This is a Savanna forest with something of the appearance of English coppiced woodland on clay soil. The chief trees are *Tectona hamiltoniana* (dahat) and *Terminalia oliveri* (than); *Acacia catechu* is also common. *Tectona hamiltoniana* usually branches near the ground and does not reach an average height of more than 30 ft. (9 m.); *Terminalia oliveri* is here usually a straggling and ill-formed tree. *Dendrocalamus strictus* is common and is the only bamboo. The undergrowth is almost entirely grass, especially *Andropogon apricus*. Climbers are often abundant.

Than-Dahat forest covers large areas on stiff clays with a rainfall of 35–39 in. (89–99 cm.); it grows side by side with the considerably less xerophilous Indaing (which, however, is found on sandy soils).

12. Euphorbia semi-desert. Here thorny euphorbias such as *Euphorbia antiquorum* are characteristic. Various bushy species occur and occasionally a few trees of *Bombax insigne*. The bushes are often far apart and the ground between bare or partly covered with grass. This formation occurs in the centre of the 'Dry belt', but only on alkali soils (especially those with gypsum in the subsoil) and on clay.

There is thus in Burma a graded series of formations ranging from Evergreen Dipterocarp forest (closely approximating to typical Rain forest of the Malayan region) at one extreme to Semi-desert scrub at the other. The nature of the climax vegetation in a given area shows a clear relation to the total annual rainfall, but doubtless the real relationship is with the length and severity of the seasonal drought, rather than directly with the annual total; the amount falling during the wet season is likely to be of relatively small importance.

The work of Stamp & Lord (1923) has demonstrated clearly that the distribution of the formations is greatly dependent not only on climate but on soil. The characteristics of the soil, especially its depth and porosity (which probably means in effect its water-supplying power during the dry period) profoundly modify the effects of rainfall, so that climate and soil compensate for one another to a considerable extent. In the same climate, but on different soils, different formations may be found; for instance, Than-Dahat Savanna forest, Sha Thorn forest and Sha Thorn scrub may be found side by side on different soils in the same locality enjoying practically identical climatic conditions. Similarly, the same community may be found on different soils in different climates. The further from the rain-forest end of the series, i.e. the drier the soil-climate complex, the more marked is the effect of soil differences; thus Evergreen Dipterocarp, Semi-Evergreen (Pyinkado) and Moist Teak forest appear to be comparatively independent of the nature of the soil, but Dry Teak forest (at a rainfall of about 100–90 cm.) already varies greatly on different soil types and at still lower rainfalls the formation found differs entirely on different soils.

A comparison of this series of vegetation types with those of Trinidad shows many resemblances. In both series all the climax types are dominated by woody plants; in Burma grassy undergrowth becomes more and more prominent with decreasing rainfall, but in Trinidad grasses are not important constituents of any of the climax formations. In both Burma and Trinidad with increasing dryness of the environment the vegetation becomes lower and less dense and deciduous tree species become more and more common, first in the upper, later in the lower, stories of the vegetation. With only the very scanty information we possess about the structure and physiognomic features of the Burmese formations, it is impossible to go further than this, and clearly, in the present state of our knowledge, it would be unwise to try to apply a common terminology to the climax formations of Burma and Trinidad.

A series of climatic climax formations closely resembling those of Burma is found in India (see Champion, 1936), and doubtless also in Siam etc. The relationship of climax vegetation to climate and soil in India is less clear owing to the large amount of modification of the forests due to human activities. The names for the Indian equivalents of the various formations found in Burma, according to the nomenclature of Champion (1936), are included in Table 34.

Savanna in south-eastern Asia

The forest formations of Burma have been described in some detail because they show very clearly the ecotone or gradual transition of climax vegetation from evergreen forest at one extreme to semi-desert at the other. The change in the vegetation is consequent on the increase in the length and severity of the annual dry season and in Burma it is uncomplicated by the occurrence of grassland, other than secondary grasslands which are clearly the result of deflected successions. A comparative scarcity of grassland and open savanna, as compared with the regions of tropical Africa and America with comparable climates, is characteristic of most of the Indo-Malayan region, though in some parts, for instance in the highlands of Ceylon (the patanas), in the Lesser Sunda Islands and in south-eastern New Guinea, fairly large areas of grassland exist. It is not proposed to discuss the status of these grassland areas in detail. Many of them, including probably the areas dominated by the alang-alang grass, *Imperata cylindrica*, are demonstrably secondary grasslands arising after the destruction of Rain forest or deciduous forest as a result of shifting cultivation and burning. Thus the small savanna-like areas in south-eastern Borneo (Winkler, 1914) and the grasslands of the Ramu and Markham valleys in New Guinea have certainly arisen in this way. The patanas or upland grasslands of Ceylon (Pearson, 1899) seem more closely comparable with the savannas of northern South America and may perhaps owe their existence to edaphic causes as well as to fire. There is no evidence that any of the grasslands of south-eastern Asia is a climatically determined climax community.

THE CLIMATIC ECOTONE IN AFRICA

The consideration of the relation of Tropical Rain forest to deciduous forest and savanna in Africa has been left to the last, partly because of the immense area covered by the two latter formations, and partly because man has modified the vegetation of Africa to such an extent that it is much more difficult to disentangle the complex interactions of biotic factors, climate and soil, than in either tropical America or south-eastern Asia.

In Africa, as was seen in Chapter 1, there is a central mass of 'closed forest', occupying much of the Congo basin, French Equatorial Africa, the Gaboon and the Cameroons, with a narrow western prolongation which follows the Gulf of Guinea till it eventually disappears in French Guinea at about long. 12° W. and lat. 8° N. Even at its centre the 'Closed forest' is not entirely unbroken, but is almost everywhere interrupted by occasional stretches of savanna, varying in size. At its margin it passes, often abruptly, into the wide zone of savanna or 'parkland' which surrounds it.

The term Closed forest (which it will be convenient to use in the present discussion) is equivalent to the *forêt dense* of modern French writers and the *Hylaea* of the Germans and includes all types of lowland forest in which the trees grow in closed canopy (top story seldom less than 120 ft. (37 m.) high in mature communities) and the ground is not covered with grass. In the central parts of its area the Closed forest consists mainly of evergreen trees, but towards the periphery the proportion of deciduous species gradually increases and the floristic composition changes; in consequence a distinction is usually drawn between two types of Closed forest, the Evergreen forest or Tropical Rain forest proper, and the Mixed Deciduous forest (*subxerophiler Tropenwald* of Mildbraed, 1922).

In tropical Africa west of the Cameroons there is a regular gradient from the climate of the wet coastal belt, with a comparatively short dry season during which the humidity of the air remains relatively high, to the arid and strongly seasonal climate of the interior, culminating in the desert climate of the Sahara. In consequence there is a series of well-defined zones of vegetation running approximately parallel to the coast (and to the equator). Because of this regular zonation, and because the vegetation is better known than that of East Africa (where the natural zonation is complicated by the less simple relief) and Central Africa, it is chiefly the West African region which will be considered here. A useful survey of the grasslands of southern, eastern and central tropical Africa is given by Michelmore (1939).

Zonation in West Africa

The successive zones of vegetation have been variously named and classified. The following scheme following the well-known classification of Chevalier (1900, 1938) seems the best at present available (highland areas omitted):

(i) The Closed forest zone. This embraces the area of the Closed forest, as defined above, including both the tall evergreen forest near the coast and the Mixed Deciduous forest further inland. It forms a belt of varying width. From a point slightly to the east of Lagos in Nigeria to as far west as a point north of Accra in the Gold Coast it is interrupted by a long stretch of savanna country which here reaches the sea. Most of the Closed forest belt is, or has at one time been, cultivated and the remaining forest is practically all secondary. Over-intensive shifting cultivation has led locally to a deflected succession (see p. 390); stages in the secondary succession have been invaded by grasses and have been converted by the action of recurrent fires into what A. P. D. Jones (1945) has termed Derived Savanna woodland. This is similar physiognomically, and to some extent floristically, to the savanna of the next zone, though it betrays its origin in the presence of occasional oil palms (*Elaeis*) and other trees characteristic of the Closed forest zone.

(ii) The Guinea zone. Here, on sites with unimpeded drainage, the vegetation of uncultivated areas varies from open savanna with scattered, mainly deciduous, fire-resistant trees to Savanna woodland with grassy undergrowth. Where the water-supplying power of the soil is above the average, outliers of Closed forest are found (Keay, 1947). Many, but not all of these outliers are found by watercourses; thus only some of them can be correctly termed Gallery or fringing forests. Most of this zone, with the exception of very small areas, is subject to frequent fires and the vegetation is to be regarded as a fire-climax rather than as a climatic climax. It is convenient to divide the Guinea zone into two subdivisions, though the boundary between them is not always sharp:

(*a*) The Southern Guinea sub-zone. The common savanna trees here include such species as *Daniella oliveri*, *Lophira alata*, *Terminalia glaucescens*, etc., most of which have close allies in the Closed forest. The outliers of high forest within this sub-zone resemble the Mixed Deciduous type of Closed forest in floristic composition (the species include *Khaya grandifoliola*, *Triplochiton scleroxylon*, etc.), but are usually less tall and have two rather than three tree stories. The existing Southern Guinea savannas are mostly open, but in sparsely populated districts there are patches of Savanna woodland in which not all the trees are fire-tolerant species; such patches probably represent approximately the climatic climax.

(*b*) The Northern Guinea sub-zone. Here the dominant trees include *Isoberlinia doka*, *I. tomentosa*, *Monotes kerstingii*, *Uapaca somon*, etc. These are identical with or closely related to species of the 'miombo' (*Brachystegia-Isoberlinia*) woodland[1] of southern and eastern tropical Africa; species related to those of the Closed forest are much less common than in the Southern Guinea sub-zone. As in the latter, nearly all the vegetation has been much modified by fire, but small areas of Savanna woodland exist which are probably not very different from the climatic climax. Forest outliers occur chiefly as fringing forest in which the undergrowth consists mostly of Closed forest species, but the upper tree story of species dominant in the Southern Guinea savannas (*Terminalia glaucescens*, *Anogeissus schimperi*, *Malacantha alnifolia*, etc.), together with a few found in both Closed forest and savanna, e.g. *Khaya senegalensis*, *Afzelia africana*, *Diospyros mespiliformis*.

(iii) The Sudan zone. Evergreen forest outliers are almost entirely absent. The vegetation, which is highly modified by man, consists of open savanna. The trees are partly fine-leaved thorny species such as are dominant in the Sahel zone and partly broad-leaved thornless species which also occur in the Northern Guinea savannas. Common trees include *Adansonia digitata* (baobab), *Lannea microcarpa*, *Balanites aegyptiaca*, *Combretum glutinosum*, *Sclerocarya birrea*, etc.

(iv) The Sahel zone. The climax formation of this zone is probably Thorn woodland, dominated by spiny trees with finely divided leaves, such as species of *Acacia*. The existing woodland is open, usually so open that a car can drive between the trees (Aubréville, 1938); a ground-cover of grasses or herbaceous plants is absent or discontinuous. The vegetation has been greatly modified by man and grazing animals, as in the previous zones, but owing to the openness of the vegetation fire is not an important factor. The flora is small and largely different from that of the Sudan zone; near water Sudanian trees such as *Adansonia*, *Diospyros mespiliformis* and *Anogeissus schimperi* are found, but scarcely form Gallery forests.

(v) Sahara zone. Here there is only a poor and highly specialized desert flora.

---

[1] Mildbraed (1933 *a*) has applied the term *Miombowald* to similar Savanna woodland in the northern Cameroons.

In the present discussion we shall confine our attention to the Closed forest zone and the boundary between it and the savanna ('Derived' and Guinea savannas).

The ecotone within the Closed forest

Within the Closed forest there is manifestly a transition or ecotone from the luxuriant, almost entirely evergreen, forests of the wettest areas to the less luxuriant, more markedly deciduous forests bordering the savannas of the Guinea zone, which have usually been referred to as 'Mixed Deciduous forest' (*Triplochiton-Piptadenia* pre-climax of Chipp (1927)). The climatic gradient responsible for the ecotone has already been considered (Chapter 6). The change in the vegetation is chiefly in floristic composition; the structure and physiognomy of the Closed forest remains very similar throughout. The term Mixed Deciduous forest has been used for a variety of communities, some probably of approximately climax status, others seral, growing under a wide range of climatic conditions and as the meaning of 'Mixed Deciduous forest' has thus become confused, many wish to abandon the term. Jones & Keay (unpubl.) divide the Closed forest of Nigeria into three 'regions': Rain forest, Wet (Evergreen) forest and Dry (Evergreen) forest. This terminology is perhaps an improvement on any previously suggested and may be adopted here, at least provisionally, though it should be understood that Wet Evergreen forest, and probably also Dry Evergreen, as well as the 'true Rain forest' of Jones and Keay all come within the definition of Rain forest used in this book.

Most of what has been described as Rain forest in West Africa, including the Nigerian forest described above (Chapter 2), is Wet forest in this sense. It has been described in detail by Chipp (1927), Aubréville (1938) and Richards (1939).[1] The primary forest[2] is apparently always mixed in floristic composition and may probably be regarded as a single association. While in most features of structure and physiognomy it resembles the Rain forest of tropical Asia and America, it has a number of characteristics which are probably related to the comparatively severe seasonal drought of its climate (up to five consecutive months with less than 10 cm. rainfall). Thus the vegetation is rather less tall and luxuriant than the finest American and Asiatic Rain forests and the A and B stories tend to be more open (though it is not easy to be certain that this is not due to past disturbance by man). Trees which become bare of leaves for several weeks annually are common, though they do not form such a large proportion as in young secondary forest and clearings. Many other species lose all their leaves at one time and simultaneously produce a new crop. It may thus be said that seasonal activity in the vegetation is in general more pronounced than in typical rain forests. The presence of geophytes which perennate during the dry season

---

[1] The Gold Coast forest described by Foggie (1947) is probably best classified as Wet forest, though it is perhaps intermediate in character between Wet forest and Dry forest.
[2] See footnote, p. 29.

as underground corms or tubers has been mentioned previously (p. 98). Much of the forest of the Congo basin is probably similar to this West African Wet forest, which has close analogies, further afield, with Beard's Evergreen Seasonal forest in Trinidad.

In Nigeria 'true Rain forest', as understood by Jones and Keay, occupies only a small area west of the Cameroons and has nowhere been studied in detail. Some of the forests of the Oban-Calabar region are probably to be included here and the type extends (or formerly extended) to the Niger delta and the southern border of Benin province; part of the forest of Liberia and the adjoining parts of Sierra Leone may be similar. East of the Cameroons 'true Rain forest' is probably found in Spanish Guinea, Equatorial Africa and the central part of the Congo basin. True Rain forest is richer in species than the Wet forest and possesses a number of characteristic species which are not found in the Wet or the Dry Evergreen forests. Apart from floristic composition it differs very little from Wet forest, though the A and B stories are probably more continuous, the proportion of truly deciduous trees still smaller and other 'seasonal' plants, such as geophytic herbs, scarce or absent.

The Dry Evergreen forest of West Africa is mainly found forming the northern fringe of the Closed forest, though it is not in any sense a transition to the savannas of the Sudanian zone. A complete description of the structure and composition of Dry Evergreen forest has not been published, but a considerable amount of information about forests of this type is given by Chipp (1927), Mackay (1936), MacGregor (1934, 1937), Aubréville (1938), and Richards (1939). It is difficult to speak definitely about its characteristics as so much of it has been destroyed or modified by shifting cultivation that it is rarely seen in a primary or even mature condition.

The trees are probably arranged in three stories, but the upper stories are less dense and their average height is perhaps lower, than in Wet forest. The most important difference from Wet forest, as indicated by the name Mixed Deciduous forest, is the larger proportion of trees deciduous during the dry season, but evergreen species are numerous and the bare periods of the deciduous species are not completely coincident; thus at no time does the forest have the leafless appearance of a temperate deciduous forest. The shrub and lowest tree stories are almost entirely evergreen. Lianes are frequent, but epiphytes, as might be expected, are few both as species and individuals. Many species of trees are common to the Dry and the Wet forests, but the former are floristically poorer, many species characteristic of the true Rain forests and the Wet forests being lacking, e.g. *Lophira procera*. Other species are confined to the Dry forests or reach their greatest abundance there; among these may be mentioned especially *Triplochiton scleroxylon*, which is very common and may form locally as much as 20% of the stand, *Chlorophora excelsa*, *Mansonia altissima*, *Sterculia rhinopetala* and other Sterculiaceae. Some trees common in the Dry forest are found in the Wet forest chiefly in clearings and secondary forests. Along streams and in other edaphically favourable habitats some Wet forest species extend northwards into the Dry forest, just as Closed forest species form outliers in favourable habitats within the Guinea savannas. The herbaceous under-growth of the Dry forest, though appearing just as luxuriant in the wet season as that of the Wet forest, tends to dry up during the dry season and species which die down completely are more abundant.

As Aubréville says (1938, pp. 123–4), the difference between the Dry and the Wet Closed forest is chiefly floristic; the floristic differences do not produce a fundamentally different physiognomy. The transition from one community to the other is usually gradual, as in the Shasha Forest Reserve in Nigeria where one passes into the other imperceptibly, but where a change of soil and climate coincide it may be more abrupt.

Various views have been held as to the status of what we are now terming Dry Evergreen forest. The proportion of deciduous trees is hardly large enough to class it as a Monsoon forest, as is done, for example, by MacGregor (1937); indeed, according to Mildbraed (1922) there are no tall closed leaf-shedding forests in Africa strictly comparable with the tropical deciduous forests of India and Burma. The resemblance to the Semi-evergreen Seasonal forest of Trinidad is closer. Since the structural differences between Dry forest and Wet forest, and between Wet forest and true Rain forest, are very slight and the floristic differences mainly in degree, it scarcely gives a true picture of the situation to rank the three types of Closed forest as separate formations, and as there are real, though small, differences in climatic requirements between Dry forest and Wet forest, Dry forest cannot be logically regarded as a pre-climax of Wet forest, as is done by Chipp (1927). In Clements' recent terminology (1936, pp. 274–5) the three types of Closed forest could be regarded as faciations of one climax or formation, or perhaps the term subformation might be used.

The Closed-forest–savanna boundary

In most places Dry Evergreen forest forms the northern edge of the Closed forest and the boundary between Closed forest and savanna, whether Derived (see p. 336 above) or true Guinea savanna, is generally quite sharp, often as sharp as that between woodland and cultivated land in Europe. In the immedi-ate neighbourhood of the forest boundary the savanna consists of open grassland with scattered clumps of trees, or, more frequently, of very open low woodland with an undergrowth of tall grasses.

There is a great contrast in floristic composition, as well as in structure and physiognomy, between the Closed forest and the savanna. The great majority of the species, both dominant and subordinate, are different in the two, but several genera are represented in the Closed forest and savanna by 'vicarious' species strongly contrasting in physiognomy, but closely related taxonomically. Thus *Khaya* is represented by *K. ivorensis* and *K. anthotheca* in the Wet Evergreen forest, by *K. grandifoliola* in the Dry forest and by *K. senegalensis* in the savannas. The small fire-resistant tree *Lophira alata*, so characteristic of the Derived and Southern Guinea savannas, is so similar in its taxonomic characters to the tall Wet forest *L. procera* that many authorities have regarded the two as conspecific (the relationship is discussed by Aubréville, 1936). This last example is peculiar in that usually neither species is found in the intervening Dry forest. These 'vicarious' species should perhaps be regarded as ecotypes.

The great differences between the Closed forest and savanna vegetation are of course dependent on great differences in the environment and one of the chief environmental factors is certainly fire.[1] The Closed forest will not readily burn and is practically immune from fire; almost all West African savannas, on the other hand, are subject to bush fires yearly or at more frequent intervals. These fires, according to their severity, destroy the above-ground parts of the herbaceous vegetation and most of the other undergrowth, including the tree seedlings; the trees themselves, though usually not killed, are severely damaged and this damage is responsible for their gnarled and deformed appearance. It is obvious that the savanna flora must be subject to severe selection; only species resistant to periodic fires can maintain themselves. Practically all African savanna vegetation, where not forming stages in secondary successions, is thus a biotic climax determined by fire (fire climax). The important subject of bush fires cannot be discussed fully here; reference may be made to the account of Lebrun (1947), in which the effects of fire on the chief plant formations of tropical Africa, and on the various physiognomic categories of savanna plants, are fully described.

What relation does this fire climax bear to the natural climatic climaxes of the region and to what extent is the sharp boundary between Closed forest and savanna natural? Grass burning is a practice so widespread and so ancient in West Africa (it was recorded at the time of Hanno's voyage, before 480 B.C.) that both these questions are very difficult to answer. It is probable that no certain solution of either problem is possible with the data at present available.

It should be realized that the numerous more or less isolated outliers of evergreen or semi-evergreen Closed forest within savanna areas cannot be regarded as relics of the original climatic climax, except in the Derived savannas, which lie topographically and climatically within the Closed forest belt (in some cases in Wet forest regions). Closed forest outliers in the Guinea savannas consist of Gallery forests along streams and of a certain number of forest patches not near water. The latter, where not artificially planted by the natives, can mostly be shown to occur where edaphic or topographic factors are locally favourable; some were formerly connected to the Gallery forests by strips of forest now destroyed. All these outliers are dependent on special local habitat conditions.

---

[1] Reference should be made here to an important paper by Morison, Hoyle and Hope-Simpson (1948), which deals very fully with the ecological factors, including fire, affecting the savanna in the south-western Sudan.

An experimental approach to the problem of the climax in West African savannas was made by W. D. MacGregor (1937). Three sample plots were laid out in the savanna at Olokemeji in Nigeria, close to the margin of the Closed forest. Plot A was burnt fiercely every year in March, B was burnt lightly every year in December and C was completely protected from fire. After six years A showed little change, but in C there was a considerable increase in the height of the vegetation and a significant invasion of Closed forest species, though no such species had been present at the beginning of the experiment. B had also been invaded by Closed forest species, but not to the same extent as C. From the results of this experiment and from observation of the surrounding area MacGregor concluded that complete fire protection would cause a succession from savanna to Closed forest to take place; a forest community would become established and the grass cover would be eliminated. Closed forest (Dry Evergreen forest) should therefore be regarded as the climatic climax of the area.

Since the Olokemeji savannas are certainly derived and lie within the climatic limits of the Closed forest, these results contribute nothing to the problem of the climatic climax in the area of the Guinea savannas, which differ in flora and climate from the Derived savannas. Similar experiments in the Southern and Northern Guinea sub-zones are much to be desired, but the results might be inconclusive, because it seems likely that the natural vegetation in these areas has been so extensively destroyed that the seed-parents necessary for the establishment of the climax may no longer be available. It should also be remembered that long continued firing has probably had a profound effect on the soil, which is now probably very different from that of the climatic climax communities.

On the evidence at present available we may accept, at least for the Southern Guinea sub-zone, the view of Aubréville (1936; 1937, pp. 96–8) that, if by some miracle recurrent fires were eliminated, and at the same time the soil restored to its original condition, the open savanna and savanna woodland of the Guinea zone would become replaced by a deciduous forest, relatively open, but with mainly woody undergrowth, and scattered patches rather than a continuous cover of grasses (*forêt claire tropicale* of Aubréville; Closed Savanna woodland of Jones & Keay, 1946). The dominance of grasses in the existing vegetation is unquestionably a result of burning. The recent work of Keay (1947) emphasizes the floristic and ecological differences between the Southern and Northern sub-zones of the Guinea zone and it therefore seems probable that within this hypothetical climax there would be a gradual transition from south to north. In the Southern sub-zone a *forêt claire*, such as Aubréville describes, may well be the natural climax; farther north, in a more arid climate, the climax may change to a type of savanna woodland more similar to the miombo woodlands of southern and eastern tropical Africa, and not unlike the least degraded existing woodlands of the area.

Thus under the purely imaginary condition of complete fire protection, it may be supposed that there would be a quite gradual transition from the tall ever-green and semi-evergreen Closed forest to the lower, mainly deciduous, *forêt claire* and miombo; there would be no open grassy savanna. Fire is responsible for the sharp boundary often seen between forest and open savanna; the abrupt edge of the forest is largely due to the very different biological equipment and ecological requirements of the Closed forest and savanna floras, the latter of which consists mainly of species selected for fire-resistance.

There is abundant evidence that in most of West Africa, as well as in other parts of the continent, the Closed forest has retreated and the savanna advanced during recent times. The process is indeed still continuing; Scaëtta (quoted by De Wildeman, 1934) states that in Ruanda-Urundi (eastern Belgian Congo) it is retreating at a rate of about 1 km. per year. In a few places, on the other hand, an invasion of the savanna by the forest has been reported, e.g. in the southern Belgian Congo (Lebrun, 1936) and in Uganda (Eggeling, 1947). As was said in Chapter 1, it is certain that the rapid shrinkage of the African forest at the present day is chiefly due to shifting cultivation, savanna burning and other human activities, and there is little positive evidence that the 'desiccation of Africa' or the 'advance of the Sahara' is responsible. If a secular change of climate is still in progress, its effects would be very much slower than the changes in the forest boundary now being witnessed.

Savannas of the Congo forest region

Though we are dealing here chiefly with the relations of forest and savanna in the western part of tropical Africa, brief reference should be made to the savanna inliers of the central basin of the Congo, of which Robyns (1936) has given an interesting account. Over most of the Central Forest Area (*District Forestier Central*) of the Belgian Congo the climatic climax is undoubtedly Closed forest, of Rain forest, Wet Evergreen or Dry Evergreen type, but grassland areas of several kinds are found enclosed within the forest. These Robyns classifies as:

I. Terrestrial formations:
   A. Edaphic savannas.
   B. Climatic savannas.
   C. Secondary savannas.
II. Aquatic formations.

The Aquatic formations are stages in the succession leading to Evergreen Swamp forest as an edaphic climax; they are therefore comparable to the South American 'water savannas', etc. Of the Terrestrial formations, the Secondary savannas, as their name implies, are also seral stages; they are produced by human activities, or occasionally by elephants, and, if left to themselves, revert to forest. The Climatic savannas occur on the northern and southern boundaries of the forest and are found only in the western part of the central basin. Their status would appear to be similar to that of the savannas of the Guinea sub-zone in West Africa.

The Edaphic savannas are of particular interest, as they seem to form the closest parallel to the South American savannas discussed earlier in this chapter. Two types of Edaphic savanna are recognized: those on 'laterite' and *ésobes*. Little is known of the former, which are found only in the Ubangi region and are dominated by the grass *Ctenium newtonii*. *Ésobes* are pockets of savanna of small extent—a typical example near Coquilhatville occupies about 3 ha.—situated in the most low-lying part of the central basin, usually near rivers, and never on plateaux or watersheds. The flora of these *ésobes* is very uniform; the dominant plant is the grass *Hyparrhenia diplandra*, with which are associated various other herbaceous plants, bushes and stunted trees. Usually the *ésobes* occupy old river sandbanks which have been left dry and become attached to the land by the shifting of the stream; a belt of fringing forest is commonly found between them and the river bank. The soil varies from sand to clay in texture; in some cases it is a 'pure white sand' (like the soil of the Berbice savannas of Guiana). The *ésobes* are not burnt and Robyns regards some of them as 'natural' savannas in which edaphic factors prevent the successful growth of trees; others, however, are being invaded by the surrounding forest. It may be that they should be regarded as stages in a succession in which tree colonization is delayed rather than permanently prevented by unfavourable edaphic conditions. Whatever their true status the *ésobes* show in several respects a remarkable resemblance to the pockets of savanna and muri in the Guiana Rain forest (p. 324).

CONCLUSIONS

This very incomplete survey of the relation of evergreen forest to deciduous forest and savanna in various parts of the tropics has shown the bewildering complexity of the problem. The broad generalizations of Schimper are open to criticism, but in the present state of our knowledge it is not possible to put forward any very clear or definite conclusions in their place.

On the whole it seems likely that where soil conditions are not unfavourable to the growth of trees and the natural vegetation is not subject to frequent fires, the Tropical Rain forest always gives place at its climatic limit to Deciduous forest and this to Savanna woodland, while with further increase in the dryness of the climate Savanna woodland in its turn is replaced by Thorn woodland and desert vegetation. The transition from Rain forest to Deciduous forest is usually quite gradual and is characterized by a change of floristic composition, involving a decrease in the number of species and a shift of dominance from evergreen to more or less regularly deciduous trees, rather than by any striking or sudden change of physiognomy. There is also a decrease in the number of strata due to the A story becoming gradually more and more open and eventually disappearing.

Savanna is a name applied to plant communities of varied physiognomy and status found over a wide range of climatic conditions; some are seral stages, others are certainly stable climaxes. Savannas in which trees are dominant (with or without a continuous ground-cover of grasses) may be a climatic climax, but many types of savanna should be regarded as fire-climaxes; these often arise from Deciduous, Semi-deciduous, or, even, Evergreen forest as a result of deflected successions brought about by cultivation and repeated burning. Open savannas, with trees growing scattered or in occasional clumps, and treeless grasslands may arise by the degradation of forest or Savanna woodland by excessive cultivation or burning, but in some cases they are probably edaphic climaxes due to local soil conditions unfavourable to the growth of trees. The nature of the factors responsible is uncertain, but one which probably operates in some cases is seasonal waterlogging alternating with dry conditions during the rest of the year. There is little support for the view that lowland tropical grasslands are ever a climatic climax in equilibrium with a 'tropical grassland climate'; grassland therefore should not be regarded as occupying a place in the natural climatic ecotone from Tropical Rain forest to desert.

CHAPTER 16

THE TROPICAL RAIN FOREST AT ITS ALTITUDINAL AND LATITUDINAL LIMITS

The changes in the vegetation with increasing altitude are no less striking in the wet tropics than in temperate regions. As a forested tropical mountain is climbed, both the physiognomy of the dominant species and the structure of the vegetation change. The tall luxuriant Tropical Rain forest of the lowlands gives place to other communities, also evergreen, but lower in stature, simpler in structure and floristically poorer. The purely tropical flora is left behind and is replaced by a montane flora in which many of the genera, or even the species, are temperate. The Tropical Rain forest gives way to a formation-type which it is proposed to call the Submontane Rain forest; this at a still higher altitude is succeeded by another formation-type, the Montane Rain forest. In many places, especially on exposed ridges and isolated peaks, the Montane Rain forest consists of dwarf crooked trees smothered with an overwhelming abundance of epiphytes, especially hepaticae and mosses; this very characteristic type of vegetation is often called Mossy forest or Elfin woodland. Submontane and Montane Rain forest are not always clearly distinguished and both are sometimes termed Temperate Rain forest, but it is preferable to reserve this name for formations outside the tropics.

On the highest mountain ranges yet another forest belt may lie above the Montane forest, extending up to the climatic tree limit; beyond this there are only communities of low-growing alpine plants, bare rock and permanent snowfields. From the lowland tropical forest to the snow-line there are thus five major zones of vegetation in all.

This altitudinal zonation, which in its general features appears to be similar in the Old and New Worlds, depends on a gradient of climate, but the relation of climate to altitude is by no means simple. The variation of some of the climatic elements with altitude was considered in Chapter 6; it may be recalled that in the tropics the mean temperature decreases by about 0.4–$0.7°$ C. per 100 m. increase of elevation. It is, however, important to realize that, besides temperature, almost every other climatic factor affecting plant growth changes with altitude, including rainfall, atmospheric humidity, wind velocity and sunshine. These factors as a rule do not vary consistently with height; thus on high tropical mountains rainfall often increases from sea-level up to a zone of maximum precipitation and decreases above this zone. Since vegetation, fauna, climate and soil are parts of a single inter-related ecosystem, there are also changes in the character of the soil with increasing altitude; these were briefly mentioned in Chapter 9.

Because the rate of change of climatic factors with height varies from place to place, the actual altitudinal limits of the vegetation zones are different on different mountain ranges and on different parts of the same mountain. On small isolated peaks and ridges the zones are lower than on extensive ranges; they also tend to be lower on coastal mountains than on those further from the sea. This is the so-called *Massenerhebung* effect, which is well known in the Alps and other European mountains. The lowering of the upper limit of the Tropical Rain forest and of the Montane zone on tropical mountains is exactly parallel to the lowering of the tree limit on the fring:..g *Voralpen* compared with the great central ranges of the Alps. Differences in exposure to wind are at least partly responsible for this *Massenerhebung* effect on tropical mountains. Beard (1946 b, p. 41) has shown that on the Northern Range of Trinidad the zones of vegetation are depressed towards the eastern end, which faces the prevailing trade wind (Fig. 40).

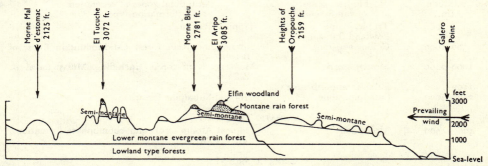

Fig. 40. Section from east to west through the Northern Range, Trinidad, British West Indies, showing the variation in the altitudinal limits of the climatic plant formations. After Beard (1946 b). Vertical scale 2000 ft. to 1 in.; horizontal scale 4 miles to 1 in.

ZONATION ON THE MOUNTAINS OF MALAYSIA

A large part of the Indo-Malayan and Australasian rain-forest area is mountainous and the altitudinal zonation of the vegetation has been more studied here than in any other part of the tropics. Few of the mountains in this part of the world exceed 3500 m. and most of them do not reach the climatic tree limit. Only the isolated peak of Kinabalu in North Borneo and the high ranges of New Guinea are over 4000 m. high; the latter alone have permanent snow. Many of the volcanoes of the Sunda Islands exceed 3000 m. and are partly or completely treeless at the summit, but the absence of trees is probably due to fires or to local edaphic conditions; it is doubtful whether any of the Sunda Islands reach the climatic tree limit.

Van Steenis (1935 b) has published a comprehensive review, mainly from the floristic point of view, of the altitudinal zonation of species on the mountains of

Malaysia. Using a method previously adopted by Sendtner for the European Alps, he attempts to determine objectively the natural delimitation of the zones. The upper and lower altitudinal limits of about 800 Malaysian mountain plants were tabulated and it was found that the majority tended to reach their upper or lower limits at certain critical levels or within narrow zones of altitude rather than in the intervening height intervals. These critical levels may be regarded as the natural boundaries between one zone and the next.

From the table it may be concluded that there are five zones of vegetation in Malaysia (the lowest two each subdivided into two sub-zones) and that the critical altitudinal limits are 1000, 2400, 4000 and 4500 m. respectively. Though arrived at by a floristic rather than a phytosociological or ecological method, the zones recognized by van Steenis agree well with those which, as has already been pointed out, seem to be general throughout the wet tropics; this is shown in Table 35.

TABLE 35. *Altitudinal zones of vegetation in Malaysia*

Altitude (m.)	Floristic zones of van Steenis	Altitudinal zones of vegetation (formation-types)
0–1000	Tropical zone	Tropical Rain forest
0–500	Lowland Sub-zone	
500–1000	Colline sub-zone	Submontane Rain forest (Mid-mountain forest of Brown)
1000–2400	Montane zone	Montane Rain forest (including Mossy forest or Elfin woodland)
1000–1500	Submontane sub-zone	
2400–4000	Subalpine zone	Tropical Subalpine forest
	CLIMATIC TREE LIMIT	
4000–4500	Alpine zone	Tropical Alpine scrub and chomophyte formations, etc.
	CLIMATIC SNOW-LINE	
4500 upwards	Nival zone	Nival chomophyte vegetation

The critical heights calculated by van Steenis are necessarily only very rough averages and the altitudinal limits of the zones vary greatly in different parts of Malaysia. These variations do not depend on latitude, but chiefly on the *Massenerhebung* effect referred to above (see van Steenis, 1932, p. 176 and 1935b, pp. 298–300; Richards, 1935, p. 345). For example, on the small isolated Gunong Santubong on the coast of Borneo the transition to mossy Montane forest takes place as low as 750 m., on G. Belumut in Johore at 840 m. (Holttum, 1924), but on the main range of West Java a similar type of forest is not met with till about 1650 m. (Seifriz, 1923) and in eastern New Guinea not till about 1500–2400 m. (Lane-Poole, 1925a, b). The limits of the other zones show a similar variation.

A characteristic feature of Malaysian mountain vegetation is the strong contrast in geographical affinities between the lowland and the mountain flora. In the lowland Rain forest the bulk of the flora belongs to Indo-Malayan genera, and this is also largely true of the Submontane zone, but in the Montane forest there is a considerable proportion of genera such as *Leptospermum*, *Tristania*, *Phyllocladus*, etc. which are clearly of Australasian origin, though the montane flora also includes Indo-Malayan and North Temperate elements. Thus the Australasian element does not increase gradually with increasing altitude, as might be expected, but increases suddenly when the boundary of the Montane forest is reached. In Malaysia, the Indo-Malayan and Australasian floras are to a large extent ecologically segregated, as they are in subtropical Australia (see below, p. 369, and Richards, 1943). The origin and history of the Malaysian mountain flora has been dealt with exhaustively by van Steenis (1934, 1935, 1936).

Zonation on the high ranges of New Guinea

The mountains of New Guinea are the most favourable region in the eastern tropics for studying the altitudinal zonation of vegetation. Since they are high enough to reach the limits of permanent snow, the complete sequence of zones can be observed from sea-level upwards; they are also non-volcanic, so that the development of climax vegetation is not prevented by eruptions or unfavourable soil conditions. New Guinea has the further advantage that its vegetation has been relatively little affected by man and in some places perhaps not at all.

Though New Guinea is very inadequately explored both geographically and botanically, there is a considerable amount of information about the vegetation of the great mountain ranges, thanks chiefly to the work of Gibbs (1917) on the Arfak Mountains, Lam (1945) and van Steenis (1937 *b*) on the Central Range of Dutch New Guinea, and Lane-Poole (1925 *a*, *b*) on Papua and eastern New Guinea.

In the eastern mountains Lane-Poole distinguishes the following zones:

(i) Lowland forest—0–1000 ft. (300 m.). This is the primary Tropical Rain forest which covers the greater part of the surface of the island, except in the 'Dry belts' which have a rainfall of less than 180 cm. It is mixed in composition, very tall and luxuriant, and has the normal characteristics of Tropical Rain forest.

(ii) Foothills forest—1000–5500 ft. (300–1650 m.). This is not as tall as the Lowland forest and consists mainly of trees of small girth. The few trees with plank buttresses, such as *Ficus* and *Alstonia* spp., can be regarded as stragglers from the Lowland zone. The dominant trees are dicotyledons. One of the most abundant and characteristic species is an oak, *Quercus junghuhnii*, which sometimes forms pure stands up to 1 acre (0·4 ha.) in extent; other common trees are *Cedrela toona*, *Vatica papuana*, species of *Albizzia*, *Elaeocarpus*, *Eugenia*, *Archidendron*, etc. Conifers are absent except towards the upper limit of the zone. The ground-cover consists chiefly of ferns and species of *Elatostemma*; lianes and epiphytes are less plentiful than in the Lowland forest.

(iii) The Mid-mountain forest—5500–*c*. 7500 ft. (1650–2250 m.). The lower limit of this zone coincides with the cloud belt. The forest here is an oak-conifer association. Species of *Quercus* (including *Lithocarpus*?) are among the commonest dicotyledonous trees and the conifer *Araucaria cunninghamii* is characteristic, especially on ridges; other conifers include species of *Podocarpus* and *Phyllocladus*. The average leaf size is smaller than in the lower zones and compound leaves are rarer; buttressed trees are absent. The ground-cover consists of *Elatostemma*, filmy ferns and mosses, including the very large *Polytrichum*-like *Dawsonia*. Lianes are few and small; an epiphytic lichen (*Usnea* sp.?) is common in the crowns of the trees.

(iv) Mossy forest—lower limit 5000–8000 ft. (1500–2500 m.) (average about 7500 ft. (2250 m.)); upper limit 10,000–11,000 ft. (3000–3350 m.). This is the Montane Rain forest which is a characteristic feature of mountains in all parts of Malaysia. It consists of a single layer of gnarled crooked trees which are only about 6 m. high; their trunks and branches are hidden by an enormously thick covering consisting chiefly of leafy liverworts and mosses which almost doubles their diameter. The conifer *Podocarpus thevetiifolia* in some places forms 80 % of the stand; two species of *Eugenia* are also common. Bamboos and tree ferns are present, but rattans are absent and other lianes very scarce. The average temperature in this zone is about 10° C., but the chief characteristics of the climate are the continual dampness and persistent mist.

(v) High Mountain forest—lower limit 10,000–11,000 ft. (3000–3350 m.). This is taller than the Mossy forest; it consists of fair-sized trees, the majority being conifers of the genera *Dacrydium*, *Libocedrus*, *Phyllocladus* and *Podocarpus*. A few dicotyledonous trees such as species of *Eugenia* and *Calophyllum* are found. The High Mountain forest does not usually form a continuous belt; but forms blocks or patches separated by grassland with scattered shrubs and tree ferns. It lies above the mist belt and enjoys a sunnier and probably drier climate than the Mossy forest.

Above 12,000 ft. (3600 m.) are no more trees and the vegetation consists of grassland, bogs and, on rocky areas, dwarf scrub. The extensive high mountain grasslands above the apparent tree limit in eastern New Guinea appear at first sight to be true alpine grassland, that is to say a climatic climax above the climatic tree line, but Lane-Poole found good reason for believing that they are in fact a biotic climax due to fires started by the natives when hunting the wallaby. Observations suggested that tree growth is possible on these mountains up to at least 4050 m., except on very steep slopes and in boggy places. The term Mountain savanna used by Lane-Poole for these high altitude grasslands is thus appropriate.

The observations of Lam (1945) on Doormantop (3580 m.) in western New Guinea indicate a similar zonation to that found by Lane-Poole on the eastern ranges. The transition from tall Rain forest to Mossy forest of dwarf trees here takes place at about 1420 m. The uppermost forest zone, immediately below the tree limit, here also consists chiefly of conifers; there is a gradual increase in the abundance of conifers with altitude (Fig. 41). Towards 2600 m. many species reach their upper limit and various herbs and shrubs characteristic of the zone above the tree limit begin to appear. Above the forest limit (which is fairly

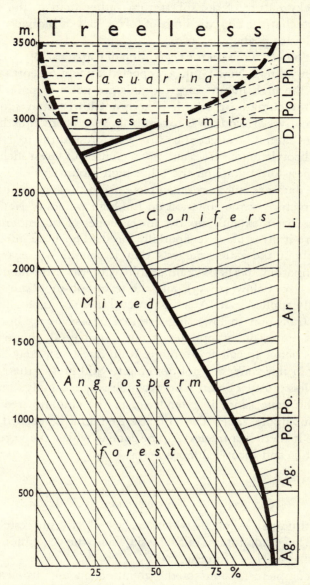

Fig. 41. Altitudinal distribution of conifers in forests of New Guinea. After Lam (1945). Ag., *Agathis alba*; Ar., *Araucaria*; D., *Dacrydium*; L., *Libocedrus*; Ph. *Phyllocladus*; Po., *Podocarpus*.

sharply defined) there are still small isolated trees of *Libocedrus* and a species of *Casuarina* in sheltered places, but most of the vegetation here consists of heath and bog-like communities and dwarf plants growing in crevices between bare rocks. There is no Mountain savanna on Doormantop and no evidence that the tree limit is not natural, though the absence of trees near the summit may be due to the rocky nature of the substratum rather than to the climate. Lam himself suggests that the climatic limit of forest (not of individual trees) in this part of New Guinea is about 3000 m.

Some recent observations on the vegetation of Carstensztop (5040 m.), a snow-capped peak in western New Guinea, reported by van Steenis (1937), confirm the view that the high mountain grasslands are not a climatic formation and indicate that the climatic tree limit is here not lower than 4200 m. On this mountain the natives are said never to climb above 3200 m., thus the vegetation at higher altitudes cannot be affected by fire. The forest at the tree limit is a mixed woodland of dicotyledonous trees and not a coniferous forest such as apparently forms the highest forest zone on the other high mountains of New Guinea.

Though the data are still very incomplete, from these accounts we can form a fairly clear picture of the zonation on the high ranges of New Guinea. The Tropical Rain forest extends up the lower slopes of the mountains to about 1000 m., where it gives place to Foothills forest. The little information which is available about the latter suggests that it should be regarded as an independent formation, perhaps comparable with the Mid-mountain forest of the Philippines (see below, p. 353), and the term Sub-montane Rain forest can be provisionally applied to it. At a higher level there is, as on all wet Malaysian mountains, a transition to low, single-storied Mossy forest but since a great abundance of epiphytic bryophytes is not a universal characteristic of this formation it is better to call it Montane Rain forest. In New Guinea conifers are abundant in this zone, but not dominant. Still higher there is a zone with a drier and less misty climate; here conifers are usually, but not always, dominant. This zone, since it extends up to the climatic tree limit, may be termed the Subalpine zone. The climatic tree limit is reached at 4200 m. or higher; above it there is a treeless Alpine zone in which the vegetation consists mainly of shrubs and dicotyledonous herbs.

Zonation on the lower mountains of Malaysia

The lower altitudinal zones, in which we are chiefly interested, can be well studied on the lower mountains of Malaysia where the vegetation is much better known than that of New Guinea.

The most complete account of the zonation on any tropical mountain is undoubtedly Brown's (1919) study of Mt Maquiling (1140 m.) in the Philippines. This work is remarkable not only for the wealth of precise data on the composition and structure of the vegetation at different altitudes, but also for its very complete analysis of the climate and for an attempt to relate the climate to the physiognomy of the vegetation on physiological lines. Only a very brief summary can be given here.

Four zones are recognized: (i) the Parang zone, 0–200 m.; (ii) the Dipterocarp forest zone, 200–600 m.; (iii) the Mid-mountain forest zone, 600–900 m.; and (iv) the Mossy forest zone (900–1140 m.).

The Parang zone was originally covered with Rain forest, similar to the Dipterocarp forest of the second zone, but all the original forest has been cleared and, where it is not under cultivation, this region is occupied by a mixture of secondary communities (seral stages) varying from open grassland dominated by *Imperata* (alang-alang) and *Saccharum spontaneum* to closed secondary Rain forest.

The Dipterocarp forest occupying the next higher zone is a mixed evergreen forest association in which three species of Dipterocarpaceae form a large proportion of the first story; it is a type of Mixed Dipterocarp forest (see Chapters 10 and 11) which is the climatic climax over a large part of the Malayan region. The structure is normal for Tropical Rain forest. There are three stories of trees, the average height of which is about 27, 16 and 10 m. respectively; the tallest trees are about 36 m. high. The ground covering consists largely of young trees and rattans, but, above 500 m. and in ravines, ferns and other herbaceous plants are common. Lianes, especially rattans, are very abundant, but epiphytes are comparatively rare and are chiefly confined to the large branches of the tallest trees. Mosses and liverworts are not abundant on the tree-trunks. Physiognomic features characteristic of Tropical Rain forest, such as buttressing and cauliflory, are well developed. Of plants over 2 m. in height, the great majority have leaves belonging to Raunkiaer's 'mesophyll' size-class (see fig. 15, p. 83) and there are a few 'macrophylls' or 'microphylls'.

Floristically, all the Dipterocarp forest on Mt Maquiling belongs to a single association with numerous dominants; it is a typical Mixed Rain forest. On a sample plot (0·25 ha.) there were ninety-two woody species over 2 m. high. Since the most abundant species in the first story is *Parashorea malaanonan* (Dipterocarpaceae), and in the second story *Diplodiscus paniculatus* (Tiliaceae), Brown terms it the *Parashorea-Diplodiscus* association.

The transition from the Dipterocarp to the Mid-mountain zone is gradual. The Mid-mountain forest is also a mixed evergreen community. There are two associations, the *Quercus-Neolitsea* association and the *Astronia rolfei* association, both of which differ from the *Parashorea-Diplodiscus* association in having a two-storied structure.

In the *Quercus-Neolitsea* association the average height of the trees diminishes with altitude; at 700 m. the first story is about 17 m. high and the second about 4 m. high; the tallest tree noted was 21·8 m. The top story is thus about equivalent in height to the second (B) story in the Dipterocarp forest; the *Quercus-Neolitsea* association is in fact very like the *Parashorea-Diplodiscus* association, except that it lacks the dominant dipterocarp layer. The canopy is closed, but much more open than in the Dipterocarp forest. The undergrowth is much less dense, but ground herbs and ferns are more abundant; herbaceous plants frequently form a thick carpet over the ground. Climbers are numerous. One of the most striking features of the *Quercus-Neolitsea* association is the abundance of epiphytes, which increases with altitude; epiphytes are here less abundant than in the Mossy forest, but considerably more so than in the Dipterocarp forest and less limited to the crowns of the trees. Foliage is mostly 'mesophyll' but as in the Dipterocarp forest there is a small proportion of 'microphylls' and 'macrophylls'. In the *Quercus-Neolitsea* association species of *Quercus* (especially *Q. soleriana*) and *Neolitsea villosa* are the commonest trees in the higher story. The number of species of woody plants over 2 m. high on a sample plot (0·25 m.) was seventy. Many of these species are also found in the Dipterocarp forest.

From 750 to 850 m. the Mid-mountain forest consists of the *Astronia rolfei* association. This is lower than the *Quercus-Neolitsea* association and has a less well-developed top story; it is probably to be regarded as a variant of the same formation ('lociation' in Clement's terminology) characteristic of steep slopes. The top of Mt Maquiling, above about 900 m., is occupied by the Mossy forest, a very characteristic community which differs more from the Dipterocarp and Mid-mountain forest than these two formations differ from each other. The number of tree stories is here reduced to one, the average height of which is only 6–10 m. The thick covering of epiphytes (chiefly mosses on the branches and liverworts and small ferns on the trunks), and the great development of aerial roots, give the dwarf trees a fantastic appearance. The average size of the leaves is much smaller than in the Mid-mountain and Dipterocarp forests; 'microphylls' are as abundant as 'mesophylls' and 'macrophylls' are absent. There is usually a fairly dense ground-cover of ferns and herbaceous flowering plants and, as in Mossy forest elsewhere in Malaysia, the distinction between ground flora and epiphytes is indistinct, species usually terrestrial sometimes growing as epiphytes and vice versa. Climbing plants, both epiphytic and rooted in the ground, are plentiful, but not numerous in species.

The Mossy forest forms a single association, the *Cyathea-Astronia* association; there is no clear dominant, the commonest species being the tree fern *Cyathea caudata* and species of *Astronia* (Melastomaceae). Of the twenty-one species of trees found in the Mossy forest, eight also occur (as understory species) in the Dipterocarp forest, but in general the flora is very different from that of the other formations.

The chief differences in structure and composition between the three altitudinal zones on Mt Maquiling are summarized in Table 36. From this table it can be seen that the chief change in the structure of the vegetation with increasing altitude is a decrease in height, accompanied by a reduction in the number of tree stories. The change from a three-storied to a single-storied structure takes place, it is interesting to note, not by a telescoping of the stories together, but by the gradual disappearance first of the top, then of the second, story. The increase of the number of woody plants per unit area (which in the Mossy forest is nearly double that in the Dipterocarp forest) is doubtless dependent on the decrease in the size of the individual trees. No figures are available of the increase in the epiphytic vegetation with altitude.

In addition to the changes in structure, there are changes in physiognomy and in floristic composition. The most striking physiognomic changes with increasing altitude are the decrease in average leaf size (Fig. 15, p. 83) and the disappearance of features such as buttressing and cauliflory. The outstanding floristic change is the great decrease in the number of species; this, however, is not accompanied by any tendency to single-species dominance (cf. Chapter 11).

TABLE 36. *Comparison of sample plots (area 0·25 ha.), Mt Maquiling, Philippine Islands.* (After Brown, 1919)

	Dipterocarp forest (450 m.)	Mid-mountain forest (700 m.)	Mossy forest (1020 m.)
Number of tree stories	3	2	1
Height of tallest tree on plot (m.)	36	22	13
Average height of stories (m.)	27, 16, 10	17, 4	6
Number of individuals of woody plants over 2 m. high	353	539	610
Number of species of woody plants over 2 m. high	92	70	21

The relationship of the altitudinal zonation of the vegetation on Mt Maquiling to climatic factors was briefly discussed in Chapter 6. The dwarfing of the trees with increasing elevation is thought by Brown to be due not to the fall in the mean temperature alone, but to the combined effects of decreased temperature and decreased illumination (due to increased cloudiness). From the data given by Brown it may be surmised that the soil in the Dipterocarp forest (and probably also in the Mid-mountain forest) is of the tropical red or yellow earth type and that in the Mossy (Montane) forest of podzol type. The Mossy forest soil is probably very acid and highly deficient in plant nutrients. Brown attributes little importance to soil differences as factors responsible for the change in the vegetation with increasing altitude, but it should be noted that the Mossy forest of Mt Maquiling, like Montane forest throughout the tropics (see Richards, 1936, pp. 354–5), has characteristics usually associated with 'physiological drought' (dwarfing, many species with small leathery leaves, etc.). Such characters are similar to those which in temperate climates are found in the vegetation of acid, mineral-deficient soils and it is perhaps tempting to look for their explanation in some chemical characteristic of the soil, such as extreme mineral deficiency or poor nitrification, rather than exclusively in climatic factors, as Brown does. Similar views have been recently put forward in explanation of the 'physiological xeromorphy' of European bog plants (see Steemann-Nielsen, 1940).

The general features of the zonation on Mt Maquiling are typical of most of the mountains of medium height in the Philippines, Borneo, the Malay Peninsula, etc. On Mt Dulit in Borneo (Richards, 1936) the forest at the foot of the mountain and on its lower slopes is Mixed Dipterocarp forest as at Mt Maquiling;

this Lowland forest has not been as extensively or completely destroyed as at Mt Maquiling, so there is no parang zone. There is little change in the structure or composition of the forest up to about 450 m., but from this height to the lower boundary of the Montane (Mossy) forest at 970–1100 m. there is a zone of forest differing from the Mixed Dipterocarp forest in composition and probably in structure; this was not closely examined, but it appeared to be similar to Brown's Mid-mountain forest. The Montane forest is marked here by a remarkably sharp boundary and is extremely different in general appearance from both the lowland and the Mid-mountain forest. The trees were seldom more than 15 m. high and were much lower than that in exposed situations; they formed two not very distinct stories, rather than one, as on Mt Maquiling. Buttressing and other physiognomic features characteristic of Tropical Rain forest were lacking and the foliage was small and coriaceous, sometimes even ericoid. The outstanding feature, as in other Mossy forests, was the extraordinary development of epiphytic vegetation; the majority of these epiphytes were leafy liverworts, but mosses, ferns and flowering plants were also abundant.

The altitudinal zonation in the Malay Peninsula has been described by Holttum (1924), Symington (1936, 1943) and others. The Dipterocarpaceae, the dominant family in the lowland Rain forests, reach their upper limit, according to Symington, at about 4000 ft. (1200 m.) on the main ranges and at about half this height on isolated mountains. The upper limit of the dipterocarps may be taken as the approximate dividing line between the lowland Tropical Rain forest and the 'mountain forests'. Three altitudinal sub-zones of the Dipterocarp forest are recognized by Symington: (i) The Lowland Dipterocarp forests, characterized by the preponderance of typically 'lowland' species of dipterocarps (chiefly *Shorea* spp. of the 'red meranti' group) in the top story. (ii) The Hill Dipterocarp forests in which 'hill' species, such as *Shorea curtisii* are the dominants; this sub-zone is not usually reached till 1000 ft. (300 m.) or higher. (iii) High Hill Dipterocarp forests. Here *Shorea curtisii* becomes scarce and exclusive 'high hill' Dipterocarps such as *Shorea platyclados* and *S. ciliata* appear. Two zones of 'mountain forests', lying above the normal upper limit of the Dipterocarpaceae, are distinguished: the Mountain Oak forests and the Mountain Ericaceous forests. Dipterocarps are not usually present in either of these zones. The Mountain Oak forests extend up to 5000–6000 ft. (1500–1800 m.) on the higher mountains; they are tall, and two-storied and are characterized by the abundance of oaks (*Quercus* and *Lithocarpus* spp.). The Mountain Ericaceous forests, which occur on the highest mountains above the Oak Forests, are dwarf and mossy; species of oak are few and ericaceous trees such as species of *Pieris*, *Vaccinium* and *Rhododendron* are prevalent. Symington's three sub-zones of Dipterocarp forest clearly belong to the Tropical Rain forest formation; the Mountain Oak forests, which are similar to the Foothills forest of New Guinea and the Mid-mountain forest of Mt Maquiling, represent the formation here termed Submontane Rain forest. The Mountain Ericaceous forest is similar to the Montane Rain forest of other Malaysian mountains. On the extensive mountain ranges of the Peninsula, owing to the *Massenerhebung* effect, the altitudinal limits of the respective zones are usually much higher than on mountains such as Mt Maquiling and Mt Dulit.

The vegetation of the high volcanoes of Java has often been described (see Schimper, 1935; Docters van Leeuwen, 1933; Seifriz, 1923 and 1924). Three forest zones are usually recognized in West Java: (i) the Tropical Rain forest, which at about 1500–2200 m. passes into (ii) the Temperate Rain forest; this is poorer in species and differs from the Tropical Rain forest in the absence of buttressed trees, the smaller size of the foliage, the thicker trunks and the greater abundance of epiphytes, as well as floristically; the commonest trees are here oaks, *Castanopsis* spp., *Litsea javanica* and *Podocarpus imbricatus*. Above this is (iii) the Elfin Woodland or Mist forest; this extends to the highest summits and is a typical dwarf Mossy forest. On the crater floors the forest is replaced by grassland or by scrub in which the woolly shrub *Anaphalis javanica* is characteristic; these communities probably owe their existence to fires or special edaphic conditions.

In the drier climate of East Java the zonation is somewhat different. Tropical Rain forest, of a drier and less luxuriant type than that of West Java, is found in the lowest zone, but the zone occupied in West Java by Temperate Rain forest is here occupied by a community dominated by *Casuarina junghuhniana*; in appearance it resembles Savanna woodland. The highest summits are covered with grassland and dwarf scrub. There is evidence that the *Casuarina* woodland is, at least in part, a fire-climax; perhaps the grassland and scrub of the highest zone also do not represent a true climatic climax.

THE TROPICAL RAIN FOREST AT ITS ALTITUDINAL LIMITS IN AFRICA

The high mountain vegetation of Africa, with its spectacular tree-like senecios, lobelias and Ericaceae, is fairly well known, but it cannot be said that the altitudinal zonation on many tropical African mountains has been adequately studied or described.

For studying the changes in the vegetation at the upper altitudinal limits of the Tropical Rain forest, Africa is not a favourable region. Most of the higher African mountains lie in areas of relatively dry climate, and savanna and similar more or less xerophytic communities, rather than tropical forest, cover their lowest slopes. Even where the Rain forest reaches the base of the mountains, as at the eastern margin of the Congo basin, belts of dry and wet climate are intricately intermingled, so that the sequence of plant formations dependent on the altitudinal gradation of climate is not easily disentangled from that due to local variations in rainfall. Further, since much of the mountain vegetation of

Africa grows on young volcanic soils where it has scarcely had time to complete its development, and much has been modified by fire, grazing and shifting cultivation, even the recognition of climax types offers considerable difficulty.

Conditions at the altitudinal limit of the African Closed forest (Tropical Rain forest, cf. Chapter 15) are best known on the western slopes of the high volcanoes which divide the Congo basin from the Great Rift valley, thanks especially to the work of Lebrun (1935, 1936a).

Here the great lowland Rain forest of Central Africa approaches close to the mountains in many places, though much of it has been destroyed and its eastern margin is receding westwards. At middle altitudes on the mountains there is a zone of evergreen non-tropical rain forest in which an appreciable proportion of the flora is subtropical or temperate rather than tropical in its affinities; similar forests, belonging to the same formation and variously termed Subtropical, Warm Temperate, Temperate or Montane Rain forests (for full synonymy, see Pitt-Schenkel, 1938) occur scattered throughout the East African mountains; provisionally they may all be regarded as belonging to the same formation type as the Montane Rain forests of Malaysia.

According to the useful map given by Lebrun (1935, Pl. 1), there is actual continuity between these upland forests and the lowland Tropical Rain forest of the Congo in a few places, notably to the north-west of Ruwenzori and to the west of Lake Kivu. More often, however, the lowland and the upland forest are separated by a narrow band of xerophilous savanna and grassland; this belt is due, according to Lebrun, to *Foehn* winds descending from the mountains which produce a localized belt of dry climate. Where the regular altitudinal gradient of climate is not interrupted in this way, the mean temperature decreases by about 0·6°/100 m. increase of altitude and the annual rainfall increases to a maximum of 225 cm. at about 2400 m. and then decreases at higher altitudes. In consequence of the increased cloudiness, insolation decreases with altitude to a minimum in the zone of maximum precipitation.

Where the lowland and the upland rain forest are continuous, the following zones may be distinguished:

(i) Tropical Rain forest zone (*Forêt équatoriale* of Lebrun)—upper limit 1100–1300 m.
(ii) Transition forest zone (*Forêt de transition*)—lower limit 1100–1300 m., upper limit 1650–1750 m.
(iii) Montane Rain forest zone (*Forêt mésophile de montagne*)—lower limit 1650–1750 m., upper limit 2300–3400 m.

The upper limit of the Montane Rain forest, which coincides with the level of maximum precipitation, is marked by the beginning of the very characteristic Bamboo zone (lower limit 2200–2400 m., upper limit 2600 m.), formed by a very dense consociation of *Arundinaria alpina*, which is met with on most of the higher East African mountains. Above 2600 m. the Bamboo zone is succeeded by the Ericaceae zone (lower limit 2600–3100 m., upper limit 3700–3800 m.), dominated by arborescent species of *Erica* and *Philippia*, or (especially in the eastern group of the Virunga Mountains) by a *Hagenia* zone (2600–3100 m.), a type of dwarf open woodland; finally come the Alpine zone, the chief home of the arborescent senecios and lobelias (lower limit 3700–3800 m., upper limit 4600 m.), and the zone of permanent snow.[1]

The Transition forest was first recognized as a definite community by Lebrun (1934), but is still very little known. It is less luxuriant than the lowland Tropical Rain forest and differs from it in the smaller height and diameter of the trees; the volume of timber per unit area is lower (see below). Floristically it is a mixed association very rich in species, many of them endemic; some of the species are also abundant in the lowland forest (e.g. *Carapa grandiflora*, *Cynometra alexandri*, *Symphonia gabonensis*, *Uapaca guineensis*, etc.), others are strays from the Montane zone above.

Where it is not succeeded (at its upper limit) by a dry *Foehn* belt, the Transition forest passes gradually into the Montane Rain forest, a tall closed community consisting of two tree stories (*strate dominante* and *strate dominée*), one, or perhaps two, shrub-layers, and a layer of ground herbs. The average height of the top tree-story is about 25 m.; the crowns of the trees are compact and separate from one another, not interwoven with lianes as in the Tropical Rain forest. The following figures show some of the more striking differences between the Montane forest, the Transition forest and the Tropical Rain forest:

TABLE 37. *Comparison of the evergreen forest formations on the mountains of the eastern Belgian Congo.* (After Lebrun, 1936a, p. 151)

	Tropical Rain forest	Transition forest	Montane Rain forest
No. of trunks (over 20 cm. diameter) per hectare	115	180	220
Mean height of boles (m.)	13	12	10
Mean height of crowns (m.)	30	25	20
Mean diameter of trunks (cm.)	60	40	35
Timber vol. per hectare (cu.m.)	400–600	300	200

In the Montane Rain forest the great majority of the trees are evergreen and very few are buttressed. An obvious difference from the Tropical Rain forest is the absence or scarcity of lianes. The abundance of epiphytes (mosses, lichens, ferns, orchids, Loranthaceae) is characteristic and increases with altitude, but the growth of moss on the tree-trunks does not reach the same dimensions as in the Ericaceae zone at much higher altitudes. The shrub-layer is very dense, but the herb-layer is quite open except in clearings and includes many suffrutescent species. As in the Tropical Rain forest there is no moss stratum on the ground, though mosses are found at the foot of trees and on boulders.

[1] The zonation on Mts Kenya and Aberdare (Fries & Fries, 1948) is similar, except that as these mountains rise from a 'basal plateau' with a dry climate, zones (i) and (ii) are replaced by grassland ('*Themeda* steppe').

Floristically the Montane forest is very different from the Tropical Rain forest; although many of its components are generically or even specifically identical with lowland species, some are of temperate affinities. The forest is a mixed association with numerous dominants.

Lebrun recognizes three sub-zones (*horizons*) of the primary Montane Rain forest: Lower (1600–1900 m.), Middle (1900–2100 m.) and Upper (2100–2400 m.). The Lower sub-zone closely resembles the Transition forest. The trees are taller and the undergrowth denser than in the two other sub-zones. Of the trees, 30% have compound leaves and a few species are deciduous. Among the chief dominants of the first story are *Entandrophragma speciosum*, *Ficalhoa laurifolia*, *Lebrunia bushaie*, *Ocotea usambarensis* and *Strombosia grandifolia*. Tree ferns are frequent. The herb-layer includes some temperate species, e.g. *Dryopteris filix-mas* and *Sanicula europaea*. A few lianes are characteristic of this sub-zone. Epiphytes, though abundant, are not as luxuriant as at higher altitudes and lichens of the genus *Usnea* are not conspicuous. In the middle sub-zone the trees are taller than in the upper sub-zone, but not as tall as in the lower (the data for the Montane forest given in Table 37 refer to this sub-zone). The percentage of trees with compound leaves here falls to 25%. The chief first story dominants include *Entandrophragma speciosum*, *Ficalhoa laurifolia*, *Ocotea viridis* and *O. usambarensis*, *Parinari mildbraedii*, *Podocarpus milanjianus* and *P. usambarensis*, *Strombosia grandifolia* and *Symphonia globulifera*. Tree ferns, often associated with *Ensete edule*, are common in ravines. The herbaceous layer is very open and includes representatives of temperate genera. There are very few large lianes. Epiphytes are abundant, but *Usnea* is not conspicuous. The forest of the upper sub-zone has something of the character of Elfin woodland. The trees do not exceed 10–15 m. in height and their trunks and branches are twisted and irregular. The foliage of the trees is mostly small, coriaceous and entire; only 15% of the species have compound leaves. The characteristic first story trees of the sub-zone are: *Ekebergia ruppeliana*, *Ficalhoa laurifolia*, *Olea hochstetteri*, *Parinari mildbraedii*, *Podocarpus milanjianus*, *Sideroxylon adolfi-friederici* and *Symphonia gabonensis*. Shrubby undergrowth is well developed, but the herbaceous ground flora is not well represented.

It is instructive to compare this zonation in the eastern Belgian Congo with that described by Moreau (1935b)[1] for the Usambara mountains in Tanganyika. These consist of two small isolated massifs, the Eastern Usambaras reaching a height of about 4900 ft. (1500 m.), and the Western over 7000 ft. (2000 m.); both are islands of wet climate in a dry region where types of 'deciduous parkland' (savanna) are the characteristic vegetation. Though much of the original forest has been replaced by cultivation, luxuriant rain forest still remains on the eastern slopes of the mountains. In the humid areas of the Usambaras, Moreau distinguishes three forest zones: Lowland (to 2500 ft., 760 m.), Inter-

---

[1] Additional information kindly provided by Mr P. J. Greenway has been included in the following account.

mediate (2500–4500 ft., 760–1370 m.) and Highland (above 4500 ft., 1370 m.). In the Lowland forest the important species of trees are *Antiaris usambarensis*, *Ficus* spp., *Sterculia appendiculata*, *Trema guineensis*, *Chlorophora excelsa* and *Albizzia* spp. This list suggests an old secondary rather than a primary community; the original primary forest may have resembled the West African Dry Evergreen (Mixed Deciduous) forest, rather than 'true Rain forest'. The undergrowth of the Lowland forest tends to be thorny, consisting largely of *Acacia pennata* and *Harrisonia abyssinica*; the grass *Olyra latifolia* forms thickets several feet high and there is little other herbaceous ground flora. Lianes are plentiful, but there are few epiphytes except scattered plants of *Platycerium angolense*. The Intermediate forest is the most luxuriant of the three types, indeed Moreau states that it is the most luxuriant forest in East Africa. The species of trees here are very numerous and many of them are the single representative in the area of their respective genera and are endemic; *Macaranga usambarica*, *Allanblackia stuhlmannii*, *Piptadenia buchananii*, *Isoberlinia scheffleri* and *Parinari* spp. are among the important dominants. Tree ferns are common. In the more open places the ground flora is varied and abundant. Lianes and strangling figs are prominent; epiphytes are numerous in both individuals and species. The general aspect of the Highland forest, which is fully developed only in the Western Usambaras, is 'not unlike that of the Intermediate forest, but the floristic composition is different. The most important trees, dominant either together or separately, are *Podocarpus* spp. and *Ocotea usambarensis*; forests of *Juniperus procera* ('cedar') occur in the same zone. Pitt-Schenkel (1938) has given a detailed account of these Highland forests (which he terms 'warm temperate rain forest').

Further details about the structure and composition of the Usambara forests are much to be desired, but provisionally we can perhaps regard the Lowland forest as representing the *forêt équatoriale* of the eastern Belgian Congo, the Intermediate forest Lebrun's Transition forest and the Highland forest the Montane Rain forest.

On the higher mountains of West Africa a direct transition from Tropical Rain forest to upland formations can seldom be observed, owing to the large areas of cultivation on the lower slopes.

On the island of S. Tomé in the Gulf of Guinea rain forest probably originally covered the whole surface up to the highest point (2024 m.). Three zones can be recognized (Exell, 1944): the Lower Rain forest region (0–800 m.), the Mountain forest region (800–1400 m. almost everywhere) and the Mist forest region (1400–2024 m.). Little remains of the Lower Rain forest which has been replaced by cultivation. It appears to have been a mixed association of the Wet Evergreen forest type (p. 338) rather than 'true Rain forest' and was probably similar in composition to the lowland forest on Fernando Po (p. 362) and the neighbouring mainland. This community probably passed gradually into the Mountain forest and the transition seems to have been 'characterized by a change in the constituents of the forest rather than in its general aspect' (Exell, 1944, p. 18). In the Mountain forest the trees are of great height, forming a dense canopy. The undergrowth consists largely of bushes of *Chasalia doniana*, up to 1 m. high, with a great abundance of ground orchids and ferns, the latter perhaps more abundant here, both in species and individuals, than anywhere else in Africa. Some tree ferns are present. Large lianes are abundant and epiphytic mosses, lichens and flowering plants are so numerous as almost to conceal the trunks. The Mountain forest is very rich floristically; Rubiaceae and Euphorbiaceae are particularly abundant and Leguminosae absent or uncommon. The Mist forest occupies the zone of maximum rainfall where there is an almost continual mist. The trees are smaller than in the lower zones and epiphytes are even more abundant than in the Mountain forest. The flora of this zone is not well known, but many species occur which seem to be absent at lower levels and most of these are endemics. Some temperate species occur.

On the mountains of the island of Fernando Po (3291 m.), and on Cameroons Mountain (4067 m.) on the African mainland, the zonation is similar, but both these mountains are much higher than S. Tomé and the upper slopes are treeless. On Cameroons Mountain forest ascends to over 2000 m. Above the tree limit there is open grassland with occasional bushes and stunted trees (Maitland, 1932), which is burned annually by the natives. The highest part of the mountain is an 'alpine desert' of lava sparsely colonized by grasses and other plants. The forest on the lower slopes, which up to about 1000 m. was probably similar to that in the neighbouring lowlands, has mostly been cleared to make way for plantations and native farms. From about 1000 m. upwards a fairly continuous belt of forest still remains, but most of it covers lava of recent or relatively recent origin and probably does not represent a climax. The lower part of this forest is tall but differs considerably in floristic composition from the lowland rain forest; temperate species such as *Sanicula europaea* and *Viola abyssinica* occur in the undergrowth, lianes are scarce and bryophytes very abundant. Societies of the tree fern *Cyathea manniana* in disturbed areas and gaps are a conspicuous feature. Above about 1500 m. the number of tree species is much reduced and nearly pure stands of *Syzygium* sp., often with a dense undergrowth of *Mimulopsis violacea* and other Acanthaceae, cover a large area. Near the forest limit the trees become dwarf and crooked; the chief species include *Hypericum leucoptychodes*, *Lasiosiphon glaucus*, *Rapanea neurophylla*, *Pittosporum mannii* and *Schefflera hookeriana*. The forest here has much of the aspect of a Malayan Mossy forest, except that bryophytes, though abundant, are considerably less luxuriant; it is evidently a type of Montane Rain forest. The forest limit is certainly not climatic and is probably determined partly by the presence of young lava flows and partly by the annual fires which prevent the invasion of the grassland by trees.

From the inadequate data, it would appear that in Africa, where the Tropical

Rain forest is in direct contact with upland evergreen forest, that is to say, where its upper limit is truly altitudinal and not 'climatic' or biotic, the general features of the transition are not dissimilar to what is seen in Malaysia. The tall Transition forest of middle altitudes in East Africa, the Mountain forest of S. Tomé and the forest from about 1000 to 1500 m. on Cameroons Mountain seem to have much in common, physiognomically and structurally, with the Submontane Rain forest of Malaysia, while the African and the Malaysian Montane forest are also similar.

The dwarf mountain forest of East Africa (Lebrun's Upper sub-zone) seems in many respects comparable with the dwarf mossy type of Montane Rain forest of Malaysia. It may, however, be premature to apply, at present, the same terms to what appear to be analogous formations in Africa and Malaysia.

THE TROPICAL RAIN FOREST AT ITS ALTITUDINAL LIMITS IN SOUTH AMERICA AND THE WEST INDIES

The Amazonian forest extends westwards to the foot of the Andes, where the character and the composition of the vegetation begin almost at once to change. With increasing height above sea-level the Tropical Rain forest gives place to other evergreen forest formations with a different and less exclusively tropical flora. On the wet eastern slopes of the Andes in Colombia, Peru and Bolivia, the following regions are usually distinguished: (i) The basal region of Tropical Rain forest (*montaña*), differing only slightly in aspect and flora from the Rain forest of the Amazon lowlands (*montaña real*, the Hylaea of Humboldt); this extends up to about 1200–1500 m. Above this lie two regions of evergreen forest in which both the flora and general appearance of the vegetation are very different from the region below, viz. (ii) the transitional *medio yungas* region, and (iii) the *ceja de la montaña* (brow of the mountain). The *ceja* begins in Peru at 1800–2000 m. and extends up to 3400–4000 m. (Weberbauer, 1911); in Bolivia its limits are approximately 2200–2800 and 3400 m. (Cárdenas, 1945). Still higher, above the *ceja de la montaña*, lies the Alpine region of the tropical Andes, consisting mainly of open paramo, or steppe, in which the arborescent Compositae (chiefly species of *Espeletia* and *Culcitium*), strikingly similar in physiognomy to the tree senecios of the corresponding region on the African mountains, are characteristic.

The Tropical Rain forest of the basal region of the Andes differs from that of the lowlands in lacking certain species and genera. The upper altitudinal limit of many characteristically tropical elements, such as the palms *Iriartea*, *Phytelephas* and *Bactris*, coincides with the upper limit of this basal Rain forest, but some tropical elements extend higher. The *medio yungas* region thus has a mixed flora in which both temperate and tropical elements play a part. In this zone are the famous cinchona forests; bamboos (*Chusquea* spp.) and tree ferns are also striking features. The *medio yungas* forests would seem to have considerable similarity to the tall Montane (Temperate) forests of East Africa. The *ceja de la montaña* forests (*Buschwald*, *Gebüsch* and *Gesträuch* of Weberbauer, 1911) are dominated by low-growing trees with gnarled and twisted trunks abundantly overgrown with epiphytic bryophytes; the foliage is predominantly small, coriaceous and entire. They are, in fact, a typical 'mist forest' or Elfin woodland formation; it is evidently closely analogous to the Mossy forest of Malaysia, to which indeed it has a strong resemblance in floristic composition.

These evergreen forest formations of the tropical Andes have been frequently described (see especially Weberbauer, 1911, 1945; Herzog, 1923; Cuatrecasas, 1934), but since the available information is mainly floristic, they will not be further discussed here.

The mountain forests of the West Indies, especially those of Trinidad and the Lesser Antilles, are better known than those of the Andes. The forest formations and their altitudinal zonation have been dealt with by Gleason & Cook (1926), Stehlé (1935), Beard (1944a, 1946b) and others. Though the plant formations are essentially similar in their physiognomy to those displayed on a much larger scale in the Andes, owing to the great *Massenerhebung* effect of the latter, corresponding types of vegetation are found on the relatively isolated coastal mountains of the West Indies at much lower altitudes.

In Trinidad Beard (1946b) recognizes three zones of vegetation: Lower Montane Rain forest from 800 to 2500 ft. (250–760 m.), 'Montane Rain forest' from 2500 to 2900 ft. (760–880 m.) and Elfin woodland (on the summit of Aripo above 2900 ft. only). The exact altitudinal limits of these formations and the width of the transition from one to the other vary somewhat; as mentioned earlier (cf. p. 347, Fig. 40) all the zones are depressed towards the wind-exposed eastern end of the Northern Range.

Beard's Lower Montane Rain forest should be included in the Tropical Rain forest formation-type as here understood. It differs very little from typical lowland Rain forest, except that the trees are lower. The average height of a mature dominant tree is about 90 ft. (27 m.), except in sheltered places, and the highest story forms a compact canopy at 70–100 ft. (22–30 m.); below this there is an ill-defined discontinuous lower tree-story (Fig. 42). Physiognomic characters typical of rain-forest trees are not well developed; thus buttressing is not conspicuous and cauliflory is rare. Neither lianes nor epiphytes are very abundant. Palms and tree ferns are not important. Only one association, the *Byrsonima spicata-Licania ternatensis*, is present. The estimated number of species per 100 acres (40 ha.) is eighty-seven, which is slightly less than are usually found in the lowland 'Evergreen Seasonal forest' of Trinidad (p. 319). About 28% of the stand is formed by *Licania ternatensis*.

The 'Montane Rain forest' occupies only a limited area and is confined to the Aripo massif. The forest is definitely two-storied (Fig. 42b) and its average height is still lower than that of the Lower Montane forest. The upper tree-story forms a closed canopy at 50–65 ft. (15–20 m.) and no tree exceeds 75 ft. (23 m.). The trees mostly lack buttresses and other characteristic rain-forest features, but both lianes and epiphytes are very abundant and luxuriant, the epiphytes including many mosses. The foliage is 'mesophyll' to 'microphyll' in size and tends to be leathery. Beneath the lower tree story there is a layer composed chiefly of palms and tree ferns. The two most abundant species in the single association are *Richeria grandis* and *Eschweilera trinitensis*, both of which are characteristic of the zone. The flora is very restricted, including only twenty-eight tree species.

The Elfin woodland formation is represented by a consociation of the stilt-rooted tree *Clusia intertexta* which is only on the actual summit of Aripo. The dominant tree forms a discontinuous upper story 20–25 ft. (6–7·5 m.) high; below this there is a very dense layer of tree ferns and small palms about 10 ft. (3 m.) high. The epiphytic vegetation is extremely luxuriant and consists entirely of bryophytes and lichens which cover all the trunks and branches. In addition to the dominant there are only six other dicotyledonous trees; the total woody flora (including tree ferns and palms) amounts to only eleven species.

The three climax formations of the Trinidad mountains recognized by Beard seem to be quite closely analogous to the corresponding zones in the eastern tropics. The Lower Montane Rain forest, essentially tropical in flora and not very different in structure from lowland Tropical Rain forest, is clearly equivalent to the somewhat modified variants of the Mixed Dipterocarp forest found on the lower slopes of the Malaysian mountains, such as Symington's Hill Dipterocarp forests and High Hill Dipterocarp forests. The Trinidad formation which Beard terms 'Montane Rain forest' is analogous, both in structure and in its floristic relations, to the zones above and below, to the Malaysian Mid-mountain forest, Mountain Oak forest, etc. for which we are proposing the name Submontane Rain forest formation. The resemblance in structure and physiognomy between the Elfin woodland of Trinidad and the 'mossy' type of Montane Rain forest in Malaysia is obvious, though floristically, as might be expected, there is little similarity.

In the other West Indian islands the zonation is not different in any fundamental respect from that in Trinidad. The corresponding formations differ floristically to some extent from those of Trinidad; thus in the Lesser Antilles the *Byrsonima spicata-Licania ternatensis* association of the Lower Montane Rain forest zone is replaced by a *Dacryodes excelsa-Sloanea* association, but without any great change of physiognomy or structure. Similarly the Elfin woodland formation is often represented in the Lesser Antilles by a *Didymopanax-Charianthus* association instead of a Clusietum (Beard, 1945, etc.). On the higher islands the zones are broader owing to *Massenerhebung* and treeless formations are present at high altitudes. In Guadeloupe (Stehlé, 1935) for example, 'hygrophytic forest' extends up to about 1000 m. In its lower part this community is like the Lower Montane forest of Trinidad; above it appears to resemble the Submontane Rain forest ('Montane Rain Forest' of Beard). From about 1000 to 1120 m. there is Elfin woodland, dominated, as in Trinidad, by a species of *Clusia*

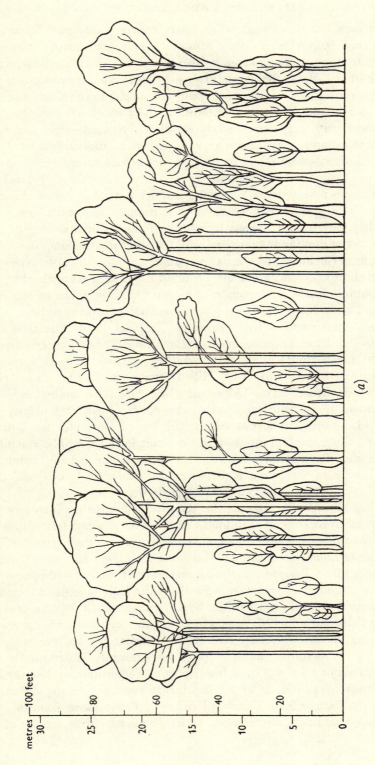

metres ─ 100 feet

30
25 — 80
20
15 — 60
10 — 40
5 — 20
0

(a)

Tropical Rain forest (Lower Montane Rain forest of Beard).

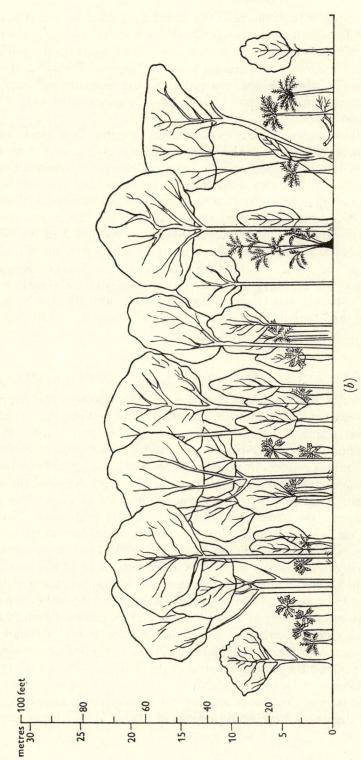

(b)

Submontane Rain forest (Montane Rain forest of Beard).

Fig. 42 *a* and *b*. Profile diagrams of Tropical Rain forest and Submontane Rain forest, Northern Range, Trinidad, British West Indies. After Beard (1946*b*).

(*C. venosa*). At its upper limit the Elfin woodland is replaced by a savanna-like community, the Lobelietum guadeloupense, which in physiognomy seems to be similar to the paramo of the Andes. Still higher there is a Pitcairnietum, consisting of Bromeliaceae, ferns, etc. and a ground cover of *Sphagnum, Breutelia* and other bryophytes. Finally, there is the vegetation of the volcanic summits and plateaux, dominated by *Sphagnum,* mosses and hepatics with only a few flowering plants.

In some of the Antilles communities dominated by palms form a zone above the Submontane Rain forest, e.g. the stands of *Euterpe globosa* on Puerto Rico at 650–750 m. (Gleason & Cook, 1926). Some of these palm communities are probably seral, but their status needs further investigation. The status and ecological relationships of the *Pinus caribaea* forests of the Greater Antilles are also not clear. On some of the West Indian islands, e.g. St Vincent (Beard, 1945), hurricanes and volcanic eruptions have destroyed much of the climax vegetation, which has been replaced by seral communities (see pp. 281–2).

Shreve (1914*a*) has made a detailed study of the physiological ecology of the forest at 4500 to 7420 ft. (1373–2262 m.) on the Blue Mountains of Jamaica. The community studied was termed Montane Rain forest by Shreve, but in the terminology used here it seems to be in part Submontane, and in part true Montane, Rain forest.

TROPICAL RAIN FOREST AT ITS LATITUDINAL LIMITS

In eastern Australia, south-east Asia (southern China, northern Burma, etc.), and on the east coast of South America, evergreen rain forest reaches and passes the geographical tropic, extending northwards or southwards far into the subtropical regions. These extra-tropical extensions of the tropical evergreen forest differ in various respects from the true Tropical Rain forest and should probably be regarded as a separate formation-type which we shall term the Subtropical Rain forest, applying the name Temperate to formations such as the Subantarctic forests of south-eastern Australia and New Zealand, southern Chile, etc. which, though evergreen, are dominated by temperate trees such as *Nothofagus* spp. and have a flora which is not tropical (whatever may have been its ultimate origin). Apart from these large extensions of the evergreen forest outside the tropics, which are in more or less direct continuity with the main rain-forest belt, here and there in favourable situations within the sub-tropics there are small outliers of broad-leaved evergreen forest which at least in their floristic composition recall the Tropical Rain forest. Examples of these are the 'hammock forests' of Florida (Davis, 1943) and an interesting relict patch of fringing forest on the banks of the Rio de la Plata (34° 47′ S.) described by Cabrera & Dawson (1944).

The available information about Subtropical Rain forests is mainly floristic and a detailed discussion here would be out of place. It will be of interest, however, to consider briefly the Subtropical Rain forest of Australia which is linked by a direct transition with the true Tropical Rain forest to the north.

Though much of it has been destroyed, it is still possible to trace the Australian rain forest as a narrow discontinuous strip running parallel to the east coast from Queensland southwards to New South Wales and Victoria (for the detailed distribution, see Francis, 1929). In tropical Queensland this forest closely resembles the rain forest of Indo-Malaya in general aspect and, except for the complete absence of Dipterocarpaceae, in floristic composition. As the belt is followed southwards the vegetation changes: the number of species (especially of trees) diminishes, sub-tropical and temperate elements mingle in increasing numbers with the Indo-Malayan flora, and characteristic tropical physiognomic features such as buttressing, cauliflory, etc. disappear or become less evident. According to Francis (1929, p. 4), a tendency to single-species dominance is concomitant with the reduction in number of species; thus in the Dorrigo highlands (30–31° S.) *Ceratopetalum apetalum* 'sometimes almost monopolizes the stand'. These changes in the character of the rain forest occur gradually and first become apparent at about the latitude of the Richmond River in northern New South Wales (29° S.) (Francis, *loc. cit.*). The prevailing type of foliage remains evergreen and 'mesophyll' throughout the rain-forest belt and lianes and epiphytes continue to be abundant. Thus, even at its southern limits, the Subtropical Rain forest retains much of the aspect of a tropical forest, though it is less luxuriant.

The contrast in flora and physiognomy between the Subtropical Rain forest and the adjacent formations—Sclerophyll forest dominated by species of *Eucalyptus* and Subantarctic Rain forest dominated by *Nothofagus*—is everywhere extremely sharp. Even where one formation is invading another the boundaries are not hard to trace. Towards its southern limit the Subtropical Rain forest depends increasingly on favourable conditions of soil and climate. Thus, as mentioned in Chapter 6, basaltic soils favour the development of Subtropical Rain forest, but soils of lower water-holding capacity, such as those overlying sandstone, favour the sclerophyll communities (Fraser & Vickery, 1938). In some areas the local microclimate is of equal or greater importance (Osborn & Robertson, 1939). The Subantarctic Rain forest is adapted to cooler conditions than the Subtropical Rain forest and occurs at higher altitudes and further south. In subtropical Australia there is thus a clear ecological segregation of the Indo-Malayan rain-forest flora from the mainly endemic Australian sclerophyll flora; this is closely parallel to the situation in Malaysia where, as was said above, the Australian flora is restricted mainly to the Montane forest and to certain special lowland habitats.

The excellent account by Fraser & Vickery (1937, 1938, 1939) of the forests of the Upper Williams River and Barrington Tops district in New South Wales (32° S.) enables a fairly detailed comparison to be made between the Subtropical Rain forest of Australia and rain forests in the tropics. The area studied is mountainous, rising to heights of some 5000 ft. (1525 m.). The average rainfall of the district is about 150 cm. At the lower altitudes the average minimum temperature for the coldest month (July) is rather under 40° F. (4° C.) and the average mean annual maximum between 80° and 90° F. (26–32° C.). In the highest part of the area temperatures are much lower and snow falls in winter.

Three climax formations occur: mixed Subtropical Rain forest, Subantarctic Rain forest and Eucalyptus (Sclerophyll) forest. The Subtropical Rain forest covers the floor and sides of the upper part of valleys facing southeast, east and north-east, from about 300 m. to an upper limit varying from 1000 to 3000 m. above sea-level. The Subantarctic forest, a consociation of *Nothofagus moorei*, occurs above the upper limit of the Subtropical forest; the downward extension of the former is perhaps limited by high summer temperatures and the upward extension of the latter by low winter temperatures. The Eucalyptus forest occupies the valleys below the lower limit of the Subtropical Rain forest, the crests and upper slopes of the ridges and the plateau.

As everywhere in this region, the three formations remain everywhere very distinct, though many species are found in more than one formation. The great differences in their physiognomy and composition are well shown in their biological spectra (Fig. 43). In all this area the Subtropical Rain forest is advancing on the Eucalyptus forest, except at its upper limit, and in some places it is rapidly invading it. Fraser & Vickery regard this expansion as part of a recovery from a retrogression in the late Tertiary, due to orogenic movements. Some of the sclerophyll elements in the rain forest, such as the two species of *Eucalyptus*, are probably relics left behind in this advance.

The Subtropical Rain forest of the Upper Williams River district is a tall, dense, evergreen forest composed of trees of many families, genera and species. The main tree stratum forms a closed canopy at 60–80 ft. (18–24 m.); the upper surface of the canopy is very uneven and occasional emergent individuals rise to about 100 ft. (30 m.). This layer consists of a mixture of species of *Dysoxylum*, *Ficus*, *Cedrela* and other genera with typical laurel-like rain-forest foliage; all but three of these species are evergreen. Above the main (B) canopy there is a very discontinuous A story of two much taller species, *Eucalyptus saligna* and *Syncarpia laurifolia*, the former of which reaches a height of 130–170 ft. (40–52 m.); both of these trees are abundant in the adjacent Eucalyptus forest and physiognomically have more affinity with the sclerophyll flora than with that of the rain forest. In having this tall A story formed of trees of a different physiognomic type to the main canopy, this forest resembles the 'kamerere forest' of New Britain (p. 38) which has a similar A story of *Eucalyptus deglupta* overtopping a dense mixed canopy of typical rain-forest species. Below the B story the Williams River forest has a C story of trees 15–20 ft. (5–6 m.) high (species of *Diospyros*, *Drimys*, *Eupomatia*, etc. and the tree-fern *Alsophila leichhardtiana*), a shrub-layer and a ground-layer consisting of tree seedlings, ferns and a few perennial herbs.

The dominant trees (B stratum) have many of the usual physiognomic features of rain-forest trees. The trunks are straight and slender and sometimes

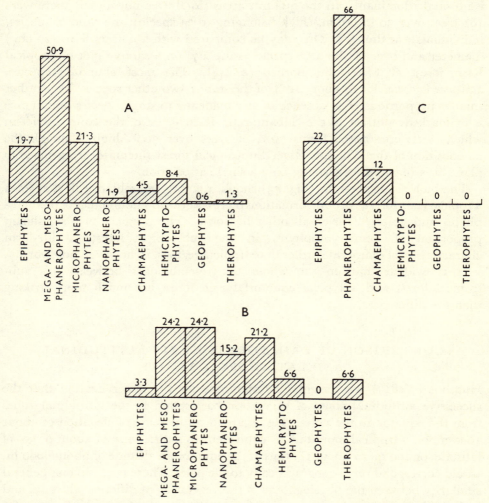

Fig. 43. Raunkiaer biological spectra for Subtropical Rain forest, Subantarctic Rain forest and Tropical Rain forest. A. Subtropical Rain forest, Upper Williams River district, New South Wales. B. Subantarctic Rain forest, Upper Williams River district, New South Wales. C. Tropical Rain forest, Moraballi Creek, British Guiana. A and B from data of Fraser & Vickery (1938), C as in Table 2, Chapter 1. Figures for percentage of species in total flora, not including ferns.

buttressed, though buttressing is not well developed except in hollows where the soil is very wet. The leaves of most species belong to the 'mesophyll' size-class and many have drip-tips; pulvini are almost universal (see p. 89). Some tropical features, such as cauliflory, do not occur.

Lianes of many species are common and are sometimes of large dimensions; they seldom reach the A story. The epiphytic vegetation is abundant and consists of orchids, ferns, bryophytes, etc.

The floristic composition of this Subtropical Rain forest is mixed and there is no single dominant. In the quadrat areas (total area not stated) there were 181 trees over 30 ft. (10 m.) high belonging to 34 species; the ratio of species/individuals was thus 5·3. This may be compared with 261 trees 8 in. (20 cm.) diameter and over of about 98 species (ratio 2·7) on a sample plot of Tropical Rain forest at Mt Dulit, Borneo (p. 244). The most abundant species, *Schizomeria ovata*, forms about 10% of the stand; two other species form rather smaller proportions. The absence of any tendency to single-species dominance is in marked contrast to the Subantarctic Rain forest of the same district in which *Nothofagus moorei* forms 76% of trees over 30 ft. high. Though the composition of the Williams River Subtropical forest fluctuates from place to place, it is regarded as forming only a single association.

The biological spectrum (Fig. 43) shows a strong resemblance to that for a Tropical Rain forest. In both formations there is a great predominance of phanerophytes and epiphytes over all other life-forms. The presence of a significant proportion of hemicryptophytes in the subtropical forest, a life-form characteristic of temperate rather than tropical regions, is, however, noteworthy.

It is evident that on the whole the resemblances between this subtropical forest and a typical equatorial rain forest are much more striking than the differences.

COMPARISON OF RAIN FOREST AT ITS ALTITUDINAL AND LATITUDINAL LIMITS

Humboldt (1817) was originally responsible for the generalization that the successive altitudinal zones of vegetation correspond to the latitudinal zones from the equator to the poles. He regarded an increase of elevation of about 1000 m. on a tropical mountain as equivalent to a difference of about 9° 30′ of latitude, or 100 m. to about 1°; thus at the equator an altitude of about 8800 m. would correspond phytogeographically to the poles. There is an obvious general similarity between the physiognomy of the vegetation at different altitudes and at latitudes with the same mean temperature, but the correspondence is far from exact and, as Lam (1945, p. 82) has pointed out, Humboldt's conception is too simple. The climate of the equivalent altitudinal and latitudinal zones is never the same, since in a given altitudinal zone at the equator the length of day and the seasonal changes, to mention only two factors, are different from those in higher latitudes.

C. Troll (1948) has also stressed the differences between the climate of tropical mountains and that of the temperate lowlands; the former (*Tageszeitenklima*) is characterized by a small seasonal and a relatively large daily temperature range, the latter (*Jahreszeitenklima*) by a large seasonal and a relatively small daily range. Troll points out, however, that in several respects, notably the frequency of frost and the small range of the mean monthly temperature, tropical mountain climates are much more similar to that of cool subantarctic regions than to any northern temperate climate. To this fact, he suggests, may be traced the much larger representation of south temperate (including subantarctic) than of north temperate elements in the tropical mountain floras. Even the plant communities of tropical mountains resemble those of the southern hemisphere; the Montane Rain forest of the tropics, for example, shows many similarities to the cool Temperate Rain forest of New Zealand, south Chile etc., but practically none to any north temperate forest community. This fact gives rise, as Troll puts it, to an asymmetry between the vegetation zones of the northern and southern hemispheres.

The data presented in this chapter are too incomplete to allow the theory of Humboldt or its modification by Troll to be fully tested, but a brief comparison between the communities adjoining the Tropical Rain forest at its altitudinal and latitudinal limits may be of interest.

With increasing altitude, as we pass from the lowland to the Montane Rain forest, the climax vegetation undergoes both structural and floristic changes. The average height of the trees decreases with altitude, the rate of decrease being much affected by wind-exposure. The three tree stories of the lowland forest are reduced to two, probably by a gradual thinning and ultimate disappearance of the A story. The average leaf size becomes smaller and when the Montane forest is reached microphylls and even smaller leaves become common. Drip-tips disappear. Buttressing and cauliflory rapidly become less frequent. Large lianes become rarer and eventually disappear. Epiphytes become much more abundant (at least up to the level of the Montane forest), but in the epiphytic communities flowering plants play a decreasing part, and ferns and bryophytes an increasing part. The main floristic changes are the increasing proportion of temperate species and a reduction in the total number of species, especially of trees. The impoverishment of the flora is not usually accompanied, as it is in edaphically unfavourable lowland habitats, by any marked tendency to single-species dominance.

Many of the differences between the Australian Subtropical forest and a Tropical forest are similar to these. There is little difference of height between the Subtropical and Tropical forest; this confirms the view that the decreased height of the Montane forest is due to low light intensity owing to cloudiness (as Brown suggests), and to wind exposure, rather than to decreased temperature alone. The Subtropical forest of the Williams River has three tree stories like a Tropical forest, but the highest story is sparsely developed and most subtropical forests are perhaps two-storied. Buttressing and cauliflory decrease with increasing latitude as with increasing altitude. No precise information about leaf sizes in the Subtropical forest is available, but since the prevailing leaf size is mesophyll, there is probably no marked difference from the tropical forest. Drip-tips do not disappear, as with increasing altitude. The floristic differences between the Subtropical forest of Australia and a Tropical forest are closely comparable with those due to increasing altitude. In both cases the number of species is reduced and the tropical flora becomes mixed with a temperate flora. In the Subtropical forest, however, there is sometimes a tendency to single-species dominance.

Part VI

MAN AND THE TROPICAL RAIN FOREST

CHAPTER 17

SECONDARY AND DEFLECTED SUCCESSIONS

Wherever the rain forest is the climatic climax, particularly in thickly populated regions, large areas of the 'virgin' or primary forest have been destroyed and replaced by cultivation or by secondary communities varying from stands of herbs and grasses to dense forest not very dissimilar in appearance to the original primary vegetation. These communities, excepting those dominated mainly by grasses (which usually result from deflected successions) are generally called 'second-growth' or 'secondary forest'. It has been objected that many so-called secondary communities follow the repeated destruction of the primary natural vegetation and are in fact tertiary, quaternary, etc., but the term secondary forest is convenient and too long established to be replaced by a new term such as 'substitution forest' (*forêt de remplacement*) suggested by Lebrun. In many countries local names are used for secondary communities; thus in British Guiana secondary forest is termed 'low bush' or *mainap*, in eastern Brazil *capoeira*, in the Philippines *parang*, throughout Malaysia *belukar* or *bluka* (sometimes also *utan muda*).

Secondary communities derived from Tropical Rain forest, whether they have the aspect of forest, scrub, savanna or (as is often the case) of a chaotic wilderness of trees, shrubs, herbs and climbers, are always more or less unstable and are thus stages in secondary successions. Left to themselves and protected from felling, burning and grazing, they are gradually invaded by primary forest species and there is little doubt that the climatic climax would ultimately re-establish itself, though probably only after a very long time. Where they are subjected to grazing or recurrent fires deflected successions set in, leading to apparently stable biotic climaxes.

Though many types of secondary vegetation derived from Tropical Rain forest have been described, they have seldom been closely studied and very few systematic observations have been made on the successions of which they are stages. It is unfortunate that this should be so, because no aspect of the ecology of the Tropical Rain forest is of greater practical value or promises results of more theoretical importance. From the fragmentary observations available it is evident that these secondary and deflected successions are complex and vary greatly from place to place, depending on differences in the habitat and in its previous history. The details can only be fully elucidated by careful quantitative observations continued over a long period of years. Such observations would help to fill the most serious gap in our present knowledge of the Rain forest.

The destruction of the primary forest which gives rise to secondary successions may take place in different ways and for various reasons. By far the most important cause of destruction up to the present has been the system of shifting cultivation (the ladang system of Malaysia and the taunggya system of Burma) which is practised by nearly all the native peoples of the tropics, except in a few regions such as tropical India and the lowlands of Java, where intensive sedentary systems of cultivation have been developed. The exact procedure in shifting cultivation varies somewhat from place to place, but the general principle is always the same. A patch of forest is felled, though some of the larger trees are often spared, either because of the labour involved in cutting them or for some superstitious reason. The fallen trees are burnt and then crops such as cassava, maize, yams, bananas or hill rice are planted. Since humus is rapidly destroyed by exposure to the sun and the ground generally receives little cultivation and no manure, the soil soon becomes impoverished and infertile. After one or more crops have been harvested, the number depending on the inherent fertility of the soil and the intensity of leaching and erosion, the plot is abandoned and the development of second-growth begins. In the course of the succession soil fertility is at least partially restored and the secondary forest may be cleared and cultivated again after some years. Whether the native farmer makes his fields on the sites of former cultivation or always in virgin forest will of course depend on the density of the population and the extent of suitable land. Lane-Poole (1925 a) says that in New Guinea the establishment of peaceful conditions under civilized government has led to a shortening of the cycle of cultivation so that the natives tend to clear the secondary forest before the fertility of the soil has been sufficiently restored. Whether the fields are made on old, or always on new land, the area of 'fallow' (or secondary vegetation) is generally very large compared with that actually in cultivation and thus from the point of view of forest conservation and land usage the system is exceedingly wasteful.

Plantation agriculture in the tropics is, or aims at being, permanent, but in fact it has often been little less temporary than the shifting cultivation of the natives and has led to a reckless expenditure of soil fertility. When a plantation ceases to be profitable, owing to soil exhaustion, erosion, plant diseases or economic causes, it is abandoned and a secondary succession begins as on an old native field. Long-lived crops such as coffee or rubber may survive for a time in competition with the second-growth vegetation, but the course of the succession is essentially the same as that due to shifting cultivation. In most tropical countries there are large areas of abandoned plantations showing various stages of secondary successions.

Forest destruction is not always due to cultivation. There has been some controversy as to whether unfelled rain forest (in contrast to drier types of tropical forest) will in fact burn. It is probable that fires in the rain forest are always local and can take place only under exceptional weather conditions, but fires, not as a rule leading to the complete destruction of the trees, are certainly not uncommon in the rain forest of British Guiana and doubtless also occur in other parts of the tropics. Patches of forest are sometimes felled by strong winds, such as tornadoes and hurricanes, and secondary successions may be started in this way, e.g. in the West Indies.

The treatment an area has received before secondary succession begins may thus vary very much. The subsequent course of events will depend on the size of the cleared area, its distance from uncleared areas, the local conditions of climate, topography and soil type, and on the degree of humus destruction, soil erosion and soil impoverishment which has taken place. Soil impoverishment will in turn depend on several factors, especially how long and in what way (if at all) the area has been cultivated, whether it has been burnt and the quantity of debris and 'slash' left on the ground after clearing. Soil erosion is affected by the slope of the ground, the amount and intensity of the rainfall and many other factors. It is not surprising that the course of the secondary succession and the composition and physiognomy of the communities which may develop also vary very widely.

Areas of primary rain forest subjected to selective exploitation of timber (selection felling) may be termed depleted forest. The gaps left by the removal of the timber trees become colonized by secondary forest species and the community thus comes to consist of a patchwork of primary and secondary forest of very irregular structure. Depleted forest of this kind occupies large areas, for instance in the mahogany-producing districts of Nigeria (Richards, 1939, p. 12) and in the greenheart region of British Guiana; a similar community in Borneo (*gelichteter Urwald*) is described by Winkler (1914, p. 205). Sometimes the effect of depletion is to cause a vigorous outburst of growth in the undergrowth already present, without much invasion of species from elsewhere.

GENERAL FEATURES OF SECONDARY RAIN FOREST

Before considering the successions themselves, it will be worth noting briefly some of the general characteristics of 'typical' secondary forest, i.e. the earlier seral stages found on areas which have been cultivated or exploited for timber, but not subsequently grazed or burnt. Though, as has been emphasized, secondary communities on rain-forest sites are extremely varied, secondary forest in this sense has well-marked features and is as a rule easily recognized.

In the first place secondary forest is lower and consists of trees of smaller average dimensions than those of primary forest (cf. the term 'low bush'). It would be useless to give definite figures, but it is obvious that a forest consisting mainly of smaller trees than in forests on similar sites in the same region must always be suspected of being secondary. Since it is comparatively seldom that an area of forest is clear-felled or completely destroyed by fire, occasional trees much larger than the average are usually found scattered through secondary forest. Seen from above it has a more level 'surface' than primary forest. For example, the young secondary forest 'fallow' in the cultivated area of the Tinjar valley in Borneo looks like a smooth lawn when seen from Mt Dulit 1200 m. above; in the more darkly coloured primary forest the crowns of individual trees can be distinguished at a much greater distance.

Very young secondary forest is often remarkably regular and uniform in structure, though the abundance of small climbers and young saplings gives it a dense and tangled appearance, unlike that of undisturbed primary forest, and makes it difficult to penetrate. At a somewhat later stage in the succession an extremely irregular structure is characteristic; tangles of lianes, razor grass (*Scleria* spp.) and similar plants are scattered haphazard among tall stands of trees with little undergrowth and casting a deep shade. Irregular structure, as has been mentioned, is one of the distinguishing features of depleted forest and it might be expected that where secondary succession starts on an area more or less completely cleared of its original vegetation the stand would always be even-aged and regular. In the earliest stages of succession even-aged stands of one or a few woody species are in fact quite common. Thus in south Sumatra van Steenis (1933) met with a secondary forest with a remarkably regular two-layered structure, the upper story consisting of trees of *Macaranga* sp. (*bancana*?) about 10 m. high, the lower of a dense thicket of *Lantana camara* 2–3 m. high; this community was presumably a stand of uniform age which had grown up on an area completely cleared of its previous vegetation.

In course of time these even-aged stands of pioneer species become senescent and often they are unable to regenerate under the new ecological conditions they have created. In this way most of the trees over a considerable area may become simultaneously liable to wind-throw or death from some other cause and large gaps may then arise which become temporarily occupied by climbers and only gradually filled by the slower-growing trees which dominate the next phase of succession. This is one of the causes of the abundance of lianes which is characteristic of secondary forest. Where the pioneers include a variety of species with different natural spans of life the gaps arise at different times and thus produce the patchy and irregular structure characteristic of later stages of the succession. The even mixture of age classes found in mature forest is only attained after a long period of development.

Secondary forest can generally be even more readily recognized by its floristic composition than by its structure or physiognomy. It is sometimes stated that the dominants characteristic of secondary forest are absent in primary forest. This is untrue; there are probably few if any species entirely restricted to secondary forest, though many primary forest species are absent from young secondary forest. Secondary species occur in small numbers in natural openings in primary forest, such as the 'holes' made by the death of large trees; they are, however, far more abundant in secondary than in primary forest, so that the general 'floristic picture' is different. The composition of secondary forest will be considered in detail when the succession itself is discussed, but meanwhile it may be noted that in tropical America the abundance of such trees as *Cecropia* spp., *Vismia guianensis*, *Miconia* spp., certain species of *Inga* and *Byrsonima* and many others is an indication of secondary forest. In West Africa *Musanga cecropioides*, *Trema guineense*, *Fagara macrophylla*, *Harungana madagascariensis*, *Macaranga barteri*, *Pycnanthus angolensis*, *Sterculia tragacantha*, etc., and in Malaysia species of *Macaranga, Mallotus, Glochidion, Trema, Rhodamnia, Elaeocarpus*, etc. have a similar significance. Burkill's (1919) analysis of the composition of secondary forest about thirty years old at Singapore shows that there is a very different representation of families as compared with primary forest. In the secondary forest the families represented by the greatest number of species were Euphorbiaceae and Urticaceae; those represented by the greatest number of individuals the Myrtaceae, Rhizophoraceae, Lauraceae and Urticaceae (in this order). None of these, except the Myrtaceae and Lauraceae, is among the most abundant families in primary forest in Malaya. The Dipterocarpaceae, so characteristic of the Malayan climax forest, were entirely absent in this secondary forest and the Rubiaceae, which are abundantly represented in the primary forest, were only tenth in order of number of individuals.

There are no suitable figures for comparing the number of species per unit area in secondary and primary forest, but there is no doubt that secondary forest is much poorer in species than primary, unless very small areas are considered. A rough index of the difference is Chevalier's statement (1909) that the secondary forests of the Ivory Coast consist of about thirty species of trees and the 'virgin' forest of 250–300 (the latter figure is a serious under-estimate, but the former is probably not far from correct). Secondary forest is sometimes, but by no means always, dominated by a single, or a small number of species. As was mentioned in Chapter 11 many of the types of Rain forest with single dominants which have been described may be stages in secondary or other successions, or fire-climaxes.

Secondary forest (in the restricted sense which we are now considering) thus has well-marked characteristics by which it may be recognized. Old secondary forest is difficult or impossible to distinguish from undisturbed virgin forest, but since secondary forest consists of stages in a succession leading ultimately to the re-establishment of the climatic climax, this is only to be expected. Biotic climaxes resulting from deflected successions may closely resemble other climatic climaxes, though not the rain forest from which they are derived. The diagnosis of a community as secondary by criteria such as those just considered may sometimes be confirmed by other evidence, such as the remains of buildings, native fortifications, buried fragments of pottery and other evidences of human occupation. In the Mazaruni district of British Guiana, Wood (1926) found that 'low bush' was sometimes found in places where the Indians said there had never been cultivation. It was found that the creeks giving access to such areas always showed signs of having been cleared for navigation; it could therefore be assumed that all the 'low bush' occupied sites where primary forest had been cleared or interfered with by human activities. The extensive tracts of heath-like vegetation (see below, p. 396) near Lake Toba in Sumatra and on the Dijëng plateau in Java, though at first sight looking like a stable climatic climax, betray their secondary origin (from Rain forest, following cultivation and burning) by the presence of stumps of former forest trees (van Steenis & Ruttner, 1932, p. 245).

CHARACTERISTICS OF SECONDARY FOREST TREES

The dominant trees of the earlier stages of the secondary succession differ markedly in various respects from the dominants of primary forest. Secondary forest trees in fact share a number of common characteristics which are independent of their taxonomic affinities; thus, like primary forest dominants (see Chapter 4), they form a natural biological group. Some of these characteristics are clearly adaptive, in the sense that the species possessing them could not play their special role in the vegetation without them. There are perhaps also other common characteristics the significance of which is less obvious.

In the first place, almost all secondary forest trees are light-demanding and intolerant of shade. They grow well in any opening or clearing of sufficient size, but ultimately they become suppressed by shade-tolerant species belonging to primary forest and the later stages of the secondary succession. Many secondary species, as will be seen later, cannot regenerate in their own shade, so that a community dominated by them necessarily lasts for only a single generation.

There are two other features of secondary forest trees, both of which are an essential part of their biological equipment, namely efficient means of seed dispersal and rapid growth. Efficient dispersal enables them to colonize openings and clearings as soon as they are made. Before large areas of primary forest were destroyed by man (a very recent event), the only habitats available for secondary species were openings, bare spots on riverbanks, etc., mostly small and temporary; rapid colonization was therefore necessary for their survival. Quick growth enables the species of the earlier stages of the succession to establish themselves before they are shaded out by the slower-growing secondary forest species.

Data for an accurate comparison of the distribution of different types of dispersal mechanism among primary and secondary forest species are unfortunately lacking. A careful statistical study of the subject would be of great interest and might help considerably in determining the age of late stages in secondary successions. It is, however, a matter of common observation that the majority of the trees characteristic of young secondary forest have seeds or fruits well adapted for transport by wind or animals. Thus, among common tropical American secondary species, *Vismia guianensis* is dispersed by birds and fruit-bats, *Didymopanax morototoni*, *Guazuma ulmifolia*, *Miconia* spp., *Byrsonima* spp., etc. by birds, *Ochroma* spp. by wind. Among common West African secondary forest trees *Musanga cecropioides* (like its American counterpart *Cecropia*) is very efficiently dispersed by birds and fruit-bats, *Trema* has very small light seeds, *Pycnanthus angolensis* and probably *Macaranga barteri* are bird-dispersed. Among common Asiatic species *Alstonia* and *Anthocephalus* are wind-dispersed; *Melastcma malabathricum* and most of the species of *Macaranga*, *Mallotus*, *Trema*, *Rhodamnia* and *Rhodomyrtus* are bird-dispersed. The dominants of primary forest on the other hand, though they often have winged or light seeds (see Chapter 4), are comparatively seldom animal-dispersed (except perhaps by monkeys) and frequently have heavy fruits or seeds with no special mechanism for dispersal; as has already been noted (p. 94), their means of dispersal are sometimes so poor as actually to limit their spread into suitable habitats.

Though the efficiency of their dispersal is doubtless the fundamental reason why secondary forest species are able so quickly to colonize any opening in the primary forest, however distant from the seed-parents, Symington's (1933) demonstration that dormant seeds of belukar (secondary forest) species are present in the surface soil of primary forest which has never, as far as known, been previously cleared or disturbed, is of great interest. In this connexion the large seed output of secondary forest species, and the fact that most of them seem to fruit the whole year round, may be important. Salisbury (1942, p. 209) has shown for the British flora that species of 'intermittently available habitats' have a strikingly and significantly larger seed output than those of permanently available habitats, especially those which are shaded. An analogous difference between the species of primary and secondary tropical forests would almost certainly be found. The ecesis of secondary species is always effected by seeds or fruits but their rapid spread once they have arrived is sometimes due to vegetative reproduction, e.g. *Musanga* spreads by runners (Chipp, 1913).

The quick growth of secondary forest has often been commented on. There are many scattered observations on the extraordinarily rapid growth of certain species. Thus Docters van Leeuwen (1936, p. 72) records that *Macaranga tanarius*, one of the commonest Malayan second-growth trees, grew to a height of 8 m. in 2 years at Buitenzorg. Symington (1933) found that eight individuals of the same species in Malaya grew to an average height of 11 m. in 3 years; *Trema* 'amboinense' grew to 7 m. in the same period. Chipp (1927, p. 58) saw trees of *Ceiba pentandra* 12 m. high on farmland in the Gold Coast which had been abandoned for only 3 years. In Southern Nigeria Ross (unpubl.) found that *Musanga cecropioides* reached a height of 24 m. and a girth of 2–3 m. in 15–20 years. The tropical American balsa (*Ochroma* spp.) grows equally fast (Cox, 1939). The average height of the 30-year-old secondary forest at Singapore studied by Burkill (1919) was 14–15 m. An accurate comparison of the growth-rates of secondary and primary forest trees is possible from the admirably thorough measurements of the growth of trees in the climax Dipterocarp forest and the neighbouring parang (secondary forest) made in the Philippines by Brown & Matthews (1914) and Brown (1919). The results are summarized in Table 38. These data show that the growth-rate of secondary forest species, especially those which when mature reach a diameter of over 20 cm., is consistently higher than that of Dipterocarp forest species; the average rate for the former is about two or three times as large as for the latter. According to Brown, an average secondary forest tree in the area he studied reaches 30 cm. diam. in 14½ years; this is about four times as fast as the average for common North American timber trees. The rapid growth of secondary forest trees is possible because they are unshaded during the early stages of their development; on the other hand typical primary rain-forest trees do not grow as fast even when conditions are favourable.

TABLE 38. *Average annual increase of diameter of secondary and primary rain-forest trees at Mt Maquiling, Philippine Islands.*
(After Brown, 1919)

	0–5	5–10	10–20	20–30	30–40	40–50	50–60	60–70	70–80
Parang (Secondary forest) species: (a) Not reaching a diameter of 20 cm.									
Dysoxylum decandrum	—	—	—	—	—	—	—	—	—
Ficus hauili	1·76	2·01	2·28	—	—	—	—	—	—
F. ulmifolia	0·36	1·89	—	—	—	—	—	—	—
F. sp.	0·28	—	—	—	—	—	—	—	—
Leea manilensis	0·58	0·44	—	—	—	—	—	—	—
Leucaena glauca	—	1·29	—	—	—	—	—	—	—
Melanolepis multi-glandulosa	—	0·82	—	—	—	—	—	—	—
Melochia umbellata	—	3·27	—	—	—	—	—	—	—
Mussaenda philippica	0·77	—	—	—	—	—	—	—	—
Pipturus arborescens	—	1·81	—	—	—	—	—	—	—
Premna cumingiana	0·50	0·92	—	—	—	—	—	—	—
Psidium guajava	0·22	0·77	0·92	—	—	—	—	—	—
Voacanga globosa	0·66	1·60	—	—	—	—	—	—	—
Average ...	0·64	1·40	1·60	—	—	—	—	—	—
(b) Reaching a diameter of more than 20 cm.									
Alstonia scholaris	—	3·38	—	—	—	—	—	—	—
Antidesma ghaesembilla	—	—	1·97	—	—	—	—	—	—
Artocarpus cumingiana	1·00	1·55	1·77	—	—	—	—	—	—
Bauhinia malabarica	—	2·25	2·40	2·81*	—	—	—	—	—
Canarium villosum	—	1·93	—	—	—	—	—	—	—
Columbia serratifolia	2·09	1·91	2·05	2·80*	—	—	—	—	—
Cordia myxa	—	—	1·76	—	—	—	—	—	—
Ficus nota	—	—	2·37	—	—	—	—	—	—
Glochidion philippicum	—	1·96	—	—	—	—	—	—	—
Litsea glutinosa	0·76	1·86	—	—	—	—	—	—	—
L. perrottetii	—	—	2·15	—	—	—	—	—	—
Mallotus philippensis	—	—	1·43	—	—	—	—	—	—
M. ricinoides	1·70	1·89	1·79	3·59*	—	—	—	—	—
Polyscias nodosa	—	1·70	—	—	—	—	—	—	—
Premna odorata	—	––	2·12	—	—	—	—	—	—
Sapindus saponaria	—	1·68	—	—	—	—	—	—	—
Trema orientalis	—	—	4·03	1·71*	—	—	—	—	—
Average ...	1·39	2·08	2·19	2·59	—	—	—	—	—
Mixed Dipterocarp (primary) forest species									
Dillenia philippinensis	0·46		0·55	0·39	0·24	—	—	—	—
Diplodiscus paniculatus	0·15		0·31	0·44	—	—	—	—	—
Parashorea malaanonan	0·07	0·27	0·38	0·49	0·74	0·82	0·94	0·75	0·84
Parashorea malaanonan (in open)	0·42	0·55	0·73	—	—	—	—	—	—

* These measurements refer to the '20 to 30 or 35' class.

The small maximum height and diameter reached by secondary forest trees is in no way connected with their rate of growth, but is a consequence of their short life. Average secondary forest trees in the Philippines, according to Brown (1919, p. 138), attain their full size when the stem is 30–35 cm. diam. and then, or soon after, begin to lose vitality. The exceptionally fast-growing *Melochia umbellata*, one of the smaller trees of the secondary forest, reaches its greatest diameter and dies probably within 3–4 years (ibid. p. 148). On this point further information is much to be desired, but it is evident that in general secondary forest trees are short-lived, maturing and reproducing early. In secondary Rain forest in Nigeria, Ross (unpubl., see p. 387) found that the number of individuals of *Musanga cecropioides* on sample plots $17\frac{1}{2}$ years old was much less than on plots $14\frac{1}{2}$ years old; this was because the tree begins to die (not from disease or any obvious cause) between 15 and 20 years of age and cannot regenerate in its own shade.

The soft texture and low density of the timber which is a well-known character of the majority of secondary forest trees is clearly a consequence of their rapid growth. In many secondary forest trees the wood is exceedingly soft and light; balsa (*Ochroma* spp.), which is common in the secondary forest of tropical America, has one of the lightest woods known. It is often supposed that secondary forest timbers, owing to their softness, are of little economic value; there are, however, many exceptions, including, in addition to balsa, okoumé, *Aucoumea klaineana*, a tree which forms extensive pure stands on the sites of abandoned cultivation in the Gaboon and has a valuable timber used on a large scale for veneers. In the Dry Evergreen forest region of south-western Nigeria three of the commonest species colonizing abandoned farmland are *Chlorophora excelsa*, *Terminalia superba* and *Triplochiton scleroxylon*, all of which provide useful timber.

Besides the characteristics just discussed—intolerance of shade, efficient seed dispersal and rapid growth—secondary forest trees as a group have other characteristics which are less obviously adaptive. Little attention has been given to these features, but general observation suggests, for instance, that there are differences in the nature of the foliage in primary and secondary forest. The leaves of primary forest trees, as has been seen (Chapter 4), are remarkably uniform in size, shape and texture. Those of secondary trees are less uniform; they are paler in colour, a fact noted long ago by Spruce (1908, **1**, p. 355), and a statistical study would probably show that divided or non-entire leaves are more frequent than in primary forest. A comparison of leaf sizes in primary and secondary forest might also give interesting results. Brown's studies in the Philippines (Brown, 1919, pp. 378–90) showed that there was a greater percentage of large-leaved plants in the parang (secondary forest) than in the virgin Dipterocarp forest. On the other hand, Burkill's (1919, p. 152) measurements of the sizes of the leaves of twenty woody species 2 ft. (61 cm.) high and over in secondary forest 30 years old at Singapore showed that the majority of the species fell into Raunkiaer's 'microphyll' class; two to four species had 'mesophylls' and one species 'nanophylls'. In primary forest the great majority of the species have 'mesophylls' (p. 81). No definite conclusions can be drawn from these observations, but it may well be that among secondary forest trees there is a larger proportion of species both larger and smaller than the 'mesophyll' class as compared with primary forest trees, i.e. leaf size is less uniform.

It is an interesting fact that species characteristic of 'drier' (i.e. more seasonal) types of forest are sometimes common in seral stages in rain-forest areas. In West Africa, for example, *Chlorophora excelsa* and several other common dominants of Dry Evergreen forest (see p. 338) are frequent in clearings and secondary forest in Rain forest of the Wet Evergreen type. In south India *Kydia calycina*, a tree usually found in Dry Deciduous forest, is common in clearings in Wet Evergreen forest (E. C. Mobbs).

ORIGIN OF THE SECONDARY FOREST FLORA

A thorough study of the affinities and geographical origin of the secondary forest flora is much needed. In general the constituents of secondary forest are more widely distributed than those of primary forest. Some are pan-tropical or range through the whole neotropical or palaeotropical region; very few have an endemic or localized distribution. The species of primary forest on the other hand are often very local in distribution; very few of them are found in both the Old and the New Worlds and practically none is pan-tropical. In many parts of the tropics some of the most important species in the earlier stages of the secondary succession are cultivated species which have become naturalized or aliens accidentally introduced from abroad. Thus in Malaysia three of the most abundant plants in young secondary communities are the herbs *Eupatorium pallescens* and *Ageratum conyzoides*, and the shrub *Lantana camara*; all three are quite recent introductions from tropical America. On many of the Pacific islands the original lowland Rain forest has disappeared or been greatly reduced in area; the site of the forest, where uncultivated, is largely dominated by introduced plants, many of them of American origin. In Fiji, for instance, large areas of former forest are covered with a scrub of *Psidium guajava*, *Lantana*, *Melastoma malabathricum* and other aliens.

The wide distribution of secondary forest species is no doubt chiefly due to their good dispersal and efficiency in colonizing disturbed ground. The biological equipment which made them able, before the Rain forest was extensively cleared, to take advantage of any opening or temporary habitat available, has also enabled them, under modern conditions, to cover enormous areas and often to attain a world-wide distribution. Van Steenis (1937) regards the flora of all 'anthropogenic' plant communities in Malaysia (including in this term the dominants of fire-climaxes such as *Melaleuca leucadendron*, *Pinus merkusii* and *Tectona grandis*, as well as the belukar flora) as selected by the environment from the general mass of species inhabiting mixed forest. On this view all secondary forest species originally existed in small numbers in primary Rain forest; the opportunities provided by large-scale clearing and burning have merely selected out those species able to take advantage of them. This interpretation may well be correct for the majority of secondary forest species, but some, especially the alien and pan-tropical species, are very probably invaders from other plant formations and not originally natives of the rain forest at all.

OBSERVATIONS ON SECONDARY SUCCESSIONS

As has already been pointed out, few detailed observations on secondary successions leading to Tropical Rain forest have been published. There are various incomplete observations from which, *faute de mieux*, a picture of the general course of events must be built up. Among these probably the most instructive are those made by Mr R. Ross in Nigeria in 1935, which have not so far been published.

Secondary successions in Nigeria

The work of Ross was carried out in the Shasha Forest Reserve, a low-lying area forming part of what was once a continuous belt of Rain forest (at this point some 60 km. wide), running parallel to the coast immediately inland from the outer fringe of lagoons, mangrove and Fresh-water Swamp forest. The topography, climate, soils and primary vegetation of the area have been fully described elsewhere (Richards, 1939) and some details of the structure of the mature forest will be found earlier in this book (see p. 29 n.). There is a relatively severe seasonal drought (annual rainfall 208 cm., with five successive months with less than 10 cm.) and the primary forest is of the Wet Evergreen rather than 'true Rain forest' type (see p. 338); there are, however, no reasons for believing that the secondary successions are different in any important respect from those which would take place in a less markedly seasonal climate where the climatic climax is 'true Rain forest'.

In the Shasha Reserve there are scattered villages and hamlets dating from the period before the reserve was constituted. Each village is surrounded by an enclave in which cultivation is permitted. In the immediate neighbourhood of the houses there are permanent plantations of cola and fruit trees, elsewhere the land is under a typical system of shifting cultivation. Since the enclaves are not large in relation to the population, new land brought into cultivation has always been previously farmed at a fairly recent date; practically all the uncultivated land of the enclaves is under secondary forest of various ages. When the land is cleared for cultivation a few large trees, particularly hard-wooded species such as *Lophira procera*, are left unfelled; these may be subsequently killed by fire. All the debris except the larger tree trunks is removed by burning. The chief crops are maize, yams, cassava, bananas, plantains and coco-yams

(*Colocasia*); small quantities of other species are also grown. After a few seasons the ground becomes unprofitable, a new field is made and the secondary succession begins on the old one.

By studying a series of sample areas on comparable sites where the date when cultivation was abandoned is known, it is possible to reconstruct the early stages of the succession with some accuracy. In Ross's work plots measuring 200 × 100 ft. (60 × 30 m.), which had been left uncultivated for $5\frac{1}{2}$, $14\frac{1}{2}$ and $17\frac{1}{2}$ years respectively, were examined, and one of unknown, but probably greater, age. The following account of the succession is based on these results, together with less detailed observations made elsewhere in the area.

When cultivation is abandoned the ground is bare, except for a few surviving forest trees and perennial crop plants such as bananas which persist for a while, but in a few weeks it becomes covered with a dense mass of low vegetation; after a few years it is covered with secondary forest about 16 m. high. While the ground is bare any species within the dispersal range may invade the area, provided its seedlings can establish and maintain themselves. The conditions for seedling establishment, it should be remembered, are very unlike those in the primary forest. Competition from other plants is minimal, but the ground is exposed freely to sun, wind and rain; illumination and the daily range of temperature and humidity are consequently greater. In addition, during the period of cultivation the soil will have been directly and indirectly modified in various ways. These environmental differences will be considered more fully later.

The primary invaders fall into three groups: (*a*) herbaceous weeds of cultivated land, e.g. *Phyllanthus* and *Solanum* spp.; (*b*) species, mainly woody, characteristic of secondary habitats, but also found in small openings in high forest made by the felling or death of large trees, e.g. *Musanga cecropioides*, *Trema guineense*, *Vernonia conferta* and *V. frondosa*, *Fagara macrophylla*. These are the 'typical secondary forest trees' described on pp. 382–6; (*c*) light-demanding high forest species whose seedlings are able to establish themselves under open conditions, e.g. *Erythrophleum ivorense*, *Khaya ivorensis*, *Lophira procera*.

Group (*a*), the herbaceous weeds, may form a closed stand very rapidly, but their dominance is transient and the part they play in the succession is a very minor one. The species of group (*b*), if not dominant from the first, very soon assume dominance and maintain it for 15 to 20 years. This they can do because they are tall enough to suppress the herbaceous weeds and because their growth is much more rapid than that of the high forest species in group (*c*). *Vernonia conferta* and *V. frondosa*, woody composites with large leaves, are most commonly the first dominants, the former in particular often forming pure stands. At other times *Trema guineense*, and perhaps other species, are the first dominants; one small area cleared of all vegetation in March 1935 had by May become covered with a dense growth of *Trema* seedlings about 1 m. high, mixed with a few smaller seedlings of *Albizzia gummifera*, *Carica papaya* and *Solanum* sp. Which species first attains dominance during the invasion phase is probably to some extent a matter of chance, depending on which species happen to be fruiting most abundantly at the time the site becomes available for colonization (though in view of the work of Symington referred to above, it is possible that an appreciable proportion of the invaders spring from dormant seeds in the soil). As soon as a closed community has formed, the factor of chance presumably decreases and the floristic composition is then determined mainly by competition.

In every area studied *Musanga cecropioides* had become dominant after about 3 years. This tree, the most abundant and characteristic secondary rain-forest species of tropical Africa, is apparently unable to colonize bare ground, but once a cover of vegetation has been formed, its extraordinarily rapid growth gives it a great advantage, in spite of its comparatively open canopy. During the *Musanga* phase of the succession the subdominant tree is *Macaranga barteri*, another group (*b*) species which does not grow as tall or as fast as *Musanga*. It is possibly significant that both *Musanga* and *Macaranga* are stilt-rooted trees. Beneath the shade of the tree-layer there is a dense second story of shrubs and young trees, including *Conopharyngia penduliflora*, *Discoglypremna caloneura* (on yellow earth soils), *Rinorea* spp. and *Rauwolfia vomitoria*. After 5 years the second story is already 4–5 m. high. During the early years of the *Musanga* phase, small herbs (*Geophila* spp., *Leea guineensis*, Commelinaceae, etc.) are much more abundant than in the primary forest and form a closed carpet. Later, owing to the increased shade (and perhaps owing to increased root competition) the herbaceous ground vegetation becomes more and more sparse.

After 15–20 years the *Musanga* trees die; thus on the $17\frac{1}{2}$-year-old plot the number of individuals was already considerably less than on the $14\frac{1}{2}$-year-old plot. The cause of death seems to be merely senescence; just before they die the trees show a scaling of the bark on the stilt roots, but no other symptoms of disease. Dead and dying trees are commonly blown over during the tornado season (March to May) and at this time of year prostrate trees can be seen everywhere in secondary forest about 15 years old. The dead trees have no successors of their own species, since *Musanga* is apparently unable to regenerate in its own shade. The *Musanga* consocies is thus a single-generation community.

By now, however, various other species have reached a height of 20–25 m.; among these are *Albizzia gummifera*, *Anthocleista vogelii*, *Diospyros* 'confertiflora', *Discoglypremna caloneura*, *Funtumia elastica* and *Sarcocephalus diderrichii*. Some of these are group (*b*) species and are normal constituents of the high forest. Young individuals of *Diospyros insculpta*, *Erythrophleum ivorense*, *Lophira procera* and *Scottellia* sp. are also present. On the $17\frac{1}{2}$-year-old plot there were thirty-seven group (*c*) species and twenty-seven group (*b*) species; on the $14\frac{1}{2}$-year-old plot there were only twenty-five group (*c*) to twenty-nine group (*b*) species.

Thus the secondary succession in the clearings of the Shasha Forest Reserve clearly tends towards the re-establishment of the climatic climax. The further progress of development after the first 20 years has not been traced. Its next phase would appear to be a decrease in the abundance of group (*b*) species and a corresponding increase in those of group (*c*), leading to the development of high forest. The young high forest will still be unstable, as it includes an abnormally high proportion of light-demanding species which established themselves under the open conditions at the beginning of the succession; these cannot reproduce themselves in a closed community except in gaps. Recent observations by the author in another part of Nigeria suggest that a further series of changes continues over a long period of years before a stable equilibrium is reached.

The secondary succession thus consists here of three stages: (i) the short invasion stage during which a closed covering of vegetation is established; (ii) the *Musanga* stage, during which a single species holds undisputed dominance for a single generation; (iii) the stage of slow replacement of secondary by primary forest trees. The third stage is certainly much longer than both the others, but its actual duration is unknown. The seral communities, it should be noted, are dominated by woody plants almost from the start; if there is a herbaceous phase at all, it is very brief. Grasses (e.g. shade-tolerant species such as *Leptaspis cochleata*, which also occur in the primary forest) may take a part in the succession, but only a very unimportant one; there is never even a temporary dominance of tall tussock-forming grasses such as *Imperata*, which are so important a feature of secondary and deflected successions in some parts of the tropics. Each stage of the succession prepares the way for the next. Thus the closed cover of vegetation developed during stage (i) provides the necessary environmental conditions for the establishment of *Musanga*. The deep shade of the *Musanga* consocies provides a suitable microclimate for the species which become dominant during stage (iii). Stages (i) and (ii) are characterized by rapidly growing, highly light-demanding species, typical of secondary forest. Stage (iii) witnesses the gradual return to dominance of the slower growing species of the primary forest.

Secondary succession in the Congo

Fragmentary observations in many parts of the evergreen forest region of tropical Africa show that so long as the course of the succession is undisturbed, the secondary succession following temporary cultivation is very similar, in its main features, to that described above for Nigeria. If, however, the secondary communities are subjected to fire or grazing, or if a second period of cultivation intervenes only a few years after the first, deflected successions are initiated in which tussock-forming grasses, especially *Imperata cylindrica* var. *africana* (varietally distinct from the lalang grass of Malaya), usually play the principal part. These deflected successions may culminate in the replacement of the forest by open grassland, with or without scattered trees and bushes. Such grassland, which may be called secondary savanna,[1] can maintain itself indefinitely as a biotic climax if it is grazed or periodically burnt.

A clear picture of the general course of these successions in the Congo is given

[1] Equivalent to the Derived savanna of A. P. D. Jones (see p. 336).

by Vermoesen (quoted by Lebrun (1936a)). Two divergent 'series' of successions, 'progressive' and 'regressive', are distinguished. The floristic composition of the various stages of both 'series' is very variable, depending on local differences of soil and other causes. The 'progressive series' is found where there is a long interval between successive periods of cultivation, the 'regressive series' where there is a short one. The former is a straightforward secondary succession leading to the re-establishment of a forest climax, the latter a deflected succession leading to the more or less complete dominance of grasses or of the fern *Pteridium*.

In the 'progressive' successions the abandoned cultivated ground is at first invaded by herbaceous plants of rapid growth; these complete their life-cycle in a few weeks. This herbaceous stage is followed by one of perennial and 'suffrutescent' herbs; in this herbaceous and woody climbing plants are abundant. Young trees, growing both from seeds and from stumps which have survived in the soil, are also common and some of the short-lived herbs of the first stage are still present. After some months bushes and young trees become dominant and a tree canopy is soon formed, thus beginning the third or tree stage of the succession. Vermoesen recognizes three phases in the tree stage: the first dominated by *Musanga cecropioides*, *Trema guineensis*, *Harungana madagascariensis* and *Pycnanthus angolensis*, the second dominated by species of *Bosqueia*, *Conopharyngia*, *Alstonia*, *Funtumia*, *Albizzia*, *Pentaclethra*, *Sterculia*, *Ricinodendron*, *Fagara*, *Ficus*, etc., and the third marked by the gradual return to dominance of species characteristic of primary forest. Vermoesen estimates the optimum period and duration of these phases as follows: phase 1, optimum at 10–20 years, total duration 20–30 years; phase 2, optimum 20–30 years, duration 50 years, primary forest trees becoming dominant (stage 3) after 60–100 years. Lebrun believes these figures to be an over-estimate. A very similar account is given by Chevalier (1917) of the succession in the Gaboon.

On the Ubangi the typical course of events leading to a 'regressive' succession is as follows: An area of forest is clear-felled and burnt. Several harvests of maize are gathered, but when the yield becomes insufficient the land is abandoned. The 'progressive' succession just described now takes place, but when it has reached the first phase of the tree stage (*Trema* is here the usual dominant) the ground is cleared once more and another crop is grown. During the herbaceous stage of the succession following the first cultivation the grass *Imperata* is found as isolated tufts which do not flower; during the succeeding bush and tree stages these are suppressed, but not killed, and benefit from the second period of cultivation. Immediately after the land is abandoned for the second time *Imperata* becomes dominant and may form an almost pure consocies and this community persists more or less indefinitely as a biotic climax, appearing quite stable, though established and maintained artificially. On moist clayey soils *Imperata* is replaced by *Pennisetum* savanna and on sandy ground the secondary grassland is replaced by extensive stands of *Pteridium*.

As will be seen below, the method of origin of these secondary communities of grasses or *Pteridium* in the Congo differs hardly at all from the method of origin of some of the almost identical communities in Malaya, New Guinea and other parts of the tropics.[1]

Secondary and deflected successions in the Malayan region

In the Malayan region there are enormous areas of secondary forest (belukar), and in some places scrub and grassland, which have originated from rain forest, but few systematic studies have been made of the successions in which these secondary communities are stages. There is little reason to doubt, however, that in this part of the world, as in tropical Africa, when the primary forest is destroyed, the succession leads to the re-establishment of the climatic climax, whether a period of cultivation follows or not. When the secondary communities are subjected to grazing or repeated burning, or when periods of cultivation succeed one another after a very short interval, the succession is deflected and apparently stable biotic climaxes, dominated usually by grasses, make their appearance. The natives often set fire to the secondary grasslands at regular intervals because the young shoots of the grasses which come up after burning are specially valuable for grazing.

A very full description of the early stages in a secondary succession is given by Jochems (1928) in his account of the vegetation colonizing fallow tobacco land at Deli (Sumatra). This district is noted for the fine quality and high yield of its tobacco and an extremely elaborate system of cultivation is practised. Each plot of land bears a crop of tobacco not more often than once in 8 years. The tobacco crop is usually followed by a single native crop of hill (unirrigated) rice or maize, but apart from this the land is left entirely fallow until the next tobacco crop is planted; the secondary succession thus follows its natural course for 7 years completely undisturbed. On some plantations attempts have been made to substitute 'artificial belukar' of planted trees useful as small timber or as soil improvers (Leguminosae) for the useless belukar (secondary forest) which develops spontaneously during the fallow period. The tobacco-planting season is about January and some months before planting begins the plot is cleared of belukar and the trees are stacked and burned. The soil is now cultivated to a depth of about 50 cm. and a luxuriant growth of weeds springs up or a cover crop of *Mimosa invisa* is sown. Before planting of the tobacco actually starts the weeds or cover crop are removed and the soil is thoroughly cultivated once more. The tobacco now remains on the ground till about July; during the early part of this period the land is kept clean, but towards the harvest weeds are

[1] It may be added that on the borders of the Budongo Rain forest in Uganda (Eggeling, 1947), where the forest is expanding at the expense of the surrounding grassland, the succession from grassland to forest differs considerably from the African successions described above. It is not certain whether the Budongo succession is really to be regarded as secondary, as it is not known whether the Rain forest is reoccupying areas from which it was formerly cleared or whether, owing to climatic change or some other reason, it is invading entirely fresh ground.

allowed to come up. If a rice or maize crop is to follow the tobacco, the land is cleared, leaving only scattered individuals of young secondary forest trees. The natives harvest their rice or maize in December. During the long fallow period which now ensues a succession occurs in which three stages can be recognized: (i) the herbaceous stage, a floristically rich community consisting mainly of short-lived herbs and young trees; (ii) the *Trema-Blumea* stage, in which the chief plants are the small trees *Trema* sp., *Blumea balsamifera* and *Abroma angustum*; (iii) the tree stage which is typical belukar. When no rice or maize crop is grown stage (i) is less fully developed than otherwise and stage (ii) is eliminated, so that stage (i) passes directly into stage (iii). Stage (iii) is dominated by one or more of a small number of species; according to the floristic composition the following five types of belukar are distinguished: *Trema* (*Trema* sp.), *Macaranga* (*M. tanarius* or *M. denticulata*), *Callicarpa* (*C. tomentosa*), *Melochia* (*M. umbellata*), mixed (no single dominant). The following are the chief subordinate tree species in all five types: *Commersonia bartramia, Ficus fistulosa, F. toxicaria, Millettia atropurpurea, Pithecellobium lobatum*. The chief shrubs are *Desmodium polycarpum* and *Lantana camara*, the chief herbaceous species the fern *Dryopteris arida* and lalang grass, *Imperata cylindrica* var. *major*. Herbaceous lianes play a large part in these communities, the commonest species being *Argyreia capitata, Lygodium scandens, Merremia vitifolia, M. umbellata* and *Pericampylus glaucus*. Jochems gives full details of the composition of a number of sample plots.

In some parts of the tobacco land seedlings of the large tree *Koompassia excelsa*, a constituent of the adjoining primary forest, are abundant. This depends on the proximity of seed-parents, but Jochems regards the presence of these seedlings as evidence that the succession would lead ultimately to the re-establishment of primary forest if it were not cut short by the next period of cultivation. It is noteworthy that though the grass *Imperata* is common on the Deli tobacco land, it rarely becomes dominant; according to Jochems it only does so if the secondary vegetation is burned. The underground stems of this grass are resistant to fire and after a fire it thus gains an advantage over its competitors.

The succession on the tobacco land of Deli, though not untypical of Malaysian secondary successions in general, is doubtless determined to some extent by the peculiar and complex system of agriculture in which it plays a part.

A useful, but incomplete, study of secondary vegetation at Kepong in the Malay Peninsula has been published by Symington (1933). In this locality secondary communities at various stages of development occupy abandoned vegetable gardens on land which was under Rain forest not many years previously. These have been subjected to various degrees of interference by burning, grass-cutting, etc. It is inferred that the first stage of the succession was colonization by herbaceous plants, including grasses and creepers. This is followed by the dominance of shrubs, such as *Melastoma polyanthum* and the introduced *Lantana camara*, and then by typical secondary forest of *Macaranga* and similar trees which undergoes a gradual change continuing for many years until a community approximating to the high forest climax is developed. Though the general course of the succession can thus be reconstructed fairly easily, several important problems remain to be solved, especially why *Imperata* becomes dominant, arresting the succession for long periods, in some places and not in others. A series of sample plots (area 1 chain = 20 m. square) was marked out and studied over a period of 3 years in an attempt to throw light on these problems, but it was not possible to pursue the work to a definite conclusion.

Two plots were on land which had been cultivated for some years and abandoned shortly before observations began. When first marked out these plots were a mixed community of about fifty-two species, including *Trema* 'amboinensis', *Macaranga tanarius*, *Bridelia tomentosa* and several other young trees, *Musa malaccensis*, shrubs, a few surviving cultivated plants, and herbaceous plants of various kinds, including ferns, sedges and grasses. There were some small colonies of lalang (*Imperata*) and one or two patches of the grass *Brachiaria distachya*. During the first few months the plots were under observation generations of herbaceous plants succeeded one another rapidly; the grasses and creepers spread by rapid vegetative growth. After a year the dominance of herbaceous weeds was declining, partly owing to the growth of the trees, shrubs and *Musa*, and partly owing to the increase of *Imperata*, which had spread by means of its rhizomes. After 3 years, when the plots were examined for the last time, trees, sometimes with an understory of *Lantana*, formed a dense canopy with practically no herbaceous undergrowth over part of the area; over the remainder *Imperata* was almost exclusively dominant, though in one place a colony of *Brachiaria* had resisted invasion by *Imperata* for a long time. Only traces of the mixed herbaceous community remained. On these plots *Imperata* had thus established itself as a stage in the succession over part of the area, but not over the whole of it.

Another plot was chosen as the purest *Imperata* community in the neighbourhood; its previous history was unknown. Though at first sight *Imperata* seemed to exclude everything else, scattered shrubs, herbs and creepers were present. The chief change during the 3 years the plot was under observation was that *Imperata* became distinctly less vigorous and over part of the area it was replaced by the fern *Nephrolepis biserrata*. Bushes of *Melastoma* reached a height of 3 m. and now cast sufficient shade greatly to reduce the vigour of the *Imperata* growing beneath them. In this plot there was no evidence that woody species could establish themselves among the *Imperata* after it had become a closed community; the shifting of dominance which appeared to be taking place from *Imperata* to woody species was brought about by the increase of species which had established themselves at an early stage in the succession.

One more of Symington's plots deserves mention; this was laid out in a small clearing on the fringe of the primary forest. This area had never been cultivated and it was chosen in order to see whether the succession on rich forest soil was different from that on the impoverished soil of abandoned cultivated land. The vegetation before clearing consisted of primary forest species with a small admixture of belukar species, invaders from the neighbouring secondary vegetation. In the 3 years the plot was studied a typical secondary forest dominated by *Macaranga tanarius* established itself. The first stage of the succession was dominated by the herbaceous plants *Physalis minima* and *Amaranthus viridis*, beneath which masses of seedlings of *Trema, Musa* and other belukar species had germinated. *Trema* later became dominant, but eventually it was overtopped by the *Macaranga*. It was noteworthy that no *Imperata* appeared on this plot.

Secondary successions in the Philippines have been described by Whitford (1906) and by Brown (1919). A considerable part of the land area of the islands (about 40%) is covered by grassland, most of it *cogonales* or secondary grassland dominated by *Imperata* (here called cogon), or by *Saccharum spontaneum*, a tussock-forming grass of similar habit. Much of the lowlands is also occupied by parang, a mixture of grassland and secondary forest similar to the Malayan belukar. Both *cogonales* and parang probably occupy the site of former evergreen Dipterocarp forest; cultivated land and grassland quickly revert to forest if not interfered with by man and protected from fire. Several successive fires leave grasses in possession of the land and in the parang the trees tend to spread in the intervals between fires; parang thus represents a 'tension belt' between grassland and forest, persisting for years, the trees sometimes gaining and sometimes losing ground. In the grasslands scattered fire-resistant trees (*Antidesma ghaesembilla, Bauhinia malabarica* and *Acacia farnesiana*) occur as single individuals which tend to spread and form clumps. When fire is excluded small trees, shrubs and lianes invade the grassland, but before a closed canopy of trees is formed the tussock-forming *Imperata* and *Saccharum* tend to be replaced by grasses forming taller but less dense stands.

The secondary forest of the Philippines is a 'dense, heterogeneous tangle' of a great variety of small trees, shrubs and lianes. The trees include *Ficus* spp., *Litsea glutinosa, Macaranga tanarius, Melochia umbellata, Trema orientalis* and many others. Most of them are typical small, fast-growing, soft-wooded, secondary forest trees. *Bischofia javanica* is exceptional in being tall and *Psidium guajava* is fairly hard-wooded. The secondary forest varies greatly in composition for no apparent reason. It is presumed that it will develop into climax evergreen forest, but the course of the succession has not been traced. On areas which have been cleared but not cultivated the succession is much less complex than on grassland.

The other available data on secondary successions in those parts of the Malayan region where Rain forest (or some similar type of evergreen forest) is the climax agree well with the general picture given by the results so far summarized. In eastern New Guinea (Lane-Poole, 1925*a, b*) the natives practise shifting cultivation on a cycle not usually shorter than 8–10 years. Between the periods of cultivation dense secondary forest of species of *Trema, Macaranga,*

Geunsia, Pipturus, etc. grows up. Under peaceful conditions the cycle has tended to shorten and this leads to invasion of the land by *Imperata* and other grasses. When these appear the natives tend to burn the land annually to assist them in hunting game. In this way a secondary grassland is soon produced and stabilized as a biotic climax. The extensive grasslands of the Markham (Marr, 1938) and Ramu valleys have undoubtedly arisen in this way.

Abandoned cultivated land in South Bantam (Java) is usually colonized first by *Imperata* and later by shrubs such as *Melastoma malabathricum* and *Lantana* (Backer, 1913). The shrub stage is followed by the development of a typical secondary forest of *Vitex, Grewia, Dillenia,* etc.

Frey-Wyssling (1931) gives an interesting description of a heath or *maquis*-like scrub of such species as *Leptospermum flavescens, Melastoma decemfidum, Rhodomyrtus tomentosa,* etc. found on the plateau of Habinsaran (alt. 800 m. and over) in the Batak country of Sumatra. Similar communities cover large areas in the whole of the Lake Toba region. Frey-Wyssling compares this scrub with the grasslands found under very similar ecological conditions on the Karo plateau and concludes that both communities have taken the place of evergreen (rain) forest, scrub arising where anthropogenic influence (fire and grazing) is less, and grassland where, owing to a denser population, it is more intense.

Other useful data on secondary communities and successions in Malaysia are given by Winkler (1913, 1914), Burkill (1919), Endert (1920), van Roosendael & Thorenaar (1924), van der Laan (1925) and van Steenis (1932, 1933, 1937*a*).

From what has been said it should be clear that it would be useless at present to generalize about secondary successions in the Malayan region, but it is evident that there are considerable similarities between the successions here and in tropical Africa. The secondary forests of the Malayan region are much richer in species than those of Africa and consequently more varied in composition; this doubtless depends on the much greater richness of the Malayan flora as a whole. It may be an important difference between the successions in Malaysia and those of Africa that in the former a stage of dominance by lalang grass (*Imperata cylindrica* var. *major*), sometimes long continued, is apparently often a normal feature of the straightforward undeflected secondary succession, while in tropical Africa *I. cylindrica* var. *africana* would appear to take part only in the deflected successions following over-cultivation, etc. The part played in succession by this species, which is of little economic value and is in fact a serious pest, is one of the most important ecological problems of the tropics. Further investigation may throw more light on Symington's (1933) interesting suggestion that the undesirable *Imperata* phase of the secondary succession might be eliminated by sowing the seeds of suitable secondary forest trees on bare ground. Space will not permit a general discussion of the ecology of *Imperata*, of which a useful review has recently been published (Gray, 1944).

The successions which have been described above are those found on the 'normal' soils of the Malayan region, that is to say on those of tropical red earth or tropical yellow earth type. On podzols where the climax is Heath forest (see Chapter 9, pp. 211–14 and Chapters 10 and 11) secondary communities of different physiognomy and very different composition are found. These have never been fully described, but they are extensively developed in some parts of Borneo, e.g. about Miri and Claudetown in Sarawak. The secondary 'xerophilous Myrtaceae' forest on white sand in East Borneo, referred to by Endert (1925), is evidently a community of this kind; *Tristania obovata* and *Eugenia bankensis* are here the commonest trees. The heath-like padang vegetation of sandy soils in Borneo, Bangka, etc., mentioned in an earlier chapter (p. 211), may possibly be sometimes a biotic climax resulting from a deflected succession.

Secondary successions in tropical America

In clearings and on old cultivated land in the American tropics secondary Rain forest develops which is very similar in structure to secondary Rain forest in the Old World, though of course quite different in its floristic composition. The successions concerned have been little studied, though it is certain that they differ considerably on different soils.

Throughout the wetter parts of tropical America the very rapidly growing, soft-wooded trees of the genus *Cecropia* are a characteristic feature of young secondary forest. These trees may form pure stands which, like the *Musanga cecropioides* communities of tropical Africa, are probably single-generation communities (cf. Bouillienne, 1926, p. 82). The cecropias in fact very closely resemble *Musanga* (to which they are taxonomically related), not only in the shape of their leaves and in their general physiognomy, but also in their ecological role. The grass *Imperata cylindrica*, so important in secondary and deflected successions in the Old World tropics, does not occur in South America except in Chile, but *I. brasiliensis* and other grasses play an analogous part. Scrambling sedges of the genus *Scleria*, known as razor-grass because of their sharp-edged leaves, often form impenetrable tangles in secondary forest in Guiana.

Kenoyer (1929) has reconstructed the course of the succession on old clearings on Barro Colorado Island (Panama). In the first stage of colonization the chief plants are grasses (about twenty species) and sedges (species of *Cyperus* and *Scleria*), but dicotyledonous herbs (Amaranthaceae, Euphorbiaceae, Compositae, Solanaceae, Mimosoideae, etc.) are also abundant. Most of these pioneers are short-lived. After a year large-leaved monocotyledons, especially the gaudy heliconias and the panama-hat palm, *Carludovica palmata*, appear, together with numerous seedlings of the trees which dominate the next stage (species of *Trema, Cecropia, Apeiba, Ochroma, Cordia*, etc.) and various herbaceous plants not present in the first year. Lianes help to make this vegetation, now about man-high, into an impenetrable tangle.

After 2 years, trees, of which the most conspicuous are *Cecropia mexicana* and *C. longipes*, become dominant and a young secondary forest is formed. Other trees found at this stage are *Ochroma limonense* (balsa) and the palm *Attalea gomphococca*. After 15 years this community still persists, but has become much richer in species. *Ficus* spp., eight to ten species of *Inga*, many small melastomaceous trees and shrubs, *Protium* spp. and very many others are now present. Lianes are numerous, including the conspicuous ribbon-like *Bauhinia excisa*. There is an abundant shrubby and herbaceous undergrowth. By this time a number of primary forest species have established themselves, but the community is still much denser and more difficult to penetrate than primary forest. The indications are that the succession would eventually lead to the development of climax forest.

The secondary succession in forest clearings in Trinidad (Marshall, 1934, pp. 38–9) and in the Guianas is probably very similar to that in Panama; the succession in French Guiana has been briefly sketched by Benoist (1925).

In the West Indian islands secondary communities of all ages are very common, but little is known about the successions. Stehlé (1935) says that in Guadeloupe the first colonists of open spaces in forests at low and medium altitudes (similar to the Lower Montane Rain forest of Beard, see p. 364) are weeds such as *Ageratum conyzoides*, *Ipomoea nil*, *Hyptis* spp., *Lantana aculeata*, etc. Two types of older secondary communities are recognized in the 'middle region': (*a*) a secondary forest chiefly composed of *Cecropia peltata*, *Hibiscus tulipiflorus*, *Ochroma pyramidale* and *Oreopanax dussii*, found where the original humus layer still exists, (*b*) a scrub dominated by *Croton corylifolius*, *C. flavens* and species of *Miconia*, on sites of old cultivation where the humus layer has been destroyed, at low altitudes. In (*a*) tree ferns (*Cyathea* spp.) and *Gleichenia* sp. are present; the latter is a common feature of secondary communities in many damp tropical regions. The importance of pteridophytes in secondary vegetation on impoverished soils in damp climates has been emphasized by Wardlaw (1931), who found that in abandoned banana plantations in St Lucia the dominant trees after 2 years were a tree fern, *Cyathea* sp.,[1] *Cecropia palmata* and *Schefflera* sp., accompanied by bushes such as species of *Miconia* and *Peperomia*.

Freise (1938) has given a detailed account of the *capoeira* or secondary forest on abandoned cultivated ground in the coastal region of Brazil (Pernambuco to S. Catharina). The secondary succession here seems to be different in some respects from that in the rain-forest region of Central and northern South America. The first phase is the colonization of the ground by grasses (e.g. *Andropogon* spp., *Chloris bahiensis*, *Brachiaria reptans*) and herbs such as *Thalia* and *Alternanthera* spp. The pioneers are soon joined by climbers of extremely rapid growth (*Canavalia*, *Pachyrrhizus* and *Phaseolus* spp.). Some shrubs survive from the previous cultivation and these together with invading light-demanding trees form islands which tend to spread and shade out the grasses. In the young secondary forest which now develops several species of *Cecropia*, each charac-

---

[1] The abundance of tree ferns in secondary communities at relatively low altitudes is characteristic of the Lesser Antilles (Beard).

teristic of different soil conditions, occur; these live for 8–12 years. Of the remaining secondary forest trees some (e.g. Euphorbiaceae, Sterculiaceae, etc.) reach an age of 25–30 years, others live longer; most are soft-wooded.

Some of the secondary forest species are said never to occur in primary forest and some primary forest species never in secondary. In Freise's view the succession would ultimately lead to a climax of different composition to the original primary forest. If this is so, it may be because the primary forest has been so extensively destroyed that few suitable seed-parents now remain. The greatest age of the secondary forest is estimated to be 150–200 years; in this time there would be five to eight generations of the shorter-lived, and three to five generations of the longer-lived, tree species.

The secondary forest of this region is often interfered with by felling, burning, etc. In this way deflected successions are started, leading to the development of grassland or *Pteridium* communities or to the denudation of the bare rock.

GENERAL FEATURES OF SECONDARY SUCCESSIONS IN THE RAIN FOREST

The detailed observations on secondary and deflected successions which have been described above, although incomplete, allow a few general conclusions to be drawn.

In every part of the Tropical Rain forest the tendency of the secondary successions is towards the restoration of the climatic climax, though fire, grazing and soil deterioration due to over-cultivation may deflect the succession and lead to the stabilization of biotic climaxes.

The first phase of the succession is dominated by weeds, including grasses, which are usually short-lived and are ephemerals rather than annuals since their life-cycle is often much shorter than a year. The next phase may be dominated by shrubs, but dominance often passes almost directly from herbaceous plants to trees. The trees may form a canopy either by mass invasion or by gradual spread from isolated pioneers (the latter seems to occur when grasses such as *Imperata* have previously formed a dense cover). Thus in one way or another a secondary forest of trees predominantly short-lived, fast-growing and wind- or animal-dispersed grows up and this by slow changes develops into a community similar to the climatic climax of the region. Young secondary forest tends to be even-aged and is often dominated by a single species; the first dominants may form a single-generation community which dies without reproducing itself and is followed by communities with other dominants. In time the secondary forest becomes more mixed in age-structure and floristic composition till it approximates to the condition of mature forest.

The secondary successions thus reproduce on a larger scale the changes which occur in the normal regeneration of a primary forest, in which the gaps formed by the death of large old trees are temporarily colonized by the same fast-growing, easily dispersed species as dominate secondary forest. In 'depleted forest' where large numbers of trees have been removed by selective felling, the situation is intermediate between a clear-felled area and a primary forest with naturally formed gaps; the changes which occur are presumably like a telescoped secondary succession.

The course of a secondary succession is to some extent dependent on the intrinsic characteristics of the soil and is different, for example, on tropical red earths and tropical podzols. The history of the ground between the destruction of the original forest and the beginning of the secondary succession is of great importance; if there is an intervening period of cultivation long enough to alter the structure of the soil and destroy its reserves of humus, or if there is considerable erosion, the subsequent succession will be different from that on ground where the secondary communities have begun to develop immediately after the removal of the forest.

A comparison of the secondary successions with the primary xeroseres described in Chapter 12 will show that the two are essentially similar. The greater the soil deterioration during the bare or cultivated period following the destruction of the primary forest, the more closely will the subsequent succession resemble a primary xerosere.

Deflecting factors alter the course of the succession by giving an advantage to certain species over others. This often leads to the replacement of forest by grassland closely simulating an edaphic or climatic climax. Partial destruction of the forest by fire may also give a selective advantage to certain species at the expense of others and so lead to large changes in the floristic composition of the community.

There is of course a considerable general similarity between the secondary and deflected successions in all regions where forest is the climatic climax. In England a sequence of herb, shrub and tree stages in the secondary succession is found similar to that in the tropics. The dominants of seral forest communities in temperate countries share many of the characteristics of secondary rain-forest dominants, such as light-demanding temperament, rapid growth, short life and good means of dispersal. The birches (*Betula* spp.), for example, have much in common with tropical secondary forest trees. The origin by deflected succession of grassland and certain types of pine woods in temperate regions is closely analogous to the corresponding processes in the Rain forest.

Little is known of the time-scale of secondary successions in the tropics. Chevalier (1948) states that the forest on the site of the ancient temples of Angkor Vat in Cambodia, destroyed probably some five or six centuries ago, now resembles the virgin tropical forest of the district, but still shows certain differences. In general it seems clear that the longer the period between the destruction of the primary forest and the onset of the secondary succession and the greater the modification of the soil and the environment in general during this period, the longer the time needed for the re-establishment of the climax.

ENVIRONMENTAL CHANGES DURING SECONDARY
AND DEFLECTED SUCCESSIONS

As has been repeatedly emphasized, the Tropical Rain forest is an excellent example of an ecosystem in which climate, soil, vegetation and fauna are components in an extremely complex equilibrium. It follows that when one of these components, the primary forest, is destroyed, whether partially or completely, by felling and burning, the other components undergo an equally violent change. During the subsequent secondary successions microclimate, soil and fauna undergo development parallel to that of the vegetation until a balance is once more attained. Secondary successions are in fact the evolution of an unstable into a stable ecosystem.

The removal of the forest cover at once changes the illumination at ground-level from a small fraction to full daylight. The temperature range greatly increases and the average and minimum humidity of the air become much lower. There is a change from the complicated system of microclimates characteristic of high forest (Chapter 7) to conditions closely approximating to the standard climate of the locality. Exposure to sun and rain very quickly alters the properties of the soil. Where the slope is sufficient, erosion will begin to remove the surface layers or their finer fractions. The rise in soil temperature leads to a rapid disappearance of humus. According to Corbet (1935) the destruction of humus when the soil is exposed is a chemical rather than a microbiological process; the number of bacteria in the soil is not significantly different before and after the clearing of the forest (see p. 217). Associated with the loss of humus is a very considerable loss of soil nitrogen, mainly in the form of gas (Corbet, 1935).

The decrease in combined nitrogen following forest clearance is only part of the general impoverishment of the soil. As was seen in an earlier chapter, there is a delicate balance in the primary forest between soil and vegetation, plant nutrients circulating in an almost closed cycle between the two, the small amount of soluble nutrients lost by leaching being replaced from unweathered soil minerals. When the forest is destroyed this cycle is broken; if the vegetation is left to rot on the surface of the ground much of the mineral matter contained in it will probably be lost in the drainage water.

If the secondary succession is allowed to start immediately after the forest has been cleared, there is little opportunity for soil erosion and a closed cover of vegetation is formed in a few weeks; the process of soil impoverishment is arrested before it can go very far and soil and microclimate begin at once to return to their original condition. If on the other hand there is a long period of cultivation after clearing, particularly if the crop does not provide an adequate ground-cover, the long exposure of the soil, together with the losses of plant nutrients in the harvested crops, leads to such large changes in the structure, humus-content and nutrient status of the soil that the time needed to restore the equilibrium between soil and vegetation, even when there is no large-scale erosion, becomes much greater. The importance of the length of the cultivation period and the extent of humus destruction to the course and length of the subsequent succession are thus easily understood.

The general nature of the environmental changes outlined above is fairly well known, but little precise information about them is available. From the ecological point of view it would be particularly valuable to have more knowledge of the very complex changes taking place in the soil immediately after the destruction of the forest and during the subsequent successional changes. Almost the only data available on the subject are those of Duthie, Hardy & Rodriguez (1936) in their study of the soils of the Arena Forest Reserve in Trinidad.

The climax vegetation is here a *Carapa guianensis-Eschweilera subglandulosa-Maximiliana caribaea* Mixed forest (a facies of the Evergreen Seasonal forest of Beard, see p. 319). The soil ('caroni sand' type) is loose and sandy in texture. A comparison was made of the soils on four sample plots over a period of 2 years, the plots being as follows: (a) natural forest; (b) forest cleared, but not burned or cultivated; (c) forest cleared and burned, but not cultivated; (d) forest cleared, burned and planted with 'provision crops'. The soil properties chiefly investigated were: the organic matter content, the total nitrogen content, the carbon/nitrogen ratio and the amount of available plant nutrients (determined by measurement of the electrical conductivity of soil extracts made under standard conditions).

It was found that clearing alone probably caused an immediate reduction in the total amount of organic carbon and nitrogen in the soil and a decrease in the carbon/nitrogen ratio. These changes were evident in this particular soil to a depth of 2 ft. (62 cm.). As the felling of the trees was carried out at the beginning of the wet season, the vegetation soon recovered by natural regeneration and the carbon and nitrogen content and the carbon/nitrogen ratio of the soil began to increase. This was perhaps due to the incorporation in the soil of organic matter from the leaves shed by the felled trees. This process continued more rapidly during the following dry season and in the second wet season biochemical changes were mainly confined to the lower soil layers, where the carbon/nitrogen ratio continued to decrease. By the beginning of the third wet season (last date of sampling) the increase in the organic carbon and nitrogen contents was considerable, but the carbon/nitrogen ratio had not reached its original level. It was found that mineral nutrients were liberated abundantly during periods of rapid organic decomposition, but tended to be lost by leaching during the wet seasons.

On the plot which had been cleared and burned, but not cultivated, the amount of organic carbon and nitrogen and the carbon/nitrogen ratio of the soil all remained at a lower level than on the forest which had been cleared without burning. This was shown to a depth of at least 1 ft. (30 cm.). Burning liberated salts with an alkaline reaction and considerably diminished the acidity of the soil, though the latter effect disappeared after 2 years; it did not affect the normal liberation of mineral nutrients from the felled trees.

The soil changes on the plot which had been cleared, burned and cultivated were similar to those on the two cleared but uncultivated plots, except that the recovery of the organic matter status of the soil was appreciably arrested, owing to the much smaller amount of humus added to the soil by the crop compared with that added by a natural cover of vegetation.

The results of this investigation are summarized in Table 39.

TABLE 39. *Soil changes following felling of forest, burning and cultivation at Arena Forest Reserve, Trinidad.*
(Data from Duthie, Hardy & Rodriguez, 1936)

The plots were felled on 28 May 1933 and plots 2 and 4 were burned on 23 August 1933.

	Organic matter (%)			Total nitrogen content (%)		
	0–6 in. (0–15 cm.)	6–12 in. (15–30 cm.)	12–24 in. (30–60 cm.)	0–6 in. (0–15 cm.)	6–12 in. (15–30 cm.)	12–24 in. (30–60 cm.)
Plot 1. Natural forest:						
26 June 1933	2·37	0·97	0·64	0·13	0·07	0·05
18 July 1935	2·19	1·07	0·79	0·09	0·06	0·05
Plot 2. Burned:						
26 June 1933	1·41	1·35	0·87	0·05	0·05	0·04
18 July 1935	0·93	0·93	0·77	0·05	0·04	0·03
Plot 3. Not burned:						
26 June 1933	1·31	0·73	0·52	0·05	0·03	0·02
18 July 1935	1·64	0·99	0·52	0·08	0·05	0·03
Plot 4. Burned and cultivated:						
26 June 1933	1·57	0·81	0·53	0·06	0·03	0·02
18 July 1935	1·15	0·83	0·59	0·05	0·04	0·03

	Carbon/nitrogen ratio			Available nutrients (electrical conductivity values)		
	0–6 in. (0–15 cm.)	6–12 in. (15–30 cm.)	12–24 in. (30–60 cm.)	0–6 in. (0–15 cm.)	6–12 in. (15–30 cm.)	12–24 in. (30–60 cm.)
Plot 1. Natural forest:						
26 June 1933	11·2	9·1	7·7	41	28	25
18 July 1935	13·7	11·2	9·7	52	35	37
Plot 2. Burned:						
26 June 1933	14·5	14·8	12·7	42	43	36
18 July 1935	13·7	14·0	14·3	59	56	31
Plot 3. Not burned:						
26 June 1933	15·1	14·0	13·1	35	30	56
18 July 1935	11·7	11·6	11·1	51	43	34
Plot 4. Burned and cultivated:						
26 June 1933	15·6	17·6	18·3	31	27	30
18 July 1935	13·1	11·6	11·7	25	35	39

From this work it is clear that in the conditions of the area investigated the effects of any deterioration due to forest clearance alone disappear very quickly. The effects of prolonged cultivation are much more lasting.

THE FUTURE OF THE TROPICAL RAIN FOREST

The Tropical Rain forests of the world consist, as we have seen, of a number of plant formations differing in floristic composition, but very similar in structure and physiognomy; for this reason they are regarded as a single formation type or pan-climax. Mature or primary Rain forests are extremely complex plant communities and exist in a delicate but stable equilibrium with the other components of the ecosystem, the climate, soil and animals.

Extremely little is known of the past history of the rain-forest formations, but there is reason to believe that they have persisted, unchanged or changing exceedingly slowly, from a very remote period. Secular climatic changes have caused the boundaries of the rain forest sometimes to expand and somtimes to retreat, but the formations have probably seldom been forced to adapt themselves to sudden or violent changes in the environment. Until the most recent period in its history man has had little effect on the Tropical Rain forest; large areas of it have been altogether uninhabited or inhabited only by food-gathering peoples with no more influence on the vegetation than any of the other animal inhabitants.

At length, after existing thus for millions of years, the rain forest ecosystem has very recently—in most of its area only within the last 100 years or so—been rudely disturbed by the spread of western civilization to the tropics. This has involved not only the widespread destruction of the forest to make way for plantation agriculture—the cultivation of rubber, coffee, cocoa and similar crops for export—but the increase of the native populations under settled government has given rise to even more widespread clearing of the forest in the interests of subsistence agriculture, mainly in the form of shifting cultivation. Thus, within a very short space of time the primaeval climax forest communities have been replaced over immense areas by cultivation, ruderal communities and seral stages. Till this change took place the Tropical Rain forest was in a stage of development not unlike the forests of Europe in the Mesolithic period, when habitation was limited to the forest fringes and restricted sites accessible by river. The subsequent destruction of the forest all over the tropics is comparable with the clearing of the European forest by agricultural peoples beginning in the Neolithic period, except that the one process has been accomplished in a few decades, while the other lasted for thousands of years.

Though it has been estimated (Heske, 1938–9) that the Tropical Rain forest still forms about half the world's forest area, the destruction of what remains is continuing at increasing speed. Unless determined efforts are made to halt the destruction, the whole of the Tropical Rain forest may disappear within the lifetime of those now living, except for a few inaccessible areas and small 'forest reserves' artificially maintained mainly as sources of timber. At best but a small fraction of the original forest will survive and the greater part of a formation-type which has clothed a large fraction of the earth's surface for many millions of years is doomed to disappear. What consequences are likely to follow?

The most far-reaching is the probable effect on the course of plant evolution in the world as a whole. The rain-forest flora with its immense wealth of species belonging to thousands of genera and scores of families has acted in the past as a reservoir of genetical diversity and potential variability. During at least the more recent epochs in the earth's history it has been a centre of evolutionary activity from which the rest of the world's flora has been recruited. It is not intended to imply that there have not been other active foci of plant evolution, but as we saw in Chapter 1 there is clear evidence that much of the flora of the temperate regions is derived directly or indirectly from the tropics; other centres such as South Africa and Australia have had much less influence on the regions outside their limits. The tropical forest has thus played a part different from that of any of the other major plant formations, perhaps because evolution has not there been interrupted by seasonal checks to plant activity or by secular climatic changes such as glacial periods. It is also possible that the high and constant temperatures of the Rain forest are specially favourable to the survival of mutations.

This role, of a source of supply of genetical material for evolution and of new forms of plant life, the Tropical Rain forest will play no longer, or only in a much diminished degree. The area of the Rain forest itself has been so much reduced and the adjacent formations which served as the path of plant migration from the wet tropics to other regions have been so much modified or destroyed, that the invasion of the subtropical and temperate regions by plant lineages evolved originally in the tropics must, if it has not already ceased completely, become of small importance. It is therefore likely that the destruction of the Tropical Rain forest accomplished during the past 100 years has changed fundamentally the future course of plant evolution and closed many avenues of evolutionary development.

Another consequence of the destruction of the rain forest possibly of great importance is its effect on climate; this is a controversial subject and it is not proposed to deal with it here. The influence of forests in general on climate has been discussed by Zon (1912), Brooks (1927b) and Nicholson (1936). The influence on precipitation of Tropical Rain forest in particular has been discussed by Bernard (1945).

The more obvious and immediate results of the disappearance of the Tropical Rain forest concern its direct economic value. This is twofold: first, it is a source of raw materials, especially timber, fuel, cellulose and 'minor forest products' such as resins, gums, camphor, rattans, etc. and secondly, it protects the soil, especially in hilly districts, against erosion and excessive run-off after heavy rain.

The Tropical Rain forest is the only source of a number of valuable timbers such as greenheart, many kinds of mahogany, etc., but its importance as a supplier of timber is less than might be expected. This is partly because the greater part of the world demand for timber is for coniferous softwoods, which are usually absent in the tropical lowlands (Heske, 1938–9, estimated that in 1937 coniferous wood provided 92·5% of world timber exports and 88% of imports), and also because Tropical Rain forests, as we have seen, are usually mixed associations with a very large number of dominants, no one of which contributes a large share to the whole stand. The dominant species are not only varied in their technological value but only a small proportion are at present regarded as economically useful. Appel (1938–9) states that of the 500 timber species in the Cameroons Rain forest, only some 30% are commercially valuable. In the future, uses are likely to be found for a much larger proportion of rain-forest species and the development of plywood manufacture in the tropics is already making it possible to utilize a far larger proportion of the timbers than formerly. Yet it may be surmised that the future development of timber production in the tropics will tend towards the creation of artificial communities composed of relatively few valuable species either by planting or by sylvicultural treatment of existing primary, depleted or secondary mixed forests. In the formation and maintenance of such communities the ecological study of natural single-dominant Rain forests may be expected to give some guidance.

To a botanist the policy of concentrating attention on a small number of 'economically valuable' species suggests certain doubts. Among the many thousands of species of rain-forest trees there is only a small fraction of which the properties, mechanical, technical, biochemical, etc. are even approximately known. Among the rapidly disappearing majority of 'useless' species there may be many of unsuspected value, some as timbers for special purposes, others for their chemical by-products, and possibly some for producing new types of tree by hybridization. The reservoir of natural material represented by the rain-forest flora is in danger of disappearing before its value has been adequately explored.

The economic value, known and potential, of the Tropical Rain forest is alone a sufficent reason for conserving the remaining areas of primary forest on a much larger scale than has yet been attempted, but there are also aesthetic and scientific reasons. The tropical forest is a field for biological research which has no substitute and in it must lie the key to a vast amount of scientific know-ledge which could not be obtained elsewhere. Corner (1946) has made an eloquent plea for a great expansion of research in tropical botany. It is certain that existing botanical ideas are based far too largely on the specialized and impoverished floras of temperate regions, and the wealth of material for observation and experiment in the evergreen forests of the tropics is still not generally realized. 'I fear', says Corner, 'lest all the virgin lowland forest of the tropics may be destroyed before botany awakes.' What the situation requires is a system of nature reserves under national or international administration, in which the natural vegetation is completely protected from disturbance by timber exploiter or by native cultivator. Such reserves must be extensive and must be provided with facilities for botanical and zoological research. The Albert National Park in the Belgian Congo is an example of nature preservation which should be copied; if samples of the ancient plant communities of the tropics are to be preserved for the study and admiration of future generations, relatively small 'forest reserves', in which exploitation continues under control, are not enough.

REFERENCES

ADAMSON, R. S. (1910). Note on the roots of *Terminalia Arjuna* Bedd. *New Phytol.* **9**, 150–6.

AIYAR, T. V. V. (1932). The Sholas of the Palghat Division. Parts I and II. *Indian For.* **58**, 414–32, 473–86.

ALLEE, W. C. (1926a). Measurement of environmental factors in the tropical rain-forest of Panama. *Ecology*, **7**, 273–302.

—— (1926b). Distribution of animals in a tropical rain-forest with relation to environmental factors. *Ecology*, **7**, 445–68.

ALLORGE, V. & P. (1939). Sur la répartition et l'écologie des hépatiques épiphylles aux Açores. *Bol. Soc. Broter.* (ser. 2), **13**, 211–31.

ANGOT, A. (1928). *Traité élémentaire de météorologie* (ed. 4, revised by C. E. Brazier). Paris.

APPEL, E. (1938–9). Waldnutzung im westafrikanischen Urwald. *Z. Weltforstwirtsch.* **6**, 95–101.

ARBER, A. (1934). *The Gramineae, a study of cereal, bamboo and grass.* Cambridge.

AUBRÉVILLE, A. (1933). La forêt de la Côte d'Ivoire. *Bull. Com. Afr. occid. franç.* **15**, 205–61.

—— (1936). *La flora forestière de la Côte d'Ivoire.* Paris.

—— (1938). La forêt coloniale: les forêts de l'Afrique occidentale française. *Ann. Acad. Sci. colon., Paris*, **9**, 1–245.

BACKER, C. A. (1909). De flora van het eiland Krakatau. *Jversl. topogr. Dienst Ned.-Ind.* Batavia (Java).

—— (1913). Een lastige vreemdeling. *Trop. Natuur*, **2**, 27–31, 33–6.

—— (1929). *The problem of Krakatao as seen by a botanist.* Weltevreden (Java) and The Hague.

BAIN, F. M. (1934). *The Rainfall of Trinidad.* Dept. Agric. Trinidad and Tobago.

BAKER, J. R. (1938). Rain-forest in Ceylon. *Kew Bull.* (1938), 9–16.

BAKER, J. R. & I. (1936). The seasons in a tropical rain-forest (New Hebrides). Part 2. Botany. *J. Linn. Soc. (Zool.)*, **39**, 507–19.

BAKER, J. R. & HARRISSON, T. H. (1936). The seasons in a tropical rain-forest. Part 1. Meteorology. *J. Linn. Soc. (Zool.)*, **39**, 443–63.

BAREN, J. VAN (1931). Properties and constitution of a volcanic soil, built in 50 years in the East-Indian Archipelago. *Med. Landbouwhoogesch. Wageningen*, **35** (Verh. 6), 1–29.

BEARD, J. S. (1942). The use of the term 'deciduous' as applied to forest types in Trinidad, B.W.I. *Emp. For. J.* **21**, 12–17.

—— (1944a). Climax vegetation in tropical America. *Ecology*, **25**, 127–58.

—— (1944b). The natural vegetation of the island of Tobago, British West Indies. *Ecol. Monogr.* **14**, 135–63.

—— (1944c). Forestry in the Windward Islands. *Development and welfare in the West Indies, Bull.* no. **11**.

—— (1945). The progress of plant succession on the Soufriere of St Vincent. *J. Ecol.* **33**, 1–9.

—— (1946a). The Mora forests of Trinidad, British West Indies. *J. Ecol.* **33**, 173–92.

—— (1946b). The natural vegetation of Trinidad. *Oxf. For. Mem.* no. **20**.

—— (1949). The natural vegetation of the Windward and Leeward Islands. *Oxf. For. Mem.* no. **21**.

BECCARI, O. (1904). *Wanderings in the great forests of Borneo.* Ed. and transl. by F. H. H. Guillemard. London.

BENNETT, H. H. & ALLISON, R. V. (1928). *The soils of Cuba.* Trop. Res. Foundn., Washington.

BENOIST, R. (1925). La végétation de la Guyane française (suite). *Bull. Soc. bot. Fr.* **72**, 1066–78.

BERNARD, E. (1945). *Le climate écologique de la Cuvette Centrale Congolaise.* Brussels: Publ. Inst. Nat. Agron. Congo Belge.

BERRY, E. W. (1925). Tertiary flora of Trinidad. *Studies in Geology*, **6**, 71–161.

BEWS, J. W. (1927). Studies in the ecological evolution of the Angiosperms. *New Phytol.* **26**, 1–21 *et seq.*

BILHAM, E. G. (1938). *The Climate of the British Isles.* London.

BLANFORD, H. R. (1929). Regeneration of evergreen forests in Malaya. *Indian For.* **55**, 333–9, 383–95.

BLUM, G. (1933). Osmotische Untersuchungen in Java. I. *Ber. schweiz. bot. Ges.* **42**, 550–680.

BOEREMA, J. (1919). Intensiteit der zonnestraling. *Hand. eerste ned.-ind. natuurw. Congr.* 99–101.

BOOBERG, G. (1932). Grondvormen, etages, en phytocoenosen van Java's vegetatie. *Hand. 6de ned.-ind. natuurw. Congr.* 1931, 329–46.

BØRGESEN, F. (1909). Notes on the shore vegetation of the Danish West Indies. *Bot. Tidsskr.* **29**, 201–17.

BOUILLIENNE, R. (1926). Savanes équatoriales en Amérique du Sud. *Bull. Soc. bot. Belg.* **58**, 217–23.

—— (1930). Un voyage botanique dans le Bas-Amazone. *Arch. Inst. bot. Univ. Liége*, **8**, 1–185.

BRAAK, C. (1924). Het klimaat van Nederlandsch Indië. Deel 1. *Verh. magn. met. Obs. Batavia*, **8**.

—— (1931). Klimakunde von Hinterindien und Insulinde, in W. Köppen & R. Geiger, *Handbuch der Klimatologie*, **4**. Berlin.

BRAUN, E. L. (1935). The undifferentiated deciduous forest climax and the association-segregate. *Ecology*, **16**, 514–19.

—— (1941). The differentiation of the Deciduous Forest of the eastern United States. *Ohio J. Sci.* **41**, 235–41.

BRENNER, W. (1902). Klima und Blatt bei der Gattung *Quercus*. *Flora*, **90**, 114–60.

BROCKMANN-JEROSCH, H. & M. (1925). Jamaika. *Vegetationsbilder*, Reihe **16**, Hft. 5–6.

BROOKS, C. E. P. (1927 a). The mean cloudiness over the earth. *Mem. Roy. Met. Soc.* **1**, 127–38.

—— (1927 b). The influence of forests on rainfall and run-off. *Quart. J. R. Met. Soc.* **54**, 1–17.

—— (1932). *Climate; A Handbook for Business Men, Students and Travellers*, 2nd ed. London.

BROWN, W. H. (1919). *Vegetation of Philippine mountains*. Manila.

BROWN, W. H. & MATTHEWS, D. M. (1914). Philippine Dipterocarp forests. *Philipp. J. Sci.* **9** (sect. A), 413–561.

BÜNNING, E. (1947). *In den Wäldern Nord-Sumatras*. Bonn.

—— (1948). Entwicklungs- und Bewegungsphysiologie der Pflanze. *Lehrbuch der Pflanzen-physiologie*, **2**, **3**. Berlin, Göttingen und Heidelberg.

BURGER, H. (1933). Waldklimafragen. II, III. *Mitt. schweiz. ZentAnst. forstl. Versuchsw.* **18**, 7–54, 153–92.

BURKILL, I. H. (1919). The composition of a piece of well-drained Singapore secondary jungle thirty years old. *Gdns' Bull. Singapore*, **2**, 145–57.

—— (1928). The main features of the vegetation of Pahang. *Malay. Nat.* **2**, 11–21.

BURTT DAVY, J. (1938). The classification of tropical woody vegetation types. *Imp. For. Inst. Paper*, no. **13**.

BÜSGEN, M. (1903). Einige Wachstumsbeobachtungen aus den Tropen. *Ber. dtsch. bot. Ges.* **21**, 435–40.

BÜSGEN, M. & MÜNCH, E. (1929). *The structure and life of forest trees*. Transl. by T. Thomson. London.

BUSSE, W. (1905). Ueber das Auftreten epiphyllischer Kryptogamen im Regenwaldgebiet von Kamerun. *Ber. dtsch. bot. Ges.* **23**, 164–72.

BUXTON, P. A. & LEWIS, D. J. (1934). Climate and tsetse flies; laboratory studies upon *Glossina submorsitans* and *tachinoides*. *Philos. Trans.* ser. B, **224**, 175–240.

BYERS, H. G., KELLOGG, C. E., ANDERSON, M. S. & THORP, J. (1938). Soil formation, in *Soils and Men*. *Yearb. U.S. Dep. Agric.* pp. 948–78.

CABRERA, A. & DAWSON, G. (1944). La selva marginal de Punta Lara. *Rev. Mus. La Plata*, (N.S.), Sec. Bot. **5**, 267–382.

CALDWELL, J. (1930). The movements of food materials in plants. *New Phytol.* **29**, 27–43.

CÁRDENAS, M. (1945). Aspecto general de la vegetación de Bolivia, in F. Verdoorn. *Plants and plant science in Latin America*. Waltham, Mass., 312–13.

CARTER, G. S. (1934). Reports of the Cambridge Expedition to British Guiana, 1933. Illumination in the rain forest at ground level. *J. Linn. Soc. (Zool.)*, **38**, 579–89.

CHAMPION, H. G. (1936). A preliminary survey of the forest types of India and Burma. *Indian For. Rec.* (N.S.), **1**, 1–286.

CHAPIN, J. P. (1932). Birds of the Belgian Congo. Part 1. *Bull. Amer. Mus. Nat. Hist.* **65**.

CHAPMAN, V. J. (1944). 1939 Cambridge University Expedition to Jamaica. Part I. A study of the botanical processes concerned in the development of the Jamaican shore-line. Part II. A study of the environment of *Avicennia nitida* Jacq. in Jamaica. Part III. The morphology of *Avicennia nitida* Jacq. and function of the pneumatophores. *J. Linn. Soc.* (*Bot.*), **52**, 487–533.

CHARTÈR, C. F. (1941). *A reconnaissance survey of the soils of British Honduras.* Government of British Honduras, Trinidad.

CHEESEMAN, T. F. (1903). The flora of Rarotonga. *Trans. Linn. Soc. Lond.* (*Bot.*), **6**, 261–313.

CHENERY, E. M. (1948). Aluminium in plants and its relation to plant pigments. *Ann. Bot., Oxford,* N.S., **12**, 121–36.

CHENERY, E. M. & HARDY, F. (1945). The moisture profile in some Trinidad forest and cacao soils. *Trop. Agriculture, Trin.,* **22**, 100–15.

CHENGAPA, B. S. (1934). Andaman forests and their reproduction. *Indian For.* **60**, 54–64, 117–29, 185–98.

CHEVALIER, A. (1900). Les zones et les provinces botaniques de l'Afrique Occidentale française. *C.R. Acad. Sci., Paris,* **130**, 1205–8.

—— (1909). L'extension et la régression de la forêt vierge de l'Afrique tropicale. *C.R. Acad. Sci., Paris,* **149**, 458–61.

—— (1912). Carte botanique, forestière et pastorale. *Géographie,* **26**, pl. 1 (explanation by Hulot, pp. 276–7).

—— (1917). La forêt et les bois du Gabon. *Les végétaux utiles d'Afrique tropicale française,* Fasc. **9**, Paris.

—— (1924). Sur la forêt primitive tropicale et la forêt secondaire. *C.R. Soc. Biogéogr.* **1**, 39–40.

—— (1938a). *Flore vivante de l'Afrique Occidentale française,* **1**. Paris.

—— (1938b). Sur la présence d'une Broméliacée spontanée en Guinée française. *Bull. Soc. bot. fr.* **85**, 489–90.

—— (1948). Biogéographie et écologie de la forêt dense ombrophile de la Côte d'Ivoire. *Rev. Bot. appl.* **28**, 101–15.

CHIPP, T. F. (1913). The reproduction of *Musanga Smithii. Kew Bull.* (1913), **96**.

—— (1922). Buttresses as an assistance to identification. *Kew Bull.* (1922), 265–8.

—— (1927). The Gold Coast Forest. A study in synecology. *Oxf. For. Mem.* **7**.

CHRISTOPHERSEN, E. (1927). Vegetation of Pacific Equatorial Islands. *Bishop Mus. Bull.* **44**.

CLEMENTS, F. E. (1936). Nature and structure of the climax. *J. Ecol.* **24**, 252–84.

COPELAND, E. B. (1907). Comparative ecology of the San Ramon Polypodiaceae. *Philipp. J. Sci.* (sect. C), **2**, 1–76.

CORBET, A. S. (1935). *Biological processes in tropical soils, with special reference to Malaya.* Cambridge.

CORNER, E. J. H. (1933). A revision of the Malayan species of *Ficus*: Covellia and Neomorphe. *J. Malay Br. Asiat. Soc.* **11**, 1–65.

—— (1935). The seasonal fruiting of agarics in Malaya. *Gdns' Bull. Singapore,* **9**, 79–88.

—— (1938). The systematic value of the colour of withering leaves. *Chronica Bot.* **4**, 119–21.

—— (1939). A revision of *Ficus,* subgenus *Synoecia. Gdns' Bull. Singapore,* **10**, 82–161.

—— (1940). *Wayside trees of Malaya* (2 vols.). Singapore.

—— (1946). Suggestions for botanical progress. *New Phytol.* **45**, 185–92.

COSTER, C. (1923). Lauberneuerung und andere periodische Lebensprozesse in dem trockenen Monsun-gebiet Ost-Javas. *Ann. Jard. bot. Buitenz.* **33**, 117–89.

—— (1925). Die Fettumwandlung in Baumkörper in den Tropen. *Ann. Jard. bot. Buitenz.* **35**, 71–104.

—— (1926a). Periodische Blüteerscheinungen in den Tropen. *Ann. Jard. bot. Buitenz.* **35**, 125–62.

—— (1926b). Die Buche auf dem Gipfel des Pangerango. *Ann. Jard. bot. Buitenz.* **35**, 105–19.

—— (1927–8). Zur Anatomie und Physiologie der Zuwachszonen und Jahresringbildung in den Tropen. *Ann. Jard. bot. Buitenz.* **37**, 49–160; **38**, 1–114.

—— (1932). Wortelstudiën in de Tropen. I. De jeugdontwikkeling van het wortelstelsel van een zeventigtal boomen en groenbemesters. *Tectona,* **25**, 828–72.

Coster, C. (1933). Wortelstudiën in de Tropen. III. De zuurstof behoefte van het wortelstelsel. *Tectona*, **26**, 450–97.

—— (1935*a*). Licht, ondergroei en wortelconcurrentie. *Tectona*, **28**, 612–19.

—— (1935*b*). Wortelstudiën in de Tropen. IV. Wortelconcurrentie. *Tectona*, **28**, 861–78.

Cox, H. A. (1939). *A handbook of Empire timbers*. Dep. Sci. and Industr. Res., London.

Cuatrecasas, J. (1934). Observaciones geobotanicas en Colombia. *Trab. Mus. Cienc. nat., Madr.*, Ser. Bot. **27**.

Czapek, F. (1909). Ueber die Blattentfaltung der Amherstien. *S.B. Akad. Wiss. Wien*, **118**, 201–30.

Däniker, A. U. (1929). Neu-Caledonien, Land und Vegetation. *Vjschr. naturf. Ges. Zürich*, **74**, 170–97.

Davis, J. H. (1940). The ecology and geologic rôle of mangroves in Florida. (Pap. from Tortugas Lab. no. 22.) *Publ. Carneg. Instn*, no. 517, 303–412.

—— (1943). The natural features of southern Florida, especially the vegetation and the everglades. *State of Florida, Dep. of Conservation, Geol. Bull.* no. **25**.

Davis, T. A. W. (1929). Some observations on the forests of the North West District. *Agric. J. Brit. Guiana*, **2**, 157–66.

—— (1933). Report on the Balata industry in British Guiana with special reference to the Rupununi District. *British Guiana, Second Legislative Council, Third Sess.*, 1932.

—— (1941). On the island origin of the endemic trees of the British Guiana peneplain. *J. Ecol.* **29**, 1–13.

Davis, T. A. W. & Richards, P. W. (1933–4). The vegetation of Moraballi Creek, British Guiana; an ecological study of a limited area of Tropical Rain Forest. Parts I and II. *J. Ecol.* **21**, 350–84; **22**, 106–55.

Deuerling, O. (1909). Die Pflanzenbarren der Afrikanischen Flüsse mit Berücksichtigung der Wichtigsten pflanzlichen Verlandungserscheinungen. *Münch. geogr. Studien*, no. 24. (Abstr. *Bot. Zbl.* 1910, **114**, 91–4.)

Diels, L. (1906). Die Pflanzenwelt von West-Australien. *Veget. Erde*, **7**.

—— (1918). Das Verhältnis von Rhythmik und Verbreitung bei den Perennen der europäischen Sommerwaldes. *Ber. dtsch. bot. Ges.* **36**, 337–51. (Summary in Schimper, 1935, pp. 626–9.)

Diels, L. & Hackenberg, G. (1926). Beiträge zur Vegetationskunde und Floristik von Süd-Borneo. *Bot. Jb.* **60**, 293–316.

Dingler, H. (1911). Versuche über die Periodizität einiger Holzgewächse in den Tropen. *S.B. bayer. Akad. Wiss. München, math.-naturw. Kl.* 127–43.

Dixon, W. A. (1882). On the inorganic constituents of some epiphytic ferns. *Proc. Roy. Soc. N.S. Wales*, **15**, 175–83.

Docters van Leeuwen, W. M. (1933). Biology of plants and animals occurring in the higher parts of Mount Pangrango-Gedeh in West-Java. *Verh. Akad. Wet., Amst.* (sect. 2), **31**, 1–270.

—— (1936). Krakatau, 1883 to 1933. A. Botany. *Ann. Jard. bot. Buitenz.* **46–7**, 1–506.

Doyne, H. C. (1935). Studies in tropical soils. Increase of acidity with depth. *J. Agric Sci.* **25**, 192–7.

Dugand, A. (1934). The Transition Forests of Atlántico, Colombia. *Trop. Woods*, **40**, 1–14.

Du Rietz, G. E. (1931). Life-forms of terrestrial flowering plants. I. *Acta phytogeogr. suec.* **3** (1).

Duthie, D. W., Hardy, F. & Rodriguez, G. (1936). *Soil investigations in the Arena Forest Reserve, Trinidad.* (Typescript. Imp. Coll. Trop. Agric. Trinidad. Summary by R. L. Brooks.) *Imp. For. Inst. Paper*, no. **6**, 1937.

Eggeling, W. J. (1935). The vegetation of Namanve Swamp, Uganda. *J. Ecol.* **23**, 422–35.

—— (1947). Observations on the ecology of Budongo rain forest, Uganda. *J. Ecol.* **34**, 20–87.

Elwes, H. J. & Henry, A. (1906–7). *The trees of Great Britain and Ireland*, **1** (1906); **2** (1907). Edinburgh.

Endert, F. H. (1920). De woudboomflora van Palembang. *Tectona*, **13**, 113–60.

—— (1925). *Verslag van de Midden-Oost-Borneo-Expeditie*, 1925. Batavia.

ENDERT, F. H. (1928). Geschlachtstabellen voor Nederlandsch-Indische Boomsoorten naar vegetatiev kenmerken. *Meded. Proefst. Boschw.*, *Batavia*, **20**, 1–242.

—— (1932). De proefbaanmetingen in de panglonggebieden van Bengkalis (Sumatra's Oostkust) en Riouw. *Tectona*, **25**, 713–85.

—— (1933). Eenige resultaten van de boschverkening in de Buitengewesten. *Tectona*, **26**, 391–421.

ERNST, A. (1908). *The new flora of the volcanic island of Krakatau.* Transl. by A. C. Seward. Cambridge.

—— (1934). Das biologische Krakatau Problem. *Vjschr. naturf. Ges. Zürich*, **79** (Beibl.), 1–187.

ERNST, A., BERNARD, C. & others (1910–14). Beiträge zur Kenntnis der Saprophyten Javas. *Ann. Jard. bot. Buitenz.* **23**, 20–61; **24**, 55–97; **25**, 161–88; **26**, 219–57; **28**, 99–124.

EVANS, G. C. (1939). Ecological studies on the rain forest of Southern Nigeria. II. The atmospheric environmental conditions. *J. Ecol.* **27**, 436–82.

EXELL, A. W. (1944). *Catalogue of the vascular plants of S. Tomé (with Principe and Annobon).* London.

FABER, F. C. VON (1915). Physiologische Fragmente aus einem tropischen Urwald. *Jb. wiss. Bot.* **56**, 197–220.

—— (1923). Zur Physiologie der Mangroven. *Ber. dtsch. bot. Ges.* **41**, 227–34.

—— (1927). Die Kraterpflanzen Javas in physiologischökologischer Beziehung. *Arbeit. Treub-Lab., Buitenzorg*, **1**.

FÉHER, D. (1927). Untersuchungen über die Kohlenstoffernährung des Waldes. *Flora* (N.F.), **21**, 316–33.

FITTING, H. (1910). Ueber die Beziehungen zwischen den Epiphyllen Flechten und den von ihnen bewohnten Blättern. *Ann. Jard. bot. Buitenz.* (suppl. 3), pp. 505–18.

FOGGIE, A. (1947). Some ecological observations on a tropical forest type in the Gold Coast. *J. Ecol.* **34**, 88–106.

FOLLETT-SMITH, R. R. (1930). Report of an investigation of the soils and of the mineral content of pasture grasses occurring at Waranama Ranch, Berbice River. *Agric. J. Brit. Guiana*, **3**, 142–59.

FOXWORTHY, F. W. (1927). Commercial timber trees of the Malay Peninsula. *Malay. For. Rec.* **3**

FRANCIS, W. D. (1924). The development of buttresses in Queensland trees. *Proc. Roy. Soc. Qd*, **36**, 21–37.

—— (1929). *Australian rain-forest trees.* Brisbane.

FRASER, L. & VICKERY, J. W. (1937). The ecology of the Upper Williams River and Barrington Tops Districts. I. Introduction. *Proc. Linn. Soc. N.S.W.* **62**, 269–83.

—— (1938). The ecology of the Upper Williams River and Barrington Tops Districts. II. The Rain-Forest Formations. *Proc. Linn. Soc. N.S.W.* **63**, 139–84.

—— (1939). The ecology of the Upper Williams River and Barrington Tops Districts. III. The Eucalypt Forests and general discussion. *Proc. Linn. Soc. N.S.W.* **64**, 1–33.

FREISE, F. (1936). Das Binnenklima von Urwäldern in subtropischen Brasilien. *Petermanns Mitt.* **82**, 301–7.

—— (1938). Beobachtungen in Zweitwuchsbeständen aus dem Küstenwaldgebiet Brasiliens. *Z. Weltforstwirtsch.* **6**, 281–99.

—— (1939). Einige Bemerkungen über soziologische Verhältnisse im Urwald. *Z. Weltforstwirtsch.* **6**, 603–21.

FREY-WYSSLING, A. (1931). Over de struikwildernis van Habinsaran. *Trop. Natuur*, **20**, 194–8.

FRIES, R. E. & T. C. E. (1948). Phytogeographical researches on Mt Kenya and Mt Aberdare, British East Africa. *K. svenska VetenskAkad. Handl.* ser. 3, **25**, 1–83.

FRITSCH, F. E. (1907). A general consideration of the subaërial and fresh-water algal flora of Ceylon. Part I. Subaërial algae and algae of the inland fresh-waters. *Proc. Roy. Soc.* B, **79**, 197–254.

—— (1936). The role of the terrestrial alga in nature. *Essays in geobotany in honor of William Albert Setchell*, Berkeley (California), pp. 195–217.

FUNKE, G. L. (1929). On the biology and anatomy of some tropical leaf joints, I. *Ann. Jard. bot. Buitenz.* **40**, 45–74.

—— (1931). On the biology and anatomy of some tropical leaf joints, II. *Ann. Jard. bot. Buitenz.* **41**, 33–64.

GAMS, H. (1918). Prinzipienfragen der Vegetationsforschung. *Vjschr. naturf. Ges. Zürich*, **63**, 293–493.

GEIGER, R. (1927). *Das Klima der bodenäher Luftschicht.* Braunschweig.

GENTNER, G. (1909). Ueber den Blauglanz auf Blättern und Früchten. *Flora*, **99**, 337–54.

GHESQUIÈRE, (1925). Note sur les racines tabulaires ou accotements ailés de quelques arbres congolais. *Rev. zool. afr.* **13** (suppl. 2), 1–2.

GIBBS, L. S. (1917). *A contribution to the phytogeography and flora of the Arfak Mountains, etc.* London.

GIESENHAGEN, K. (1910). Die Moostypen der Regenwälder. *Ann. Jard. bot. Buitenz.* (suppl. 3, pt. 2), pp. 711–90.

GLEASON, H. A. & COOK, M. T. (1926). Plant ecology of Porto Rico. *Scientific Survey of Porto Rico and the Virgin Islands*, **7**, Part 1. N.Y. Acad. Sci., New York.

GOEBEL, K. (1888). Morphologische und biologische Studien. I. Ueber epiphytische Farne und Muscineen. *Ann. Jard. bot. Buitenz.* **7**, 1–73.

—— (1889). *Pflanzenbiologische Schilderungen. III. Epiphyten.* Marburg.

—— (1913). *Organographie der Pflanzen. 1te Teil. Allgemeine Organographie* (ed. 2). Jena.

—— (1922). Erdwurzeln mit Velamen. *Flora* (N.F.), **15**, 1–26.

—— (1924). *Die Entfaltungsbewegungen der Pflanzen* (ed. 2). Jena.

GOEDERT, P. (1938). Les sols de l'Afrique Centrale, spécialment du Congo Belge. Introduction— Le régime pluviale au Congo Belge. *Publ. Inst. Nat. pour L'Et. agron. Congo Belge, Hors série.*

GOODING, E. G. B. (1947). Observations on the sand dunes of Barbados, British West Indies. *J. Ecol.* **34**, 111–25.

GOSSWEILER, J. & MENDONÇA, F. A. (1939). *Carta fitogeográfica de Angola.* Lisbon.

GRAY, A. P. (1944). Ecology. Chapter IV in *Imperata cylindrica, taxonomy, distribution, economic significance and control. Imp. Agric. Bur. Joint Publ.* no. **7**. Oxford and Aberystwyth.

GREENWAY, P. J. (1938). Mafia Island, its natural history, etc. Unpubl. Memoir, E. Afr. Res. Sta. Amani.

GRESSER, E. (1919). Resumeerend rapport over het voorkomen van ijzerhout op de olieterreinen Djambi. I. *Tectona*, **12**, 283–304.

GUPPY, H. B. (1906). *Observations of a naturalist in the Pacific between 1896 and 1899.* **2.** *Plant-dispersal.* London.

GUTTENBERG, H. VON (1931). Beiträge zur Kenntnis der Laubblattassimilation in den Tropen. *Ann. Jard. bot. Buitenz.* **41**, 105–84.

HABERLANDT, G. (1898). Ueber die Grösse der Transpiration in feuchten Tropenklima. *Jb. wiss. Bot.* **31**, 273-88.

—— (1899). Erwiderung. *Jb. wiss. Bot.* **33**, 166–70.

—— (1926). *Eine botanische Tropenreise* (ed. 3). Leipzig.

HADDOW, A. J., GILLETT, J. D. & HIGHTON, R. B. (1947). The mosquitoes of Bwamba County, Uganda. V. The vertical distribution and biting-cycle of mosquitoes in rain-forest, with further observations on microclimate. *Bull. Ent. Res.* **37**, 301–30.

HARDON, H. J. (1936). Podsol profiles in the tropics. *Natuurk. Tijdschr. Ned.-Ind.* **96**, 25–41.

—— (1937). Padang soil, an example of podsol in the Tropical Lowlands. *Verh. Akad. Wet. Amst.* **40**, 530–8.

HARDY, F. (1935). The chemical and ecological researches on cacao. *Trop. Agriculture, Trin.*, **12**, 175–8.

—— (1936 a). Some aspects of cacao soil fertility in Trinidad. *Trop. Agriculture, Trin.*, **13**, 315–17.

—— (1936 b). Some aspects of tropical soils. *Trans. 3rd Int. Congr. Soil Sci.* (Oxford), 1935, **2**, 150–63.

—— (1939). Soil erosion in St Vincent, B.W.I. *Trop. Agriculture, Trin.*, **16**, 58–65.

—— (1940). A provisional classification of the soils of Trinidad. *Trop. Agriculture, Trin.*, **17**, 153–8.

HARDY, F. (1945). The soils of South America, in *Plants and plant science in Latin America* (ed. by F. Verdoorn). Waltham, Mass.

HARDY, F. & FOLLETT-SMITH, R. R. (1931). Studies in tropical soils. II. Some characteristic igneous rock soil profiles in British Guiana, South America. *J. Agric. Sci.* **21**, 739–61.

HARMON, M. H. (1942). A study of some growth factors affecting symmetrical growth in trees. *Butler Univ. Stud.* **5**, 134–44. (Abstr. *Biol. Abstr.* **17**, no. 684, 1943.)

HARRIS, J. A. (1918). On the osmotic concentration of the tissue fluids of phanerogamic epiphytes. *Amer. J. Bot.* **5**, 490–506.

HERBERT, D. A. (1929). The major factors in the present distribution of the genus *Eucalyptus*. *Proc. Roy. Soc. Qd*, **40**, 165–93.

—— (1935). The climatic sifting of Australian vegetation. *Rep. Melb. Meeting, Aust. & N.Z. Ass. Adv. Sci.* pp. 349–70.

HERZOG, T. (1923). Die Pflanzenwelt der bolivischen Anden und ihres östlichen Vorlandes. *Veget. Erde*, **15**.

HESKE, F. (1938–9). Der tropische Wald als Rohstoffquelle. *Z. Weltforstwirtsch.* **6**, 413–85.

HINGSTON, R. W. G. (1932). *A naturalist in the Guiana Forest*. London.

HOLLICK, A. (1924). A review of the fossil flora of the West Indies. *Bull. N.Y. Bot. Gdn*, **12**, 259–323.

HOLTERMANN, C. (1907). *Der Einfluss des Klimas auf den Bau der Pflanzengewebe. Anatomisch-physiologische Untersuchungen in den Tropen.* Leipzig.

HOLTTUM, R. E. (1924). The vegetation of Gunong Belumut in Johore. *Gdns' Bull. Singapore*, **3**, 245–57.

—— (1931). On periodic leaf-change and flowering of trees at Singapore. *Gdns' Bull. Singapore*, **5**, 173–206.

—— (1935). The flowering of Tembusu trees (*Fagraea fragrans* Roxb.) in Singapore 1928–35. *Gdns' Bull. Singapore*, **9**, 73–8.

—— (1938*a*). The ecology of tropical Pteridophytes, in F. Verdoorn, *Manual of Pteridology*, pp. 420–50. The Hague.

—— (1938*b*). Leaf-fall in a non-seasonal climate (Singapore). *Proc. Linn. Soc. Lond.* 150th session 1937–8), pp. 78–81.

—— (1940). On periodic leaf-change and flowering of trees at Singapore. II. *Gdns' Bull. Singapore*, **11**, 119–75.

HUBER, J. (1902). Contribuiçao á geographia physica dos Furos de Breves e da parte occidental de Marajó. *Bol. Mus. Paraense Hist. Nat.* **3**, 447. (Abstr. *Bot. Zbl.* 1903, **93**, 235–7.)

HUMBOLDT, A. VON (1817). *De distributione geographica plantarum secundum coeli temperiem et altitudinem montium prolegomena.* Paris.

—— (1852). *Personal narrative of travels to the equinoctial regions of America,* **2**. Transl. T. Ross. London.

IHERING, H. VON (1923). Das periodische Blattwechsel der Bäume in tropischen und subtropischen Südamerika. *Bot. Jb.* **58**, 524–98.

JANSE, J. M. (1897). Les endophytes radicaux de quelques plantes javanaises. *Ann. Jard. bot. Buitenz.* **14**, 53–201.

JENTSCH (1911). Der Urwald Kameruns. Folgerungen aus den auf der Expedition 1908/09 gewonnenen Erfahrungen in Bezug auf den Zustand und die Nutzbarmachung des Waldes. *Beih. Tropenpfl.* **12**, 1–199.

JOCHEMS, S. C. J. (1928). De begroeiing der tabakslanden in Deli en hare beteeknis voor de tabakscultuur. *Med. Deli. Proefstat.* (2de ser.), no. **59**.

JOHOW, F. (1885). Die chlorophyllfreien Humusbewohner West-Indiens. *Jb. wiss. Bot.* **16**, 415–49.

—— (1889). Die chlorophyllfreien Humuspflanzen nach ihren biologischen und anatomisch-entwicklunsgeschichtlichen Verhältnissen. *Jb. wiss. Bot.* **20**, 475–525.

JONES, A. P. D. (1945). Notes on terms for use in vegetation description in southern Nigeria. *Farm & Forest (Nigeria)*, **6**, 130–6.

—— (1947). Botanica nigerica. *Farm and Forest*, **8**, 10–16.

JONES, A. P. D. & KEAY, R. W. J. (1946). Descriptive terms for the vegetation of the drier parts of Nigeria. *Farm and Forest*, **7**, 34–40.

JONES, E. W. (1945). The structure and reproduction of the virgin forest of the North Temperate Zone. *New Phytol.* **44**, 130–48.

JUNGNER, J. R. (1891). Anpassungen der Pflanzen an das Klima in den Gegenden der regenreichen Kamerungebirge. *Bot. Zbl.* **47**, 353–60.

KAMERLING, Z. (1912). De verdamping van epiphyte Orchideen. *Natuurk. Tijdschr. Ned.-Ind.* **71**, 54–72.

KARSTEN, G. (1891). Ueber die Mangrove-Vegetation im malayischen Archipel. *Bibliogr. bot.* Heft **22**.

—— (1895). Morphologische und biologische Untersuchungen über einige Epiphytenformen der Molukken. *Ann. Jard. bot. Buitenz.* **12**, 117–95.

—— (1924–26). Das Licht im tropischen Regenwalde. *Vegetationsbilder*, Reihe **16**, Hft. 3.

—— (1925). Ueber mantelförmige Organe bei Epiphyten und Wurzelklettern. *Flora* (N.F.), **18–19** (Goebel Festschr.), 300–11.

KEAY, R. W. J. (1947). Forest vegetation in the savannah regions of Nigeria. *Fifth Empire Forestry Conference* (1947).

KEEBLE, F. W. (1895). The hanging foliage of certain tropical trees. *Ann. Bot., Oxford*, **9**, 59–94.

KENNEDY, J. D. (1936). *Forest flora of Southern Nigeria.* Lagos.

KENOYER, L. A. (1929). General and successional ecology of the Lower Tropical Rain-Forest at Barro Colorado island, Panama. *Ecology*, **10**, 201–22.

KLEBS, G. (1911). Ueber die Rhythmik in der Entwicklung der Pflanzen. *S.B. heidelberg. Akad. Wiss., math.-nat. Kl.*, Abhandl. **23**.

—— (1912). Ueber die periodischen Erscheinungen tropischer Pflanzen. *Biol. Zbl.* **32**, 257–85.

—— (1915). Ueber Wachstum und Ruhe tropischer Baumarten. *Jb. wiss. Bot.* **56**, 734–92.

—— (1926). Ueber periodisch wachsende tropischer Baumarten. *S.B. heidelberg. Akad. Wiss., math.-nat. Kl.*, Abhandl. **2**.

KLEIN, W. C. (1914). Kalkplanten in Nederlandsch-Indie. *Trop. Natuur.* **3**, 133–5.

KOERNICKE, M. (1910). Biologische Studien an Loranthaceen. *Ann. Jard. bot. Buitenz.* Suppl. **3**, 665–97.

KOOPMAN, M. J. F. & VERHOEF, L. (1938). *Eusideroxylon zwageri* T. & B., het ijzerhout van Borneoen Sumatra. *Tectona*, **31**, 381–99.

KOORDERS, S. H. & VALETON, T. (1894). Bijdrage tot de kennis der boomsoorten van Java, Nr. 1. *Medd. s'Lands Plantentuin*, **2**. Buitenzorg.

KOPPEL, C. VAN DER (1926). Winning van copal in het Gouvernement Celebes en onderhoorigheden, de uitvoer uit Makassar, en eenige details over het gebruik van copal. *Tectona*, **19**, 525–74.

—— (1945). Forestry in the Outer Provinces of the Netherlands Indies, in F. Verdoorn, *Science and scientists in the Netherlands Indies.* New York.

KÖPPEN, W. (1918). Klassifikation der Klimate nach Temperatur, Niederschlag und Jahreslauf. *Petermanns Mitt.* **64**, 193.

—— (1936). Das geographische System der Klimate in W. Köppen and R. Geiger. *Handbuch der Klimatologie*, **1**, Teil C. Berlin.

KÖPPEN, W. & GEIGER, R. (1928). *Klimakarte der Erde.* Gotha.

KRAMER, F. (1933). De natuurlijke verjonging in het Goenoeng-Gedehcomplex. *Tectona*, **26**, 156–85.

KRAUS, G. (1895). Physiologisches aus den Tropen. I. Das Längenwachsthum der Bambusrohre. *Ann. Jard. bot. Buitenz.* **12**, 196–210.

KURZ, S. (1875). *Preliminary report on the forest and other vegetation of Pegu.* Calcutta.

LAAN, E. VAN DER (1925). De bosschen van de Zuider- en Ossterafdeeling van Borneo. *Tectona*, **18**, 925–52.

—— (1926a). Analyse der bosschen in de onderafdeeling Pleihari van de afdeeling Bandjermasin der Zuider- en Oosterafdeeling van Borneo. *Tectona*, **19**, 103–23.

—— (1926b). De bosschen van de onderafdeeling Tanah Boemboe van be afdeeling Zuidoostkust der residentie Zuider- en Ossterafdeeling van Borneo. *Tectona*, **19**, 215–38.

LAAN, E. VAN DER (1927). De analyse des bosschen van de eilanden Poeloe Laoet en Seboekoe der Residentie Zuider- en Oosterafdeeling van Borneo. *Tectona*, **20**, 19–36.

LAM, H. J. (1945). Fragmenta Papuana. (Observations of a naturalist in Netherlands New Guinea.) Transl. L. M. Perry. *Sargentia*, **5**.

LANE-POOLE, C. E. (1925a). The forests of Papua and New Guinea. *Emp. For. J.* **4**, 206–34.

—— (1925b). *The forest resources of the Territories of Papua and New Guinea*. Parliament of the Commonwealth of Australia. Canberra.

—— (1938). Kamerere. *New Guinea Agric. Gaz.* **4**, 15–17.

LANJOUW, J. (1936). Studies of the vegetation of the Surinam savannahs and swamps. *Ned. kruidk. Arch.* **46**, 823–51.

LAUSBERG, T. (1935). Quantitative Untersuchungen über die kutikulare Exkretion des Laubblattes. *Jb. wiss. Bot.* **81**, 769–806.

LAVAUDEN, L. (1937). The equatorial forest of Africa, its past, present and future. *J. Roy. Afric. Soc.* **36** (extra Suppl.).

LEBRUN, J. (1935). Les essences forestières des régions montagneuses du Congo oriental. *Publ. Inst. Agron. Congo Belge*, Ser. sci. **1**.

—— (1936a). *Répartition de la forêt équatoriale et des formations végétales limitrophes*. Brussels.

—— (1936b). La forêt équatoriale congolaise. *Bull. agric. Congo belge*, **27**, 163–92.

—— (1947). La végétation de la plaine alluviale au sud du Lac Édouard. *Exploration du Parc National Albert, Mission J. Lebrun* (1937–1938). Fasc. **1**. Brussels.

LINDMANN, C. A. M. (1900). *Vegetationen i Rio Grande do Sul (Sydbrasilien)*. Stockholm.

LIVINGSTON, B. E. (1916). Physiological temperature indices for the study of plant growth in relation to climatic conditions. *Physiol. Res.* **1**, 399–420.

LIVINGSTON, B. E. & G. J. (1913). Temperature coefficients in plant geography and climatology. *Bot. Gaz.* **56**, 349–75.

LONGMAN, H. A. & WHITE, C. T. (1917). The flora of a single tree. *Proc. Roy. Soc. Qd*, **29**, 64–9.

LUNDEGÅRDH, H. (1931). *Environment and plant development*. Trans. E. Ashby.

LÜTZELBURG, P. VON (1925–26). Estudo botanico do Nordéste. *Minist. Viação e obr. publ.*, *Publ.* **57**, Ser. 1 A, 1–3. (Abstr. in *Bot. Jb.* **61**, 1927. Literaturber. 21–31.)

MACGREGOR, W. D. (1934). Silviculture of the Mixed Deciduous Forests of Nigeria. *Oxf. For. Mem.* **18**.

—— (1937). Forest type and succession in Nigeria. *Emp. For. J.* **16**, 234–42.

MACKAY, J. H. (1936). Problems of ecology in Nigeria. *Emp. For. J.* **15**, 190–200.

McLEAN, R. C. (1919). Studies in the ecology of Tropical Rain-forest: with special reference to the forests of South Brazil. *J. Ecol.* **7**, 15–54, 121–72.

MAITLAND, T. D. (1932). The grassland vegetation of the Cameroons Mountain. *Kew Bull.* (1932), 417–25.

MARR, C. C. (1938). Agricultural survey of the Markham valley. *New Guinea Agric Gaz.* **4**, 2–12.

MARSHALL, R. C. (1934). The physiography and vegetation of Trinidad and Tobago. *Oxf. For. Mem.* **17**.

MARTIN, F. J. & DOYNE, H. C. (1927). Laterite and lateritic soils in Sierra Leone. *J. Agric. Sci.* **17**, 530–47.

MARTYN, E. B. (1931). A botanical survey of the Rupununi Development Company's ranch at Waranama, Berbice River. *Agric. J. Brit. Guiana*, **4**, 18–25.

MASSART, J. (1898). Les végétaux épiphylles. *Ann. Jard. bot. Buitenz.* Suppl. **1**, 103–8.

MEAD, J. P. (1928). The forests of the Fiji islands. *Emp. For. J.* **7**, 47–54.

MEIJER DREES, E. (1938). Plantensociologie, boschbouw en houtteelt. *Tectona*, **31**, 166–205.

MERRILL, E. D. (1915). A contribution to the bibliography of the botany of Borneo. *Sarawak Mus. J.* **2**, 99–136.

MEUSEL, H. (1935). Wuchsformen und Wuchstypen der europäischen Laubmoose. *Nova Acta Leopoldina* (N.F.), **3** (12), 123–277.

MEZ, C. (1904). Physiologische Bromeliaceen-Studien. I. Die Wasser-Oekonomie der extreme atmosphärischen Tillandsien. *Jb. wiss. Bot.* **40**, 157–229.

MICHELMORE, A. P. G. (1939). Observations on tropical African grasslands. *J. Ecol.* **27**, 282–312.

MIEHE, H. (1911). Javanische Studien—Zur Frage der mikrobiologische Vorgänge im Humus einiger humussammelnder Epiphyten. *Abh. sachs. Ges. (Akad.) Wiss., math.-phys. Kl.* **32**, 376–98.

MILDBRAED, J. (1914). Die Vegetationsverhältnisse im Sammelgebiet der Expedition. *Wiss. Ergeb. d. deutsch. Zentral-Afrika-Expedition* 1907–1908, **2**, 603–91. Leipzig.

—— (1922). *Wissenschaftliche Ergebnisse der zweiten deutschen Zentral-Afrika-Expedition* 1910–1911 *unter Führung Adolf Friedrichs, Herzogs zu Mecklenburg.* Leipzig.

—— (1923). Das Regenwaldgebiet im äquatorialen Afrika. *Notizbl. bot. Gart. Berlin*, **8**, 574–99.

—— (1930*a*). Zusammensetzung der Bestände und Verjüngung im tropischen Regenwald. *Ber. dtsch. bot. Ges.* **48** (Generalversammlungs-Heft), 50–7.

—— (1930*b*). Sample plot surveys in the Cameroons rain-forest. (Transl. by H. M. Heyder.) *Emp. For. J.* **9**, 242–66.

—— (1933*a*). Zur Kenntnis der Vegetationsverhältnisse Nord-Kameruns. *Bot. Jb.* **65**, 1–52.

—— (1933*b*). Ein Hektar Regenwald auf Fernando Po. *Notizbl. bot. Gart. Berl.* **11**, 946–50.

MILNE, G. (1937). Essays in applied pedology. I. Soil type and soil management in relation to plantation agriculture in East Usambara. *East Afr. Agric. J.* (July 1937).

—— (1940). *Report on a journey to parts of the West Indies and the United States for the study of soils.* Agric Res. Station, Amani.

MOHR, E. C. J. (1930). Tropical soil-forming processes and the development of tropical soils, with special reference to Java and Sumatra. Transl. by R. L. Pendleton. *College of Agric., Univ. Philippines, Exp. Sta. Contrib.* no. 655 (mimeographed).

—— (1944). *The soils of equatorial regions with special reference to the Netherlands East Indies.* Transl. by R. L. Pendleton. Ann Arbor, Michigan.

MOLISCH, H. (1896). Das Erfrieren von Pflanzen über dem Eispunkt. *S.B. Akad. Wiss. Wien*, **105**, 82–95.

MOREAU, R. E. (1933). Pleistocene climatic changes and the distribution of life in East Africa. *J. Ecol.* **21**, 415–35.

—— (1935*a*). Some eco-climatic data for closed Evergreen Forest in Tropical Africa. *J. Linn. Soc. (Zool.)*, **39**, 285–93.

—— (1935*b*). A synecological study of Usambara, Tanganyika Territory, with particular reference to birds. *J. Ecol.* **23**, 1–43.

—— (1938). Climatic classification from the standpoint of East African biology. *J. Ecol.* **26**, 467–96.

MORISON, C. G. T., HOYLE, A. C. & HOPE-SIMPSON, J. F. (1948). Tropical soil-vegetation catenas and mosaics. A study of the south-western part of the Anglo-Egyptian Sudan. *J. Ecol.* **36**, 1–84.

MYERS, J. G. (1930). Notes on wild cacao in Surinam and in British Guiana. *Kew. Bull.* (1930), 1–10.

—— (1933). Notes on the vegetation of the Venezuelan Llanos. *J. Ecol.* **21**, 335–49.

—— (1935). Zonation of vegetation along river courses. *J. Ecol.* **23**, 356–60.

—— (1936). Savannah and forest vegetation of the interior Guiana Plateau. *J. Ecol.* **24**, 162–84.

NAVEZ, A. (1924). La forêt équatoriale brésilienne, I. *Bull. Soc. Bot. Belg.* **57**, 7–17.

—— (1930). On the distribution of tabular roots in *Ceiba* (Bombacaceae). *Proc. Nat. Acad. Sci., Wash.*, **16**, 339–44.

NICHOLSON, J. W. (1936). The influence of forests on climate and water supply in Kenya. *Kenya For. Dep. Pamphlet*, no. 2.

OLIVER, W. R. B. (1930). New Zealand epiphytes. *J. Ecol.* **8**, 1–50.

OLSEN, C. (1917). Studies on the succession and ecology of epiphytic bryophytes on the bark of common trees in Denmark. *Bot. Tidsskr.* **34**, 313–42.

ORTH, R. (1939). Zur Kenntnis des Lichtklimas der Tropen und Subtropen sowie des tropischen Urwaldes. *Beitr. Geophys.* **55**, 52.

OSBORN, T. G. B. & ROBERTSON, R. N. (1939). A reconnaissance survey of the vegetation of the Myall Lakes. *Proc. Linn. Soc. N.S.W.* **64**, 279–96.

OYE, P. VAN (1921). Influence des facteurs climatiques sur la répartition des épiphytes à la surface des troncs d'arbres à Java. *Rev. gén. Bot.* **33**, 161–76.

PALLIS, M. (1916). The structure and history of *Plav:* the floating fen of the delta of the Danube. *J. Linn. Soc. (Bot.)*, **43**, 233–90.

PAULIAN, R. (1946). Preliminary survey of the West African rain-forest canopy. *Nature, Lond.*, **157**, 877.

—— (1947). *Observations écologiques en forêt de Basse Côte d'Ivoire.* Paris.

PEARSON, H. H. W. (1899). The botany of the Ceylon patanas. *J. Linn. Soc. (Bot.)*, **34**, 300–65.

PENDLETON, R. L. & SHARASUVANA, S. (1946). Analyses of some Siamese laterites. *Soil Sci.* **62**, 423–48.

PENZIG, O. (1902). Die Fortschritte der Flora des Krakatau. *Ann. Jard. bot. Buitenz.* **18**, 92–113.

PESSIN, L. J. (1925). An ecological study of the polypody fern *Polypodium polypodioides* as an epiphyte in Mississippi. *Ecology*, **6**, 17–38.

PETCH, T. (1924). Gregarious flowering. *Ann. R. Bot. Gdns, Peradeniya*, **9**, 101–17.

—— (1930). Buttress roots. *Ann. R. Bot. Gdns, Peradeniya*, **11**, 277–85.

PHILLIPS, J. (1931). Mortality among plants and its bearing on natural selection. *Nature, Lond.*, **127**, 851–2.

PICADO, C. (1912). Les maires aeriennes de la forêt vierge américaine: les Bromeliacées. *Biologica*, **11**, 110–15.

—— (1913). Les Broméliacées épiphytes, considerées comme milieu biologique. *Bull. sci. Fr. Belg.* Ser. 7, **47**, 215–360.

PIJL, L. VAN DER (1934). Die Mycorrhiza von *Burmannia* und *Epirrhizanthes* und die Fortpflanzung ihres Endophyten. *Rec. Trav. bot. néerland.* **31**, 761–79.

PITTENDRIGH, C. S. (1948). The Bromeliad-Anopheles-Malaria Complex in Trinidad. I. The Bromeliad flora. *Evolution*, **2**, 58–89.

PITT-SCHENKEL, C. J. W. (1938). Some important communities of Warm Temperate Rain Forest at Magamba, West Usambara, Tanganyika Territory. *J. Ecol.* **26**, 50–81.

POLAK, B. (1933). Een tocht in het zandsteen gebied bij Mandor (West Borneo). *Trop. Natuur*, **22**, 23–8.

POLAK, E. (1933). Ueber Torf und Moor in Niederlandisch Indien. *Verh. Akad. Wet., Amst.*, **30**, 1–85.

POSTHUMUS, O. (1931). Plantae in *Festbundel K. Martin, Leidsche Geol. Meded.* **5**, 485–508.

POTONIÉ, H. & KOORDERS, S. H. (1909). Sumpfflachmoornatur der Moore des Produktiven Karbons. *Jb. preuss. geol. Landesanst.* **30**, 389–443.

POTTER, M. C. (1891). Observations on the protection of buds in the tropics. *J. Linn. Soc. (Bot.)*, **28**, 343–52.

PYNAERT, L. (1933). La mangrove congolaise. *Bull. Agric. Congo Belge*, **23**, 184–207.

RAUNKIAER, C. (1934). *The life-forms of plants and statistical plant geography.* Oxford.

RECHINGER, K. (1908). Vegetationsbilder aus dem Neu-Guinea-Archipel. *Vegetationsbilder*, Reihe **6**, Hft. 2.

—— (1910). Botanische und zoologische Ergebnisse einer Forschungsreise nach den Samoa-inseln, dem Neuguinea-archipel und den Salomoninseln. III. Siphonogamen der Samoa-inseln. *Denkschr. Akad. Wiss. Wien*, **85**, 202–388.

REID, E. M. & CHANDLER, M. E. J. (1933). *The London Clay flora.* London.

RESVOLL, T. R. (1925). Beschuppte Laubknospen in den immerfeuchten Tropenwäldern. *Flora* (N.F.), **18–19** (Goebel Festschr.), 409–20.

RICHARDS, P. W. (1932). Ecology. In F. Verdoorn, *Manual of bryology*, pp. 367–95. The Hague.

—— (1936). Ecological observations on the rain forest of Mount Dulit, Sarawak. Parts I and II. *J. Ecol.* **24**, 1–37; 340–60.

RICHARDS, P. W. (1939). Ecological studies on the rain forest of Southern Nigeria. I. The structure and floristic composition of the primary forest. *J. Ecol.* **27**, 1–61.

—— (1941). Lowland Tropical Podsols and their vegetation. *Nature, Lond.*, **148**, 129–31.

—— (1943). The biogeographic division of the Indo-Australian Archipelago. 6. The ecological segregation of the Indo-Malayan and Australian elements in the vegetation of Borneo. *Proc. Linn. Soc. Lond.* (Sess. **154**, 1941–2), pp. 154–6.

—— (1945). The floristic composition of primary tropical rain forest. *Biol. Rev.* **20**, 1–13.

RICHARDS, P. W., TANSLEY, A. G. & WATT, A. S. (1940). The recording of structure, life form and flora of tropical forest comunities as a basis for their classification. *J. Ecol.* **28**, 224–39. (Also published as *Imp. For. Inst. Paper*, no. **19**, 1939.)

RIKLI, M. (1943). *Das Pflanzenkleid der Mittelmeerländer*, **1**. Bern.

ROBERTS, R. C. (1936). Soil survey of Porto Rico. *U.S. Dep. Agric.* no. 8.

ROBERTY, G. (1946). Les associations végétales de la vallée moyenne du Niger. *Veröff. geobot. Inst. Rübel*, **22**.

ROBINSON, G. W. (1949). *Soils, their origin, constitution and classification* (ed. 3). London.

ROBYNS, W. (1936). Contribution à l'étude des formations herbeuses du District Forestier Central du Congo Belge. *Mém. Inst. Roy. Colon. Belge*, **5**.

—— (1946). Statistiques de nos connaissances sur les spermatophytes du Congo Belge et du Ruanda-Urundi. *Bull. Jard. bot. Brux.* **18**, 133–44.

RODWAY, J. (1894). *In the Guiana forest*. London.

ROOSENDAEL, J. VAN & THORENAAR, A. (1924). De natuurlijke verjonging van Ngerawan (*Hopea Mengarawan* Miq.) in Zuid Sumatra. *Tectona*, **17**, 519–67.

SALISBURY, E. J. (1925). The structure of woodlands. Festschr. C. Schroeter. *Veröff. geobot. Inst. Rübel*, Hft. **3**, 334–54.

—— (1929). The biological equipment of species in relation to competition. *J. Ecol.* **17**, 197–222.

—— (1930). Mortality amongst plants and its bearing on natural selection. *Nature, Lond.*, **125**, 817.

—— (1942). *The reproductive capacity of plants*. London.

SAXTON, W. T. (1924). Phases of vegetation under monsoon conditions. *J. Ecol.* **12**, 1–38.

SCHEFFLER, G. (1901). Ueber die Beschaffenheit des Usambara-Urwaldes und über den Laubwechsel an Bäumen desselben. *Notizbl. bot. Gart. Berl.* **3**, 139–66.

SCHENCK, H. (1892–3). Beiträge zur Biologie und Anatomie der Lianen. I. Beiträge zur Biologie der Lianen. *Bot. Mitt. Trop.* Hft. **4** (1892). II. Beiträge zur Anatomie der Lianen. *Ibid.* Hft. **5** (1893).

—— (1904). Strandvegetation Brasiliens. *Vegetationsbilder*, Reihe **1**, Hft. 7.

SCHIMPER, A. F. W. (1888). Die epiphytische Vegetation Amerikas. *Bot. Mitt. Trop.* **2**.

—— (1891). Die indo-malayische Strandflora. *Bot. Mitt. Trop.* Hft. 3.

—— (1898). *Pflanzengeographie auf physiologischer Grundlage* (ed. 2). Jena.

—— (1903). *Plant-geography upon a physiological basis*. Transl. by W. R. Fisher. Edited by P. Groom & I. B. Balfour. Oxford.

—— (1935). *Pflanzengeographie auf physiologischer Grundlage* (ed. 3). Revised by F. C. von Faber. Jena.

SCHWEIZER, J. (1932). Ueber die Periodizität des Blattwechsels bei tropischen Bäumen. *Mitt. naturf. Ges. Bern* (1932), 41–6.

SEIFRIZ, W. (1923). The altitudinal distribution of plants on Mt. Gedeh, Java. *Bull. Torrey Bot. Cl.* **50**, 283–305.

—— (1924). The altitudinal distribution of lichens and mosses on Mt. Gedeh, Java. *J. Ecol.* **12**, 307–13.

SENN, G. (1913). Der osmotische Druck einiger Epiphyten und Parasiten. *Verh. naturf. Ges. Basel*, **24**, 179–83.

—— (1923). Ueber die Ursachen der Brettwurzelbildung bei der Pyramiden-Pappel. *Verh. naturf. Ges. Basel*, **35** (Festbd. H. Christ), 405–35.

SENSTIUS, M. W. (1930–1). Agro-geological studies in the tropics. *Soil Res.* **2**, 10–56.

SEWANDANO, M. (1937). Inventarisatie en inrichting van de veenmoerasbosschen in het panglonggebied van Sumatra's Oostkust. *Tectona*, **30**, 660–75.

SHANTZ, H. L. & MARBUT, C. F. (1923). The vegetation and soils of Africa. *Amer. Geogr. Soc. Res. Series*, **13**.

SHIRLEY, H. L. (1935). Light as an ecological factor and its measurement. *Bot. Rev.* **1**, 355–81.

SHREVE, F. (1914 a). A montane rain-forest. A contribution to the physiological plant geography of Jamaica. *Publ. Carneg. Instn.*, no. 199.

—— (1914 b). The direct effects of rainfall on hygrophilous vegetation. *J. Ecol.* **2**, 82–98.

SIMON, S. V. (1914). Studien über die Periodizität der Lebensprozesse der in dauernd feuchten Tropengebieten heimischen Bäume. *Jb. wiss. Bot.* **54**, 71–187.

SMITH, A. C. & JOHNSTON, I. M. (1945). A phytogeographical sketch of Latin America, in F. Verdoorn, *Plants and plant science in Latin America*. Waltham, Mass.

SMITH, A. M. (1909). On the internal temperature of leaves in tropical insolation, with special reference to the effect of their colour on the temperature; also observations on the periodicity of the appearance of young coloured leaves of trees growing in Peradeniya Gardens. *Ann. R. Bot. Gdns. Peradeniya*, **4**, 229–98.

SMITH, J. J. (1923). Periodischer Laubfall bei *Breynia cernua* Muell. Arg. *Ann. Jard. bot. Buitenz.* **32**, 97–102.

SPANNER, L. (1939). Untersuchungen über den Wärme und Wasserhaushalt von *Myrmecodia* und *Hydnophytum*. *Jb. wiss. Bot.* **88**, 243–83.

SPRUCE, R. (1908). *Notes of a botanist on the Amazon and Andes* (2 vols.) Ed. by A. R. Wallace. London.

STAHL, E. (1893). Regenfall und Blattgestalt. *Ann. Jard. bot. Buitenz.* **11**, 98–182.

—— (1896). Ueber bunte Laubblätter. Ein Beitrag zur Pflanzenbiologie. *Ann. Jard. bot. Buitenz.* **13**, 137–216.

STAMP, L. D. (1925). *The vegetation of Burma*. Calcutta.

STAMP, L. D. & LORD, L. (1923). The ecology of part of the riverine tract of Burma. *J. Ecol.* **11**, 129–59.

STEBBING, E. P. (1937 a). *The forests of West Africa and the Sahara*. London and Edinburgh.

—— (1937 b). The threat of the Sahara. *J. R. Afr. Soc.* **36** (extra Suppl.).

STEEMANN-NIELSEN, E. (1940). Ueber die Bedeutung der sogenannten xeromorphen Struktur im Blattbau der Pflanzen auf nährstoffarmen Boden. *Dansk bot. Ark.* **10**, no. 2.

STEENIS, C. G. G. J. VAN (1932). Botanical results of a trip to the Anambas and Natoena Islands. *Bull. Jard. bot. Buitenz.* (ser. 3), **12**, 151–211.

—— (1933). Report of a botanical trip to the Ranau region, south Sumatra. *Bull. Jard. bot. Buitenz.* (ser. 3), **13**, 1–56.

—— (1934). On the origin of the Malaysian mountain flora. Part 1. *Bull. Jard. bot. Buitenz.* (Ser. 3), **13**, 135–262.

—— (1935 a). Maleische Vegetatieschetsen. *Tijdschr. ned. aardrijksk. Genoot.* Reeks 2, **52**, 25–67, 171–203, 363–98. (The map also appears in *Atlas van tropisch Nederland*, 1938. Weltevreden, Java.)

—— (1935 b). On the origin of the Malaysian mountain flora. Part 2. Altitudinal zones, general considerations and renewed statement of the problem. *Bull. Jard. bot. Buitenz.* (Ser. 3) **13**, 289–417.

—— (1936). On the origin of the Malaysian mountain flora. Part 3. Analysis of floristic relationships (1st instalment). *Bull. Jard. bot. Buitenz.* (Ser. 3) **14**, 56–72.

—— (1937 a). De invloed van den mensch op het bosch. *Tectona*, **30**, 634–52.

—— (1937 b). Over de flora van den Carstensztop, in A. H. Colijn, *Naar de eeuwige sneeuw van tropische Nederland*. Amsterdam.

—— (1938). Recent progress and prospects in the study of the Malaysian flora. *Chron. Bot.* **4**, 392–7.

—— (1942). Gregarious flowering of *Strobilanthes* (Acanthaceae) in Malaysia. *Ann. R. Bot. Gdn, Calcutta* (150th Anniv. vol.), pp. 91–7.

STEENIS, C. G. G. J. VAN & RUTTNER, F. (1932). Die Pteridophyten und Phanerogamen der deutschen limnologischen Sunda-Expedition. *Arch. Hydrobiol.* (Suppl.-Bd.), XI. Tropische Binnegewässer, 3, 231–387.

STEHLÉ, H. (1935). Essai d'écologie et géographie botanique. *Flore de la Guadeloupe et dépendances*, 1. Basse Terre (Guadeloupe).

STEHLÉ, H. (1945). Forest types of the Caribbean Islands. Part I. *Carib. For.* **6** (suppl.), 273–408.

STEUP, F. K. M. (1930). Bijdragen tot de kennis der bosschen van Norrd- en Midden-Celebes. I. *Tectona*, **23**, 857–73.

—— (1932). Bijdragen tot de kennis der bosschen van Noord- en Midden-Celebes. III. Het zoogenaande tjempaka-hoetan complex in de Minhasa. *Tectona*, **25**, 119–47.

STOCKER, O. (1935*a*). Assimilation und Atmung westjavanische Tropenbaüme. *Planta*, **24**, 402–45.

—— (1935*b*). Transpiration und Wasserhaushalt in verschiedenen Klimazonen. III. Ein Beitrag zur Transpirationsgrösse in javanischen Urwald. *Jb. wiss. Bot.* **81**, 464–96.

STRAELEN, V. VAN (1933). Résultats scientifiques du voyage aux Indes Neérlandaises de LL.AA.RR. la Prince et le Princesse Léopold de Belgique, 1. Introduction. *Mém. Mus. Hist. nat. Belgique*, hors série. Brussells.

STRUGNELL, E. J. & MEAD, J. P. (1937). An ascent of Gunong Tahan. *Malay. Forester*, **6** 131–40.

SWYNNERTON, C. F. M. (1917). Some factors in the replacement of the ancient East African forest by wooded pasture land. *S. Afr. J. Sci.* **14**, 493–518.

SYKES, R. A. (1930). Some notes on the Benin forests of Southern Nigeria. *Emp. For. J.* **9**, 101–6.

SYMINGTON, C. F. (1933). The study of secondary growth on rain forest sites. *Malay. Forester*, **2**, 107–17.

—— (1936). The flora of Gunong Tapis in Pahang. *J. Malay. Br. Asiat. Soc.* **14**, 333–64.

—— (1943). Foresters' manual of Dipterocarps. *Malay. For. Rec.* **16**.

TANSLEY, A. G. (1935). The use and abuse of vegetational concepts and terms. *Ecology*, **16**, 284–307.

TANSLEY, A. G. & FRITSCH, F. E. (1905). Sketches of vegetation at home and abroad. I. The flora of the Ceylon littoral. *New Phytol.* **4**, 1–17, 27–55.

THOMAS, A. S. (1932). The dry season in the Gold Coast and its relation to the cultivation of cacao. *J. Ecol.* **20**, 263–9.

THOMPSON, J. McL. (1943). A modern study of cauliflory. *Nature, Lond.*, **151**, 481–2.

—— (1946). The study of plant behaviourism; a common meeting ground for those engaged in botanical enquiry. *Proc. Linn. Soc. Lond.* Session **157** (1944–5), 72–91.

THORENAAR, A. (1927). Eigenaardige wortelvormigen in de moerasbosschen van Palembang. *Trop. Natuur*, **16**, 73–82.

THORNTHWAITE, C. W. (1931). The climates of North America according to a new classification. *Geogr. Rev.* **21**, 633–55.

THORP, J. & BALDWIN, M. (1940). Laterite in relation to soils of the tropics. *Ann. Ass. Amer. Geogr.* **30**, 163–94.

TIEMANN, H. D. (1935). What are the largest trees in the world? *J. For.* **33**, 903–15.

TREUB, M. (1883). Observations sur les plantes grimpantes du Jardin Botanique de Buitenzorg. *Ann. Jard. bot. Buitenz.* **3**, 160–83.

—— (1888). Notice sur la nouvelle fiore de Krakatau. *Ann. Jard. bot. Buitenz.* **7**, 213–23.

TROCHAIN, J. (1940). Contribution à l'étude de la végétation du Sénégal. *Mém. Inst. d'Afrique Noire*, **2**. Paris.

TROLL, C. (1948). Der asymmetrische Aufbau der Vegetationszonen und Vegetationsstufen auf der Nord- und Südhalbkugel. *Ber. geobot. Inst. Rübel f. J. 1947*, pp. 46–83.

TROLL, W. (1930). Ueber die sogenannten Atemwurzeln der Mangroven. *Ber. dtsch. bot. Ges.* **48** (Generalversammlungs-Heft), 81–99.

—— (1938). *Vergleichende Morphologie der höheren Pflanzen*, 1, Berlin.

TROLL, W. & DRAGENDORF, O. (1931). Ueber die Luftwurzeln von *Sonneratia* Linn. f. und ihre biologische Bedeutung. *Planta*, **13**, 311–473.

TURRILL, W. B. (1935). Krakatau and its problems. *New. Phytol.* **34**, 442–4.

TÜXEN, R. (1929). Ueber einige nordwestdeutsche Waldassoziationen von regionaler Verbreitung, *Jb. geogr. Ges. Hannover*, 1929, pp. 55–116.

Ulbrich, E. (1928). *Biologie der Früchte und Samen (Karpobiologie)*. Berlin.

Ule, E. (1901). Die Vegetation von Cabo Frio an der Küste von Brasilien. *Bot. Jb.* **28**, 511–28.

—— (1908*a*). Die Pflanzenformationen des Amazonas-Gebietes. *Bot. Jb.* **40**, 114–72; 398–443.

Ule, E, (1908*b*). Catinga- und Felsenformationen in Bahia. *Bot. Jb.* **40**, Beibl. 93, 39–48.

—— (1908*c*). Das Innere von Nordost-Brasilien. *Vegetationsbilder*, Reihe **6**, Hft. 3.

Vageler, P. (1933). *An introduction to tropical soils*. Transl. by E. Greene. London.

Vaughan, R. E. & Wiehe, P. O. (1937). Studies on the vegetation of Mauritius. I. A preliminary survey of the plant communities. *J. Ecol.* **25**, 289–342.

—— —— (1941). Studies on the vegetation of Mauritius. III. The structure and development of the Upland Climax Forest. *J. Ecol.* **29**, 127–60.

—— —— (1947). Studies on the vegetation of Mauritius. IV. Some notes on the internal climate of the Upland Climax Forest. *J. Ecol.* **34**, 126–36.

Vaupel, F. (1910). Die Vegetation der Samoa–Inseln. *Bot. Jb.* **44**, Beibl. 102, 47–58.

Vermoesen, C. (1931). *Les essences forestières du Congo Belge. I. Manuel des essences de la région équatoriale et du Mayombe*. Brussels.

Vesque, J. (1882). L'espèce végétale considérée au point de vue de l'anatomie comparée. *Ann. Sci. nat. (Bot.)*, Sér. 6, **13**, 5–46.

Volkens, G. (1912). *Laubfall und Lauberneuerung in den Tropen*. Berlin.

Voth, P. D. (1939). Conduction of rainfall by plant stems in a tropical rain forest. *Bot. Gaz.* **101**, 328–40.

Wallace, A. R. (1878). *Tropical nature and other essays*. London.

Walter, H. (1936). Nährstoffgehalt des Bodens und natürliche Waldbestände. *Forstl. Wschr. Silva*, **24**, 201–5, 209–13.

Walter, H. & Steiner, M. (1936–7). Die Oekologie der Ost-Afrikanischen Mangroven. *Z. Bot.* **30**, 65–193.

Wardlaw, C. W. (1931). Observations on the dominance of pteridophytes on some St Lucia soils. *J. Ecol.* **19**, 60–3.

Warming, E. (1892). Lagoa Santa. *K. danske vidensk. Selsk. Skr.* **6**.

—— (1909) *Oecology of Plants*. Transl. P. Groom & I. B. Balfour. Oxford.

Warming, E. & Graebner, P. (1933). *Lehrbuch der ökologischen Pflanzengeographie* (ed. 4). Berlin.

Watson, J. G. (1928). The mangrove swamps of the Malay Peninsula. *Malay. For. Rec.* **6**.

—— (1937). Age-class representation in virgin forest. *Malay. Forester*, **6**, 146–7.

Watt, A. S. & Fraser, G. K. (1933). Tree roots and the field layer. *J. Ecol.* **21**, 404–14.

Weberbauer, A. (1911). Die Pflanzenwelt der peruanischen Anden. *Veget. Erde*, **12**.

—— (1945). *El mundo vegetal de los Andes Peruanos*. Lima.

Went, F. W. (1940). Soziologie der Epiphyten eines tropischen Urwaldes. *Ann. Jard. bot. Buitenz.* **50**, 1–98.

Whitford, H. N. (1906). The vegetation of the Lamao Forest Reserve. *Philipp. J. Sci.* **1**, 373–431, 637–82.

Wiesner, J. (1895). Untersuchungen über den Lichtgenuss der Pflanzen mit Rücksicht auf die Vegetation von Wien, Kairo und Buitenzorg (Java). *S.B. Akad. Wiss. Wien*, Math.-naturw. Cl., **104**, 605–711.

—— (1897). Pflanzenphysiologische Mittheilungen aus Buitenzorg. VII. Zur Physiologie von *Taeniophyllum Zollingeri*. *S.B. Akad. Wiss. Wien*, Math.-naturw. Cl., **106**, Abt. 1, 77–98.

—— (1907). *Der Lichtgenuss der Pflanzen*. Leipzig.

Wildeman, E. de (1913). Documents pour l'étude de la géobotanique congolaise. *Bull. Soc. Bot. Belg.* **51** (vol. jubilaire), fasc. 3.

—— (1926). *Les forêts congolaises et leurs principales essences économiques*. Brussels.

—— (1930). Empattements, contreforts, racines-echassés. *Bull. Acad. Belg.*, Cl. Sci., sér. 5, **16**, 989–95.

—— (1934). Remarques à propos de la forêt équatoriale congolaise. *Mém. Inst. Roy. Colon. Belge*, **2**, 1–120.

WILKINSON, G. (1939). Root competition and sylviculture. *Malay. Forester*, **8**, 11–15.

WINKLER, H. (1913). Die Pflanzenwelt der Tropen, in R. H. Francé, *Das Leben der Pflanzen*. Stuttgart.

—— (1914) Die Pflanzendecke Südost-Borneos. *Bot. Jb.* **50** (Festbd.), 188–208.

WITKAMP, H. (1925). De ijzerhout als geologische indicator. *Trop. Natuur*, **14**, 97–103.

WOOD, B. R. (1926). The valuation of the forests of the Bartica-Kaburi area. *Report by the Conservator of Forests to the British Guiana Combined Court, Second Special Session*, 1926. Georgetown, British Guiana.

WRIGHT, H. (1905). Foliar periodicity of endemic and indigenous trees in Ceylon. *Ann. Bot. Gdns, Peradeniya*, **2**, 415–517.

ZEUNER, F. E. (1945). *The Pleistocene Period, its climate, chronology and faunal successions*. Ray Soc., London.

ZON, R. (1912). Forests and water in the light of scientific investigation. (Appendix 5, Final Rep. Nat. Waterways Commission) *U.S. Senate Document*, **469**, 62nd Cong., Sess. 2. Reprinted 1927, U.S. Dep. Agric. Forest Service.

ZON, P. VAN (1915). Mededeelingen omtrent den kamferboom (*Dryobalanops aromatica*). *Tectona*, **8**, 220–4.

INDEX OF PLANT NAMES

GENERAL INDEX